EINSTEIN'S OPPONENTS

This detailed account of the controversy surrounding the publication of Albert Einstein's theory of relativity explores the ferocious popular and academic opposition which at one time encircled one of the most important scientific breakthroughs of the twentieth century.

Based on extensive archival research, this fascinating discourse includes a compelling and entertaining examination of the contemporary literature created by Einstein's detractors. Exploring the arguments and strategies, social contexts, and motivations of Einstein's detractors, and providing unique insights into the dynamics of scientific controversies, this book is ideal for anyone interested in the history and philosophy of physics, popular science, and the public understanding of science.

MILENA WAZECK is a Newton International Fellow at the University of East Anglia, Norwich, UK, and a historian of science interested in the intersection of modern physics and its social context.

EINSTEIN'S OPPONENTS

The Public Controversy about the Theory of Relativity in the 1920s

MILENA WAZECK
University of East Anglia

Translated by

GEOFFREY S. KOBY
Kent State University

CAMBRIDGE
UNIVERSITY PRESS

CAMBRIDGE
UNIVERSITY PRESS

Shaftesbury Road, Cambridge CB2 8EA, United Kingdom

One Liberty Plaza, 20th Floor, New York, NY 10006, USA

477 Williamstown Road, Port Melbourne, VIC 3207, Australia

314–321, 3rd Floor, Plot 3, Splendor Forum, Jasola District Centre, New Delhi – 110025, India

103 Penang Road, #05–06/07, Visioncrest Commercial, Singapore 238467

Cambridge University Press is part of Cambridge University Press & Assessment,
a department of the University of Cambridge.

We share the University's mission to contribute to society through the pursuit of
education, learning and research at the highest international levels of excellence.

www.cambridge.org
Information on this title: www.cambridge.org/9781107017443

The translation of this work was funded by Geisteswissenschaften International – Translation Funding for Work in the Humanities and Social Sciences from Germany, a joint initiative of the Fritz Thyssen Foundation, the German Federal Foreign Office, the collecting society VG WORT and the Börsenverein des Deutschen Buchhandels (German Publishers & Booksellers Association).

First published 2014

A catalogue record for this publication is available from the British Library

ISBN 978-1-107-01744-3 Hardback

Contents

Translator's preface: making the imaginary accessible – translating *Einstein's Opponents*

GEOFFREY S. KOBY

For a professional translator and translation scholar, translating a book is a unique opportunity to reflect in practice the knowledge and positions that one has developed by studying theory. Each book is unique in its situation, embedded in one culture yet interesting to another. The translator acts as a cultural mediator, transmitting information and attitudes across the linguistic boundary to make them accessible to target-language readers who would otherwise be unable to access the content. At the same time, every translation is simultaneously an interpretation of the source-language text. In the early nineteenth century, Friedrich Schleiermacher already offered the distinction between a domesticating practice in translation and a foreignizing practice. Similarly, work in recent years by authors such as Lawrence Venuti (2008), Mona Baker (2006), and Susan Bassnett-McGuire (2002), has highlighted the theoretical position that there cannot be only one single, definitive translation. Instead, each translator brings a unique perspective to the translation at hand. At the same time, translators must follow the "translation brief" (also referred to as translation instructions, see Nord 1991); that is, the particular specifications agreed upon between the translator and publisher – or in this case, between the translator and author. In the following, I will describe my perspective on the translation and the decisions that were reached, often in consultation with the author, to create what we defined as the translation brief: standing squarely and intentionally in the fluent, "domesticating" tradition of English language translation, this work is intended to be an accessible, readable, and factually accurate translation that simultaneously reflects the spirit of the German original.[1]

[1] The text has been slightly revised in English by the author in interaction with the translator, where the initial translation made it apparent that either clarification was necessary, or that some point needed to be expanded. In addition, the references have been updated to some extent and English sources have been used as appropriate.

As discussed above, every source text is subject to a range of possible interpretations. This leads to the question of what constitutes an "accurate" translation – a question that can only be understood within the boundaries of a particular genre of translation. Although Venuti (2008, p. 2) asserts that fiction is "the most translated genre worldwide," I would argue that this statement needs to be modified; fiction may be the most translated genre that appears in traditional publishing venues, but non-fiction in its many varieties is much more likely to be the most translated genre. This includes both self-published documents (e.g. corporate reports and technical specifications), and documents such as contracts or training materials used internally in a large variety of multilingual organizations, along with literally any kind of document that may need to be translated for legal purposes (e.g. lawsuits). The size of the membership of professional translation associations indicates that professional translation extends far beyond non-fiction: the American Translators Association alone counts more than 11,000 members in 2013.

My conception of accuracy in non-fiction, then, can be explained using a metaphorical image of a coaxial cable (see figure). From the center outward, a coaxial cable is composed of a center core wire, a dielectric insulator, a metallic shield, and a plastic jacket. Let us assume that the message of the source text, as embodied in its wording, is represented by the central wire at the core of the cable. This original message is the guiding line that each translator follows in creating his or her translation. Given that multiple interpretations are possible for any source text, there is a range of solutions or renderings of any source text passage that would be considered by the community of professional translators to be acceptable or congruent with the intent of the source text author as embodied in the wording. The American Translators Association Certification Program recognizes four broad levels of translation accuracy in its Rubric (American Translators Association 2011): Strong, Acceptable, Deficient, and Minimal. Strong and Acceptable translations represent interpretations that contain few or no deviations or incongruities from the source text's factual information. A deficient translation is considered to deviate sufficiently far from the range of plausible interpretations of the source text to require editing by another translator. A minimal translation deviates so strongly from the range of interpretations considered acceptable by the majority of translators that the amount of effort required to edit the translation to an acceptable level of congruity might be more than that required to simply retranslate from the beginning.

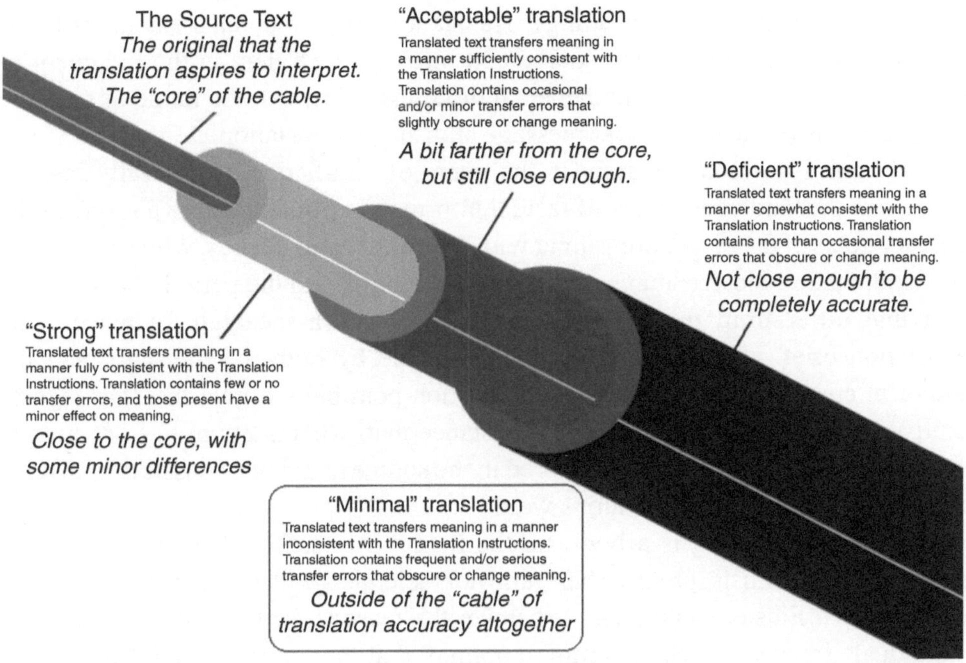

Metaphor of translation accuracy as coaxial cable. Image © 2013 Geoffrey S. Koby.

The sheathing around the coaxial cable's core represents this range of renderings. The interior of the cable represents translations with acceptable congruity, while still permitting a range of interpretation. One translator's version may be located closer to the core in terms of wording or detail of meaning, while another's may take more freedom to adapt the text or vary details for overall communicative effect – and yet still remain within the overall metaphorical cable of translational congruency with the source text. The plastic jacket of the cable represents those deficient translations that still follow the course of the cable's core, guiding the range of acceptable interpretations closely enough to be editable. Outside the cable sheathing is anything that is considered either a translation error or "not a translation." Translation errors are defined as those occurring in a document that purports to be a translation of a given original. In this case, translation errors are those renderings in a target text that suffer from a lack of congruity with the source text meaning, as interpreted by a preponderance of expert translators in that language pair. Translation errors range from minor incongruencies such as wording in the target text that, while maintaining the fundamental imagery of the source wording, does not accurately reflect its nuances, implications, or typical collocations (e.g. translating *he is trying to do X* as *he is endeavored to do X*), to target-text images that are fundamentally different from those of the source text (e.g. translating German *Haus* as *house* where in fact *company* is the source-text meaning). Translation errors may also extend as far as multiple incongruities with the source text message such that, in the judgment of expert translators, there is a fundamental disconnect between the range of possible interpretations of the source text message and such a translation.

This book is situated squarely in the genre of non-fiction, specifically history. Therefore, the kind of accuracy of factual information discussed above is one of the primary goals of the translation, along with readability and fluency, while preserving and explicitating those features of the German culture and the period under discussion that differ from the expectations of the English-speaking audience. This translation brief was used as a guiding principle by both the translator and the author in creating the most accurate translation possible – a goal that was significantly enhanced by the fortunate circumstance that, when ambiguities or multiple possible interpretations were discovered in the source text, it was possible to discuss with the author what her intentions were.

Einstein's Opponents is a book on the history of science. It focuses on those individuals – scientists, philosophers, and non-scientists with an interest in science – who opposed Einstein's theories. Whatever the reasons for their opposition, these individuals frequently held a common opinion that the modern science of the day (i.e. relativity science) as conducted by the scientists in the established institutions, was incorrect and that the scientists were (deliberately or unconsciously) suppressing the "truth" that could be found outside of the institutions.

This idea of the unreasonableness of "establishment sciences" has not changed in 100 years. It has even become a trope in the genre of science fiction, particularly science fiction humor, where the character of the "opponent of established science" has become stereotypical. This is exemplified in the fiction story *Babel* by Steven Utley, which appeared in 2004 in a prominent science fiction magazine. Utley has his characters (all of whom oppose institutionalized science) use the term "establishment scientists" in contrast with "faith-based scientists" and others. In the story, there are multiple competing "non-establishment" groups with fictional names such as the "Advocates for Biblical Creation," the "Institute for Extraterrestrialist Studies," the "Advocates for Intelligent Design." One of Utley's characters says, "By scientists, however, I mean *true* scientists, as opposed to members of the conservative uniformitarian-evolutionist establishment… That's quite a mouthful, isn't it? And I didn't even get in 'relativist' and 'indeterminist.'" Here, "The true scientist formulates a sound theory and then amasses the incontrovertible evidence that supports it." In the fictional world, a new discovery has been made and, according to the non-scientist character, "Astronomical observations will corroborate the work done in Newtonian optics, proving that the Universe is not only much younger, but much smaller than establishment science says. Then, goodbye Einstein! Relativity will join evolution in the same unmarked mass grave." This sentiment could easily have been expressed by some of the non-fictional characters in *Einstein's Opponents*.

In this translation, because the topic is the history of science, the general approach is to assume that the reader is interested in the factual breakdown of the events and in the polemics of the writers of the time who expressed their resistance to the idea of relativity. Now, resistance to the idea of relativity in science is not particularly German; therefore the translation is crafted to communicate the tone and attitude of the original writer quoted in the text *as if they were fluently expressing their ideas in English*. The translation does not attempt to make the translation sound "foreign" except in cases where cultural differences require wording or expressions that are not typical for English.

Technology contributing to accuracy

Computer-assisted translation software was used to create the translation. The specific package used was SDL Trados Studio 2011 Freelance™ as a translation environment with translation memory, along with the package's terminology management application, MultiTerm™. The advantages of using such a translation environment are many; first, the translation memory feature stores each sentence as it is translated, allowing similar sentences to be reused and adapted. Second, the translation memory can be searched using a concordance feature for words or phrases that have previously been translated. Both of these features enhance consistency across longer stretches of text and within the entire book. Third, terms, phrases, and names that recur frequently can be added to the terminology management application to ensure that they are translated identically at each occurrence. Using these features prevents drift in the use of similar words or terms across the period of time it takes to translate an entire book. In addition, they enhance translation speed by serving as an aid to memory, thus obviating the need to repeat searching for terminology that fades in the translator's mind over time.

Names and cultural phenomena

This book includes a number of names and cultural phenomena specific to the German cultural environment that are explained here for the interested reader.

Spengler, Oswald. In the discussion of Rudolf Mewes' work *Kriegs- und Geistesperioden im Völkerleben und Verkündigung des nächsten Weltkrieges*, the source text merely uses the adjective *spenglerschem*. The German reader recognizes this as a name used as an adjective referring to the work of Oswald Manuel Arnold

Gottfried Spengler (1880–1936), a German historian and philosopher best known for his large book *The Decline of the West (Der Untergang des Abendlandes)*, published in 1918 and 1922.

Kneipp. The text refers to naturopathy, including the Kneipp movement, and also frequently refers to the "hydropathic Kneipp physician" Franz Xaver Kleinschrod as a prominent universal riddle solver. Thus a brief explanation of the Kneipp movement is in order. This alternative health movement was founded in the late 1800s by Sebastian Kneipp (1821–1897) and continues to be well known in Germany. It is best known for the use of water treatments such as "hot and cold rinses, knee and thigh rinses, cold rinses, cold footbaths (15–20 seconds) and water stepping." According to www.kneipp.com, "During his life, Kneipp used knowledge of the healing powers of water and medicinal plants and combined them with his own insights to create a systematic teaching. Thus the priest and naturopath created a visionary life philosophy that sees man going about his daily habits and routines and his natural environment as an inseparable and balanced entity. He considered the elements of water, plants, exercise, diet and balance to be closely linked. The name Kneipp not only stands for scientifically based naturopathic products and treatments, but also for a holistic life philosophy."

Skeptron. The *skeptron* or ancient Greek scepter was a long staff which was a symbol of authority, sometimes understood to exemplify the idea "…that authority comes to language from outside, a fact concretely exemplified by the *skeptron* that, in Homer, is passed to the orator who is about to speak" (Bourdieu 1999, p. 109).

Semi-Kürschner. "*Sigilla Veri*, a very rare encyclopaedic work of anti-Semitism, was compiled by Philipp Stauff, who also produced a reference work called Semi-Kürschner, the title referring to Joseph Kürschner's annual German Literary Calendar, which was known as the Kürschner. The 'Semi' in the title is supposed to refer to 'Semites', the SemiKürschner being strenuously anti-Semitic" (Gale Cengage Learning 2013, p. 2).

Quotes and idioms

Most of the quotes in this book come from unpublished letters or from works in German that have never been translated. In these cases, the quotes are translated with the intent of expressing the personality of the writer as expressed in the text. However, there are four cases in which it was necessary to refer to published sources in order to link to the tradition as embodied therein. Chronologically, the oldest quotes come from Isaac Newton. Rather than re-translate a quote that was originally in Latin and translated into German, I referred back to the early published translation of *Philosophia naturalis principia mathematica* into English by Andrew Motte (Newton 1729).

The second case of quoting deals with the issue of how to refer to Kantian philosophy. In order to maintain consistency with existing Kant scholarship, the Kantian terms in this text are all taken from Waibl's *Dictionary of Philosophical Terms* (2011).

The final two cases are literary quotes, one from Goethe and one from Schiller, which are cited by individuals quoted in this book. In each case, the quotes are widely known in Germany and are cited without reference to the work in question. The Goethe quote, *Mir graut vor dir!* is cited in a longer quote from Franz Kleinschrod in Chapter 1 and comes from Goethe's *Faust*, where the character Margarethe fears Heinrich (Faust) due to his pact with the devil. In this case, the challenge was selecting from among the published translations of this widely known work. Four early to more recent translations are available on the Internet. Given the time period that this book discusses, it seemed reasonable to choose a translation more contemporary to that period. The following choices were available:

- S.T. Coleridge, *Henry, now you make me tremble*. (1821)
- Charles T. Brooks, *Henry! I dread to look on thee*. (1856)
- Bayard Taylor, *Henry! I shudder to think of thee*. (1890)
- George Madison Priest, *Henry! I shrink from you!* (1932)

The choice used, Bayard Taylor's translation, seemed to best fit both in terms of the time period and the preceding context of Kleinschrod's words, specifically the immediately preceding … *a terror and horror of this humanity seizes us*.

In Chapter 4, Ernst Gehrcke is quoted using a line from Schiller's 1796 play *Fiesco* to express his attitude towards Paul Weyland in a longer quote about the Philharmonic lectures. However, like many popularly-known phrases, the Schiller quote is actually only a near-quote. The original German is *Der Mohr hat seine Arbeit gethan, der Mohr kann gehen*, which translates as *The Moor has done his work, the Moor may go*. Gehrcke, however, writes *Der Mohr hat seine Schuldigkeit getan*. Thus it was necessary to modify the quote as well, to *The Moor has done his duty*.

Terminology

The general approach to terminology that I have taken in this translation is to make it accessible to non-German speakers. Any concepts that are unique to the German culture are explicitated. Titles (except journals) are translated, for two reasons: first, to allow non-German speakers to follow the line of argument – the titles of many works become part of the narrative; and second, to give non-German speakers access to the meaning of the reference works, so that they may consider using them in their own research. (Note that titles of Einstein's own works are taken from published translations where available; otherwise, they are taken from the most commonly used, already existing, reasonable translation found in sources such as the World Wide Web.)

Challenges

Four terms presented challenges in the translation, requiring deliberation and discussion with the author and, finally, a decision. In all four cases, the focus was kept on the translation brief of clarity and comprehensibility. These terms are *Welträtsellöser* (*world riddle solver*), *Weltanschauung* (*world view*), *Grundsätzlichkeit* (*categorical nature*), and *Wissenschaftlichkeit* (*scientific approach/scientific nature/scientific claims*). The challenge in the first two terms was to choose between competing terms and arrive at one that was transparent and acceptable to both author and translator, while remaining in the traditional usage of the time period and the discipline of the history of science. For the latter two terms, the challenge is found in their nature as the German words cover a wide range of meanings.

The initial proposed translation for *Welträtsellöser* was *universal mystery solver.* Based on a modern understanding of the word *Rätsel* (*riddle, puzzle, mystery*) and the idea of *Welträtsel* being those thorny problems of origins that have mystified mankind for centuries, this solution seemed at first reasonable. However, the context of the early 1900s rendered this impossible. The tradition at the time, including some published works, used the term *world riddle*, so the translation follows this practice; if the term were newly coined, the translator would have selected the initial solution. Incidentally, one of the English language reviews of the German original, *Einsteins Gegner*, avoids this issue by using the German term: *…those that had found their own private solution to the riddles of the universe, based on their own newly found principles; in German, the Welträtsellöser* (van Dongen 2010, p. 78). However, the mandate of clarity in the translation brief for this translation precluded the use of German terms, no matter how clearly defined, unless absolutely necessary.

This also applies to the second challenge, *Weltanschauung*. Of course, we are perfectly aware that *Weltanschauung* has been borrowed into English. However, it is well known that words, once borrowed, take on a life of their own in the borrowing language.[2] Thus the English *Weltanschauung* has a particular meaning that does not coincide with the German *Weltanschauung*, which is broader. In addition, the text uses the adjectival form *weltanschaulich*, which English does not have. The initial proposed solution was *ideology/ideological*, which seemed neat and easy. However,

[2] For instance, the English adjective *handy* has been recently borrowed into German, where it has become a noun meaning *cellular telephone*. Germans are puzzled when they use the word in English and are not understood.

discussions with the author focused on the fact that *ideology* is a narrower term than *Weltanschauung*, which encompasses more than just a political or social ideology; we agreed on *world view* as a general solution to this issue, where it is understood to encompass both ideology and a broader conception by an individual of the world as a whole. This did cause challenges for the translator, as there is no neatly corresponding adjective – *worldviewical* simply wouldn't work – so the adjectival uses are adapted on a case-by-case basis.

The latter two challenges are both terms that use the German endings *–lichkeit*, which break down into *–lich* (*–ly*) and *–keit* (*–ness*). Thus *–lichkeit* can roughly be translated as *–ly-ness*, which unfortunately is not an acceptable form in English. Thus *Grundsätzlichkeit* could be crudely translated as *fundamental-ly-ness* – a coinage that, in addition to suffering from ugliness, is intransparent to the reader. The translation *categorical nature* was arrived at using the following thought process. The dictionary meaning of *Grundsätzlichkeit* is *fundamental nature*, but that does not work well in the collocation *Vehemenz und Grundsätzlichkeit*. Therefore, working from *grundsätzlich* (*in principle, absolutely, fundamentally*), but not wanting to call them *fundamentalists*, the translation chosen (or derived) was *categorical nature* – by evolution from the term *fundamental* to its synonym *absolute* to *categorical*. Similarly, *Wissenschaftlichkeit* could be crudely rendered as *scientific-ly-ness*, where the dictionary meaning is *scientific nature* or *academic nature* depending on context. In the positive sense, it is sometimes rendered as *scientific claims*, while in the negative, if a paper were to lack *Wissenschaftlichkeit*, it would fail to have a *scientific approach*. Thus this term was rendered variously depending on context.

Word play

One of the issues in translating any language is how to handle plays on words, puns, and the like. While humor may sometimes be lost in translation, the general approach here is to show all of the meanings embedded in a German word or phrase to the extent possible. For instance, because of the way German words are derived, there are often roots in German that can be used variously as nouns, verbs, adjectives, etc., which permit very vivid wordplay using these roots. In every instance I have endeavored to transmit the flavor of the German wordplay. An example from the text is the sentence *Der den Raum oder die Stofflosigkeit überall erfüllende stofflose Stoff, genannt Weltäther, ist ein unbegreiflicher Begriff, und alle Lehren, welche auf ihm beruhen, sind genau ebenso unvollkommen und trügerisch, wie die Grundlage.*[3] Here, the wordplay revolves around the two basic roots *Stoff*

[3] "The immaterial matter filling space or materiallessness everywhere, which is called universal ether, is an inconceivable concept, and all theories that are based on it are precisely as imperfect and deceptive as their basis."

(*matter*) and *Begriff* (*concept*). *Stoff* is represented in this sentence by the words *Stofflosigkeit* (*matter-less-ness or material-less-ness*), *stofflose* (*material-less or immaterial*), and *Stoff* (*matter*). *Begriff*, whose root *greif-/griff-* is related to the English word *grasp*, means *that which is grasped*. In this sentence, it shows up in the phrase *un-begreif-licher Begriff*, literally *un-grasp-able thing that is grasped* – or as I rendered it in English, the *inconceivable concept*. Two other examples also illustrate this point. One is the title of a newspaper article, *Die flüchtige Relativität*, which I have rendered as *The flee(t)ing relativity*. The word *flüchtig* can mean either *fleeing* or *fleeting*, and the word play is, I hope, neatly expressed with the parenthetical letter. The last example involves a visual picture in German that had to be converted for the English audience. The author points out that *Bereits der Begriff »Gleichzeitigkeit« beinhalte die absolute Gleichzeitigkeit, so wie der Begriff Schimmel ein weißes Pferd beinhalte.*[4] The problematic word here is *Schimmel*, which in German means a *white horse*. Since English lacks such a word, an over-literally close translation would read *the term "white horse" contains the idea of a white horse*, which is clearly circular and nonsensical. I therefore chose the word *stallion*, which contains the idea of a male horse in order to preserve both the equine metaphor and the idea of words that include concepts, without confusing the reader.

Another wordplay issue that occurred in a few passages was the *Verdeutschung* (*Germanizing*) of ideas that were expressed in what the German language considers a foreign word, i.e. a word not composed of German linguistic elements. The German word *Relativität* (*relativity*) itself is etymologically foreign to the German language, despite the fact that the word *relativ* is relatively widely used. Thus there was discussion in the press of replacing *relativity theory* with a German compound word – *Bezügigkeitsanschauungstheorie*[5] – which, as a translation of *relativity theory*, could obviously be translated right back into English with that term; but in doing so one would lose the point that the Germans who were coining this term were struggling to represent the idea in comprehensible terms in their own language.

A similar phenomenon occurred with the German word-formation element *empor-* (*upward*), one example being in Willy Schlüter's 1919 book *Empor-Menschlichung (Empormenschlichung). Einführung in das deutsche Tat-Denken, sowie Deutsches Tat-Denken (Tatdenken)*. The challenge here is translating a title (much less a text!) that is both largely incomprehensible and simultaneously plays with the compounding possibilities of the German language. It can be roughly translated as *Upwardly-Humanizing (Upwardlyhumanizing). Introduction to Germanic Deed-Thinking, as well as German Deed-Thinking (Deedthinking)*, which I hope transmits to the reader

[4] "The term 'simultaneity' itself was said to already contain absolute simultaneity, just as the term stallion contains the idea of a male horse."
[5] Rendered in Chapter 2 as "relational observation theory."

the idea that the German title is largely incomprehensible and incoherent.[6] Despite this opacity, however, this terminology also had an influence that was reflected in the press, as quoted on p. 225, where *Empormenschlichungsdeutschtum* is translated as *upwardly humanizing Germanness* (in opposition to *progress Jewry* and *obtrusive Jewry*), while *emporlügen* (literally *upwardly lying*) required a phrasal interpretation: *... to lie ... in order to elevate.*

Conclusion

The act of translating a book is a complex undertaking, comprising interaction with the text, the subject matter of the text, the foreign culture, the translation brief, and the translator's interpretation of the text in the act of embedding it into the context of other English language texts. In this case, fortunately, it was also possible to interact with the author, which I believe has led to a translation that is clear, readable, and useful to its English language readers. It now goes forth from the translator to find its way in the world on its own. May it be as effective in English as it was in German.

Kent, Ohio
July, 2013

Bibliography

American Translators Association. (2011). *Rubric for Grading. Version 2011*. www.atanet.org/certification/aboutexams_rubric.pdf.

American Translators Association. (2013). *About Us: History*. www.atanet.org/aboutus/history.php.

Baker, Mona. (2006). *Translation and conflict: a narrative account*. London & New York: Routledge.

Bassnett-McGuire, Susan. (2002). *Translation Studies*. 3rd edn. London & New York: Routledge.

Bourdieu, Pierre. (1999). *Language and Symbolic Power*, ed. John Thompson, trans. Gino Raymond and Matthew Adamson. Cambridge, MA: Harvard University Press.

Gale Cengage Learning. (2013). *Testaments to the Holocaust: Propaganda Materials*. www.galeuk.com/testaments/pdfs/Testaments%20to%20the%20Holocaust%20Content%20Propaganda%20Materials.pdf.

Goethe, Johann Wolfgang von. (2007 [1821]). *Faustus*, trans. Samuel Taylor Coleridge; edited by Frederick Burwick, and James C. McKusick. Oxford: Oxford University Press.

[6] Graf 2008, p. 279 found him similarly opaque: "In unverständlichen, von höchstem Identifizierungsrausch geprägten Texten beschwor Schlüter den Gedanken, dass alles Denken Tat und jede Tat Denken sei. Die Verbreitung seiner Lehre sollte letztlich zu einer 'Emportüchtigung', 'Emporgeistigung' und 'Empormenschlichung' der Massen führen." ["In incomprehensible texts characterized by the greatest possible intoxication of identification, Schlüter conjured up the thought that all thinking would be deeds and that every deed would be thinking. The dissemination of his theory was ultimately intended to lead to an 'upward competencizing', 'upward intellectualizing' and 'upward humanizing' of the masses."]

Goethe, Johann Wolfgang von. (1856). *Faust: a tragedy*, trans. and notes by Charles T. Brooks. Boston: Ticknor and Fields.

Goethe, Johann Wolfgang von. (2005 [1890]). *Faust*, trans. Bayard Taylor. State College, PA: Penn State Electronic Classics Series. www2.hn.psu.edu/faculty/jmanis/goethe/Goethe-Faust.pdf

Goethe, Johann Wolfgang von. (1932). *Faust*, trans. George Madison Priest. New York: Covici, Friede.

Graf, Rüdiger. (2008). *Die Zukunft der Weimarer Republik: Krisen und Zukunftsaneignungen in Deutschland, 1918–1933 [The future of the Weimar Republic: Crises and Appropriations of the Future in Germany, 1918–1933]*. Munich: Oldenbourg.

Newton, Isaac. (1729). *The Mathematical Principles of Natural Philosophy*, trans. Andrew Motte. www.en.wikisource.org/wiki/The_Mathematical_Principles_of_Natural_Philosophy_(1729)

Nord, Christiane. (1991). *Text analysis in translation: Theory, methodology, and didactic application of a model for translation-oriented text analysis*, 1st edn. (C. Nord & P. Sparrow, trans.). Amsterdam: Rodopi.

Schiller, Friedrich. (1796). *Fiesco; or the Genoese Conspiracy: a Tragedy*, trans. Georg Heinrich Noehden and John Stoddart.

Schleiermacher, Friedrich. (2004). On the different methods of translating, trans. S. Bernofsky, in *The Translation Studies Reader*, ed. Lawrence Venuti, 2nd edn. London/New York: Routledge, 44–63.

Schlüter, Willy. (1919). *Empor-Menschlichung (Empormenschlichung). Einführung in das deutsche Tat-Denken, sowie Deutsches Tat-Denken (Tatdenken) [Upwardly-Humanizing (Upwardlyhumanizing). Introduction to Germanic Deed-Thinking, as well as German Deed-Thinking (Deedthinking)]*, 1st edn., Dresden Laube.

Utley, Steven. (2004). Babel. *Analog Science Fiction and Fact*, **124**(3), 75–81.

van Dongen, Jeroen. (2010). On Einstein's Opponents, and Other Crackpots. Essay review of "Einsteins Gegner. Die öffentliche Kontroverse um die Relativitätstheorie in den 1920er Jahren" by Milena Wazeck. *Studies in History and Philosophy of Modern Physics*, **41**, 78–80.

Venuti, Lawrence. (2008). *The Translator's Invisibility*, 2nd edn. London/New York: Routledge.

Waibl, Elmar. (2011). *Dictionary of Philosophical Terms, German–English, English–German/Wörterbuch philosophischer Fachbegriffe, Deutsch–Englisch, Englisch–Deutsch*. Vienna: Facultas. Also available online: www.uibk.ac.at/philosophie/institut/mitarbeiter/waibl_ordner/philwoeb/philwb.htm.

Preface

I spent the summer of 2002 in an archive room with no windows. Stored there, in moving boxes and banana crates, were the papers of the physicist and Einstein opponent Ernst Gehrcke, which had just been acquired by the Max Planck Institute for the History of Science (whose German abbreviation is MPIWG). My initial sorting of the material was followed by countless weekends of reading. Fiery pamphlets against the theory of relativity accompanied by agitative correspondence and solemnly worded alternative universal theories set against modern physics cast light in this small room on a shadowy world at the beginning of the twentieth century. Who were all these people who protested so vehemently against Einstein's theory? What provoked them to consider one of the most important physics theories of the twentieth century to be one of the greatest threats of their time? In order to answer these questions, this book brings together two strands of research where, until now, few connections have been apparent – the history of the popularization of science and the history of modern physics. It will show that the critics of the theory of relativity came from a non-academic social background whose roots extend farther back than the 1920s – in fact their roots developed in the course of the expansion of the sciences and their popularization in the nineteenth century.

In order to analyze the historical phenomenon of non-academic criticism of the theory of relativity, a discourse analysis approach was applied to an area that has hardly been touched by historical discourse analysis, namely the history of modern physics. The heated public controversy about Einstein's theory of relativity is understood here as a discourse of marginalized knowledge and thus subjected to a new analytical perspective that is, I hope, instructive to both the history of science and the sociology of knowledge.

This book developed from an interest in the role of science for the concept of reality in modern societies that I developed during my studies of political science. It is thanks to Jürgen Renn that I received the intellectual and institutional flexibility that was necessary for this work. Our collaboration in the context of various research projects in his department at the Max Planck Institute for the History of Science encouraged me to understand the history of knowledge as a comprehensive research field that must also deal with bodies of knowledge that extend beyond established academic sciences. The linking of source-oriented research with theoretical reflection, as is practiced in these research projects, decisively shaped my approach to research questions.

The present work was accepted as a dissertation by the Faculty of Humanities (I) at Humboldt University of Berlin, and has been slightly revised for publication. I would like to thank Jürgen Renn and Rüdiger vom Bruch for supervising this work as first and second readers; my thanks also go to Gabriele Metzler for her stimulating third reader's report. I must also thank my colleagues at the Max Planck Institute for the History of Science, Katja Bödeker, Jochen Büttner, Peter Damerow, Lindy Divarci, Carmen Hammer, Dieter Hoffmann, Christoph Lehner, Simone Rieger, Matthias Schemmel, Matteo Valleriani, and many others, for countless helpful, encouraging, and stimulating conversations. Safia Azzouni, Katja Bödeker, Dieter Hoffmann and Matthias Schemmel were careful readers of earlier versions of this work or portions thereof. I always benefitted from their critical comments and friendly pointers. I would like to express my particular thanks to Matthias Schemmel for always being willing to clarify physics issues.

This book refers to source material that has been newly discovered in some cases. The support of the library at the Max Planck Institute for the History of Science under its director, Urs Schoepflin, made many things easier and some things possible in the first place; my heartfelt thanks go to Bianca Henn for her archival cataloging of Gehrcke's papers. Additionally, I would like to thank Ann Kenne of the Department of Special Collections of the University of St. Thomas for her help in providing archival material.

Many thanks to Simon Capelin and his team from Cambridge University Press for their help producing this book. Thanks also go to Geoffrey S. Koby for his excellent and fast translation of the manuscript from the German. The translation of this book, originally published in German in 2009, was funded by Geisteswissenschaften International – Translation Funding for Work in the Humanities and Social Sciences. I am very grateful to have received this grant. The manuscript was slightly revised for the English edition.

This book is dedicated to my sister Jana and my father Jürgen. Thank you for always being there for me.

New York,
July 2013

Abbreviations

AoN	Academy of Nations
DGTP	Deutsche Gesellschaft für technische Physik [German Society for Technical Physics]
DGWAP	Deutsche Gesellschaft für Weltätherforschung und anschauliche Physik (DGWAP) [German Society for Universal Ether Research and Comprehensible Physics]
DMB	Deutscher Monistenbund [German Monist League]
DPG	Deutsche Physikalische Gesellschaft [German Physical Society]
GDNÄ	Gesellschaft Deutscher Naturforscher und Ärzte [Society of German Natural Scientists and Physicians]
GN	Ernst Gehrcke Papers, Archival Collection, Max Planck Institute for the History of Science, Berlin
MIT	Massachusetts Institute of Technology
MPIWG	Max-Planck-Institut für Wissenschaftsgeschichte [Max Planck Institute for the History of Science]
RP	Arvid Reuterdahl Papers, Department of Special Collections, University of St. Thomas, St. Paul, MN
SNG	Schweizerische Naturforschende Gesellschaft [Swiss Natural Science Society]

Introduction

Prague, October 1913. Oskar Kraus, an associate professor of philosophy at the German University, sends alarming letters to Ernst Gehrcke, a physicist at the Reich Institute of Physics and Technology in Berlin:

People are suffering from extreme fatigue, and an irritability that is due not least to the absurd theories of the relativists. I have a burning desire to see the *source of error* revealed for all of the absurdities that you yourself, honored sir, have accurately characterized. I also see that you have already revealed internal contradictions and absurd consequences multiple times. But where is the *source of error*? Because despite my calculation errors, I am still able to recognize the fact that the theory of relativity is false.[1]

Kraus excuses himself for literally bombarding Gehrcke with letters in the span of only a few days, in which one letter often revokes the statements made in the preceding one, but he confesses in his despair, "[I] would not know […] anyone else but you who as a specialist would not reject the intervention of a philosopher from the start."[2] Oskar Kraus was not an isolated case. A large number of people who were just as disturbed – including some philosophers and physicists, but many more scientific laypersons – turned to Gehrcke, who had taken a position early on as an opponent of the theory of relativity. Gehrcke's papers include a large number of letters whose authors critically discuss modern physics, which, in addition to the content of the correspondence itself, is also expressed in the pamphlets against the theory of relativity that are often enclosed.[3] Two things stand out about these pamphlets and

1 Kraus to Gehrcke, October 11, 1913, GN 72-A-2. In the German original of this book, spelling, punctuation, and emphases in the source texts were left as in the original. In the translation, as in the original, emphases are uniformly shown in italics. Comments and supplements are shown in square brackets.

2 Kraus to Gehrcke, October 12, 1913, GN 72-A-3.

3 For a quantitative overview of the publications on the theory of relativity in the German-speaking area, see Goenner, Hubert. (1992). The reception of the theory of relativity in Germany as reflected by books published between 1908 and 1945. In *Studies in the History of General Relativity*, ed. Jean Eisenstaed and Anne J. Kox. Boston [*et al.*] (Einstein Studies 3), 15–38.

their accompanying correspondence: first, the vehemence and the fundamental approach that was used in discussing the theory of relativity, and second, the vigor with which the authors – physicians and lawyers, but above all engineers, that is, citizens pursuing successful careers who had never before been heard of in scientific circles – pointed out the scientific nature of their refutation of the theory of relativity. This fact allows us to derive the following three central questions:

- Why did a highly abstract theory such as the theory of relativity, which was not directly relevant to everyday life, provoke such an intense reaction in many scientific laypeople?
- What arguments did they raise against modern physics? Which scientific concepts and knowledge were appealed to?
- Why did an academically respected physicist such as Ernst Gehrcke allow himself to become involved with this non-academic opposition to the theory of relativity?

In this book, I would like to provide answers to these questions and, by doing so, focus on aspects of the controversy surrounding the theory of relativity that have received little attention until now.

The current status of research

The controversial reception of the theory of relativity in the public sphere of the 1920s is a topic that touches on many areas and is thus mentioned correspondingly often in the secondary literature. The spectrum of focus areas already discussed is broad, ranging from the question of the possibilities and limits of popular communication of the theory of relativity,[4] through the adaptation of the theory of relativity in art and literature or the relationships between modern physics and modern art,[5] to the political attacks, particularly nationalistic and anti-Semitically motivated attacks on the theory of relativity as "Jewish physics" and on Einstein as a person.[6]

[4] Cf. particularly the differentiation among "primary literature" down to "quaternary literature" in Hentschel, Klaus. (1990). *Interpretationen und Fehlinterpretationen der speziellen und der allgemeinen Relativitätstheorie durch Zeitgenossen Albert Einsteins [Interpretations and Misinterpretations of the Special and General Theory of Relativity by Albert Einstein's Contemporaries]*. Basel [*et al.*], pp. 55ff.

[5] Cf. particularly Donley, Carol C. and Friedman, Alan J. (1985). *Einstein as Myth and Muse*. Cambridge [*et al.*]; Könneker, Carsten. (2001). *Auflösung der Natur, Auflösung der Geschichte. Moderner Roman und NS-›Weltanschauung‹ im Zeichen der theoretischen Physik [Dissolving Nature, Dissolving History. The Modern Novel and the National Socialist 'World View' Under the Sign of Theoretical Physics]*. Stuttgart/Weimar; Miller, Arthur I. (2001). *Einstein, Picasso. Space, Time, and the Beauty that Causes Havoc*. New York, NY.

[6] Cf. Goenner, Hubert. (1993a). The reaction to relativity theory I: The anti-Einstein campaign in Germany in 1920. *Science in Context*, No. 6, 107–13; Hermann, Armin. (1994). *Einstein: der Weltweise und sein Jahrhundert. Eine Biographie [Einstein: The Wise Man of the World and his Century. A Biography]*. Munich [*et al.*], pp. 238ff.; Rowe, David. (2002). Editorial Note: Einstein's encounters with German anti-relativists. In *The Collected Papers of Albert Einstein*, Vol. 7: *The Berlin Years: Writings, 1918–1921*, ed. Michel Janssen *et al.* (2002a). Princeton,

The public criticism by physicists[7] and philosophers[8] was also investigated as part of the public controversy[9] about the theory of relativity.

In the biographically oriented Einstein research, as well as Einstein research focusing on cultural and political contexts, Einstein's opponents of the 1920s, that is, those people who rejected either the special or the general theory of relativity (or both), usually appear in connection with a few spectacular events. Attention has focused above all on the public series of anti-Einstein events organized by the anti-Semitic agitator Paul Weyland. Along with Weyland himself, Ernst Gehrcke also participated as a speaker at the opening event on August 24, 1920, at the Berlin Philharmonic.[10] The discussion between Einstein and Nobel Prize winner Philipp Lenard, possibly the best-known of Einstein's opponents, at the annual meeting of the Society of German Natural Scientists and Physicians (GDNÄ) in Bad Nauheim in fall of the same year, has attracted just as much attention.[11] The public protest by

101–13. On "Jewish physics," cf. particularly Beyerchen, Alan D. (1977). *Scientists Under Hitler: Politics and the Physics Community in the Third Reich*, New Haven, and Litten, Freddy. (2000). *Mechanik und Antisemitismus. Wilhelm Müller (1880–1968) [Mechanics and Anti-Semitism. Wilhelm Müller (1880–1968)]*. Munich.

7 Cf. this chapter, Note 6, as well as Hentschel, 1990; Kleinert, Andreas. (2005). Philipp Lenard and Johannes Stark: Two Nobel laureates against Einstein. In *Albert Einstein. Chief Engineer of the Universe*, ed. Jürgen Renn (2005a), 3 vols. Weinheim, Vol. 1: *One Hundred Authors for Einstein*, 226–29; Schönbeck, Charlotte. (2000). *Albert Einstein und Philipp Lenard: Antipoden im Spannungsfeld von Physik und Zeitgeschichte [Albert Einstein und Philipp Lenard: Adversaries in the Conflict of Physics and Contemporary History]*. Berlin [*et al.*] (Schriften der Mathematisch-Naturwissenschaftlichen Klasse der Heidelberger Akademie der Wissenschaften No. 8); Rowe, David. (2006). Einstein's allies and enemies: Debating relativity in Germany, 1916–1920. In *Interactions: Mathematics, Physics and Philosophy, 1860–1930*, ed. Vincent F. Hendricks *et al.* Dordrecht (Boston Studies in the Philosophy of Science **251**), 231–80.

8 Cf. Hentschel 1990, for a comprehensive treatment.

9 The reception of the theory of relativity that occurred primarily within a scientific community is not the topic of this study. On that topic, cf. particularly Staley, Richard (2008b). *Einstein's Generation: The Origins of the Relativity Revolution*. Chicago and the essays in Glick, Thomas F. (ed.). (1987). *The Comparative Reception of Relativity*. Dordrecht [*et al.*] (Boston Studies in the Philosophy of Science **103**). On the reception of the general theory of relativity, cf. particularly the contributions in Howard, Don and Stachel, John (eds.). (1989). *Einstein and the History of General Relativity*. Boston [*et al.*] (Einstein Studies **1**) and in Eisenstaed, Jean and Kox, Anne (eds.). (1988). *Studies in the History of General Relativity*. Boston [*et al.*] (Einstein Studies **3**). On the reception among astronomers, cf. Crelinsten, Jeffrey. (2006). *Einstein's Jury: The Race to Test Relativity*. Princeton [*et al.*].

10 Cf. van Dongen, Jeroen. (2007). Reactionaries and Einstein's fame: "German Scientists for the Preservation of Pure Science," relativity and the Bad Nauheim conference. *Physics in Perspective*, **9**, 212–30; Fölsing, Albrecht. (1997). *Albert Einstein. A Biography*. New York, NY, 460–65; Goenner 1993a; ibid. (2005). *Einstein in Berlin, 1914–1933*. Munich, 179–85; Grundmann, Siegfried. (1967). Das moralische Antlitz der Anti-Einstein-Liga [The moral countenance of the Anti-Einstein-League]. *Wissenschaftliche Zeitschrift der TU Dresden [Scientific Journal of the Technical University of Dresden]*, **16**(5), 1623–26; ibid. (2005). *The Einstein Dossiers: Science and Politics – Einstein's Berlin Period*. Berlin [*et al.*], 98–106; Hermann, Armin. (1977). Der Kampf um die Relativitätstheorie [The battle about the theory of relativity]. *Bild der Wissenschaft* 109–16; Hermann 1994, 240–47; Kleinert, Andreas. (1993). Paul Weyland, der Berliner Einstein-Töter [Paul Weyland, the Einstein Killer of Berlin]. In *Naturwissenschaft und Technik in der Geschichte. 25 Jahre Lehrstuhl für Geschichte der Naturwissenschaft und Technik am Historischen Institut der Universität Stuttgart [Science and Technology in History. 25 Years of a Chair for the History of Science and Technology at the Historical Institute of the University of Stuttgart]*, ed. Helmuth Albrecht. Stuttgart, 198–232; Rowe 1993, pp. 105–8; Rowe 2006, pp. 251–57.

11 Cf. van Dongen 2007; Fölsing 1997, pp. 466–68; Goenner 2005, pp. 185–88; Hermann 1994, pp. 247–49; Kleinert, Andreas and Schönbeck, Charlotte. (1978). Lenard und Einstein. Ihr Briefwechsel und ihr Verhältnis vor der Nauheimer Diskussion von 1920 [Lenard and Einstein. Their correspondence and their relationship before the Nauheim discussion of 1920]. *Gesnerus*, **35**(3–4), 318–33; Rowe 1993, pp. 108–11; Rowe 2006, pp. 257–63; Schönbeck 2000.

Einstein's opponents at the annual meeting of the GDNÄ in Leipzig in 1922,[12] and the appearance of the collection *100 Autoren gegen Einstein [100 Authors Against Einstein]* in 1931,[13] have also been investigated.

The research to date on the opponents of the theory of relativity is characterized by a strong focus on individual protagonists, particularly Lenard, Stark, Gehrcke, and Weyland, and specific events, particularly the presentations at the Philharmonic and the dispute in Bad Nauheim. In addition, this phenomenon is discussed primarily from the perspective of what it meant for Einstein to be confronted with attacks on his science and on himself as a person. In this book, however, the central question is what it meant for the persons who classified themselves as Einstein's opponents to be confronted with the theory of relativity. A new perspective arises when the countermovement to the theory of relativity is not investigated as a movement that originated in the existence of the theory of relativity or due to Einstein. Instead, a broader historical framework is opened up that takes the social contexts of the opponents of modern physics into account and deals with the dynamics of the dispute about the theory of relativity.

Instead of approaching Einstein's opponents with fixed categories and assigning their motives and arguments to science-related (content-based) categories on the one hand and non-science-related (anti-Semitic, nationalistic) categories on the other, as is particularly the case in works by Hubert Goenner and Klaus Hentschel,[14] the analysis presented here intends to make comprehensible the manner in which the dispute about the content of science includes genuine political dimensions and is subject to processes of politicization that develop their own momentum. The significance of the presentations at the Philharmonic and the discussion in Bad Nauheim, which are practically synonymous with the opposition to the theory of relativity in previous Einstein research, becomes relative in light of the reconstruction of a broader and longer-lasting countermovement that even took on institutionalized forms. The existence of this institutionalized network of Einstein opponents has been overlooked by the Einstein research until now. This is due not only to the availability of sources, which has only improved in the past few years (cf. pp. 9ff.), but also particularly to the

[12] Cf. Fölsing 1997, pp. 523–24; Goenner 2005, pp. 188–93; Hermann 1994, pp. 279–83; Wazeck, Milena. (2005). 'Einstein on the murder list!' The attacks on Einstein and the theory of relativity in 1922. In Renn 2005a, 222–25.

[13] Cf. particularly Goenner, Hubert. (1993b). The reaction to relativity theory in Germany III: "A hundred authors against Einstein." In *The Attraction of Gravitation: New Studies in the History of General Relativity*, ed. John Earman, Michel Janssen, and John Norton. Boston [*et al.*] (Einstein Studies **5**), 248–73.

[14] Cf. Hentschel 1990, pp. 74ff.; Goenner 1993a; Goenner 1993b; Rowe 2006, pp. 263ff. is also critical of this approach.

one-sided orientation of the research about the opponents of the theory of relativity that has just been outlined.

Investigations of the controversial reception of the theory of relativity always raise the question of why the theory of relativity was able to achieve a public effect and polarization in the Weimar Republic that was unknown for a scientific theory up until then. Many positions on this issue focus on different areas falling in the continuum between the view that Einstein was a colorful personality, politically polarizing, and a media product, and the finding that the confirmation of the theory of relativity at the 1919 solar eclipse was the starting gun for reception of the theory of relativity in the public that developed its own momentum.[15] However, the "public" in these cases is usually a diffuse public about which little is known.

Newer work, on the other hand, emphasizes the fact that explanations of the unique popularity of Einstein and the theory of relativity must include multiple factors: from Einstein's personality and political involvement to the fascination with a new "great theory" and its incompatibility with everyday concepts of space and time, from the ideological environment and the specific cultural and political contexts of the Weimar Republic to the role of the popular press, which had just come into being.[16]

In *Einstein and Our World*[17], David Cassidy emphasizes the heterogeneity of the public response to the theory of relativity, pointing out that the reception of the theory of relativity in the public was both conditioned by the incompatibility of modern physics with the understanding of science that was widespread in the population, and related to the role of physics as a substitute world view in the post-war period. In *Einstein and Our World*, Cassidy covers a broad range of reception types and therefore only deals marginally with the adverse public reception that is the focus of this book.

In his book *Auflösung der Natur, Auflösung der Geschichte [Dissolving Nature, Dissolving History]*, Carsten Könneker has situated the Einstein controversy in the cultural context of the Weimar Republic and particularly demonstrated the connection with a broader debate about relativity linked to Spengler and Nietzsche.[18]

15 Cf. e.g. Elton, Lewis. (1986). Einstein, general relativity, and the German press, 1919–1920. *Isis*, **77**, 95–103; Fischer, Ernst Peter. (1996). *Einstein. Ein Genie und sein überfordertes Publikum [Einstein. A Genius and his Overburdened Public]*. Berlin [*et al.*]; Fölsing 1997; Pais, Abraham. (1994). *Einstein lived here*. Oxford [*et al.*].

16 Pais 1994, pp. 194f. also calls for this, although he resorts primarily to Einstein's (political) biography and anecdotes about Einstein.

17 Cassidy, David. (2004). *Einstein and Our World*, 2nd edn. New York, particularly pp. 93–110.

18 Paul Forman had already asserted a connection between the cultural environment and modern physics with regard to the development of quantum mechanics. Cf. Forman, Paul. (1971). Weimar culture, causality, and quantum theory: adaption by German physicists and mathematicians to a hostile environment. *Historical Studies in the Physical Sciences*, **3**, 1–115 and ibid. (1984). Kausalität, Anschaulichkeit and Individualität: How cultural

Könneker's goal is "to include the development of modern physics and its extensive tendentious vulgarization in the debate about the background of the origin and rise of National Socialism."[19] This approach leads Könneker to conclusions and assignments of guilt that must be evaluated critically; for instance in his concluding remarks on "the guilt of physics": "With his contributions to the development of modern physics, and as a highly controversial person politically, Einstein had made a decisive contribution to the increasing hardening of the ideological fronts in the Weimar Republic or to the fact that they developed at all in the form observed."[20]

In contrast, the assumption in this book is that the debate about the theory of relativity originated neither in the development of modern physics nor in Einstein himself and his political statements.[21] I rather argue that the ideological use and politicization of the theory of relativity must be interpreted instead as a phenomenon of specific sociopolitical and epistemological contexts of the long turn of the century.

The popular objections to the theory of relativity have been examined particularly by Goenner[22], and most extensively by Hentschel. In his comprehensive work on the controversial reception of the theory of relativity, *Interpretationen und Fehlinterpretationen der speziellen und der allgemeinen Relativitätstheorie durch Zeitgenossen Albert Einsteins [Interpretations and Misinterpretations of the Special and General Theory of Relativity by Albert Einstein's Contemporaries]*,[23] Hentschel investigates the popular literature from the aspect of (successful) popularization on the one hand and vulgarization (leading to misunderstandings) on the other. Using this perspective, he attributes the popular content-based objections in the opposing publications to incorrect interpretations, arrogance, or inadequate examination of the serious popular literature. However, Hentschel's primary interest is not the contexts of popular criticism, but rather the philosophical reception of the theory of relativity and the development of a standard for the adequacy of philosophical interpretations of the theory of relativity. Therefore, it is only in passing that he takes note of a phenomenon

values prescribed the character and lessons ascribed to quantum mechanics. In Nico Stehr and Volker Meja (eds.). *Society and Knowledge: Contemporary Perspectives in the Sociology of Knowledge*. New Brunswick, 333–47.

19 Könneker 2001, p. 6. 20 Ibid., p. 359.

21 Cf. Könneker's contrasting assessment: "The reason that the tone of the discussion between advocates and opponents [...] became increasingly aggravated was due to the public appearances by Einstein, who made no secret of his sympathies for pacifists, leftists, and Zionists, and also to the vulgarized content of the theory itself." Ibid., p. 3. Here, however, Könneker does not ask the decisive question of the political and social context that would make it possible for the political convictions of a scientist such as Einstein to achieve this public reaction in any way.

22 Goenner 1993a. 23 Hentschel 1990.

that is placed in the center of my work: "What is actually astonishing about the innumerable 'elementary refutations' of the T[heory of]R[elativity] in the popular literature about the T[heory of]R[elativity] in the Twenties is the insolence with which lesser and least intellects meddle in the affairs of the best mathematicians of their time."[24]

This raises the question of where this self-assurance came from, which appears as "insolence" to a modern historian of science. "Boundless overestimation of ability"[25] as a motivation for this group of people is unsatisfactory as an explanation for a historical phenomenon. Or does this actually represent a pathological development that would rather be the domain of psychology than the history of science? There are good reasons to deny this. Instead, the fact that the examination of the theory of relativity in the pamphlets took place based on a level of knowledge that was usually below the level of a graduate physicist and certainly below the level of a mathematically trained theoretical physicist – which often resulted in quite unconventional interpretations of the theory of relativity – indicates that it is necessary to deal with the knowledge content and conceptions of science of these scientific laypeople and to place oneself at the level of "what people know."[26] Here, the concept of "what people know" does not refer to what is called common sense,[27] but rather "a whole series of knowledges that have been disqualified as nonconceptual knowledges, as insufficiently elaborated knowledges: naive knowledges, hierarchically inferior knowledges, knowledges that are below the required level of erudition or scientificity."[28]

This book will show that the theory of relativity was fought so vehemently because it threatened other bodies of knowledge or was perceived as such a threat. One can describe the confrontation between incompatible bodies of knowledge existing in parallel as a conflict of paradigms or as a conflict resulting from contradictory styles of thinking.[29] The crucial point is that conflicts of this type are of a fundamental nature and therefore cannot be resolved within the scope of an established science – after all, what is the "right" science and what is the "true" knowledge are precisely what is being disputed. This book shows that these conflicts about the theory of relativity occurred not only in the context of

[24] Ibid., p. 556. [25] Ibid.

[26] Foucault, Michel. (2003). *"Society must be defended": Lectures at the Collège de France, 1975–1976.* New York, p. 7.

[27] The concept of common sense is not used in this book for the simple reason that its content is vague and it is therefore analytically useless, unless one specifies what is called "common sense" as a specific conception within a particular manner of thinking that must be contextualized historically and culturally, which has established itself as the self-evident one and from which a contrast to "nonsensical" thinking is perceived.

[28] Foucault 2003, p. 7.

[29] Cf. Kuhn, Thomas S. (2012 [1962]). *The Structure of Scientific Revolutions*, 4th edn. Chicago; cf. Fleck, Ludwik. (1979 [1935]). *Genesis and Development of a Scientific Fact.* Chicago.

academic science, but also involved non-academic bodies of knowledge. This was not just about a dispute between physicists such as Gehrcke who adhered to classical physics or philosophers such as Kraus who were committed to specific philosophical systems on the one side, and modern physicists on the other; instead, it was specifically about a conflict between representatives of non-academic bodies of knowledge and modern physics. As the first chapter will show, these non-academic researchers were frequently already in conflict with academic science prior to the confrontation with the theory of relativity. They felt forced to defend their bodies of knowledge against modern physics due to the scientific acceptance and the overwhelming presence of the theory of relativity in the public.

From this perspective, I will show that the rupture in knowledge caused by the transition to modern physics ran deeper and differently than previous studies suggest. The criticism in the non-academic sphere in no way represented a genuinely different opposition than that of experimental physicists such as Lenard or Gehrcke, or philosophers such as Oscar Kraus; instead, the opposition of the experimental physicists and the philosophers can be described using the same analytical tools as the rebellion of the non-academic critics – that is, as a counter-discourse of marginalized, disqualified, and subjugated bodies of knowledge against modern physics. This counter-discourse questioned the axioms of modern physics and simultaneously asserted specific demands on what constitutes science. It denied the status of a theory of physics to the theory of relativity and provided niches for devalued knowledge and alternative systems of recognition. This counter-discourse was apparent not only in a content dimension in the narrow sense in the form of the existence of other bodies of knowledge, but also in a strategic dimension – that is, in amalgamations, networks, and counter-measures against the marginalization of these bodies of knowledge.

This book's approach of analyzing the opposition to the theory of relativity as an "insurrection against the centralizing power-effects that are bound up with the institutionalization and workings of any scientific discourse organized in a society such as ours"[30] is determined by a discourse analysis perspective. The following discussion will not deal with the question of the correctness or reasonableness of objections to modern physics, but will rather deal with claims to being scientific that were asserted and fought for, with tactics and strategies of this battle, and with the alliances between academic physicists and philosophers and the non-academic opposition which arose predominantly from a shared feeling of

[30] Foucault, 2003, p. 9.

being threatened and from unexpected commonalities in the understanding of what constitutes a theory of physics.

Structure of the book

This book has four chapters. The first chapter explains that in the context of the popularization of science in the nineteenth century, a space developed for non-academic "free natural science," which had a strained relationship to academic science in terms of the content of its knowledge, but also habitually.

The second chapter shows that these non-academic natural scientists had a mental block against reception of the theory of relativity (as did some physicists and philosophers), and that the theory of relativity was perceived as a threat to other bodies of knowledge. The reaction to this perceived threat was both content-based and strategic.

The third chapter focuses first on the content-based criticism of the theory of relativity, with an emphasis on the alternative bodies of knowledge that constituted the starting point for the criticism. In addition, it shows that the status of the theory of relativity as a theory of physics was attacked based on a conception of science that differed from the conception of science of modern physics.

In the fourth chapter, the social networks and protest campaigns come to the fore with their strategic reaction to the theory of relativity. The danger for various other bodies of knowledge was perceived as so great that it was considered necessary not only to develop argumentation strategies against the theory of relativity, but also to forge unusual alliances and joint campaigns. This is shown in the network against the theory of relativity comprising both academic and non-academic opponents of Einstein. This chapter shows that the content-based criticism of the theory of relativity and the public protest campaigns against it resulted from a marginalization process that had both epistemological and social dimensions in both the non-academic and academic spaces.

Sources

For the most part, this book makes use of sources from Ernst Gehrcke's papers held at the Max Planck Institute for the History of Science that have been developed since 2004. Since Gehrcke functioned as a contact person for many of Einstein's opponents, his papers include a comprehensive collection of pamphlets, manuscripts, and correspondence from both academic and non-academic opponents.

Ernst Gehrcke (1878–1960)

Photograph courtesy of the Max Planck Institute for the History of Science, Archival Collection.

Ernst Gehrcke was born in Berlin on July 1, 1878. In 1897, he commenced his studies of physics, mathematics, and chemistry at the University of Berlin. He attended lectures by van t'Hoff and Planck, among others, but he was most strongly influenced by Emil Warburg, at whose institute he worked from 1899 to 1901 and where he completed his doctorate in 1901. In 1904, he completed his post-doctoral thesis at the University of Berlin, and in 1921 he was appointed there as an honorary professor. Gehrcke spent almost his entire professional life at the Reich Institute of Physics and Technology in Berlin, where he was a member from 1901 to 1946, and director of the optics department from 1926. After 1946, he worked at the University of Jena, and then in the German Office of Weights and Measures. He died in 1960 in Birkenwerder.

Sources: Gehrcke 1901; Gehrcke 1946

Arvid Reuterdahl (1876–1933)

Source: Reuterdahl papers; courtesy of the Department of Special Collections, University of St. Thomas, St. Paul, MN.

Arvid Reuterdahl was born in Karlstadt, Sweden, on February 15, 1876. His family emigrated to the United States in 1882. In 1899, he completed his MA at Brown University and began his professional career as a mathematics and physics teacher at Providence Technical High School. From 1904 to 1905, he was a professor of physics at Colby College in Maine. He later founded a company and worked in Spokane, Washington, as Assistant City Engineer, Water Commissioner, and President of the Board of Public Works. This was followed by professional appointments as a consulting engineer in Boise, Idaho (1913–1915) and Kansas City, Missouri. From 1915 to 1917, he was a professor for theoretical and applied mechanics at the Polytechnical Institute in Kansas City. From 1918 to 1922, Reuterdahl chaired the Department of Engineering and Architecture at the University of St. Thomas in St. Paul. He left this post in 1922 to become independent in the Ramsey Institute of Technology.

Source: RP, Series 1, Box 1

At the beginning of the 1920s, Gehrcke started a comprehensive collection of newspaper articles about Einstein and the theory of relativity. Because clipping services were the primary source, this collection is representative. The collection originally comprised about 5,000 articles, but some folders were lost in World War II. The (roughly) 2,700 articles that were preserved are included in the papers and represent an important source on the public discussion of the theory of relativity.[31] Furthermore, material from the papers of Arvid Reuterdahl, particularly correspondence, was used. Reuterdahl was the driving force behind the creation of an international network of Einstein's opponents in the early 1920s. A few years ago, his papers were presented to the University of St. Thomas, St. Paul, Minnesota, by his descendants and, like Gehrcke's papers, analyzed for the first time with regard to the materials on Reuterdahl's opposition to the theory of relativity for the present book.

The correspondence among Einstein's opponents was a particularly informative source for this investigation. Outside of Gehrcke's and Reuterdahl's papers, the availability of sources on Einstein's non-academic opponents is sketchy; there are hardly any collections of papers and the standard reference works generally do not include any biographical information. The paths of Einstein's non-academic opponents partially crossed with those of the lesser and greater academic luminaries, and they have left traces in their papers. In these cases, it goes without saying that it was impossible to carry out a systematic examination; instead, I limited my efforts to pursuing specific references in individual cases.

The table on p. 13 provides a summary of those opponents of Einstein who are the focus of this book. The individuals who played a particularly significant role in the network of Einstein's opponents are additionally introduced in biographical sketches. In addition, this book includes three case studies on opponents of Einstein who can be assigned to various non-academic contexts. Exemplary case studies were chosen as a method of presentation in order to comprehensibly show with adequate depth not only the connection between non-academic arguments and individual biographies, but also between non-academic arguments and the social and epistemological frameworks of specific world views.

Einstein's opponents were an extremely heterogeneous group that carried on both content-based and social conflicts with modern physics. Depending on the

[31] The newspaper article collection has been completely digitized by the MPIWG; all of the newspaper articles cited in this book with GN as the source can be found online at http://echo.mpiwg-berlin.mpg.de/content/modernphysics/gehrcke/newspaper_articles as part of the Echo Project.

Overview of Einstein's opponents

Name	Dates	Education/Profession	Basis for criticism
Browne, Robert T.	born 1882	Civil servant	Own theory
Fricke, Hermann	1876–1949	Physicist, patent office employee	Own theory
Gehrcke, Ernst	1878–1960	Physicist	Classical physics
Geissler, Kurt	born 1859	Senior primary school teacher	Philosophy
Gilbert, Leo	born 1862	Engineer, journalist	Own theory
Glaser, Ludwig	born 1889	Physicist	Classical physics
Gleich, Gerold von	born 1869	Brigadier general (retired)	Classical physics
Guillaume, Edouard	1881–1959	Physicist, patent office employee	Classical physics
Heidenreich, Eyvind L.	born 1860	Engineer	Own theory
Kleinschrod, Franz	1860–1934	Physician	Own theory
Kraus, Oskar	1872–1942	Philosopher	Philosophy
Kühn, Lenore	1878–1955	Philosopher, journalist	Philosophy
Lenard, Philipp	1862–1947	Physicist	Classical physics
Mewes, Rudolf	born 1858	Engineer	Own theory
Mohorovičić, Stjepan	1890–1980	Teacher, geophysicist	Classical physics
Orthner, Rudolf	1884–1961	Chemist	Own theory
Palagyi, Melchior	1859–1924	Teacher, philosopher	Philosophy
Patschke, Arthur	1865–1934	Engineer	Own theory
Reuterdahl, Arvid	1876–1933	Engineer, physicist	Own theory
Riem, Johannes	born 1868	Astronomer	Classical physics
Ruckhaber, Erich	1876–1956	Businessman, philosopher	Own theory
See, Thomas J. J.	1866–1962	Astronomer	Own theory
Vogtherr, Karl	born 1882	Physician	Philosophy
Weinmann, Rudolf	born 1870	Philosopher, author, actor	Philosophy
Zacharias, Johannes	born 1850	Engineer	Own theory
Ziegler, Johann H.	1857–1936	Chemist	Own theory

type of conflict, Einstein's opponents are therefore classified into various categories in this book. With regard to content-based criticism, a differentiation is thus appropriate among the opponents of Einstein who argue primarily based on classical physics, on philosophy, or on their "own theory."

In contrast, a differentiation by social status is helpful when considering the social marginalization processes. On this level, the relationship between institutionalized academic science and Einstein's opponents was determined less by content-based argumentation and more by education and profession practiced – that is, in terms of whether Einstein's opponents did or did not belong to academic physics or philosophy. Therefore, a distinction is made in this book between academic physicists and academic philosophers, as well as between academic and non-academic opponents of Einstein. Academic training corresponds to a great degree with the nature of the content-based criticism (physicists generally argue in terms of physics, philosophers philosophically, and engineers in the framework of their own mechanical theories, for instance), but the correspondence is not complete (physicists argue based on their own theories, while engineers argue in terms of physics).[32]

[32] Thus Hermann Fricke was a physicist with a doctoral degree, although he did not hold an academic position. However, he falls within the category of "world riddle solvers" based on his argumentation. Brigadier General von Gleich, on the other hand, did not have any academic physics training, yet argued purely on the basis of classical physics, corresponded with physicists, and wanted to have nothing to do with the "world riddle solving" criticism. The categorization undertaken here is intended to describe the primary basis of the content-based criticism and should not be seen as dogmatic; in fact, there were many borderline cases, which are pointed out at the appropriate places in this book.

1
The world riddle solvers

Around the turn of the century, an "independent natural science" movement existed separately from academic science. Its practitioners did not conceive of it as a hobby or as amateur science. This independent natural science movement arose in the course of the popularization of science in the nineteenth century and was a phenomenon of the world views at the turn of the century. The independent researchers shared a holistic conception of science that was incompatible with the academic conception of science. Their attitude was openly anti-academic and they did not aim at coexistence with academic science, but rather at confrontation. It was these independent researchers, these "world riddle solvers," who appeared in the 1920s as non-academic critics of the theory of relativity.

1.1 The dark side of the popularization of science

The nineteenth century is considered the "middle-class century."[1] A middle-class public that had become increasingly self-confident also extended its demand for participation to the area of the sciences. The "age of science"[2] not only brought with it the formation and differentiation of the disciplines[3] and successes in their technical application, but also the need on the part of the public to participate in scientific knowledge. The founding of book clubs, museums, and collections and the organization of natural history walks and lectures were an expression of this interest. Not

[1] A comprehensive overview of middle-class research is provided by the four volumes of *Bildungsbürgertum im 19. Jahrhundert [The Educated Middle Class in the 19th Century]*, Stuttgart 1985–92.

[2] This is the title of a lecture presented by Werner Siemens at the 1886 annual meeting of the German Society of Natural Scientists and Physicians in Berlin. Cf. Siemens, Werner. (1997). Das naturwissenschaftliche Zeitalter [The age of science], reprinted in *Forschung und Fortschritt [Research and Progress]*, ed. Dietrich von Engelhardt, *Festschrift zum 175jährigen Jubiläum der Gesellschaft Deutscher Naturforscher und Ärzte [Festschrift for the 175th Anniversary of the Society of German Scientists and Physicians]*. Stuttgart, 167–74.

[3] Cf. Stichweh, Rudolf. (1984). *Zur Entstehung des modernen Systems wissenschaftlicher Disziplinen. Physik in Deutschland 1740–1890 [The Development of the Modern System of Scientific Disciplines. Physics in Germany 1740–1890]*. Frankfurt (Main).

least of these, a broad spectrum of scientific associations offered varied opportunities for amateur scientific activity to everyone from coleopterists to hobby astronomers.[4]

In his comprehensive study of the popularization of science in nineteenth-century Germany, Andreas Daum not only examines the history of popular science associations and societies and the popular science publication media, but also particularly describes the social spaces that had in part only opened up with the popularization of science – spaces in which various forms of the reception and use of science became possible, and in which various protagonists in the richly populated border zone between "expert" and "layperson" could find and invent their roles, and acquire science.

But the acquisition of science and the transformation of knowledge in various social and epistemological contexts still represent an area of research that has not been examined extensively until now. This applies particularly to a discriminating examination of the public as a non-homogeneous construct in its role as the addressee, consumer, and producer of knowledge.[5] In fact, the public continues largely to be seen as the more passive participant in the popularization of science. This corresponds with the dominant understanding of the relationship between science and the public.[6] This assumes a hierarchy, a rationality gradient, between scientific knowledge and non-scientific knowledge, as Roger Cooter and Stephen Pumfrey express it in their programmatic article "Separate spheres and public places: Reflections on the history of science popularization and science in popular culture": "On this reading, 'popular' knowledge and practice of science had to be a diminished simulacrum – simpler, weaker or distorted in proportion to the distance between the learned and the lay communities."[7]

This understanding of popularization is based on the assumption of a principled separability between scientific knowledge and non-scientific knowledge using specific criteria. In this conception, laypeople are supposed to consume the knowledge prepared and simplified by science. They acquire certain pieces of knowledge and educate themselves, and in return they will grant acceptance and trust to science.

[4] Cf. the table of natural history associations in Germany in Daum, Andreas. (2002). *Wissenschaftspopularisierung im 19. Jahrhundert. Bürgerliche Kultur, naturwissenschaftliche Bildung und die deutsche Öffentlichkeit 1848–1914 [Popularization of Science in the 19th Century. Middle-class Culture, Scientific Education, and the German Public 1848–1914]*. 2nd edn, Munich, pp. 91–95.

[5] Gregory and Miller point out the fuzziness of the lay-expert distinction. Everyone is an expert in a few areas and a layperson in many. But for Gregory and Miller, the layperson is primarily considered as an individual with his or her individually specific personal knowledge, meaning that socio-epistemological contexts and non-academic bodies of knowledge do not play a role here. Cf. Gregory, Jane and Miller, Steve. (1998). *Science in Public: Communication, Culture, and Credibility*. New York, pp. 95–99.

[6] Cf. Hilgartner, Stephen. (1990). The dominant view of popularization: Conceptual problems, political uses. *Social Studies of Science*, **20**(3), 519–39.

[7] Cooter, Roger and Pumfrey, Stephen. (1994). Separate spheres and public places: Reflections on the history of science popularization and science in popular culture. *History of Science*, **32**(3), 237–67 (see p. 240).

According to this model, criticism or opposing standpoints are primarily attributable to irrationality, ignorance, and a lack of insight.[8]

Various studies have shown that this "one-way street model" does not accurately describe the practice of popularization, and emphatically pointed out that the reception of science is an open process since every form of acquisition is simultaneously a form of production.[9] Whether the knowledge is "correct," that is, understood in the sense intended by the popularizer, can be neither foreseen nor entirely controlled. Cooter and Pumfrey emphasize that, "'Popular science' may diverge from 'learned science' not because the latter is poorly understood, but because it is developed by its recipients for different purposes."[10] In addition, they argue for taking the opposition and resistance to science popularization into account as well and thus to move towards a side of popularization that is essentially conflicted and where investigations are largely lacking. Using the example of non-academic criticism of the theory of relativity, this book makes a contribution to the investigation of this conflicted side of the process of science popularization. In the following pages, it will be demonstrated that Einstein's non-academic opponents must in large part be placed on the dark side of the popularization of science. I would first like to clarify what I am designating as the "dark side" of science popularization and which aspect of popularization I am investigating using the non-academic criticism.

In the narrower sense, popularization of science means the planned imparting of science. In the broader sense, popularization of science describes a complex process not just of imparting, but also of acquisition of knowledge. The *popularization* of the theory of relativity was carried out by scientists, professionalized popularizers, and self-appointed popularizers (from teachers to priests) who attempted, more or less successfully, to impart the insights of academic science to the broad public. This popularization of the theory of relativity in the narrower sense, which was reflected in countless generally comprehensible introductions, is not the subject of this book. What is less clear at first glance is that this book does not deal with *amateur science* either. Even though the group of non-academic researchers examined in the present book can certainly be seen as a special group of amateur scientists, it is useful to assign a separate term that emphasizes the specific feature of this group and avoids misunderstandings. Particularly in Anglo-American research, the term "amateur

8 See, e.g., Cooter and Pumfrey 1994 and Hilgartner 1990 on the criticism of this conception of popularization.

9 The following studies must be cited as groundbreaking, without making any claims as to completeness: Allen, David Elliston. (1994). *The Naturalist in Britain*. Princeton; Cooter, Roger. (1984). *The Cultural Meaning of Popular Science: Phrenology and the Organization of Consent in Nineteenth-Century Britain*, Cambridge [*et al.*]; Daum 2002; Kohlstedt, Sally Gregory. (1976). The nineteenth-century amateur tradition: The case of the Boston Society of Natural History. In *Science and Its Public: The Changing Relationship*, eds. Gerald Holton and William A. Blanpied. Dordrecht/Boston, 173–90; Shapin, Steven and Schaffer, Simon. (1985). *Leviathan and the Air-Pump: Hobbes, Boyle, and the Experimental Life*. Princeton.

10 Cooter and Pumfrey 1994, pp. 249–50.

scientist" focuses more strongly on the participatory "Republic of Science"[11] than on the confrontational role of the outsider. In addition, a definition of "amateur" – such as was provided in one of the first studies of amateur science in the German-speaking area in the context of the "Liebhaber und Wissenschaft" [Hobbyists and Science] conference of the German Society for the History of Science – is unsuitable for the questions formulated in this book:

As a first approximation, amateurs (laypersons, hobbyists, dilettantes, autodidacts, independent scholars, non-specialists, outsiders) should [. . .] be understood as those who perform scientific work independently of institutionalized and established science that is accepted into established science. As individuals, they can either remain outside of the scientific profession, be included into it due to their accomplishments, or become the founder of a new science or discipline.[12]

This definition refers to amateurs whose accomplishments were already recognized by science as a scientific achievement. The struggle for recognition of an amateur's accomplishment, which was usually carried out beforehand, is ignored. There is no consideration whatsoever of those who definitely considered themselves scientists but whose struggle for recognition remained unsuccessful. From a perspective that investigates claims to a scientific nature, any examination of the amateur that commences only after official recognition by academic science cannot be productive; the same applies to labeling non-recognized scientific claims as pseudo-science.[13] Concepts such as "deviant sciences"[14] or "non-academic research" are more suitable, although in this case "non-academic" does not refer to formal criteria such as the researcher's employment status, designating instead scientific claims outside academia. Such non-academic claims to a scientific nature can be asserted by widely differing theories and world views. Historical examples include disputed scientific claims made by such theories as mesmerism, phrenology, or Karl von Reichenbach's Odic force. Theories such as the hollow Earth hypothesis, glacial cosmogony, and geocentrism, and not least bodies of knowledge such as homeopathy and anthroposophy that are characterized by

[11] Sheets-Pyenson, Susan. (1985). Popular science periodicals in Paris and London: The emergence of a low scientific culture, 1820–1825. *Annals of Science*, **42**, 549–72 (see p. 562).

[12] Toellner, Richard. (1986). Liebhaber und Wissenschaft. Zur Rolle des Amateurs in der Geschichte der Wissenschaften. Ansprache zur Einführung in das Symposium [Hobbyists and science. The role of the amateur in the history of the sciences. Address introducing the symposium]. *Berichte zur Wissenschaftsgeschichte*, **9**, 137–45 (see p. 143).

[13] The term "pseudo-science" is avoided in this book for the simple reason that it is predominantly used pejoratively. Cf. particularly the contributions in Wallis, Roy. (1979). *On the Margins of Science: The Social Construction of Rejected Knowledge*. Keele (Sociological Review Monograph Bd. 27); Nowotny, Helga. (1979). Science and its critics: Reflections on anti-science. In Nowotny, Helga and Rose, Hilary. *Counter-Movements in the Sciences*. Dordrecht (Sociology of Science Yearbook, Vol. 3), 1–26, particularly p. 5.

[14] Cf. Dolby, R. G. A. (1979). Reflections on deviant science. In Wallis 1979, 9–49; also cf. Ben-Yehuda on the definition of deviant science: "A deviant science is a science that by virtue of its hypotheses or methodology, is regarded by the relevant scientific establishment as deviant." Ben-Yehuda, Nachman. (1985). *Deviance and Moral Boundaries: Witchcraft, the Occult, Science Fiction, Deviant Sciences and Scientists*. Chicago, p. 106.

a world view could (and still can) show a sometimes not inconsiderable following that was (and is) firmly convinced of the scientific nature of their views and who carried on (and still carry on) sometimes vehement arguments with the eminent scientific authorities. The controversy about "intelligent design" could be added to the list as a current example. Granted, in December 2005 a court found that "I[ntelligent] D[esign] is not science"; however, since the ruling was based on specific scientific standards that were contested by the opposing parties, it was unable to end the controversy.[15]

Newer studies avoid identifying amateurs using formal criteria. Daum's pioneering work points out the vagueness of the concept of the scientific amateur, ultimately arguing for a habitus-oriented concept of amateur science:

> A strictly sociological definition of amateur science is [. . .] hardly possible. The type can only be understood as an expression of the process of differentiation of the private sphere of middle-class society and as an undeniable complement to the professionalization of science.
>
> Describing the amateurs as non-academics would be misleading. The nature associations counted numerous graduates of academic courses of study among their members. A strict contrast with the categories of professionalization would mean that Alexander von Humboldt, for instance, would be counted as an amateur.
>
> Amateur science is better defined using categories of cultural behavior. It is primarily oriented to the ideal of self-determined educational activity separate from professional duties. Its goal is the satisfaction of personal interest without necessarily disseminating one's own findings in a scientific discourse. It allows more space for the joy of conviviality and social connection than the performance ethos of a professional researcher.[16]

This concept of the amateur scientist makes it possible to describe amateur science as a complex cultural phenomenon. However, it cannot be applied to the non-academic researchers who appeared on the scene as Einstein's opponents. The non-academic opponents of the theory of relativity urgently wanted their findings to be acknowledged in the academic discussion; more than that, they wanted to fundamentally change this discussion. They did not want reform, but rather a coup d'état. The conviviality and social connections among the non-academic researchers resulted more from their marginalization by "official science"[17] than from their enjoyment of social association life. And perhaps the most telling argument against applying the term "amateur scientist" is that they never understood themselves as amateurs. They were outsiders to academic science who were quite aware of their role as outsiders but considered themselves scientists nevertheless. In contrast to the

[15] The ruling in the case of Kitzmiller, *et al.* vs. Dover Area School District, *et al.*, Case No. 04cv2688 is available online at: http://ncse.com/files/pub/legal/kitzmiller/highlights/2005-12-20_Kitzmiller_decision.pdf. The fact that such rulings cannot end the debate does not mean that they are not useful in order to maintain a certain standard of science in teaching.

[16] Daum 2002, p. 108.

[17] Ziegler, Johann Heinrich. (1908). *Konstitution und Komplementät der Elemente [Constitution and Complementarity of the Elements]*. Bern, p. 4.

academic scientists – "experts," "specialists," "government scientists" – they regarded themselves as the "true natural scientists" and were convinced that they had solved the riddles of the universe. Therefore, this book uses the term *world riddle solver* for those persons who conducted research separate from established science but were not amateurs according to their own self-conception. Instead, they saw themselves as unacknowledged colleagues of the eminent scientific authorities, often disputing that these authorities represented real science, which they instead claimed for their own comprehensive theories.

1.2 The phenomenon of world riddle solving

The view that science should solve "world riddles" became widespread at the end of the nineteenth century, particularly due to Ernst Haeckel's 1899 book *Die Weltraetsel [The Riddle of the Universe]*.[18] However, the term "world riddles" (German *Welträtsel*) had already been used earlier in the nineteenth century for the great questions of science, such as the nature of force and matter or the origin of life. It particularly became known due to Emil du Bois-Reymond's presentation entitled "The Seven World Riddles."[19] What is less well known is the fact that many individuals had already begun working on solving the world riddles at the beginning of the twentieth century, often without any direct connection with Haeckel's world riddles, although monistic views certainly played an important role in non-academic research (see pp. 35ff.). In 1912, Carl Beckenhaupt, a member of an association of non-academic researchers, the Gesellschaft Weltwissen [Society for Universal Knowledge], described the world riddle solvers as a phenomenon:

> It is well known that, in our time, more and more men from practical scientific professions are not merely expressing an extraordinary interest in universal theoretical questions, but also turning it into independent work. Over the course of the past ten years, about 50 documents written by physicians, chemists, engineers, military officers, farmers, and even businessmen have come into our hands, and the actual number certainly amounts to hundreds. These documents either deal with the indefensibility of previously proposed theories, or they seek to establish new principles; any value that these documents may have is not under discussion here, but rather simply the fact that these men, standing in the difficulties of working life, are seized by such dedicated striving for clear natural laws, and then the question of whether this striving for theoretical systems should be granted scientific significance![20]

[18] Haeckel, Ernst, (1901 [1899]). *The Riddle of the Universe at the Close of the Nineteenth Century*. New York [*et al.*].

[19] Cf. Du Bois-Reymond, Emil. (1882). *Über die Grenzen des Naturerkennens. Die sieben Welträthsel. Zwei Vorträge [On the Limits to Understanding Nature. The Seven World Riddles. Two Lectures]*. Leipzig.

[20] Beckenhaupt, Carl. (1914). Über die Notwendigkeit geordneter Ausgleiche der Wissenschaft und Lebenspraxis. Ein Versuch zur Organisation der freien Forschung [On the necessity of an orderly balancing of science and life experience. An attempt at organizing independent research]. Part 1 in *Weltwissen [Universal Knowledge]*, **2**(42), 205–07 (see p. 205).

In fact, it was particularly members of the working middle class with a scientific educational background who began developing their own "universal theories" about the world. These efforts did not relate to individual scientific questions; instead, they represented comprehensive new theories that were often in conflict with the theories accepted by academic science. Among these world riddle solvers, engineers represented by far the largest group.[21] Ernst Gehrcke's papers include a large number of such "universal theoretical" outlines by an extremely wide variety of authors. From the booklet *Kann die Erde erkalten? Die Gestirne als Kraftquelle und die Ursachen der Schwerkraft [Can the Earth Go Cold? The Stars as a Source of Power and the Causes of Gravity]* by a pharmacy owner to the booklet *Gegen die Wahnvorstellung vom heißen Erdinnern [Against the Delusion of a Hot Interior of the Earth]* by a printing company owner; from the *Rätsel des Lebens [Mystery of Life]*, solved by a hydropathic Kneipp physician, to the *Rätsel der Schwerkraft [Mystery of Gravity]*, solved by a senior military medical officer; from the *Universellen Weltformel und ihrer Bedeutung für die wahre Erkenntnis aller Dinge [Universal Unified Theory and its Significance for True Insight into All Things]* by a chemist to the *Lösung der Welträtsel durch das einheitliche Weltgesetz der Kraft [Solution of the World Riddles by the Uniform Universal Law of Force]* by an engineer – new theories were inundating the non-academic scientific market: *Eine neue Sonnentheorie [A New Solar Theory]*, *Die neue Physik [The New Physics]*, *Neue Energien im Sonnensystem [New Energies in the Solar System]*, *Neues vom Weltall [New Information from the Cosmos]*, *Neue Wege zur Lösung der Welträtsel [New Paths to Solving the World Riddles]*, *Eine neue mechanische Universal-Theorie [A New Mechanical Universal Theory]*, *Neue Energetik [New Energetics]* or *Neueste Resultate über Weltkraft und Radialströmung [Latest Results about Universal Force and Radial Current]*, to name only a few titles.[22]

An individual document about an "ethical theory of gravitation" could perhaps still be dismissed as a curiosity. Crates full of new theories of light, ether, and gravitation indicate by their sheer numbers that this is a historical phenomenon. But what do these world riddle solver documents have to do with the theory of relativity? The answer is, a great deal, as a closer look at the publication list of the world riddle solvers shows. It was primarily these authors and the authors of many similar documents who appeared in the 1920s as non-academic opponents of the theory of relativity – and that is one of the central discoveries of this book. The

[21] This can be explained by, among other things, the engineers' understanding of nature, which was strongly influenced by their professional practice of interacting with nature. Cf. Dienel, Hans-Liudger. (1992). *Herrschaft über die Natur? Naturvorstellungen deutscher Ingenieure 1871–1914 [Control Over Nature? German Engineers' Conceptions of Nature, 1871–1914]*. Stuttgart.

[22] Wutke 1909; Goldzier 1906; Kleinschrod 1912; Haedicke 1902; Ziegler 1902; Patschke 1905a; Fricke 1933; Lamberty 1926; Kniepf 1904; Müller 1913; Feigl 1932; Hack 1930; Gilbert 1912; Holz 1904.

Fig. 1.1 The world riddle solvers' own theories: Kleinschrod 1928; Ziegler 1902; Gilbert 1912; Goldzier 1906; Patschke 1905a; Fricke 1919.

chemist announced *Das Ende der sog. Relativitätstheorie [The End of the So-Called Theory of Relativity]* with his universal unified theory in 1923; the engineer announced the *Umsturz der Einsteinschen Relativitätstheorie [Overthrow of the Einsteinian Theory of Relativity]* with his universal law of force in 1920; the editor of *Neue Energetik [New Energetics]* declared *Das Relativitätsprinzip, die jüngste Modenarrheit der Wissenschaft [The Relativity Principle, the Latest Fashionable Folly of Science]* as early as 1914; the Patent Office employee discovered the *Fehler in Einsteins Relativitätstheorie [Error in Einstein's Theory of Relativity]* with his new theory of gravity in 1920; the hydropathic Kneipp physician made a statement on *Das Lebensproblem und das Positivitätsprinzip in Zeit und Raum und das Einsteinsche Relativitätsprinzip in Raum und Zeit [The Problem of Life and the Positivity Principle in Time and Space and the Einsteinian Relativity Principle in Space and Time]* in 1920, as did the senior military medical officer on *Die physikalische Unhaltbarkeit der Relativitätstheorie Einsteins [The Physical Indefensibility of Einstein's Theory of Relativity]* in 1932.[23]

A central point is to examine the theories developed before, alongside, and after this criticism of the theory of relativity and to investigate them for their significance

[23] Ziegler 1923; Patschke 1920; Gilbert 1914; Fricke 1920a; Kleinschrod 1920; Haedicke 1932.

Fig. 1.2 Anti-Einstein writings of the world riddle solvers: Kleinschrod 1920; Goldzier 1921; Gilbert 1914; Ziegler 1923; Fricke 1920a; Patschke 1920.

to the motivation and nature of the attacks on Einstein's theory. This analysis will show whether the criticism of the theory of relativity was only an episode for these non-academic researchers, or whether the theory of relativity represented a catalyst that brought the confrontation between non-academic and academic science to the boiling point.

1.3 Contexts of world riddle solving

World riddle solving was a phenomenon of cultural contexts of the "long turn of the century" from 1880 to 1930. The following section sketches the three central contexts of the phenomenon of world riddle solving; each is illustrated using a case study. In this work, it must remain only a brief overview. A differentiation is made among the contexts of occultism, life reform, and monism. Although intersections do exist between the occultist and monistic movements and the life reform movement, such a differentiation is useful because the forms of social organization of these contexts on the one hand, and their individual conflict zones with academic science on the other, which are the primary focus of this book, differ significantly.

1.3.1 Occultism

The rise of the sciences and the accompanying decline in the significance of the Christian churches in the nineteenth century caused new forms of religiosity to develop. Non-denominational movements arose; in Europe, oriental religions such as Buddhism attracted many intellectuals in particular, while others sought religious fulfillment in turning to pre-Christian "primitive religions" or in occult movements.[24] Occultism is understood here as a generic term for all types of "esoteric lore" that claim to deal with supernatural phenomena that cannot be or have not yet been explained scientifically.[25] Among the various expressions of modern occultism, it was particularly theosophy and spiritism, and in Germany also anthroposophy (which emerged from theosophy) that took up the questions of the meaning of life that had been excluded from science.

The Theosophical Society founded by Helena Blavatsky in the US in 1875 and the Anthroposophical Society founded by Rudolf Steiner out of the German section of the Theosophical Society in 1912/13 established an institutional foundation for these movements and drew large numbers particularly from intellectual circles, including many artists but also scientists. In fact, discoveries such as X-rays and radioactive waves reinforced opinions that there could be additional types of rays that would only be detectable by sensitive persons. The interest in occultism was particularly widespread among French scientists, while in Germany in the 1890s, the astronomer Karl Friedrich Zöllner organized meetings with a medium in order to research the fourth dimension.[26] The occult enjoyed undiminished popularity even in the 1920s. On the one hand, periodical titles such as *Zeitschrift für wissenschaftliche Xenologie [Journal of Scientific Xenology]* bear witness to the occult movements' claims to having a scientific nature; among established scientists, on the other hand, there was a prevailing and open interest in getting to the root cause of

[24] For a brief overview, cf. Ulbricht, Justus H. (1998). Religiosität und Spiritualität [Religiosity and Spirituality]. In *Handbuch der deutschen Reformbewegungen 1880–1933 [Handbook of German Reform Movements, 1880–1933]*, ed. Diethard Kerbs and Jürgen Reulecke, Wuppertal, 495–98.

[25] On the various approaches to defining the occult, cf. Truzzi, Marcello. (1974). Definition and dimensions of the occult: Towards a sociological perspective. In *On the margin of the visible: sociology, the esoteric, and the occult*, ed. Edward A. Tiryakian, New York, NY [*et al.*], 243–55.

[26] On the role of occultism among French scientists, cf. Bensaude-Vincent, Bernadette and Blondel, Christine. (2002). *Des savants face à l'occulte, 1870–1940 [Scholars Face the Occult, 1870–1940]*. Paris. Zöllner describes his research particularly in Zöllner, Karl Friedrich. (1879). *Wissenschaftliche Abhandlungen [Scientific Transactions]*, Vol. 3: *Die transcendentale Physik und die sogenannte Philosophie: eine deutsche Antwort auf eine ›sogenannte wissenschaftliche Frage‹ [Transcendental Physics and So-Called Philosophy: A German Answer to a 'So-Called Scientific Question']*. Leipzig. Cf. also Meinel, Christoph. (1991). *Karl Friedrich Zöllner und die Wissenschaftskultur der Gründerzeit: Eine Fallstudie zur Genese konservativer Zivilisationskritik [Karl Friedrich Zöllner and the Scientific Culture of the Founders' Years: A Case Study on the Genesis of Conservative Criticism of Civilization]*. Berlin (Berliner Beiträge zur Geschichte der Naturwissenschaften und der Technik [Berlin Contributions to the History of Science and Technology], Vol. 13). On the public reception of X-rays, cf. inter alia Hessenbruch, Arne. (2000). Science as public sphere. X-rays between spiritualism and physics. In *Wissenschaft und Öffentlichkeit in Berlin, 1870–1930 [Science and Publicity in Berlin, 1870–1930]*, ed. Constantin Goschler, Stuttgart, 89–126.

allegedly supernatural phenomena.[27] It is true that many occult movements claimed to be sciences,[28] but the majority of their protagonists were not interested in a detailed critique of natural science. Instead, they were generally interested in confronting natural science, which in their understanding was proceeding one-sidedly reductionalistically and materialistically, with a new science that reconciled spirit and religion with science. The conflicts with academic science therefore arose in connection with the question of whether the "supernatural" would be justified as an object of science, and whether the observations and meetings with so-called mediums carried out scientifically (according to the occult understanding) fulfilled generally accepted standards of science.[29] The claims of a scientific approach asserted by the occult movements were based on a new definition of scientific criteria. Fundamental principles of the modern understanding of science, such as reproducibility or publishing of results for which scientific claims are made, were pushed aside by proclaiming that something was "higher knowledge" that was not accessible to everyone.

Individual points of conflict with academic science as to content consisted, above all, of the question of what the fourth dimension was[30] with regard to the nature of radiation. The occult conceptions of radiation referred back to "animal magnetism" (mesmerism) and Reichenbach's "Odic force," both of which were non-academic theoretical developments at the end of the eighteenth and the beginning of the

[27] The "Deutsche Okkultistische Gesellschaft" (D.O.G.) [German Occult Society], among others, devoted itself to this purpose. Cf. the excerpt from the D.O.G. program [leaflet], GN 83-F-18. Even skeptical Einstein could not avoid making himself available as an object of investigation at a graphological meeting. Cf. Ein graphologisches Portrait Einsteins [A graphological portrait of Einstein]. *Vossische Zeitung* newspaper, February 18, 1922, GN 4, XVI, 4r. On the presence of spiritism in Berlin around 1900, cf. Teitel, Corinna. (2000). The culture of knowledge in the metropolis of science. Spiritualism and liberalism in fin-de-siècle Berlin. In Goschler 2000, 127–54.

[28] Cf. e.g. the statement in the occult periodical *Die Übersinnliche Welt. Monatsschrift für wissenschaftliche Begründung des Okkultismus [The Supernatural World. Monthly Journal for Scientific Justification of Occultism]*: "Since occultism is now a branch of natural science [. . .]," Schögel, W. (1922). Philosophie und Okkultismus [Philosophy and occultism]. *Die Übersinnliche Welt [The Supernatural World]*, **2**, 17–25 (see p. 21). Cf. also the statement by the physicist Oliver Lodge: "The obscure communications and strange movements which are now studied or experienced in spiritualistic circles, and which by some are thought to be miraculous or impossible – so impossible that the reports of them usually excite ridicule – will gradually take their place in the orderly scheme of recognised science. Such things as travelling clairvoyance and reciprocal dream experiences may become intelligible." Lodge, Oliver. (1933). *My Philosophy. Representing my Views on the Many Functions of the Ether of Space*. London, p. 238.

[29] Cf. e.g. the questions that were asked by a committee of the Psychological Society of Berlin: "1) Does [. . .] clairvoyance exist? 2) Does telepathy exist [. . .] 3) Does telekinesis exist [. . .] 4) Do materializations of the deceased or other materializations exist [. . .]?" A call to submit experimental proof to the committee appeared in an occult periodical. Kleine Mitteilungen [Brief Announcements]. *Die Übersinnliche Welt [The Supernatural World]*, **3**–4, 1922, p. 61.

[30] The theory of relativity was sometimes seen as confirmation of the existence of the occult fourth dimension and used to justify resumption of the Zöllner investigations. Cf. e.g. Pick, Leopold. (1922). *Die vierte Dimension als Grundlage des transzendentalen Idealismus [The Fourth Dimension as the Foundation of Transcendental Idealism]*. Leipzig, p. 3: "However, a bending of space, which is the prerequisite for Einstein's theory of relativity, is inconceivable without assuming a fourth dimension, so that Zöllner's long-derided hypothesis was finally deemed worthy of a scientific discussion."

nineteenth centuries.[31] Developments in academic physics were reflected in this context to the extent that, for instance, the electromagnetic ether was interpreted in an occult manner in the sense of theosophy, and the discoveries of X-rays and radioactivity were integrated into comprehensive occult theories of radiation. Occult thinking placed physical radiation into completely different contexts than did academic science. Theosophy, for instance, assumed that X-rays, which shine through the body, were "astral light," and its adherents saw radioactive radiation as providing scientific proof of the existence of other "incorporeal" effects that would explain such things as clairvoyance and telepathy "scientifically."[32]

World riddle solvers from the occultism context are characterized by their regard of natural phenomena as supernatural phenomena. In these theories, the existence of a God (or of the Divine) and the existence of a Spirit that was not comprehensible by traditional natural science were considered to be unquestioned facts that were integrated into the line of argumentation. Among Einstein's opponents examined in this book, Arvid Reuterdahl in particular, who was involved in the theistic movement in America, and the chemist Johann Heinrich Ziegler, who will be introduced in an exemplary case study, should be assigned to this group.

Case study: Johann Heinrich Ziegler. The chemist and the primordial light

On New Year's Eve, 1900, Johann Heinrich Ziegler handed a sealed envelope over to his lawyer. It contained a manuscript entitled "The Universal Unified Theory [Die universelle Weltformel]." Who was this world riddle solver who was so convinced of the significance of his theory? And what did the universal unified theory look like? The industrialist Ziegler family was documented to have been in Winterthur since 1500, and had probably resided there since as early as 1300. As factory owners, the Zieglers were members of the mercantile middle class in Switzerland, which comprised a fairly limited group.[33] Johann Heinrich Ziegler initially started a

[31] On mesmerism, cf. Darnton, Robert. (1968). *Mesmerism and the End of the Enlightenment in France*. Cambridge, MA. [*et al.*]; Winter, Alison. (1998). *Mesmerized. Powers of Mind in Victorian Britain*. Chicago; and the contributions in *Franz Anton Mesmer und der Mesmerismus. Wissenschaft, Scharlatanerie, Poesie [Franz Anton Mesmer and Mesmerism. Science, Charlatanry, Poetry]*, ed. Gereon Wolters. Konstanz 1988. No comprehensive study of Odic force has yet been published.

[32] Cf. Hoormann, Anne. (2003). *Lichtspiele. Zur Medienreflexion der Avantgarde in der Weimarer Republik [Moving Pictures. The Media Reflection of the Avantgarde in the Weimar Republic]*. Munich, pp. 41f. and the references there. The positions reported here come from the book written by the founder of modern theosophy, Helena Blavatsky, *Isis Unveiled: A Master-Key to the Mysteries of Ancient and Modern Science and Theology*, Vol. 1: *Science*. New York 1877.

[33] Cf. Tanner, Albert. (1995). *Arbeitsame Patrioten – wohlanständige Damen. Bürgertum und Bürgerlichkeit in der Schweiz 1830–1914 [Industrious Patriots – Respectable Ladies. Bourgeoisie and the Middle-Class Way of Life in Switzerland, 1830–1914]*. Zürich. Tanner states that the number of individuals belonging to the mercantile middle class in Switzerland was just under 30,000. Cf. ibid. p. 47. Ziegler's biographical information comes from various autobiographical remarks in his works as well as from Saager, Adolf. (1930). *Der Winterthurer Naturphilosoph Johann Heinrich Ziegler [Johann Heinrich Ziegler, the Natural Philosopher of Winterthur]*. Winterthur.

Fig. 1.3 Johann Heinrich Ziegler. Source: Reuterdahl papers; courtesy of the Department of Special Collections, University of St. Thomas, St. Paul, MN.

course of business education, dropped out of subsequent university studies of economics, was registered for three semesters at the newly-founded Winterthur College of Technology, then completed a one year voluntary placement at his father's factory without any great motivation, performed his military service, and registered for two semesters in Berlin, where he studied chemistry but primarily enjoyed the social life.

Ziegler's life, which had been quite aimless up to this point, changed when it finally became possible to persuade his father to sell the business. Freed from the burden of the dilapidated factory, Ziegler now began studying chemistry in earnest with Emil Fischer, Carl Graebe, and Adolf von Baeyer in Munich and Erlangen, finishing with a doctoral degree that he received in 1883 at the age of 26. He then worked for several years in industrial research at the companies Geigy in Basel and Poirier in Paris, among others, as well as at the universities of Geneva and Zurich.

A real gold-rush mood had been prevailing in organic chemistry since the middle of the nineteenth century. Chemists working in industry participated financially in the commercial exploitation of syntheses. It was even frequently possible to earn a fortune with them: "One chemist, one synthesis, one mansion" – such a career was not unusual.[34] However, Ziegler was a member of the successor generation and the claims had largely been staked. Nevertheless, he achieved a certain fame in color chemistry.[35] In the 1890s, Ziegler was at the pinnacle of his career. A large German dye factory offered him the directorship of its scientific laboratory, but Ziegler allowed the contract to fall through for reasons that are unclear. The fact that he had already developed a strong interest in philosophical questions in the 1890s may have played a role, as he would have wanted to avoid the commitment that work in an industrial laboratory would have brought. In order to ensure that he could earn a living nevertheless, he founded an antipyrine factory, but had similar difficulties to his father as a businessman. After only three years, in fall 1899, Ziegler's factory had gone bankrupt. The liquidation of the company was still underway when the bankrupt Ziegler became entirely absorbed in theoretical questions in 1900. Within three months, he wrote down the broad outline for his system of universal unified theory.

[34] Cf. Simon, Christian. (1997). Chemie in der Schweiz [Chemistry in Switzerland]. In *Chemie in der Schweiz: Geschichte der Forschung und der Industrie [Chemistry in Switzerland: History of Research and the Industry]*, ed. Thomas Busset, Andrea Rosenbusch, and Christian Simon. Basel, 25–57, particularly pp. 42f.; Studer, Tobias. (1997). Wirtschaftliche Rahmenbedingungen als Determinanten des Berufsbildes des Chemikers [Structural economic factors as determinants for the professional profile of the chemist]. In the same volume. 233–52, particularly p. 240.

[35] In particular, he discovered tartrazine, an artificial yellow azo dye that is known today as E102 (FD&C Yellow 5) and is still used in some countries as a food coloring. Fierz-David gave particular prominence to Ziegler's contributions in his textbook on artificial organic dyes, which, however, is possibly less attributable to Ziegler's actual significance than to the personal acquaintance between him and Fierz-David. Cf. Fierz-David, Hans Eduard. (1926). *Technologie der Textilfasern [Technology of Textile Fibers]*, 3 vols, Vol. 3: *Künstliche organische Farbstoffe [Artifical Organic Dyes]*. Berlin 1926.

Ziegler explains why he tried his hand at being a theoretician at all by looking back to the direct connection between his practical and theoretical work:

As a color chemist, I had already felt an active desire for a more correct understanding of those relationships [between light and matter] years ago in technology and endeavored, like many others, to derive regularities and to discover the laws that govern the nature of the colors [. . .]

The longer I carried on this activity, the more I became conscious of the complete inadequacy of the prevailing views on the nature of matter as well as that of light [. . .] I said to myself that since color, this special form of light, appears as a child of matter and light, and since only similar things can enter into productive interaction according to their natures, then those parents of such apparently so different natures would have to be of one and the same nature, and if one wished to tackle the problem of body color with some prospects for success, one would, above all, have to discern the true unity, or the true character of light and matter.

All at once, the color problem thus transformed itself into the ultimate scientific or philosophical problem, into the problem of the true essence of nature. From that time on, it was synonymous with the universal problem and as such was concerned with nothing less than the knowledge of absolute truth.[36]

In his *Universelle Weltformel [Universal Unified Theory]*, Ziegler developed an ontology based on the premise that the first step would have to be to know about the simplest thing of all, the "primordial thing [*Urding*] itself,"[37] and this was light. Ziegler did not further critically examine the fact that light occupied this exposed position as the primordial thing. He stated that light was the "most obvious" and "simplest" perceivable manifestation of nature and thus sufficiently qualified as the "primordial thing."[38]

Following the two unified theory essays, the color problem ceased to play a central role in Ziegler's writings, but from mundane light Ziegler inferred the existence of primordial light, which gave his primordial light theory its name. Primordial light came into play because in the design of his theory, Ziegler refers to the epistemological premise that oppositeness is the structural principle of the world. Since the things in the world as people know and experience them are permutable, ponderable, and perceptible – like mundane light – there must be a complementary opposite to the mundane thing that is eternal, concealed, supernatural, not mundane, and ultimately religious: the supernatural primordial light. Ziegler derives all physical phenomena from primordial light. In his view, the physical atoms accessible to science, that is, the "normal" atoms that a physicist or chemist deals with, are formed from the primordial light atoms

[36] Ziegler 1902, pp. 6f.

[37] The introduction of the German "Ur-" ("primordial") prefix is typical of the attempts by the world riddle solvers to put science on a metaphysically secure basis (see also pp. 154ff.).

[38] All quotes in this paragraph are from Ziegler 1902, p. 8. A second unified theory essay appeared in 1903, cf. Ziegler 1903.

that are in principle inaccessible to humans. Likewise, forces such as gravitation are seen as merely secondary manifestations of the battle of the opposites.[39]

A unified theory with a claim to universality extends not only to inanimate nature, but also to the living world, and not least to humans as biological and social creatures (cf. pp. 54f.). Ziegler was convinced that he had discovered the essence of the universe in his metaphysical primordial light atoms and spent the next few decades demanding support and recognition for his unified theory from academic science. In the years following 1901, he published more than 20 booklets and articles; his last work, *Das Geheimnis des Selbstverständlichen: der Schlüssel zum Welträtsel [The Secret of the Self-Evident: The Key to the World Riddle]*, appeared in 1935. He died one year later at the age of 79.

1.3.2 Life reform

The materialistic conception of nature that began to be generally accepted in the mid-nineteenth century was faced with competition at the end of that century. In science, new developments in physics such as thermodynamics and electrodynamics led to a revision of the mechanical conception of the world, while in society, materialistic natural science was confronted by a reform movement that viewed it skeptically.[40] Although the life reform movement conceived of itself as apolitical, it manifested itself as a political factor in the German imperial period and the Weimar Republic simply due to its critical approach to culture and civilization. The slogan "back to nature" expressed its longing for comprehensive renewal and its seeking for refuge in "the natural." In fact, this topic had been part of the discourse on the alienation of modernity ever since the Enlightenment,[41] but at the beginning of the twentieth century it was ideologized, religionized, and politicized in a specific way: Nudism and physical culture, organization of leisure time (hiking, cycling, swimming, etc.), home and workplace décor, aversion to cities, and country communes were new forms for appropriation of nature. Yet the nature cult says nothing about "the natural." Instead, it says something about the relationship between humans and nature and about conceptions regarding the boundary lines between culture and

[39] Cf. Ziegler, Johann Heinrich. (1904b). Über das wahre Wesen der sogenannten Schwerkraft [On the true nature of so-called gravity]. In *Die wahre Einheit von Religion und Wissenschaft. Vier Abhandlungen [The True Unity of Religion and Science. Four Treatises]*. Zürich, 60–85, particularly pp. 64f.

[40] Paul Forman attributed later developments in physics, particularly quantum mechanics, in large part to scientists' reactions to the challenges of this scientifically skeptical context. However, he pays too little attention to the fact that science could continue to count on a high level of agreement in society. His study also neglects the internal scientific development processes that led to the development of quantum mechanics. Cf. Forman 1971, 1984.

[41] What must particularly be mentioned here is Rousseau's critique of civilization and his conception of the "natural state." Cf. particularly Rousseau, Jean Jacques. (2009 [1755]). *Discourse on the Origin of Inequality*. Oxford [*et al.*]; ibid. (1993 [1762]) *Émile*. London [*et al.*].

nature. Here, spontaneity, naturalness, and experiencing instead of dissecting are set in opposition to the mechanistic thinking of science.

Points of friction between life reform and academic science primarily came to light in the relationship between naturopathy and orthodox medicine. Naturopathy, which is linked to names such as Priessnitz (1799–1851), Hahn (1824–1883), Schroth (1798–1856), and Kneipp (1821–1897), had its roots in folk medicine, but only achieved the status of a specific social movement when it became organized and associations were founded in the mid-nineteenth century. As such, it represented an important part of, and indeed the origin of, the life reform movement,[42] sharing its theoretical points of reference, particularly the proximity to the natural philosophy of the Romantic period and to the more recent cultural and life philosophy. This heterogeneous movement united an affinity for natural life, rejection of orthodox medicine, and a turn towards natural therapies such as light, air, or water therapy, physiotherapy, and diet. Estimates assume ten million organized members of – and sympathizers with – life reform in Germany around 1930.[43]

As part of the life reform movement, naturopathy must be understood as a movement that comprised a strong social organization along with an epistemological level, and whose topic areas extended far beyond therapeutic approaches and medical issues. When considering approaches to healing an illness, naturopathy's holistic approach in particular made it necessary to take into account not just the diseased body part, but rather the entire person and his or her social circumstances, including the areas of housing, nutrition, clothing, "physical discipline," spiritual welfare, and spirituality.[44] The significance of this ideological factor becomes apparent, for instance, in the fact that there was resistance in the naturopathy movement to orthodox medicine merely accepting the therapeutic possibilities of alternative medicine – as long as orthodox medicine did not share the accompanying theoretical framework.[45]

[42] Cf. e.g. Rothschuh, Karl. (1983). *Naturheilbewegung, Reformbewegung, Alternativbewegung [Naturopathic Movement, Reform Movement, Alternative Movement]*. Stuttgart.

[43] Cf. Bothe, Detlef. (1991). *Neue Deutsche Heilkunde 1933–1945 [New German Medicine 1933–1945]*. Husum, p. 26.

[44] On the general topic of life reform, cf. Barlösius, Efa. (1997). *Naturgemäße Lebensführung. Zur Geschichte der Lebensreform um die Jahrhundertwende [Natural Conduct of Life. On the History of Life Reform around the Turn of the Century]*. Frankfurt (Main)/New York; Bothe 1991; Krabbe, Wolfgang. (1998a). Lebensreform/Selbstreform [Life Reform/Self-Reform]. In Kerbs and Reulecke 1998, 73–75; and particularly on naturopathy, Brauchle, Alfred. (1951). *Die Geschichte der Naturheilkunde in Lebensbildern [The History of Naturopathy in Life Vignettes]*. Stuttgart; Krabbe, Wolfgang (1998b). Naturheilbewegung [Naturopathic movement]. In Kerbs and Reulecke 1998, 77–85.

[45] "A naturopathic physician is not someone who occasionally uses a natural treatment when he has reached the limit of his mechanical preparations, just as little as a naturopathic physician becomes a conventional medical practitioner because he occasionally uses a medication; instead, a natural physician is someone who regards and treats the illness physiatrically, that is, not purely pathologically-mechanistically." Kleinschrod, Franz. (1911a). *Wissenschaftliche Begründung der Naturheilkunde. Eine prinzipielle Untersuchung [Scientific Justification of Naturopathy. An Investigation of Principles]*. Wörishofen, pp. 163f.

What makes world riddle solvers from the context of life reform particularly interesting is the balancing act that their relationship to science represented. On the one hand, it often stood as a synonym for empty materialism and was therefore rejected. On the other hand, scientific evidence, whose power of legitimation was not doubted, was supposed to be used to enhance the status of the goals of life reform. In contrast to other theories from this context that remained focused on the biological or medical areas, the world riddle solvers from the life reform movement, mostly physicians or biologists by training, also dealt expressly with physics. This also made it necessary for them to react to the theory of relativity, as becomes apparent in the example of the hydropathic Kneipp physician Franz Xaver Kleinschrod presented in the following case study.

Case study: Franz Kleinschrod. The hydropathic Kneipp physician and meta-mechanics[46]

Franz Xaver Kleinschrod came from lower middle-class circumstances. He was born in Ochsenfurt on May 31, 1860, the son of a master baker. Kleinschrod was the only member of his family to attend secondary school and beyond. He first studied law in Würzburg, but soon dropped this course of study in favor of studying medicine with Max von Pettenkofer (1818–1901), Hugo von Ziemssen (1829–1902), and Carl von Voit (1831–1908) in Würzburg and Munich from 1882 to 1887.

In 1887, after passing his medical examinations and starting a family, he opened a general practice in the Upper Palatinate. Since the practice did not go well, the family soon moved to Schlingen, one of Wörishofen's neighboring villages, where Father Sebastian Kneipp had already achieved great fame with his alternative cures. A personal encounter with Kneipp had left Kleinschrod deeply impressed; in fall 1889, he was already Kneipp's first spa physician and moved into the "Doctor House" across from the monastery in Wörishofen. However, he only stayed there until August 1891, after which he worked as the head of various spa medical practices. In 1909 he settled definitively in Wörishofen with a seasonal practice, but still kept a Munich residence. Since managing a sanatorium did not fulfill him intellectually, he made a decision to lead a double life: on the one hand as a physician in Wörishofen during the summer, and on the other as an "independent scientist" in Munich in the winter.[47]

[46] This portrayal is based on Mariafai, Judith M. (1982). *Das Leben des Sanitätsrats Dr. med. Franz Kleinschrod [The Life of Medical Counselor Dr. Franz Kleinschrod]*. Bad Wörishofen. This biography is written from the perspective of the Kneipp movement and is structured as a personal biography to emphasize the significance of the "divinely blessed physician" as a practitioner and theoretician of the Kneipp movement. The author, a librarian and commissioner for Kneipp history research at the non-profit Kneipp association Kneipp-Bund e.V., states that all of Kleinschrod's papers were lost in World War II.

[47] Mariafai 1982, p. 24. Kleinschrod had attended philosophical lectures in Munich during the winter months already in the 1890s.

Fig. 1.4 Franz Xaver Kleinschrod. Courtesy of Mariafai 1982.

Kleinschrod's dispute with academic science and his motivation for developing his own universal theory were set in the context of the "crisis of medicine" and the ubiquitous fierce dispute between alternative and orthodox medicine. According to the Trade Regulations of the German Empire, medicine was opened up to practice like any other trade, and since the naturopathic movement was growing and large portions of the population had lost confidence in orthodox medicine, the confrontation with orthodox medicine was inevitable.[48] "Quackery" (German *Kurpfuscherei*) became a fighting word that referred in general to treatments that deviated from orthodox medicine and thus also applied to certified physicians who used alternative procedures. Kleinschrod's former teacher, pathology professor Hugo von Ziemssen stated: "We must most deeply regret the fact that physicians can degrade themselves to such a degree that they become accomplices to Kneipp's hocus-pocus. We expel such charlatans from the threshold of sacred science."[49]

Although hardly any theoreticians originated in the naturopathic movement, naturopathy was strongly shaped by specific assumptions about the nature of the world. Since these assumptions had rarely been put into words, however, Kleinschrod felt challenged to remedy naturopathy's theoretical deficit. Ultimately, after all, their treatments were often based only on the collected experience of the lay practitioners, that is, non-scientists, and academic medicine quickly used this to reproach them. As a physician trained in orthodox medicine, Kleinschrod felt that he was in a position to give naturopathy a scientific foundation that it urgently needed in order to avoid orthodox medicine's accusation of quackery.

He understood this activity – combating scientific materialism as a physician, as an ethical obligation above all. He saw scientific materialism embodied not just in orthodox medicine, but rather in every science that rejected vitalism and the existence of a "life force." It was this motivation that caused him to write his magnum opus, *Übermechanik des Lebens [Meta-Mechanics of Life]*, which was published in 1928. There, he argues that illness can only be understood once the essence of life has been discovered, and in order to do that, it is first necessary to solve the body–soul problem, that is, the problem of the connection between what is alive and spiritual and what is lifeless and physical. According to Kleinschrod's assertions, this must not occur by philosophical means, but rather by scientific means. For this purpose, he developed an all-encompassing theory that was intended to unify physics, mathematics, and biology along with religion in one theory. The central concept of meta-mechanics was the "life force," which in turn represented the focal point of a biological number theory as was advocated at the

[48] Cf. Bothe 1991, particularly pp. 16–38.

[49] Quoted from Kleinschrod, Franz. (1921). *Sebastian Kneipps Lehre. Wissenschaftliche Begründung [Sebastian Kneipp's Theory. Scientific Justification]*. Kempten, p. 47.

time particularly by Wilhelm Fliess (1858–1928) and later became popular under the name biorhythm theory.[50]

In these number theories, numerical regularities in biological processes are made ontologically absolute. In Kleinschrod's theory, for instance, it is the *number* of wing beats that enables the life force of a bird to overcome physical gravity and allows it to climb into the air. According to this logic, in which the physical explanation that the wing beats cause lift has no place, it is always the meta-mechanical life force that causes physical effects. Expressed in Kleinschrod's terminology, it sounds like this:

> Every life force has the ability to cause a very specific arithmetic physical time–energy field to be created in a purely dynamic way according to its arithmetic, meta-mechanical time rule for the material realization of a meta-mechanical law [. . .] *The life force thus works directly with a physical energy field* that it can create itself according to its meta-mechanical regularity.[51]

In addition to the flight of birds, Kleinschrod also cited Reichenbach's Odic force and "animal magnetism," among other things, as empirical evidence for the generation of physical fields by the life force. Mind-reading, clairvoyance, and hypnosis were cited (as in occult thinking) as existing facts that would verify meta-mechanics.[52] Kleinschrod also wrote a number of specialist articles that appeared primarily in the publication media of the Kneipp movement. He died in 1934.

1.3.3 Monism

Materialistic views of nature are among the oldest and most popular perspectives due to their vividness.[53] In contrast to dualistic thinking, for instance in the context of life reform (which differentiates between spirit and matter) and occult thinking (which dissolves matter into the spiritual), materialistic thinking assumes that all natural phenomena can be explained exclusively by the characteristics of the material. The phenomena otherwise attributed to the spiritual realm (soul, consciousness, psyche) are expressly included.

[50] Cf. Hines, Terence M. (1998). Comprehensive review of biorhythm theory. *Psychological Reports*, **83**, 19–64.

[51] Kleinschrod, Franz. (1928). *Die Übermechanik des Lebens [The Meta-Mechanics of Life]*, Vol. 1*: Die Herrschaft des Lebens über die tote Welt. Das tierisch-pflanzliche Leben als Leibseeleproblem [The Power of Life over the Inanimate World. Animal and Plant Life as a Body-Soul Problem]*. Berlin, p. 387.

[52] Cf. e.g. Ibid.

[53] For a historical presentation and discussion, cf. Bayertz, Kurt, Gerhard, Myriam, and Jaeschke, Walter (eds.). (2007). *Weltanschauung, Philosophie und Naturwissenschaft im 19. Jahrhundert [Ideology, Philosophy, and Natural Science in the 19th Century]*, Vol. 1*: Der Materialismus-Streit [The Materialism Dispute]*. Hamburg; Klimke, Friedrich. (1911). *Der Monismus und seine philosophischen Grundlagen [Monism and its Philosophical Foundations]*. Freiburg im Breisgau, pp. 25ff.; Lange, Friedrich Albert. (1974 [1866]). *Geschichte des Materialismus und Kritik seiner Bedeutung in der Gegenwart [The History of Materialism and a Criticism of its Significance in the Present]*, 2 vols, ed. Alfred Schmidt. Frankfurt (Main); Wittkau-Horgby, Annette (1998). *Materialismus [Materialism]*. Göttingen.

Since the mid-nineteenth century, positivism, materialism, and the delimitation from natural philosophy characterized the world view of natural science. The popularization of science was a phenomenon of this progressive environment that wished to transmit the insights of natural science to a broad public audience.[54] Therefore, a large portion of non-academic science in Germany was characterized by an understanding of reality that was positivistic and materialistic in a broad sense. Around 1900, monism in particular exercised a strong influence on popular science and non-academic research. The German Monist League (Deutsche Monistenbund, DMB) was founded in 1906 with the goal of "dissemination of a uniform world view on a natural scientific basis."[55] This was linked with an ethically motivated optimism about progress that was affected little by criticism of technology and a life reform movement skeptical about science.[56] The goal of science, according to them, was to improve the world by dominating nature.[57]

The monism of the early twentieth century was not a uniform movement in terms of both the content that was advocated and its social organization. Commonalities existed particularly in the rejection of dualistic thinking and criticism of the Christian churches. Admittedly, a specific form of monism was propagated by the founding of the DMB and the establishment of its publication forums. However, monistic thinking was much more widespread than the number of members in the

[54] Cf. Daum 2002, for a comprehensive treatment. Cf. also the assessment by Heinrich Lübbe: "In Germany, the materialism of the second half of the century turned scientific knowledge of the cosmos, the Earth, the history of life on Earth, and the anatomical and physiological 'nature' of human beings, to the extent that it could be popularized into knowledge that was taken for granted in an educated person [. . .] In summary, it can be said that materialistic-scientific popular philosophy in the second half of the nineteenth century caused natural science to become important for the modern interest in education on a broad scale." Lübbe, Heinrich. (1974). *Politische Philosophie in Deutschland [Political Philosophy in Germany]*. Munich, pp. 128f.

[55] (July 1906). Was will der Deutsche Monistenbund? [What does the German Monist League want?]. *Blätter des Deutschen Monistenbundes [Gazette of the German Monist League]*, **1**, 7–9 (see p. 7). On monism around 1900, cf. Hillermann, Horst. (1976). *Der vereinsmäßige Zusammenschluß bürgerlich-weltanschaulicher Reformvernunft in der Monismusbewegung des 19. Jahrhunderts [The Association-based Merger of Middle-class Ideological Reform Reasoning in the Monist Movement of the 19th Century]*. Kastellaun [*et al.*] (Schriften zur Geschichte und Politischen Bildung Vol. 16); Ziche, Paul (ed.). (2000). *Monismus um 1900: Wissenschaftskultur und Weltanschauung [Monism around 1900. Scientific Culture and Ideology]*. Berlin.

[56] However, monistic ideology did have political and social goals that coincided with those of life reform. School reform, the women's movement, maternity protection and sexual reform, legal and land reform, the peace movement, and the alcohol issue went under the name of "monistic cultural activities," among other things. Cf. Monistische Kulturarbeit [Monistic Cultural Activities]. *Das Monistische Jahrhundert [The Monistic Century]*, Vol. 2.2 (1913–14), author, quoted from Drehsen, Volker and Zander, Helmut. (1996). *Rationale Weltveränderung durch naturwissenschaftliche Weltinterpretation? Der Monistenbund – eine Religion der Fortschrittsgläubigkeit [Rational change in the world through natural scientific interpretation of the world? The Monist League – A religion of faith in progress]*. In *Vom Weltbildwandel zur Weltanschauungsanalyse. Krisenwahrnehmung und Krisenbewältigung um 1900 [From a Change in World View to an Analysis of Ideology. Crisis Perception and Crisis Management around 1900]*, ed. Volker Drehsen and Walter Sparn. Berlin 1996, 217–38 (see p. 230). To this extent, it is possible to attribute monism around 1900 to the reform movement in terms of its political and social goals.

[57] Cf. Drehsen and Zander 1996.

Monist League would suggest,[58] and there were quite a few groups that disassociated themselves from the DMB and represented different varieties of monistic thinking, such as the Humboldt League (with its periodical *Neue Weltanschauung [New World View]*) led by Wilhelm Breitenbach (born 1856), a student of Haeckel's, and the Universal Knowledge Society (with its periodical *Weltwissen [Universal Knowledge]*) led by the engineer Johannes Zacharias. In contrast to many representatives of occult and vitalist thinking, the monistic movement conceived of itself as a purely scientific world view from the start. Therefore, claims of their scientific nature were asserted for all monistic content, resulting in a wide variety of potential lines of conflict with academic science. However, monism, like the other ideological contexts, was primarily occupied with its own program and had largely dissociated itself from developments in academic science,[59] which on the one hand caused conflicts to intensify, but on the other hand did not allow them to arise in the first place because no confrontation occurred.

The monistic world view around 1900 had not had a purely materialistic orientation for some time; instead, pantheistic ideas that showed parallels to occult thinking could be found, particularly in Haeckel-style monism.[60] It is true that the world riddle solvers from this context claimed to be able to explain all events on the basis of mechanics. Separately from this, the substance of the mechanistic movement was often religionized and spiritualized in the monistic–pantheistic or the romantic–idealistic sense. Nevertheless, this did not mean that any dualistic position was taken. World riddle solvers from the monistic context were predominantly engineers. They almost exclusively advocated mechanical theories, as in the case of the engineer Arthur Patschke, who is introduced in the next case study.

Case study: Arthur Patschke. The engineer and the universal law of force

Arthur Patschke was born in Böhmenhöfen near Braunsberg (known today as Braniewo, Poland) in the Ermland region of East Prussia on April 13, 1865, the son of a mill owner. He wanted to become an engineer early on and left high school prematurely in 1881 in order to commence his studies at Mittweida College of

[58] Data are only available for the years 1916 and 1917, according to which the DMB had about 4000–4500 members. Cf. Weber, Heiko. (2000). *Monistische und antimonistische Weltanschauung: Eine Auswahlbibliographie [Monistic and Anti-monistic Ideology: A Selected Bibliography]*. Berlin (Ernst-Haeckel-Haus-Studien: Monographien zur Geschichte der Biowissenschaften und der Medizin Vol. 1), pp. 28f.

[59] This is also Drehsen and Zander's assessment, according to which the Monist League increasingly lost its connection to the rapid developments in the natural sciences. Cf. Drehsen and Zander 1996, p. 226.

[60] On Haeckel's view of monism as a religion, cf. Kleeberg, Bernhard. (2005). *Theophysis: Ernst Haeckels Philosophie des Naturganzen [Theophysics: Ernst Haeckel's Philosophy of the Cosmos]*. Cologne [*et al.*], particularly pp. 31ff., and Di Gregorio, Mario A. (2005). *From Here to Eternity: Ernst Haeckel and Scientific Faith*. Göttingen.

Technology, which he concluded in 1887 as a mechanical engineer.[61] This was followed in 1912–14 with supplementary studies in electrical engineering at the Royal Technical College of Charlottenburg.

His professional career started with the construction of drawing apparatus. He then worked in various companies as a design engineer for steam engine construction starting in 1894. Around 1907 he moved to Berlin, where he was employed at the Siemens-Schuckert company, and resided in Berlin until the end of his life.[62] Patschke was a successful steam engine designer. Around 1900, preliminary work began on the design and construction of a rotating steam engine that he developed, which he presented in 1902 at the Commercial and Industrial Exposition in Düsseldorf;[63] he later developed a "transverse steam turbine."[64] Due to his innovations in turbine construction, Patschke was in contact with the leading design engineers of the time, including Nils van Dalen, the 1912 Nobel Prize winner in physics, when van Dalen was chief engineer at the De Laval company who were pioneers in steam turbine construction.

Along with his work as an engineer, Patschke had another interest that went far beyond merely designing steam engines. He wanted to track down the energy expressed in the steam, that is, the essence of the forces of nature. For Patschke, there was no difference between the domains of "being an engineer" and "being a natural scientist"; instead, in his view, the fact of dealing with and utilizing nature required an engineer to discern the essence of nature. And who could do this better than an engineer, who researches, tests, and examines it practically? In retrospect, Patschke stated that he possessed a pronounced interest in fundamental questions, which arose when he experienced a steam engine at work: What force is it that moves the machine? Asked more precisely: What is the essence of force? What is movement "in and of itself"? What is energy? What does the world ultimately

61 Technikum Mittweida [Mittweida College of Technology], *Jahresbericht [Annual Report] 1886–1888*, p. 11. The biographical information in this section comes from Patschke's works, particularly Patschke 1925, supplemented by documents from Patschke's student file in the Mittweida College Archive, Student File No. 4393.

62 This is apparent from a letter from Patschke to Gehrcke. Patschke to Gehrcke, March 19, 1922, GN 82-H-19. There are no documents in the Siemens archive about Patschke. According to information from the archive, however, this is not unusual for individuals below management level.

63 Rotating steam engines, which have never become generally accepted compared to piston steam engines because they do not work effectively enough, use the steam pressure directly to move the driving wheel and are therefore significantly more compact than piston steam engines. Cf. Mauersberger, Klaus. (2001). Kolbendampfmaschine kontra Rotationsdampfmaschine [Piston steam engine contra rotating steam engine]. In *Johann Beckmann und die Folgen: Erfindungen – Versuch der historischen, theoretischen und empirischen Annäherung an einen vielschichtigen Begriff [Johann Beckmann and the Consequences: Inventions – An Attempt at an Historical, Theoretical, and Empirical Approach to a Complex Concept]*, ed. Gerhard Banse and Hans-Peter Müller. Münster [*et al.*], 65–88.

64 The characteristic feature of a transverse steam turbine according to Patschke's system consists of the fact that the steam is guided through helical windings and divided into the finest of streams, which Patschke calls "lamella jets." These extremely fine steam jets were intended to utilize the force-absorbing surface of the rotor in a particularly effective and even way. Cf. Patschke, Arthur. (1904). *Transversal-Dampfturbinen für elastische Kraftmittel [Transverse Steam Turbines for Elastic Force Media]*. Mühlheim-Ruhr.

consist of? And what is man? Of course, these questions as such are by no means new, and did not only occupy steam engine designers, but for a mechanical engineer around the turn of the century, it was natural to answer them with a universal, mechanical explanation of world events. Patschke describes what drove him to his almost thirty-year career as a world riddle solver in an autobiographical reflection:

> In my function as a practical engineer, a specialist, a mathematician, and a natural scientist, I was always only able to accept a scientific work if it had a sufficiently logical structure, that is, if it was constructed on clear mathematics and obvious mechanics. As a result of experiments with steam turbine jets and discussions with practitioners and scholars, I was inevitably led to fathoming natural forces, that is, to solving the world riddles, to electromechanics, to the uniform explanation and mechanics of the natural forces.[65]

In his theoretical design, Patschke was strongly influenced by his professional experience. In fact, he wanted to show "that the *Earth is a universal turbine*, a universal ether turbine in large."[66] He discussed natural forces not only with his colleagues, but also with members of the Natural Scientific Association of Düsseldorf, and not least with his wife, Margarete.[67] In 1905, Patschke presented his solution to the world riddles with the "Universal Law of Force," which states: "Bodies floating in gasses (heavenly bodies, planets, atoms) can only move forward when they have received force from behind."[68] Patschke based his universal mechanical theory on this fundamental law, which was asserted in 1905 in *Universal Law of Force* and formulated particularly in *Electromechanics*, which appeared in 1921. (The title is misleading, as it deals with the mechanics of Patschke's light ether atoms.) The concept of the atom is assigned central significance, since the pressure force resulting from the movement of atoms is seen as the primordial force from which all natural forces result. The movement of the atoms, in turn, can only arise from the influence of force from other moving masses according to the universal law:

[65] Patschke, Arthur. (1925). *Weltlichttheorie. Welttheorie der Naturkräfte. Lösung der Naturkraftprobleme. Erfindungen, Entdeckungen und Ausnutzung der Naturkräfte [Universal Light Theory. A Universal Theory of Natural Forces. Solution of the Natural Force Problems. Inventions, Discoveries, and Utilization of the Natural Forces]*. Diessen near Munich, p. 13.

[66] Patschke, Arthur. (1905a). *Lösung der Welträtsel durch das einheitliche Weltgesetz der Kraft [Solution of the World Riddles by the Uniform Natural Law of Force]*. Munich, p. 33.

[67] Patschke's third wife, Margarete von Oppell-Patschke, had earned a bachelor's degree in natural science in Cape Town, South Africa. She supported her husband in all of his work, also in its content. In the year of his death, 1934, she published her work *Urkraft [Primordial Force]*, in which she set out to adapt her husband's theory to the latest scientific discoveries. She is thus the only female world riddle solver that I discovered. Cf. von Oppell-Patschke, Margarete. (1934). *Urkraft. Anpassung der vor 30 Jahren von Ing. A. Patschke aufgestellten Lichtäthermechanik an d. Ergebnisse d. heut. Experimentalphysik [Primordial Force. Adaptation of the Light Ether Mechanics set forth by A. Patschke, Engineer, 30 years ago to the Results of Modern Experimental Physics]*. Berlin.

[68] Patschke 1905a, p. 12.

Soeben erschien:

Lösung der Welträtsel

durch das einheitliche

Weltgesetz

der

Kraft.

Von

Ingenieur A. Patschke

Düsseldorf-Oberkassel.

Verlag Seitz & Schauer, München.

Näheres umstehend!

Fig. 1.5 Advertising leaflet for Patschke's 1905 work *Solution of the World Riddles by the Uniform Universal Law of Force.* Source: Hochschularchiv Mittweida, Studentenakte No. 4393.

> Masses can only be moved by masses, whereby the masses driving each other must come into contact with each other. [. . .] [This results in] the fundamental theorem of force: All forces can be traced back to the movements of masses.[69]

Gravitation, cohesion, electricity, magnetism, heat, light, and electrolytic and chemical processes are explained in Patschke's *Electromechanics* as mechanical phenomena. In order to describe physical effects even in apparently empty space as impact effects in filled space, Patschke introduces the "primordial force mass," which is ultimately an ether that causes all of the forces just named, but which is accessible to science up to the present only by logical thinking, but not by direct evidence.[70] This universal law as such is not limited to the physical world; instead, it extends

> from the stars to the atoms, from the atoms to the machines, to all bodies, plants, animals, and humans, from the simplest mechanical forces to the feeling from which, by means of suitable, more perfect bodies provided with corresponding ether resistance, living beings were able to develop, nay, had to develop reason in the human being, which towers above all things, by the driving force of light ether, of universal ether.[71]

The growth of plants or human emotional life – everything is an effect of the universal ether, which in this way acquires a religious dimension in the theory.[72]

Compared to other world riddle solvers, this very busy engineer wrote only a few works. Along with some papers from 1905 and 1906 that describe the universal law in more detail, and his magnum opus, the *Electromechanics* of 1921, only a single work appeared before his death in 1934, his *Universal Light Theory*[73] of 1925.

1.4 The world riddle solvers' conception of science

Despite the heterogeneity of their social and cultural backgrounds, two important factors united Einstein's non-academic opponents: they shared a particular view of what the object and purpose of science is, and they positioned themselves as "independent researchers" contra academic science. Before I discuss the anti-academic self-concept of the world riddle solvers in the following section, it is

[69] Patschke, Arthur. (1921). *Elektromechanik: Einheitliche Erklärung und Mechanik der Naturkräfte [Electromechanics: Uniform Explanation and Mechanics of Natural Forces]*. Berlin, p. 14.

[70] Ibid., pp. 14f.

[71] Patschke, Arthur. (1905b). *Der Sturz des Anziehungsgesetzes und die Entdeckung des einheitlichen Weltgesetzes der Kraft [The Fall of the Law of Attraction and the Discovery of the Uniform Universal Law of Force]*. Munich, p. 31.

[72] On the religionization of science pp. 57ff.

[73] Patschke 1905a; Patschke 1905b; Patschke, Arthur. (1906). *Vom Stern zum Atom. Beitrag zum Weltgesetz [From the Star to the Atom. A Contribution to the Universal Law]*. Munich; Patschke 1921; Patschke 1925.

first necessary to present the world riddle solvers' conception of science, which is characterized by the demand for absolute truth and recognition of the substance of reality as well as by far-reaching ideological claims.

1.4.1 The demand for absolute truth

One conflict between the world riddle solvers and academic science existed at the level of differing concepts of truth and its ontological status. Unaffected by contemporary philosophical discussions about the concept of truth[74] and statements written by some scientists who no longer attached any significance to the concept of truth in science, the world riddle solvers demanded an absolute, objective truth taken from nature, *the* truth, *the* true nature of all things.

> Man wants to dive down to the *foundation* of the thing, to understand the *essence* of the thing [. . .] Nothing is too high for him any more! [. . .] The voices calling for a *truth* that, founded on science, would be in a position by itself to clear away all of our doubts and our distrust, as well as to satisfy our searching reason completely, are becoming ever louder. The knowledge of such truth would have to be *uniform*, after all, and [. . .] include everything *universally.*[75]

It was with this fiery summation that the world riddle solver Grubić introduced his monistic natural philosophy in 1917, which was intended to establish this truth. The world riddle solvers did not reflect upon the concept of truth; instead, it stood for absolute certainty about the essence of natural phenomena. The truth was whatever the world riddle solver had identified in his theory as the primordial cause of events.

While academic science had become restrained at the end of the nineteenth century with regard to the demand for absolute and eternal truth as a goal, or as the starting point for their research, and had largely distanced itself from it,[76] the situation was quite clear for the world riddle solvers. They knew what the truth was; after all, they knew the unified theory. Academic science, on the other hand, was not even seeking absolute truth. This was an unacceptable situation for a world riddle solver like Zacharias: "Science should consist of certainties. To date, however, it is a 'theory of complete unknowability.'"[77] At the same time, after all, he had already

[74] On the concepts of truth in the philosophical discussion of the twentieth century, cf. Skirbekk, Gunnar. (1996). Einleitung [Introduction]. In ibid. *Wahrheitstheorien. Eine Auswahl aus den Diskussionen über Wahrheit im 20. Jahrhundert [Theories of Truth. A Selection from the Discussions about Truth in the 20th Century]*. 7th ed., Frankfurt (Main), 8–34, as well as the source texts published there.

[75] Grubić, Dušan. (1917). *Universal-Kausalprozess als unser oberstes Naturgesetz [The Universal-Causal Process as our Highest Natural Law]*. Zagreb, pp. 2f.

[76] On the replacement of truth as the goal of science with the more modest goal of objectivity, cf. Daston, Lorraine and Galison, Peter. (2007). *Objectivity*. New York.

[77] Both quotes from: Zacharias, Johannes. (1920). *Rätsel der Natur und Totengräber der Wahrheit [Mysteries of Nature and Gravediggers of Truth]*. Munich, p. 117.

long ago determined the certainties of science once and for all: "Almighty nature is the soul of the world. Thus we have the strictest of order, unalterable certainty. The fundamental riddles of nature have thus been solved. The uniform view of nature has been established."[78]

Johannes Zacharias (born 1850)

Source: Reuterdahl papers; courtesy of the Department of Special Collections, University of St. Thomas, St. Paul, MN.

Johannes Zacharias was born in Karalene near Insterburg in 1850. In 1870, he joined the army as an engineering officer, receiving his training at the Artillery and Engineering School in Berlin. He then served three years in fortification construction before being mustered out of the army in 1880 as an invalid. Zacharias was initially a chief engineer, later a director of a stock corporation, followed by work in incandescent bulb fabrication, at a battery and element factory, and in the construction of streetcars. In 1893 he took on the development and management of an electrical company in Chicago. By the end of 1893, he was again working in Germany, as an expert, consulting engineer, and patent attorney. Having become financially independent, he founded *Weltwissen. Zeitschrift für Erforschung und Erkenntnis in Natur und Technik auf einheitlicher Grundlage [Universal Knowledge. Journal for Research and Insight into Nature and Technology on a Uniform Basis]* in 1912. He espoused the view that all world events were attributable to the effects of moving primordial matter.

Source: Zacharias 1921

[78] Ibid.

Ziegler, the primordial light theorist, could only agree. He was entirely aware of the great difference between his own conception of science and that of academic science and expressly took an anti-academic epistemological position. According to Ziegler's definition, science was no such thing if it was based on hypotheses; instead, he assigned it a function that in his contemporaries' understanding would rather have been assigned to metaphysics or religion, stating that science would have to seek the "ultimate reason for all knowledge":

> The purpose of [Ziegler's] primordial light theory [. . .] [is] to enable clear, uniform science and to do away with everything hypothetical and incomprehensible. Its purpose is to give modern physics, which is riddled with errors, a solid foundation of definite laws, that is, to introduce true metaphysics into it.[79]

He was completely scandalized by the exclusion of "consideration of the essence" in academic science. "That is certainly astonishing, indeed, it seems as if no greater miracle could have occurred than that the representatives of science want to explain things without knowledge of the Eternal."[80]

At the same time, as a world riddle solver, he could do nothing else than proclaim the truth. This was so clear and plausible to him that it had to convince everyone immediately. With his solution of the world riddles by his Universal Law of Force, the engineer Patschke by no means considered himself a scientific outlaw, but rather at the pinnacle of a long historical development of the knowledge of humanity. Based on his assumption that knowledge would not only continually develop due to experience accumulated over generations, but would also and above all continually come closer to an absolute truth, he understood himself as the herald of a truth that was comprehensible to each and every contemporary of average intelligence, a truth that had literally forced itself upon him: "The power of the world is an omnipotent one and accordingly also forced us to announce this revolutionary universal law, which self-evidently recommends itself due to its clarity and truth, preliminarily in its present form."[81] To put it bluntly, it was not Arthur Patschke who proclaimed the universal law; instead, the truth of the universal law proclaimed itself through its medium, Patschke. This odd combination of the self-confidence bordering on

[79] Ziegler, Johann Heinrich. (1932). *Das Urlicht und das Hauptgerüst richtiger Weltanschauung: Ueber das Absolute und das Relative [Primordial Light and the Basic Framework of Correct Ideology: On the Absolute and the Relative]*. Zürich, p. 40.

[80] Ziegler 1908b, p. 29. [81] Patschke 1905a, p. 8.

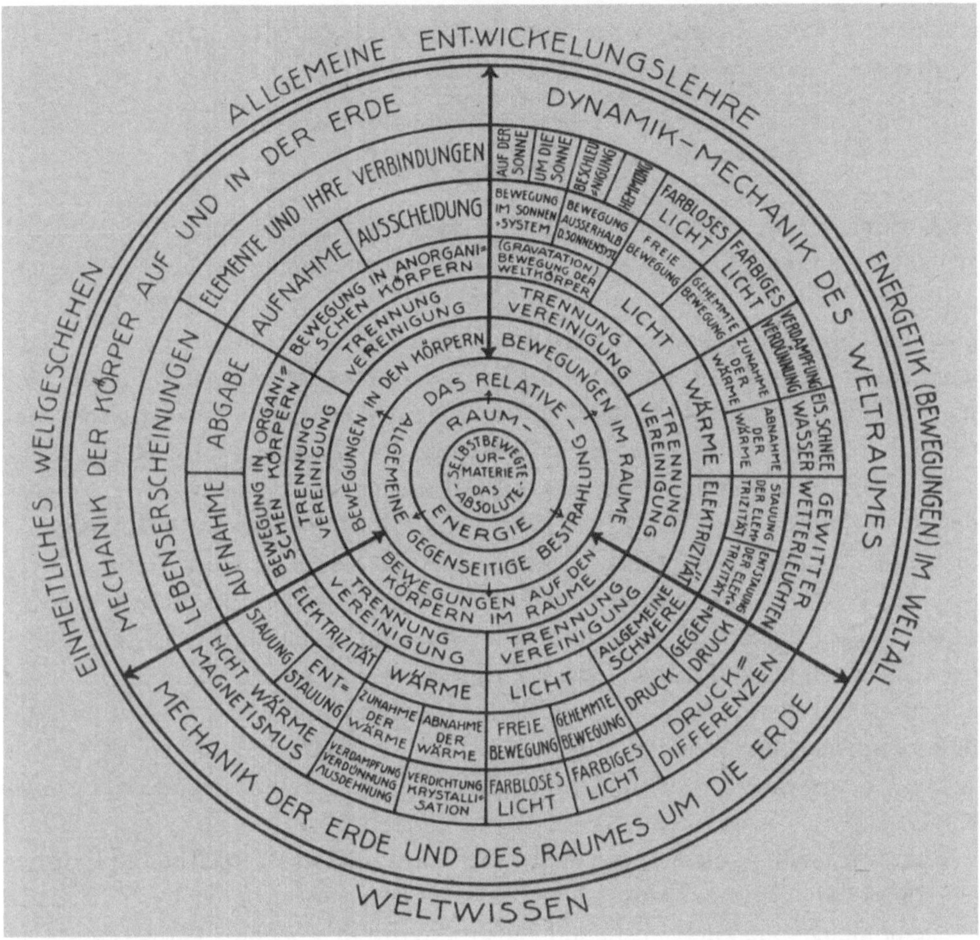

Fig. 1.6 Engineer Johannes Zacharias' unified theory traces all world events back to the effects of a primordial matter (center) that is inaccessible to humans. Source: *Weltwissen*, 5, April 1, 1921, p. 20.

presumptuousness of having discovered the universal law with the modesty and withdrawal of individuality in the face of "eternal truth" is characteristic for many world riddle solvers. This conception of a more or less self-evident universal law did not confront Patschke with the problem of how truth can be recognized; instead, like a number of other world riddle solvers, he was faced with the difficulty of explaining why this very simple truth had nevertheless not yet gained acceptance. The reason for this, according not only to Patschke, could be found in the institutionalization of the search for truth in the system of academic science, bound to tradition and fending off all innovations, which rendered it impossible for the truth to be heard at all (cf. pp. 58ff.).

1.4.2 Substance thinking

The search for an essence of reality is a form of thinking that can be labeled "substance thinking," following Ernst Cassirer's differentiation between the concepts of substance and function.[82] World riddle solvers were never professed phenomenologists, functionalists, or constructivists; as "truth-seekers," they were chiefly metaphysicists and ontologists. This had consequences for the content and line of argument of their theories. The questions about the objectivity and certainty of scientific knowledge that the world riddle solvers were asking themselves were equally being asked by most of the academic scientists. These academic scientists generally also assumed – in various epistemological versions – that reality is intersubjectively comprehensible. But while the world riddle solvers deduced their theories from an objective certainty that they postulated, modern science can be conducted without knowing the ultimate answers or assuming that they can be known in the first place. This difference in perspectives about the tasks and limits of scientific knowledge, which culminated in the *ignorabimus* dispute in academic science in the 1870s,[83] was expressed in the popular conception of science of the early twentieth century particularly in a widespread criticism of Kant.

Kant was unambiguously rejected by the world riddle solvers. Their optimistically epistemological substance thinking found itself confronted by a critically epistemological phenomenalism and constructivism that negated knowledge of an ultimate substance of reality. In the popular understanding, Kant was the person who had taken the "thing in itself" – and thus absolute truth and ultimate substance –

[82] Cf. particularly Cassirer, Ernst. (1923 [1910]). *Substance and Function*. Chicago; ibid. (1957 [1929]) *The Philosophy of Symbolic Forms*, Vol. 3: *The Phenomenology of Knowledge*. New Haven. This epistemology oriented towards certainty and absolute truth, as espoused in the academic philosophy of the time by, e.g., Dingler, was also labeled "certism" in the theory of science in contrast to science, which is oriented towards falsification. Cf. Schroeder-Heister, Peter. (2005). Certismus. In *Enzyklopädie Philosophie und Wissenschaftstheorie [Encyclopedia of Philosophy and Philosophy of Science]*, ed. Jürgen Mittelstrass, Vol. 2 (C–F), 2nd edn. Stuttgart/Weimar, 35–36, online at: http://ls.informatik.uni-tuebingen.de/psh/forschung/offprints/request/Enzy1/Certismus.php (March 20, 2013).

[83] On the contemporary context and on the effect of du Bois-Reymond's "Ignoramibus Speech," cf. Vidoni, Ferdinando. (1991). *Ignorabimus! Emil du Bois-Reymond und die Debatte über die Grenzen wissenschaftlicher Erkenntnis im 19. Jahrhundert [Ignorabimus! Emil du Bois-Reymond and the Debate about the Limits of Scientific Knowledge in the 19th Century]*. Frankfurt (Main) [*et al.*]; cf. Bayertz, Kurt, Gerhard, Myriam, and Jaeschke, Walter (eds.). (2007). *Weltanschauung, Philosophie und Naturwissenschaft im 19. Jahrhundert [Ideology, Philosophy, and Natural Science in the 19th Century]*, Vol. 3: *Der Ignorabimus-Streit [The Ignorabimus Dispute]*. Hamburg.

away from knowledge. For many world riddle solvers, his philosophy therefore constituted an antithesis to "common sense" and embodied uncertain, speculative, and ultimately "unscientific" science. The world riddle solvers' criticism of Kant operated at this elementary level and thus did not deal critically with Kantian philosophy, but rather with what it stood for in the popular conception. The "thing in itself" was understood by the world riddle solvers in the metaphysical sense, not the transcendental sense. The world riddle solvers intentionally judged what they understood as Kantian philosophy based on their own personal experience and justified their claims to epistemological competence using their practical interactions with nature. For Kleinschrod, for example, it was his experience as a physician that granted him a certain degree of authority on epistemological issues: "If Kant had been a physician, he quite certainly would have postulated a different theory of time and space. Death and the span of life alone would have certainly taught this deep thinker something different."[84] He distanced himself particularly from Kant's denial of the knowledge of the "thing in itself" – for him, the "thing in itself" was the knowledge of the essence of life – confessing:

> As a physician, I was never able to come to terms with such pessimism; my entire inner nature resisted it. In another respect as well. Every person struggles to find a world view as the cornerstone of his way of life. But we will never achieve a world view on a scientific foundation without comprehending the essence of life.[85]

Erich Ruckhaber, a businessman, also wanted to solve the "thought problem." In this respect, he found a "bright and cheerful approach that takes the bull by the horns" lacking[86] and continued:

> Kant is very much to blame for this sorry picture. With his transcendental philosophy, he elevated to a scientific dogma the supposed impossibility of human understanding ever comprehending itself. As such, it has been reigning for 150 years at almost all German university lecterns of worldly wisdom with the same broad intransigent power and complacency as the infallibility of the Pope reigns in Catholic pulpits. The orthodoxy is conspicuously similar for both of them; whoever does not believe in Kant is a heretic.[87]

Here, the fact that Ruckhaber sets up Kant as the enemy of human knowledge unencumbered by too much knowledge of transcendental philosophy is not of interest; what is interesting is that he would like to close the rift between phenomenon and the "thing in itself" using his sensualist philosophy of the "primordial principle."[88]

[84] Kleinschrod 1929, p. 7. [85] Kleinschrod 1928, p. 289.
[86] Ruckhaber, Erich. (1920a). Das Denkproblem und seine Lösung [The thought problem and its solution]. *Neue Weltanschauung [New World View]*, **9**(1), 1–9 (see p. 1).
[87] Ibid.
[88] Ruckhaber, Erich. (1959). *Das Urprinzip der Welt [The Primordial Principle of the World]*. Berlin, pp. 30f.

Erich Ruckhaber (1876–1956)

Erich Ruckhaber, 1959.

Erich Ruckhaber was born in Breslau in 1876. From 1893 to 1909, he worked as a businessman, spending the majority of this period in Africa. From 1910 to 1911, he was employed at Deutsche Bank, and from 1918 to 1929, he worked as an interpreter. He earned his doctorate in philosophy in Berlin in 1927. From about 1912, Ruckhaber regularly wrote articles in the popular science periodical *Neue Weltanschauung [New World View]*. In the 1910s and 1920s, he worked on sensualist epistemology, and in the 1930s he worked primarily on ether vortex theory and developed "biomechanics," a mechanical theory of life.

Sources: Degener's "Wer ist's" [Who's who] 1955; Ruckhaber 1959

The world riddle solvers' conviction that they had possession of the absolute truth and their adherence to substance thinking generated a potential conflict relationship between their conception of science and that of academic science. This conflict relationship was initially largely latent. Whoever had identified the "thing in itself" and knew the absolute truth had to focus primarily on obtaining recognition for the truth and not tackle the problem of other views, which could only be false at any rate – unless they were as omnipresent as the theory of relativity. The fact that these conceptions of science collided with each other across a large front in the confrontation about the theory of relativity is demonstrated later in this book (cf. pp. 155ff.).

1.4.3 Between the abyss and redemption – science and world view

The significance of science for the world riddle solvers in terms of religion and world view as discussed below was a decisive parameter in the course of the dispute with the theory of relativity that is examined later on in this book. Along with the epistemological elements of the world riddle solvers' conception of science in the narrower sense outlined above, their conception of science was additionally characterized by claims based on their world view that caused a different kind of dividing line to arise between it and academic science. In the eyes of the world riddle solvers, academic science was incomplete because, unlike the world riddle solver theories, it did not claim to be a program based on a world view that was intended to create a new society and a new human being.

For their part, the claims of the world riddle solvers based on their world view can only be understood in connection with the crisis of world view at the turn of the century and the social and political upheavals of the early twentieth century. The world riddle solvers' diagnosis of their time period is first outlined before presenting a discussion of their convictions about the treatability of the noted shortcomings with the aid of their own theories. The significance of science to world view found expression particularly in the religionization of science, whose discussion will conclude this section.

Criticism of civilization and historical philosophy from below

The fragility of the social, political, economic, and world-view contexts had certainly become particularly apparent after World War I, but the fragmentation of society by world view extended farther back. It must be placed in context with increasing industrialization and technical development and the accompanying radical changes in all areas of life in the nineteenth century.[89] Despite this continuity in

[89] Cf. e.g. the contributions in Blaschke, Olaf, and Kuhlemann, Frank-Michael (eds.). (1996). *Religion im Kaiserreich. Milieus – Mentalitäten – Krisen [Religion in the German Empire. Milieus – Mentalities – Crises]*. Gütersloh (Religiöse Kulturen der Moderne Vol. 2); Vom Bruch, Rüdiger, Graf, Friedrich W., and Hübinger, Gangolf. (1989). *Kultur und Kulturwissenschaften um 1900 [Culture and Cultural Sciences around 1900]*, Vol. 1*: Krise der Moderne und Glaube an die Wissenschaft [Crisis of Modernity and Faith in Science]*. Stuttgart; ibid., Vol. 2: *Idealismus und Positivismus [Idealism and Positivism]*. Stuttgart; Doerry, Martin. (1986). *Übergangsmenschen. Die Mentalität der Wilhelminer und die Krise des Kaiserreichs [Transitional Humans. The Mentality of the Wilhelminians and the Crisis of the German Empire]*, Vol. 1. Weinheim/Munich; Drehsen and Sparn 1996.

the transformation of world views that lasted for decades, World War I was a cesura that was understood as a departure into a new world, as redemption from antiquated structures and outdated views.[90]

Ultimately, what remained was a disillusioned Germany. According to a widespread view, not only had a war been lost, but also a world. The religious-political vacuum recorded by many contemporary witnesses was one of the principal problems of the newly founded democracy, which bore the stigma of being only a pragmatic compromise from the beginning.[91] The search for meaning in the supernatural prospered in the post-war period, as did the belief in technological progress. Prophets and leaders, comprehensiveness and clarity were demanded. The crisis situation experienced both individually and collectively in all areas – inflation, social downgrading, defeat in war, dissolution of the empire – constituted the framework for the longing for renewal both material and spiritual. The "closet religions,"[92] regardless of whether they invoked the Bible, the Bhagavad Gita, Nietzsche, nature, or reason, were unified by their promise of truth, of re-enchantment where the spell had been broken, of renewal and unity of the world.

For a broad spectrum of the population, their world view had become fragile and the meaning of history dubious; the prerogative of religion and science to interpret the essence of things was questioned and, not least for this reason, the two of them began to compete with each other. The ideological props that were offered as surrogates for meaning did not fill the vacuum of world view; instead, they were an expression of it. One thing is immediately apparent when dealing with the world riddle solver documents after World War I. As expected, the rise of modern physics caused them to be confronted by the fact that their previous understanding of the structure of nature and of the scientific theories had lost its self-evident nature. For them, however, this uncertainty was also connected with their diagnosis of a moral degeneration of society in the course of, and as a consequence of, the war. They sometimes placed themselves into social reform movements and expressed criticism

[90] Cf. e.g., Flasch, Kurt. (2000). *Die geistige Mobilmachung. Die deutschen Intellektuellen und der Erste Weltkrieg [Spiritual Mobilization. German Intellectuals and the First World War]*. Berlin; Mommsen, Wolfgang J. (ed.). (1996). *Kultur und Krieg. Die Rolle der Intellektuellen, Künstler und Schriftsteller im Ersten Weltkrieg [Culture and War. The Role of the Intellectuals, Artists, and Authors in the First World War]*. Munich; Mohler, Armin. (1999 [1950]). *Die konservative Revolution in Deutschland 1918–1932. Ein Handbuch [The Conserviative Revolution in Germany 1918–1932. A Handbook]*. 5th edn., Graz [*et al.*], pp. 32ff.

[91] On the characterization of the Weimar Republic as "practical republicanism" cf. Klueting, Harm. (1986) 'Vernunftrepublikanismus' und 'Vertrauensdiktatur': Friedrich Meinecke in der Weimarer Republik ['Practical republicanism' and a 'dictatorship of trust': Friedrich Meinecke in the Weimar Republic]. *Historische Zeitschrift*, **242**, 69–98. On the cultural situation in the Weimar Republic, cf. also the contributions in Cancik, Hubert (ed.). (1981). *Religions- und Geistesgeschichte der Weimarer Republik [Religious and Spiritual History of the Weimar Republic]*. Düsseldorf; Gay, Peter. (1968). *Weimar Culture: The Outsider as Insider*. New York; Peukert, Detlev. (1987). *Die Weimarer Republik: Krisenjahre der Klassischen Moderne [The Weimar Republic: Crisis Years of the Classic Modern Period]*. Frankfurt (Main); Ziche 2000.

[92] Cf. Bry, Carl Christian. ([1964/1924]). *Verkappte Religionen [Closet Religions]*. 3rd edn., Lochham [*et al.*].

of civilization in their theories. Therefore, they cannot merely be considered competing scientific theories from the non-academic sphere; instead, they are often expressly simultaneously a program for creating a new human being and a new social order through "pure, true knowledge." For the world riddle solver theories in the early twentieth century period, it is thus possible to see a parallel phenomenon to the religionization of politics, namely the religionization of science, and more precisely, of scientific theory.[93]

The engineer Rudolf Mewes (born 1858), a staunch supporter of the Kaiser, introduced the third part of his world riddle solver theory, *Kriegs-und Geistesperioden im Völkerleben und Verkündigung des nächsten Weltkrieges. Eine astrologisch-physiologische Skizze [War Periods and Spiritual Periods in the Life of Peoples, and Proclamation of the Next World War. An Astrological-Physiological Sketch]*, which appeared after World War I – this work is a morphological philosophy of history on a scale similar to Oswald Spengler's works – with this pessimistic diagnosis:

In the meantime, the iron dice of fate have been cast; the German nation has emasculated itself and delivered itself up to be at the mercy of the enemy alliance. It has plunged itself into ruin and cannot reverse the consequences of its ignominious betrayal of the emperor, the people, and the race. Fate is striding over the German nation with an iron boot and threatens to crush it. [. . .] Germany's situation is much more humiliating and pathetic than it was before and after the revolution of forty-eight. Deeper is the night and more hopeless the abyss of the German collapse as a consequence of the overthrow of all order and loyalty on November 9, 1918; blacker is the future that the German Empire and thus all Germandom is swaying towards after the unpardonable treason against the people under the external pressure of the rapacious alliance.[94]

For Kleinschrod, the Kneipp physician, this is what the present looked like in 1931:

Bolshevism, communism, Marxism, and also the whole Social Democratic movement have their ideological roots in this striving of humans to use their intellect only to suspend the moral-religious part of the law of humanity [discovered by Kleinschrod], simply to destroy it. The consequence of this is the false concept of the state that all of these political trends work with. Religion is a private matter, is dulling the mind of humanity, the opium of reason; family and marriage are only means to inhibit personality. Work is slavery, the state is only a feeding trough, etc., etc. When we look down on humanity from the lofty standpoint of the

[93] At the same time, the ideological significance of natural science is neither a phenomenon limited to non-academic science nor a specific feature of the 1920s. It developed particularly in the second half of the nineteenth century. At a time when ever fewer people were placing their trust in the churches' traditional knowledge of revelation and an empirically founded, uniform world view of science was able to convince more and more, the natural scientists who were able to present this unifying theory in a popular way acquired an important function. Cf. Daum 2002, pp. 200f.

[94] Mewes, Rudolf. (1922a [1896]). *Kriegs- und Geistesperioden im Völkerleben und Verkündigung des nächsten Weltkrieges [War Periods and Spiritual Periods in the Life of Peoples, and Proclamation of the Next World War]*, 3rd and 4th expanded edns., Leipzig, p. 378.

[Kleinschrod] law of humanity, then a terror and horror of this humanity seizes us: *Mir graut vor dir! (I shudder to think of thee!)* is what one would like to shout.[95]

Although these diagnoses of the present are very little elaborated, they come from individuals who conceive of themselves as natural scientists, not as historians or social philosophers. They are part of an ideological natural science, which is reflected in combinations of world view and science that seem odd at first glance. For instance, Mewes discovers a "Bolshevist characteristic"[96] in modern physics due to the assumption of probability and the disordered movement in kinetic gas theory, while Fricke, the patent office employee, remarks, "Due to his concept of 'empty' space, Newton became the founder of French materialism and Russian nihilism against his will."[97]

However, these excursions into historical philosophy would not have been anything particularly special in this period if they had not been undertaken by world riddle solvers for whom scientific theory, particularly their own theory, unequivocally represented a means to improve humanity. That is, the world-view-based dimension of the world riddle solvers' own theories was usually reflected in attempts to treat the crisis of the present day.

Natural science as therapy

The world riddle solvers were among those who wanted to catapult humanity from the catastrophic present day into a happy future with the aid of science. The significance of science to world view acquired additional momentum after World War I, also in the wider population; German science was one of the few positive aspects and something that people could still be proud of. Science was not seen as a means to economic development, but rather above all as a means to uplift the German people spiritually. To some extent, the world war was even interpreted as a crisis caused by a false scientific world view that could be treated by a correct scientific world view. In the Christian Kepler League, for instance, it was said that the war should be considered an expression of the great battle of world views between idealism and materialism.[98] As different as the details of the individual

[95] Kleinschrod, Franz. (1931). *Konnersreuth: Natürliches oder übernatürliches Geschehen? [Konnersreuth: Natural or Supernatural Event?]*. Waldsassen, pp. 133f.

[96] Mewes 1922a, p. 375.

[97] Fricke, Hermann. (1937). Neue Strahlungslehre als Weg zur Verständigung in der Physik. Schwerkraftstrahlung und Physik des ›Als ob‹ aphorisch dargestellt [New radiation theory as a path to understanding in physics. Gravity radiation and physics of the 'as if' presented aphoristically]. *Docentra*, **58**(12), 190–95 (see p. 193).

[98] After the war, the Secretary-General of the Kepler League stated, "In the causes of the war, in its conduct, and also in the revolutionary disruptions commencing after the war, we find a bursting apart of spiritually and materially oriented forces. These battles are the prelude to a battle of the ideologies that the world has never seen in this intensity. If we should succeed in coming through all of the turmoil, then the first thing to do will be to allow emotions to settle down again, and to restore the lost equilibrium of the soul to most people. Could there be anything more suitable for this purpose than a spiritual return to nature and mankind calling on her? The penetration of miraculous nature into the sciences can acquire a cultural significance in this case, but only when it

theories, intended to trigger the change to the good, the true life, ultimately were, both the diagnoses and the recipes for bringing about the redemption of humanity using the correct world view were remarkably similar. The extensive ethical significance of the engineer Patschke's universal law, for instance, resulted from the fact that:

using the technical language of nature as the universal language will contribute to drawing closer to world peace than is possible for nations using egotistical acts of war that disregard public welfare, nations that call themselves civilized, but that have taken things so far with the world war that, instead of exploiting natural forces profitably, they have abused them to a great degree in a diabolical way by making them work for mutual annihilation, and in this way made entire civilized nations poorer than the poorest of beggar nations with regard to health and prosperity for a longer period of time. Civilized nations, learn to understand the natural forces as warmongers and peacemakers, then you will achieve mutual respect and thus world peace.[99]

For Kleinschrod, the hydropathic Kneipp physician, the ethical use of his new medicine was obvious:

A science only elevates itself to significance when it emerges from its narrow framework of research and transmits its results of truth in a broad stream into the people. At this point, then, it is collaborating in elevating the people, it becomes the common property of humanity and a cultural factor. And what science would have a more social nature in its deepest essence than medicine?[100]

In contrast to the treatment plans, which generally had an international focus, Mewes, a Pan-Germanist, thought above all of Germany and the emperor, which were expected to benefit from his theory of historical-philosophical periodicity:

Particularly in the current period of decline, such a refreshment of the heart and the spirit is urgently necessary not just for individuals, but for the entire German people in the same way. Above all, however, a light-filled prospect for the future can become a certain comfort and support for our deeply stooped emperor, who has taken on the crown of thorns of abdication for his people, to await with patience and strength the mission for his people that is reserved for him.

And he emphasized "the refining and educational value of such penetration into the secrets of history, insofar as [Mewes' own] periodicity gives cause for bearing the

occurs not only on the part of professional scientists, but also by the broad mass of the people." Schöning, Max. (1920). Ziel und Zweck des Kepler-Bundes in der Gegenwart [The object and purpose of the Kepler League in the present]. *Unsere Welt [Our World]*, **12**(1), 17–28 (see pp. 17f.).

99 Patschke 1925, p. 104.

100 Kleinschrod, Franz. (1911b). *Denkschrift zur Errichtung eines Lehrstuhles für Erforschung der Gesetze und Vorgänge der Naturheilung, verbunden mit dem Lehrauftrag für sozial-ethische Volksmedizin [Memorandum on the Establishment of a Chair for Research of the Laws and Processes of Naturopathy, combined with a Special Teaching Position for Social-Ethical Folk Medicine]*. In Kleinschrod 1911a, 121–75 (see p. 173).

unavoidable with dignity and helps to strengthen hope for the future."[101] Mewes was convinced that world events ran in periods; therefore, a better time would have to dawn after passing through rock bottom.[102] Zacharias, the engineer, was just as nationalistic. After the world war, he stated, "Until now, science was international. Today, however, it is no longer, as a consequence of the war." Therefore, Germany was supposed to reestablish itself based on German knowledge and to exclude foreign science. "They can take our money and our property, but not the power of our knowledge."[103] Such nationalistic views, which arose primarily in a direct relationship with Germany's defeat in World War I, were actually rather rare among the world riddle solvers. For the vast majority, their knowledge was intended for the benefit of humanity as a whole. Despite his anti-Semitic attitude (cf. particularly pp. 233f.), Ziegler also thought about the use of his primordial light theory less in categories of nationality and race than about the redemption of humanity. "Humans can only correctly be guided by a science that is entirely certain and uniform. Anything else makes it possible to sink back into error or to fall into adversity. Pure science is therefore the sole possibility of redemption for humanity."[104] Ziegler also drew far-reaching social policy conclusions from his unified theory. His unified theory had "a sociological value that could not be denied [. . .] since it represented nothing other than the general concept of order itself. And as such, it is necessarily the worst enemy of every kind of anarchy, not just that of science."[105] Using his principle of "absolute oppositeness," he determined, among other things:

> According to the formula, society first breaks down into the class of the educated and the uneducated, or, from the point of view of material power, into rich and poor. Here, the analogy with the states of matter comes particularly to mind. The richer a rich man is, the greater his freedom of action and the less the resistance to be overcome.[106]

[101] Both quotes from: Mewes 1922a, p. 377.

[102] Such approaches, labeled "historionomy" or historical statistics, are particularly linked to the works of Friedrich von Stromer-Reichenbach, whose goal was to completely calculate the course of history. Cf. Stromer-Reichenbach, Friedrich von. (1924). *Historionomie, ihr Wesen und ihre Bedeutung. Ein wissenschaftlicher Versuch [Historionomy, its Essence and its Significance. A Scientific Experiment]*. Constance. As secondary literature, cf. Wimmer, Franz Martin. Zufall oder Notwendigkeit? Die Frage nach dem Grundgesetz des Weltlaufs [Chance or necessity? The question of the fundamental law of the course of the world]. In *Soziales und geschichtliches Verhalten des Menschen [Social and Historical Behavior of Humans]*, eds. Herbert Wendt and Norbert Loacker, Zürich (Kindlers Enzyklopädie *Der Mensch [Humans]* Vol. 5), 416–34.

[103] Both quotes from: Zacharias 1920, p. 119 (retrospective remark); cf. also Kleinschrod's statement: "May the German spirit and German research, German acumen and German diligence here first carry forward the new torch of humanity, in order to cast the scientific boycott of our hate-filled enemies like rubbish at their feet." Kleinschrod 1920, p. 58.

[104] Ziegler, Johann Heinrich. (1931). *Der grosse Welt- und Selbstbetrug der Physiker: Ein Dokument aus der Gegenwart für die Zukunft [The Physicists' Great Deception of the World and of Themselves: A Document from the Present for the Future]*. Zürich, p. 92.

[105] Ziegler, Johann Heinrich. (1914). *Die Umwälzung in den Grundanschauungen der Naturwissenschaft: Acht kritische Betrachtungen [The Upheaval in the Fundamental Views of Science: Eight Critical Reflections]*. Bern, p. 43.

[106] Ibid., p. 42.

Hermann Fricke (1876–1949)

Photograph courtesy of Ruckhaber and Hellweg, 1940.

Hermann Heinrich Karl Fricke was born in Goslar on December 16, 1876. From 1895–1901 he studied mathematics, physics, and chemistry in Freiburg (Breisgau), Berlin, Munich, and Göttingen, where he earned his doctorate in 1901 under Eduard Riecke (1845–1915). From about 1901–1904, he was a scientific assistant at the Royal Technical College of Charlottenburg and from 1904–1906, a scientific assistant at the Berlin Telegraphic Experiment Office. From 1906, he was a temporary worker at the Berlin Imperial Patent Office, where he was appointed as a permanent employee in 1914 and senior civil servant (*Regierungsrat*) in 1919. In 1928 he became a member of the Berlin Imperial Patent Office and in 1932 a senior government counselor there. Starting in 1906, Fricke had begun working out his own ideas about the essence of natural phenomena. He began with mechanical interpretations of electromagnetism, which he published in Zacharias' periodical *Weltwissen [Universal Knowledge]*, and over the next several years he developed a mechanical ether vortex theory as well as the theory of the "Earth-like sun."

Source: Ruckhaber and Hellweg 1940

For Ziegler, analogies – the simpler, the more insightful – were accepted evidence. Like colors, he also organized humans using the unified theory. It was not only biologically that lives followed the unified theory; character and abilities were also determined by the unified theory. On this basis, Ziegler designed a kind of caste system where each person was to occupy a place in society corresponding to the unified theory, and wise men were to govern as some kind of priest-kings.[107]

Sometimes, the world riddle solvers' efforts at treatment found expression not only on a theoretical level, but also on a practical one. Fricke, the patent office employee, developed a "Plan for Rapidly Overcoming the European Distress" after World War II.[108] What sounds like a precursor to the Marshall Plan is a typical example of projects with a belief in progress that were intended to improve the world using technology. Fricke wanted to solve the physical problems, above all the energy shortage resulting from the loss of coal from the Ruhr region, by using alternative energy sources, particularly water power, and to increase food production using irrigation. Since the Spree and Havel rivers near Berlin had too little water for this large-scale irrigation project, he suggested, in a sub-project entitled "Berlin on the Elbe," to reverse the flow of the Havel by routing water from the Elbe into the Havel.[109]

Patschke, the engineer, wanted to have a "temple of natural forces" erected, a "temple of peace as a symbol of world harmony . . . in order to emphatically glorify the natural forces." In large halls, "the laws of nature, which are comprehensible for anyone" were to be "clearly presented using demonstration objects for the purpose of promoting a true culture encompassing the entire world, reconciling all peoples and serving world peace."[110] He developed a "light ether turbogenerator" – at least on the drawing board – that was supposed to function like an inverse radiometer, driven by ether pressure. The ether turbogenerators were to be mounted at the peaks of temples of nature to be erected all over the world and serve as energy providers for humanity.[111]

The religionization of science

The significance of science for the world view of the world riddle solvers is particularly apparent in their striving to combine science and religion. For a long

[107] Cf. ibid., p. 44.

[108] Fricke, Hermann. (March 1, 1947). *Plan zur schnellen Überwindung der europäischen Not. Manuskript-Entwurf zur Veröffentlichung Nr. 103 zu Fricke's Urheberschriften [Plan for Rapidly Overcoming the European Distress. Manuscript Draft for Publication No. 103 of Fricke's Original Works]*. GN 86-A.

[109] Prepared in Nos. 6 and 26 of the Original Works, which did not survive: *Planung Berlin an der Elbe I* and *Planung Berlin an der Elbe II [Planning of Berlin on the Elbe I* and *II]* and in Fricke 1947.

[110] All quotes from Patschke 1925, pp. 134–36.

[111] Cf. the section on the "Lichtäther-Turbodynamo" [Light-Ether Turbodynamo] in Patschke 1921, pp. 399–401.

time, religious belief and scientific enlightenment were considered in secularization theory as a kind of zero-sum game according to which religion has been increasingly pushed aside by enlightened, rational, agnostic thinking over the course of secularization.[112] However, more recent historical and social scientific work argues convincingly that the search for security, particularly in periods of fundamental insecurity that frequently occur in the course of modernization processes, for absolute knowledge and ahistorical truth occur particularly at the level of religion, above all in the context of extra-ecclesial spiritual offerings.[113] Many world riddle solvers felt called to proclaim the truth for the benefit of humanity and thus presented themselves as "scientific prophets" in this regard.[114] For many of them, their theory had a religious dimension. For some of them, the unification of religion and science was even the top priority.

Kleinschrod, the hydropathic Kneipp physician, was one of the few who was rooted in the Christian religion. He held the view that he had combined knowledge and Christian faith into a higher unity in his meta-mechanics.[115] More often, however, the world riddle solvers were adherents of various forms of nature worship. They assumed that their own science, which guaranteed absolutely certain knowledge of the world, was the actual religion over and above the Christian denominations. Thus Ziegler, the primordial light theorist, explained:

It [the primordial light theory] is that which is positive, dependable, what matters for everything and what everything depends on, all laws and all random world events, everything whatsoever that can be an object of science or explanation. Therefore, the true theory of reality or true physics [the primordial light theory], which today still stands separate from

[112] As a classical position, cf. Max Weber's thesis of the decline of religion and religious conceptions of the world in modern times in Weber, Max. (1920). *Gesammelte Aufsätze zur Religionssoziologie [Collected Essays on the Sociology of Religion]*, Vol. 1, Tübingen, 536–73, particularly pp. 544ff.

[113] Cf. Bruce, Steve, and Wallis, Roy. (1992).Secularization: The orthodox model. In *Religion and Modernization. Sociologists and Historians Debate the Secularization Thesis*, ed. Steve Bruce, Oxford, 8–30; McLeod, Hugh. (1992). Secular cities? Berlin, London, and New York in the later nineteenth and early twentieth centuries. In Bruce 1992, 59–89; Simon-Ritz, Frank. (1996). Kulturelle Modernisierung und Krise des religiösen Bewußtseins. Freireligiöse, Freidenker und Monisten im Kaiserreich [Cultural modernization and the crisis of religious conciousness. Non-denominationalists, freethinkers, and monists in the German Empire]. In Blaschke and Kuhlemann 1996, 457–73; on the discussion of the current state of modernization theory and religion, cf. Elm, Veit. (2001). *Die Moderne und der Kirchenstaat. Aufklärung und römisch-katholische Staatlichkeit im Urteil der Geschichtsschreibung vom 18. Jahrhundert bis zur Postmoderne [Modernism and the Papal States. Enlightenment and Roman Catholic Statehood in the Judgment of Historiography from the 18th Century to the Postmodern Era]*. Berlin (Historische Forschungen Vol. 72), pp. 11–31.

[114] This would be a possible wording based on Linse's "barefoot prophets," the inflation saints of the 1920s. Cf. Linse, Ulrich. (1983). *Barfüßige Propheten. Erlöser der zwanziger Jahre [Barefoot Prophets. Redeemers of the Twenties]*. Berlin.

[115] Cf. Kleinschrod 1931, p. 125.

agnostic physics and separate from religion, has primacy over religion, meaning that henceforth, religion must be guided by it and not conversely, science guided by religion.[116]

The religious dimension of nature was often a topic for the world riddle solvers in their treatises and also in letters to the academic physicists. A Dutch engineer, Lamberty, for instance, explained the following in a letter to Max Planck in 1925: "For me, the substance of the world is God. [. . .] One may call it what one likes, world substance, atomic structure, God, nature, etc. It is always the same Being, and whoever lives in this Being and according to the true laws of this God lives according to the one true religion."[117] This dimension of world riddle solver science will be demonstrated repeatedly, with examples, in the third chapter in the presentation of the bodies of knowledge that were brought to bear against the theory of relativity (cf. particularly pp. 108ff. and pp. 145ff.).

1.5 The anti-academic attitude and dissociation from academic research

An additional line of conflict between world riddle solvers and academic physicists that was later decisive in influencing the debate about the theory of relativity arose from the world riddle solvers' anti-academic attitude. The world riddle solvers were rooted in a tradition of criticism of science as a social institution and particularly of the tenured professoriate, as was expressed at the end of the nineteenth century, particularly by Houston Stewart Chamberlain (1855–1927) and Eugen Dühring (1833–1921). Specialization and intellectual complacency were the accusations – and assigning actual scientific competence to the non-academic researchers was the consequence. Chamberlain was confronted with a specialized scientific landscape where he, as a professed polymath, had to justify taking up the skeptron,[118] which he did by taking the bull by the horns:

> Now, shouldn't an honest, openly admitted dilettantism have certain advantages over a hidden one? Isn't the situation clearer when the author immediately states, I am not a specialist in any field? Isn't it possible that comprehensive unscholarliness does better justice to a large complex of phenomena, that it will move more freely in artistic arrangement than an erudition that has engraved certain furrows in one's thinking due to intensive and lifelong specialized studies?[119]

[116] Ziegler 1931, p. 96. [117] Lamberty to Planck, May 8, 1925, [copy], GN 38-D-5.

[118] Bourdieu uses the example of the skeptron, which gives the speaker authority to speak, in his analysis of the relationship between language and power relationships. Cf. Bourdieu, Pierre. (1991). *Language and Symbolic Power.* Cambridge, MA., e.g., p. 109.

[119] Chamberlain, Houston Stewart. (1904 [1898]). *Die Grundlagen des Neunzehnten Jahrhunderts, 1. Hälfte [The Foundations of the Nineteenth Century, 1st Half]*, 5th edn., Munich, foreword to the first edition of 1898, p. IX. But cf. Chamberlain's opinion on the possibility of specialist contributions by dilettantes in the foreword to the 4th edn. of 1902: "For me, competition between experts and dilettantes is not what is important; I also doubt

Dühring went further, attacking the moral integrity of the academic scientists: "It is in the egotism and crime of the scientists, as well as in their obscurantism and mindlessly shameless insensitivity that the primary restraint on the intellectual and moral progress of the human race is to be sought from now on."[120] In their criticism of the professoriate, the world riddle solvers liked to invoke Chamberlain and Dühring; quoting these authorities likewise imbued the world riddle solver's work with authority and legitimated the avowal of non-academic research. Likewise, many expressions of sympathy and partisanship can be found for individuals who were stylized as warriors and outsiders of the history of science, and with whom people could identify, such as the physician Julius Robert Mayer (1814–1878), who asserted a priority claim to having stated the first law of thermodynamics.[121] Even after the turn of the century, treatises were published that were dedicated solely to the dispute with the professoriate. Thus the teacher and occultist Max Seiling wrote a satirical polemic *Das Professorentum, "der Stolz der Nation"? [The Professoriate, "The Pride of the Nation"?]*[122], and Ernst Westland composed a work about *Universität, Politik und Dummheit [The University, Politics, and Stupidity]* in which, in one sweeping blow against academic science, he also attacked specialization, which he said had lost its perspective of the larger whole. "Special research has become an end in itself. Since that time, it has been ridiculous, and along with it those who carry it out and take such delight in their rag collecting."[123]

whether it will be possible henceforth to achieve anything of scientific significance in any area without specialized knowledge; the layperson who succeeds in doing so is simply a scholar lacking a public appointment." Chamberlain 1904, p. XXIX.

120 Dühring, Eugen. (1905). *Logik und Wissenschaftstheorie: denkerisches Gesammtsystem verstandessouveräner Geisteshaltung [Logic and Philosophy of Science: A Total Intellectual System of Masterfully Reasonable Mentality]*, 2nd revised and expanded edn., Leipzig, pp. 475f. Dühring lost his lectureship in 1877 following vehement personal attacks on Prof. Wagner, a lecturer in the Faculty of Arts of the University of Berlin, and other scientists. Cf. Dühring's own discussion in Dühring, Eugen. (1882). *Sache, Leben und Feinde. Als Hauptwerk und Schlüssel zu seinen sämmtlichen Schriften. Mit seinem Bildniss [Matter, Life, and Enemies. As Magnum Opus and Key to his Entire Works. With a Portrait]*. Karlsruhe/Leipzig, particularly pp. 151–235, as well as the documents published by the Faculty on the affair: *Aktenstücke in der Angelegenheit des Privatdozenten Dr. Dühring [Documents on the Matter of Private Lecturer Professor Dr. Dühring]*, published by the Philosophische Fakultät (Faculty of Arts) of the Royal University of Berlin. Berlin 1877.

121 Cf. e.g. Westland, Ernst. (1901), *Universität, Politik und Dummheit [The University, Politics, and Stupidity]*. Berlin, p. 19; Ziegler, Johann Heinrich. (October 28, 1922). Relativitätstheorie und Ostwalds Farbenlehre. Ziegler oder Einstein? [The theory of relativity and Ostwald's color theory. Ziegler or Einstein?], Part 2. *Luzerner Neueste Nachrichten* newspaper (appearing under the pseudonym J. E. G. Hirzel): "[. . .] one need only recall Kepler, Galileo, and Robert Mayer," who were "envied, hated, boycotted, and whose destruction was attempted." Cf. also Ziegler to Gehrcke: "[. . .] Robert Mayer, [the] Galileo of the nineteenth century," April 26, 1923, GN 6-G-4, and Kleinschrod: "But my discovery shares the fate of all those made outside of university circles. It is simply a discovery by a mere 'general practitioner' whose voice does not easily penetrate into the hermetically sealed university rooms. Robert Mayer was [also] a simple general practitioner." Kleinschrod 1920, p. 62; but cf. also Lenard to Gehrcke, May 7, 1921, GN 3-F-10 about Dühring.

122 Seiling, Max. (1915 [1904]). *Das Professorenthum, "der Stolz der Nation"? [The Professoriate, "The Pride of the Nation"?]*. Leipzig, 3rd revised edn.

123 Westland 1901, pp. 34f.

The sophisticated and specialized landscape of science at the beginning of the twentieth century was seen as a fundamental error by the world riddle solvers, who were interested in large-scale interrelations. It was easy to declare one's allegiance to non-academic world riddle solving research, rejecting such "rag collecting" by the specialized scientists. The monist philosopher Johann Gustav Vogt (born 1849) justified the program of the periodical *Weltwissen [Universal Knowledge]*, an extremely militant publication of the world riddle solvers or "independent researchers,"[124] as the members of the Weltwissen group primarily described themselves, in this way in the first issue:

[W]e want to see this immeasurable material of knowledge [of experimental science] processed into a *systematic universe*. And for this grandiose purpose, we not only need people who observe, but *above all, people who think. Only such thinkers can be the bearers of true universal knowledge*, and *they must finally be heard*, regardless of whether they originate from the scholars' guild or from the wide circles of the educated public. Although our official science may have a monopoly on practical facilities, professorships, laboratories, etc., *they certainly do not have a monopoly on thinking, on universal knowledge.*[125]

Ziegler, the primordial light theorist, protested vehemently against the "ignorance" of the professors in 1914 due to the lack of response to his unified theory:

From the beginning, the professors ignored [. . .] as if on command, the simple chemist who had had the audacity to create order for them. The jolt with which he had overturned their eternally shaky theories had shocked them to the core. [. . .]

Because according to general [. . .] opinion, only those people are allowed to participate in the rehabilitation of science who have already lived intimately with their false theories for decades and have exploited them commercially. Other people are excluded.

These gentlemen merely forgot that thinking is fundamentally free, and [is] not only permitted to those [. . .] everywhere who have obtained a special permit from the highest authority. The true rector directing all thinking is not by any means the Rector of the University of Berlin, but rather common sense, which resides everywhere that he likes.[126]

The engineer Zacharias, an adherent of Ziegler's primordial light theory and founder of the protest periodical *Weltwissen*, expressed particularly extensive criticism of academic science. In his numerous polemics against the "monopoly of knowledge" that he asserted academic science claimed,[127] he describes the dispute between

124 The concept of the "independent thinker and researcher" as a counterpart to the academic professoriate can also be found in Dühring 1905, p. 519.

125 Vogt, Johann Gustav. (1912a). Weltwissen [Universal Knowledge]. *Weltwissen*. **1**(1), 1–4 (see p. 4).

126 Ziegler 1914, p. 142.

127 Cf. Zacharias, Johannes (1912c). Gibt es ein Monopol auf Wissen und Denken? [Is there a monopoly on knowledge and thinking?] *Weltwissen*, **1**(6), 91–93, as well as many other polemical contributions in Vols. 1–3 of *Weltwissen*.

world riddle solvers and academic science not primarily as a divergence in content, but rather as a question of power, as a struggle for the determination of truth:

> If we compare all of the blunders and encroachments by scientists [on 'independent researchers'] then we arrive at the certainty that for many of them, the matter is more about reputation, control, and egotism than about deciding disputed issues. [. . .] What is lacking is not insight, it is willingness. It is a struggle for power, not for scientific theories [. . .] [it is] a public struggle for the truth.[128]

According to Zacharias, this "public struggle for the truth" is being carried out by "hushing up, disparaging, attacking, doubting, mutually praising, research institutes, etc. [These] are the notorious and popular means of many scientific officials against any accomplishments of the independent researchers."[129] In his polemic "Professorial Wisdom," the engineer Carl Wilhelm Meyer opposed the "inflation of one's own opinion" and the "inflation of the assessment of the 'common herd'" due to the professorial title, which would by no means be an indication of outstanding intelligence, pointing out, "But those gentlemen [professors] live *from* science and are all the *more the natural opponents* of those who live *for* science."[130]

This sharp opposition does not fit into the picture of the amateur as a peaceful leisure-time scientist, a hobbyist collecting and studying plants, and a layman willing to learn. This consciously different, this professedly unacademic, this combative research fundamentally took a position in conflict with academic science, whereby the struggle was not only about the truth of theories, but also about recognition and social status – and it was entirely conscious of this. Without exception, the world riddle solvers saw themselves in a situation where they had to promote their work. They turned to two target groups. First, they sought support from the public. Followers were recruited from among their readers, while other world riddle solvers with similar theories became allies in alternative recognition systems.[131] But second and simultaneously, they also appealed to the experts.

Since the goal of the world riddle solvers was to correct the "false" academic science, they sought contact with their expert colleagues, pointed out supposed errors and false assumptions, usually in a friendly manner at first, and brought the

[128] Zacharias, Johannes. (1914c). Ein Vorschlag zur Güte (offenes Schreiben an die führenden Herren Gelehrten) [A suggestion of amicability (open letter to the leading scholars)]. *Weltwissen*, **3**(53–54), 32–33 (see p. 33).

[129] Ibid. (1914b). Akademische Politik [Academic Politics]. *Weltwissen*, **3**(51–52), 25–26 (see p. 25).

[130] Meyer, Carl Wilhelm. (1914). Professoren-Weisheit [Professorial wisdom]. *Weltwissen*, **3**(51–52), 17–21 (see p. 17). Kleinschrod asserted a claim that Kneipp's work was of a scientific nature against voices from academic medicine that denounced Kneipp as unscholarly. "Sebastian Kneipp also has a scientific side and represents a very specific scientific program. As a hydrotherapist, he has introduced valuable innovations [. . .] If a professor at the university had introduced these innovations, then his name would shine 'scientifically' today; but in reality, merely the 'unscholarly' Kneipp introduced them." Kleinschrod, Franz. (1922). Sebastian Kneipp und die Wissenschaft [Sebastian Kneipp and science], Part 1. *Kneipp-Blätter [Kneipp Pamphlets]*, **3**, 58–62 (see pp. 58f).

[131] Cf. e.g. the quotes from letters received in Patschke 1925, pp. 15–27; cf. also pp. 296f. on Fricke.

solution of the world riddles to their attention – without success. But that was only the beginning of the battle for the world riddle solvers. Dramatic phrasing such as, "The struggle is aimed at the highest and most exalted goals, at discerning the eternal in the mundane, the root of all events"[132] were more the rule than the exception.

1.6 Provisional appraisal of world riddle solving

Characteristics of world riddle solver theories

The content of world riddle solver theories may be developed in extremely different ways, but they all assume strong premises about the composition of nature and human cognitive capacity. Anyone for whom harmony and unity of the conception of the world is the highest goal, anyone who wants a theory that "blocks reason from questioning,"[133] has good reasons to assert strong premises and thus to achieve a self-sufficiency of theory that made it irrefutable, similarly to a religious belief system. This self-sufficiency often also consisted of a lack of reference to other theories and the research literature. It is true that reference to acknowledged scientific authorities provided the world riddle solvers with their urgently needed seriousness, and perhaps with access to the academic discourse. But the world riddle solvers usually saw no reason whatsoever to refer to others; the works by world riddle solvers often get by without footnotes or a bibliography and are rather more similar to a manifesto of knowledge than a scientific treatise. In the knowledge of the fact that he was conducting a totally different science, Patschke, the engineer, could self-confidently admit, for instance:

> The reason why I generally have not referred to other authors can be found simply in the fact that my writings represent original work that derives from itself due to its correctness, and which I consequently pulled out of my hat automatically, so to speak, with the aid of a single theorem, that of the [. . .] universal law of natural forces.[134]

Using sufficiently strong premises, world riddle solvers who "brightly and cheerfully"[135] approach the explanation of the universe and thus blithely eliminate any questioning of knowledge can solve world riddles using a relatively simple theory. Academic science, on the other hand, which does not acknowledge their respective premises about the fundamental essence of all things, must go astray according to this perspective. Ziegler remarked quite correctly that his theory was completely

[132] Zacharias 1920, p. 114.
[133] Fincke, Ewalt. (1913). Empiriker und Philosoph [Empiricist and Philosopher]. *Weltwissen*, **1**(7), 101–4 (see p. 103).
[134] Patschke 1925, p. 67. [135] Ruckhaber 1920a, p. 1.

immunized against scientific criticism. "But my explanations were never acknowledged by official physics, but also never refuted, because this is just absolutely impossible."[136] Insofar as reference was made to other works, this usually did not occur in the form of a discussion, but rather as situating one's own work in a particular research tradition whose questions were conclusively resolved for all time by this work.[137] Accordingly, this or that eminent authority such as Maxwell, Reichenbach, Heraclitus, or even Goethe had indeed seen the problem and attempted to solve it, but only the world riddle solver was successful, due to his completely new approach. However, some world riddle solvers, such as Kleinschrod, larded their works with all kinds of references that were even remotely possible, from the Bible to the *Annalen der Physik [Annals of Physics]*, in order to display their scientific nature.

Additional reference authors were representatives of anti-academic positions such as Dühring and Chamberlain as well as Johann Wolfgang von Goethe, who was considered a model of the "true scientist" across the various contexts of world riddle solving, more or less as a figurehead of the Romantic conception of science.[138] These were points of reference that on the one hand were motivated by strategic considerations and resulted from anti-academicism, and on the other hand represented an epistemological position. However, direct influences of scientific and popular science reading can hardly be documented for the development of their own theories. It is true that the world riddle solver theories did not develop outside of a context, specifically because they were usually developed from practical interactions with nature in professional life. The world riddle solvers based their work on knowledge that they had acquired in school, at the university, or in their training, as well as autodidactically. In particular, the engineers, who had often been trained at the Technical Colleges, and the physicians, who received basic education in physics in the context of their studies, had knowledge of classical mechanics, which is reflected in the theories.

[136] Ziegler, Johann Heinrich. (November 22, 1929). Was verstehen wir unter physikalischen und chemischen Elementen und wodurch unterscheiden sie sich hauptsächlich? [What do we understand by physical and chemical elements and how are they primarily distinguished?] *Das freie Wort [The Independent Word]*.

[137] In the context of occult thinking, reference was particularly made to non-academic theoreticians such as Karl von Reichenbach with his Odic force; the sources for vitalistic and organological thinking were vitalists such as Hans Driesch and number theorists such as Wilhelm Fliess, while mechanistic thinking referred to a large number of scientific and popular science sources of the nineteenth century, including treatises by academic scientists such as Maxwell, but also by non-academic theoreticians such as Anderssohn. Since the points of reference differed widely on an individual basis, they are explained in connection with the discussion of the individual theories.

[138] Cf. e.g, Fricke 1919, pp. 51f.; Mewes 1920b, pp. 16ff.; Ziegler 1902, p. 8. But cf. also Mohorovičić, Stjepan. (1923a). *Die Einsteinsche Relativitätstheorie und ihr mathematischer, physikalischer und philosophischer Charakter [The Einsteinian Theory of Relativity and its Mathematical, Physical, and Philosophical Character]*. Berlin/Leipzig, p. 75.

A common characteristic consisted of the fact that these theories emerged from socio-epistemological contexts of the turn of the century. The hydropathic Kneipp physician determined the essence of life, the engineer the essence of force, and the chemist the essence of light. The common factor here is not *what* was determined as the essence of the individual phenomenon, but rather *that* an individual phenomenon was seen as so fundamental that the determination of its essence would solve all of the riddles of science.

World riddle solver theories, as different as their content may be, whether materialistic or idealistic, are unified by the fact that they deduce a self-contained theory from a strong premise about the composition of reality, that they assert an absolute claim as to the truth of this knowledge, and that they are loaded with world view to a great degree.

The world riddle solvers

A glance at the social origins of the world riddle solvers shows that world riddle solving was a phenomenon of the middle class, and within that class, it occurred particularly among academically or technically trained individuals who asserted a special connection to nature due to their professional activity, for instance as an engineer or a physician.[139] It is impossible to unequivocally answer the question of whether a role was played by the fact that this often involved individuals who did not belong to the mainstream, but rather to the fringe groups of the educated middle class due to their membership in a particular occupational group (e.g. hydropathic Kneipp physician, engineer).[140] For some world riddle solvers, personal crises were the decisive factor causing them to set out on the search for absolute certainty. They were members of the "Buddenbrook Generation"[141] around the turn of the century, which was threatened by economic and social decline and therefore saw itself faced with the actual possibility of failure, as was particularly clear in Ziegler's biography. In this context, science also had partly a compensatory and therapeutic function.

[139] There were, of course, exceptions such as Hans Christiansen, who explicitly stated why he felt called as an author and artist to solve the world riddles. Cf. Christiansen, Hans. (1915). *Meine Lösung der Welträtsel [My Solution of the World Riddles]*, 2 volumes, 3rd corrected edn., Wiesbaden pp. 8ff.

[140] Thus the entire professional group of engineers, for instance, can be described as such a fringe group, because engineers did not have a traditional academic education. But only the smallest number of engineers became world riddle solvers. On the engineers as a fringe group of the educated middle class, cf. Späth, Manfred. (1992). Die Professionalisierung von Ingenieuren in Deutschland und Rußland 1800–1914 [The professionalization of engineers in Germany and Russia 1800–1914]. In *Bildungsbürgertum im 19. Jahrhundert [The Educated Middle Class in the 19th Century]*, Vol. 1*: Bildungssystem und Professionalisierung in internationalen Vergleichen [Educational Systems and Professionalization in International Comparisons]*, eds. Werner Conze and Jürgen Kocka, Stuttgart (Industrielle Welt Vol. 38), 561–88.

[141] This sudden vulnerability of the middle class at the end of the nineteenth century, the porosity of the social barriers, was described by Thomas Mann in his novel *Buddenbrooks. The Decline of a Family.* Only now was it possible for a merchant like Mr. Buddenbrook to rise to become a consul, but only now was the possible decline omnipresent as well.

However, only a few world riddle solvers had failed livelihoods. The majority of them were socially integrated citizens who regarded science as the only secure foundation in a period of world view uncertainty. In the tradition of the popular science of the nineteenth century, they were interested in "what holds the world together in its innermost part";[142] in non-academic popular science spaces such as the natural scientist associations, alternative world view movements such as life reform, or in monistic groups, they had a sounding board for their own theories and sometimes had real followings.[143] They understood themselves as citizens competent in the area of science who were able to self-confidently assert their right to have a say on an equal footing with the academic scientists. At the same time, they were so perfused with the correctness and meaningfulness of their own theories that these theories often acquired the status of a religion, and the academic world should either be converted or would be denounced as heretics.

The research of the world riddle solvers was not a supplement to academic science from the amateur sphere, but rather a non-academic parallel science on a confrontational course. And what could be more of a provocation to them than the sudden appearance of Einstein's "competing theory"?[144] A revolution occurred in the foundations of the "false" academic science, as they had always demanded – but it was not their own theory that ultimately prevailed, but rather the theory of relativity. The next chapter will show the types of reception patterns that the non-academic opposition used to react to the incomparable triumphal procession of the theory of relativity.

[142] Goethe, Johann Wolfgang von. *Faust I*, V. 282. [143] Cf. Note 131 (this chapter). [144] Ziegler 1931, p. 20.

2

The confrontation with the theory of relativity

Never before had a scientific theory so fascinated and occupied broad circles of the public as the theory of relativity. On the one hand, the new understanding of physical concepts such as space and time also called the everyday understanding of these concepts into question; on the other hand, the theory of relativity offered many points of connection for debates on world view. In the early 1920s, the theory of relativity was omnipresent: if even an average citizen could not avoid dealing with the theory of relativity, how could a world riddle solver who was competent in science? Or even a physicist or philosopher?

From this point on, this book includes academic physicists and philosophers among Einstein's opponents. This chapter shows that Einstein's opponents perceived the omnipresent theory of relativity as a threat to their bodies of knowledge and could therefore only apprehend it in a particular way and with a defensive attitude. This perceived threat gave rise to countermovements which encompassed both epistemic and socio-political aspects and are investigated later in Chapters 3 and 4.

2.1 The triumphal march of the theory of relativity in the public sphere

On December 14, 1919, the *Berliner Illustrirte* newspaper featured a full-page portrait of Einstein on the front page and stated in the caption, "A New Great Man of World History: Albert Einstein, whose discoveries mean a complete upheaval of our consideration of nature and are equal to the discoveries of Copernicus, Kepler, and Newton" (Fig. 2.1). The *New York Times* had already published an article on November 10th with these sensational headlines: "Lights All Askew in the Heavens. Men of Science More or Less Agog Over Results of Eclipse Observations. Einstein Theory Triumphs. Stars Not Where They Seemed or Were Calculated to Be, but Nobody Need Worry."

Fig. 2.1 "A new great man of world history: Albert Einstein." Source: *Berliner Illustrirte Zeitung* (Berlin Illustrated Newspaper), December 14, 1919.

What had happened? A few days earlier, the prediction of a tiny physical effect in a most complicated theory by a man largely unknown to the public had been officially confirmed. According to the general theory of relativity, at a solar eclipse it must be possible to observe that light rays passing near the Sun are deflected by an amount of 1.7 arcsec.[1] This effect results from the equivalence principle and the curvature of space. Einstein thus predicted a stronger deflection of light than would result from Newton's corpuscular theory of light, according to which a deflection of light could also be explained – the mass of the Sun attracts the mass of the light atoms – but in that case only about half as much light deflection would have been expected. Two English expeditions,[2] one under the leadership of Arthur Stanley Eddington in Principe in the Gulf of Guinea on the west coast of Africa, and the other in Sobral in Brazil,[3] captured the solar eclipse on photographic plates on May 29, 1919, and analyzed it in the following months. On November 6, 1919, it was announced at a joint meeting of the Royal Society with the Royal Astronomical Society that the average of the measurements of the two expeditions was very close to the value required by Einstein. This meant that the general theory of relativity was officially confirmed[4] and Einstein was presented worldwide to the broader public as a "new great man of world history."[5] The fact that the overwhelming public response to the theory of relativity began with the mediagenic confirmation of an experimentally testable prediction of the general theory of relativity – and not in 1905 with the publication of the special theory or in 1915 with the publication of the general theory of relativity – already shows the significance of the mass press and its

[1] Cf. Einstein, Albert. (1915b). Erklärung der Perihelbewegung des Merkur aus der allgemeinen Relativitätstheorie [Explanation of the perihelion motion of Mercury from the general theory of relativity]. Sitzungsberichte der Königlich Preußischen Akademie der Wissenschaften, pp. 831–39 (see p. 834). English translation available in *The Collected Papers of Albert Einstein*, Vol. 6: *The Berlin Years: Writings, 1914–1917*. (English translation supplement), translated by Alfred Engel, eds. A. J. Kox, Martin J. Klein, Robert Schulmann, Princeton, 1997, Doc. 24.

[2] Shortly after World War I, the question "Einstein or Newton" additionally acquired the political dimension of "Germans against Englishmen," which extended far beyond the decision about a scientific theory.

[3] This expedition was under the leadership of Charles Rundle Davidson and A. C. D. Crommelin. Cf. extensive discussion in Crelinsten 2006, pp. 123–45, particularly pp. 129ff.

[4] At the same time, objections were raised that questioned the precision of the measurement results and that turned out to be partly justified. Cf. Crelinsten 2006, pp. 125ff.; Brush, Stephen G. (1999). Why was relativity accepted? *Physics in Perspective*, **1**, 184–214; ibid., (1989). Prediction and theory evaluation: The case of light bending. *Science*, **246**(4934), 1124–29. Since the results were not conclusive, the measurements of light deflection were repeated multiple times after 1919. The expeditions that were to re-measure light deflection at the solar eclipse of 1922 likewise experienced an overwhelming media response, as is documented in numerous newspaper articles in Ernst Gehrcke's papers. On the 1922 expedition, cf. Crelinsten 2006, pp. 183ff.

[5] On the press reporting in the German-speaking area, cf. Braun, Traute. (1980). *Die Entwicklung der Physik im Spiegel der Presse: Berichterstattung des Berliner Tageblattes 1919–1923 [The Development of Physics as Reflected by the Press: The Reporting of the Berliner Tageblatt 1919–1923]*, 2 volumes, [Master's Thesis, University of Stuttgart]; Elton 1986; Könneker 2001, particularly pp. 117ff.; for a contemporary source, cf. Gehrcke, Ernst. (1924). *Die Massensuggestion der Relativitätstheorie [The Mass Suggestion of the Theory of Relativity]*, Berlin; for the U.S. cf. Pais 1994; for France, cf. Biezunski, Michel. (1991). *Einstein à Paris: temps n'est plus . . .*, Paris.

principles for this phenomenon (the other mass media, radio, film, and television, only played a relevant role at a later time).

The newspaper environment in the Weimar Republic

With more than 5,000 daily newspapers, which often were published in multiple daily editions, the number of newspapers in the Weimar Republic reached a peak that has never been exceeded since.[6] The daily newspapers of Berlin in particular achieved high circulation numbers; the *Berliner Morgenpost* from the Ullstein publishing house sold about 400,000 copies in 1914, while the *Berliner Tageblatt* from Mosse sold just under 230,000. The Mosse group also owned the *Berliner Morgen-Zeitung*, with a circulation of about 100,000 copies around 1900, and the *Berliner Lokal-Anzeiger*, founded in 1883 by August Hugo Friedrich Scherl, with a circulation of about 150,000.

The newspaper environment of the Weimar Republic was politically fragmented,[7] which encouraged the politicization of the debate about the theory of relativity. The views for or against the theory of relativity, as well as for or against Einstein as a person, corresponded conspicuously with the political orientation of the newspapers. It is possible to make a rough distinction here between the democratic-liberal press and the German nationalist and right-wing press. While the democratic-liberal press from the Mosse and Ullstein publishing houses was rather more inclined in favor of the theory of relativity, things looked different in newspapers from the corporate group of the German nationalist media entrepreneur Alfred Hugenberg. The most influential one from the right-wing political spectrum was the *Deutsche Zeitung;* others that stood out in the critical reporting on the theory of relativity were the *Staatsbürgerzeitung [Citizens' Newspaper]*, *Der Tag [The Day]* and *Die Wahrheit [The Truth]*, the latter with particularly harsh anti-Semitic articles.

The composition of the journalistic staff was also of fundamental significance for the reporting – an influencing factor that was connected to the political orientation of the newspaper, but cannot be reduced merely to that. It is true that at this time the professional profile of an editor responsible for specific departments was just developing; but smaller newspapcrs frequently consisted of a single editor covering all fields. How this editor regarded the theory of relativity varied. Smaller newspapers

[6] This discussion is based on Dussel, Konrad. (2004). *Deutsche Tagespresse im 19. und 20. Jahrhundert [The German Daily Press in the 19th and 20th Centuries]*. Münster [*et al.*], pp. 85–158. On the significance of the press to the Weimar public, cf. Moores, Karen. (1996). *Presse und Meinungsklima in der Weimarer Republik [The Press and the Climate of Opinion in the Weimar Republic]*. [Dissertation, University of Mainz].

[7] Cf. Schottenloher, Karl, and Binkowski, Johannes (eds.). (1985). *Flugblatt und Zeitung. Ein Wegweiser durch das gedruckte Tagesschrifttum [Leaflet and Newspaper. A Guide to the Printed Daily Literature]*, Vol. 2: *Von 1848 bis zur Gegenwart [From 1848 to the Present]*. Munich (Bibliothek für Kunst- und Antiquitätenfreunde Vol. 21/2), pp. 106ff.; Moores 1996.

often simply took the agency reports from "Wolffs Telegraphischem Büro," the most important German news agency, or from a larger newspaper belonging to its parent company. But even at those newspapers, science journalism was often taken care of on the side; science reports were found under "Miscellaneous" or "Art and Science." The author of the reports is usually unknown. More extensive articles, which then also carried a byline, were often written by external experts. It is noteworthy that at some newspapers, such as the nationalistic *Deutsche Zeitung*, the reporting on the theory of relativity is almost exclusively in the hands of a professed opponent of the theory of relativity who was likewise brought in from the outside as an "expert."[8]

The theory of relativity in popular culture

The theory of relativity was assimilated into popular culture[9] through the mass press. *The Times* of London had already pointed out in 1921 that the concept of "relativity" was virtually an invitation to jokes, satires, and cartoons.[10] But would the German term *Bezügigkeitsanschauungstheorie* (roughly *relational observation theory*), which was discussed in the press as a new name for the theory of relativity,[11] really have prevented the Einstein euphoria? It was certainly not just the familiar quotation "everything is relative," but rather, above all, the astonishing statements by the theory that clocks would run more slowly and objects would get shorter as a consequence of motion (cf. pp. 96f.), which challenged people to think about the effects of the theory of relativity on everyday life – because what was more in motion than the "Roaring Twenties"? *Der (schwarze) Tag [The (black) Day]* newspaper raised the following considerations about the connection of the theory of relativity to the present under the headline "Relativity theory and practice":

> Since Einstein we know that there are no absolute time and space dimensions, that they are instead relatively dependent upon one another. [. . .]
>
> Starting from these considerations, one can describe our time, measured by various dimensions, as faster than the past (a pleasant prospect, to emerge from the mess soon), since typical objects are showing a conspicuous inclination to reduce their dimensions compared to earlier. [. . .] Hats are getting smaller and smaller, skirts shorter and shorter, and a Parisian fashion designer is announcing very low ladies' heels. Any lady who thinks anything of herself will never again take a greyhound out on the promenade, but at most a toy Chinese pug.

[8] Cf. pp. 204ff., 212ff. on these niches of the counter-experts. The *Deutsche Zeitung*'s Einstein expert was the astronomer Johannes Riem.

[9] On the "popular culture," "mass culture," and also "pop culture" developing around the turn of the century, cf. e.g. Maase, Kaspar. (1997). *Grenzenloses Vergnügen. Der Aufstieg der Massenkultur 1850–1970 [Unlimited Amusement. The Rise of Mass Culture 1850–1970]*. Frankfurt (Main).

[10] "Hints on Einstein" in *The Times*, April 3, 1922, GN 4, XVII, 102r.

[11] "Einsteins – Bezügigkeitsanschauungstheorie" in: *12 Uhr Mittagszeitung [12 o'clock Midday Newspaper]*, July 8, 1921, GN 4, XI, 26r.

The Berlin Municipal Authority has ordered whole series of small cars, and the accommodation assigned to me and my wife by the housing office can only be entered with a magnifying glass before your eyes. Our politics is dominated by short-lived cabinets with a conspicuously narrow horizon. Signs enough for those who have eyes to see with.[12]

In August 1921, more than twenty German newspapers reported that it had been discovered that according to the theory of relativity, people would become thinner while dancing the waltz if they danced in the direction of the Earth's rotation, and fatter if they danced in the other direction. The press made observations as to the social complications that this would lead to, since fat people and thin people would scarcely be able to dance together now, since one of them would have to lose weight and the other gain. "According to this new theory, nothing else will be possible than for the dancing couples to pair up by weight, and it is to be feared that many a love affair will be disrupted by these unexpected effects of the theory of relativity."[13] Einstein himself was increasingly being elevated above normal mortals – worldwide. On the occasion of Einstein's visit to London in 1921, which was politically extremely relevant so shortly after World War I,[14] the *Daily Mail* reverently reported:

He came, he saw and conquered. He is a German and gives his lectures in the German language, a fact that doesn't exactly commend him to an English public. He is a Jew. He brings with him a new theory, you might say a fantastic theory, that tosses all previous views out the window, and the scientific, academic world is very conservative, as is well known. He is a revolutionary. And yet, people received him not just kindly, but rather quite enthusiastically. Where does this come from? Because Professor Einstein is a man whose genius is written on his face [. . .] Whether or not you understand his language is irrelevant. You know that you are under the spell of a captivating personality, an immense intellectual power.[15]

People were interested in Einstein's effect on women ("Einstein in Japan. A misunderstanding by the Japanese women about the theory of relativity"[16]) and in the effect of women on Einstein ("Einstein sails. U.S. women 'lovely'"[17]). People informed themselves about his telephone habits ("Einstein, the enemy of the

[12] Korff. (June 15, 1921). Relativitätstheorie und Praxis [Relativity Theory and Practice] in *Der (schwarze) Tag [The (black) Day]*, GN 4, X, 47r.

[13] Einsteins Theorie im Ballsaale. 'Walzer nach den Gesetzen der Relativitätstheorie' [Einstein's theory in the ballroom. 'Waltzes according to the laws of the theory of relativity']. *National-Zeitung*, August 24, 1921, GN 4, XI, 62r, and identical versions in at least twenty additional newspapers.

[14] Cf. e.g. Grundmann 2005, pp. 118f.; Fölsing 1993, pp. 581ff. On Einstein's role as a semi-official German diplomat, cf. particularly Grundmann 2005.

[15] Quoted according to the *Halleschen Nachrichten* newspaper, Einstein in englischem Urteil [Einstein in the English judgment], June 21, 1921, GN 4, X, 25r.

[16] Einstein in Japan. Ein Mißverständnis der Japanerinnen über die Relativitätstheorie [Einstein in Japan. A misunderstanding by Japanese women on the theory of relativity]. *Berliner Tageblatt* newspaper, 1923 [no specific date], GN 4, I, 54r.

[17] Einstein sails. U.S. women "lovely." *The Chicago Tribune*, June 1921 [no specific date], GN 4, IX, 36r.

telephone"[18]), about how he dressed ("Prof. Einstein's frock coat"[19]), and how he made music ("Einstein as a violinist"[20]). And just as the theory of relativity seemed to have a connection to everything, Einstein also had to have an opinion about everything. In the context of a single interview, he was asked about "the distress of the intellectual workers in Germany, the situation of the middle class, the connection between economic conditions and the inclination to commit political violence, the effects of association policy, the stabilization of the German mark, the possibilities of 'socialization'."[21] In the 1920s, the theory of relativity ultimately became so widespread that it turned up in recipes or had to serve as a comparative standard for the new collective bargaining law (which was incomparably more difficult to understand according to the opinion of the author of the article).[22]

The public media were the precondition for the public response to the theory of relativity (radio and film also came into play in the following years[23]). And yet the triumphal march of the theory of relativity cannot be reduced to the "agenda-setting" of a sensationalist press. The theory of relativity offered everything that could make it interesting as a public topic. In a period that was highly unstable politically and economically, a new "grand theory" offered a welcome change, certainly not just as an amusement and a diversion. Even today, astronomy and cosmology are the most fascinating areas of science for large groups of the public. Occupying oneself with the meaning of the new theory for one's scientific conception of the world and for one's world view offered an orientation towards something greater and more exalted than arduous daily life. The theory of relativity reorganized fundamental concepts such as space and time; this new understanding stood in contradiction to previous ideas which were deeply anchored and reinforced by everyday experience. At this fundamental level – What is actually time, which we experience in daily life as the constantly passing time of hours, days, weeks, and years, and as the duration of specific moments? What is space, in which we are situated and which we experience through our senses? – the theory of relativity was not an internal scientific development interesting only to specialists, but instead meaningful for everyday knowledge.

[18] Einstein, der Feind des Telephons [Einstein, the enemy of the telephone]. *Hamborner General-Anzeiger* newspaper, April 11, 1922, GN 4, XVII, 153r.

[19] Prof. Einsteins Gehrock [Prof. Einstein's frock coat]. *Leipziger Neueste Nachrichten* newspaper, July 1, 1921, morning edn., GN 4, XI, 27r.

[20] Einstein als Violinist. *Vossische Zeitung* newspaper, May 20, 1921, morning edn., GN 4, IX, 30r.

[21] Albert Einstein über die deutsche Not [Albert Einstein on German distress]. *Boersen-Courier Berlin* newspaper, October 6, 1922, GN 4, I, 68r. This was a reprint of Einstein's answers, translated into German, to a list of questions from Brailsford, the editor-in-chief of the English periodical *New Leader*.

[22] Heldt, A. (September 3, 1921). Wirtschaftliche und politische Gegenwartsfragen in Amerika [Current economic and political questions in America]. *Esslinger Zeitung* newspaper.

[23] A movie about the theory of relativity was filmed starting in 1920 and completed in 1922. Cf. Wazeck, Milena. (2010). The 1922 Einstein Film: Cinematic Innovation and Public Controversy. *Physics in Perspective*, **12**(2), 163–79.

Fig. 2.2 "The people will have their idols. . ." – Einstein and Rabindranath Tagore, winner of the Nobel Prize for Literature. Source: *Berliner Illustrirte Zeitung* (Berlin Illustrated Newspaper), August 25, 1921.

The new space and time concepts collided with previous views and challenged artists, authors, and theologians as well as the "man on the street" to deal with them seriously. In the 1920s, even fundamental questions of science were discussed in the daily newspapers. Thus, for instance, after an article about the theory of relativity by the English physicist Sir Oliver Lodge[24] appeared in *The Times* of London, a discussion arose in the letters to the editor about whether empty space could actually have physical properties, as Lodge had described in his article as a statement from the general theory of relativity. Thomas Case, a professor for moral philosophy and metaphysics and Dean of Corpus Christi College at Oxford University, questioned this:

> Sir, in a letter which you published on November 25, Sir Oliver Lodge quoted with approval a statement of Professor Einstein that 'according to the general theory of relativity space is endowed with physical qualities.' How far is this statement true? Space is an attribute of body, and to that extent Professor Einstein's statement is true. But it does not follow that space is a physical attribute of the body; on the contrary, it is without the most essential physical attributes of body. It is not impenetrable, since all finite bodies, resting or moving, penetrate space during time. It offers no resistance. It is neither passive as matter nor active as force. Therefore, though an attribute of body, space is not 'endowed with physical qualities'.[25]

If the theory of relativity had not been so omnipresent in the media, Case would hardly have felt cause to present his view of the essence of physical space in a letter to the editor of a daily newspaper, and thus to take a position on the theory of relativity.

Another focus of the public discussion was a debate about the significance of the theory of relativity to world views. In a period when world views had become fragile and had forfeited their self-evidence, and even their very justification had been called into question, the theory of relativity was usually negatively charged from the perspective of world view. For the author Hans Natonek, the theory of relativity was symptomatic of the "disunity" and "nihilism" of the time[26], and in 1921, the General Secretary of the Catholic Bonifatius League stated, "The dangers of the present can be put into three key words: Einstein's theory of relativity, Spengler's decline of the West, and Count Keyserling's travel journal."[27] Only a few people judged the theory

[24] On Lodge, cf. e.g. Lunteren, Frans Herbert van. (1991). *Framing Hypotheses: Conceptions of Gravity in the 18th and 19th Centuries*. Utrecht, pp. 232ff.

[25] Case, Thomas. (December 3, 1921). From Euclid to Einstein. Is Space Physical? The Test of Relativity. To the Editor of The Times, *The Times* (London), GN 4, XV, 24r; and Lodge, Oliver. (December 10, 1921). From Euclid to Einstein. Sir O. Lodge and Relativity. The 'Interim Tool of Exploration'. To the Editor of The Times, Ibid., GN 4, XII, 32r; Case, Thomas. (December 24, 1921). From Euclid to Einstein. Infinite Body of the Universe. Time in Connexion With Space. To the Editor of The Times, Ibid., GN 4, XV, 22r; Fraser, Bodie A. (December 27, 1921). From Euclid to Einstein. To the Editor of The Times, Ibid., GN XV, 23r.

[26] Natonek, Hans. (July 22, 1921). Relativismus [Relativism]. *Weser-Zeitung* newspaper, GN 4, XI, 40r.

[27] Quoted from Die Berliner Bonifatiustagung. Minister Dr. Brauns über die nationalen Ideale des Katholizismus [The Berlin Bonifatius Conference. Minister Dr. Brauns about the national ideals of Catholicism]. *Kölnische Volkszeitung* newspaper, June 18, 1921, evening edn., GN 4, IX, 9r. On the contemporary discussion of links between Einstein and Spengler's *Untergang des Abendlandes [Decline of the West]*, cf. Könneker 2001, pp. 145ff.; on the link between modern physics and *Untergang des Abendlandes* also cf., in general, Forman 1994.

of relativity positively from the perspective of their world view and saw in it a departure to "a healthy and useful relativistic world view."[28]

At any rate, the theory of relativity was predominantly perceived as significant from a world-view perspective[29] and, precisely for this reason, it was felt by various world-view contexts to be a provocation and an uncertainty. Waldorf school teachers just as cardinals, vitalists just as monists, all had to assess the significance of modern physics for their world view and to defend it if necessary. Occultists had to take a position on four-dimensional spacetime. After all, their very own topic, the fourth dimension, had suddenly become the favorite topic of academic physics.[30] At the same time, the theory of relativity was certainly not always perceived as a threat. Some people saw in it a confirmation of their own views, such as the Baltic botanist Karl Reinhold Kupffer (1872–1935), who considered the Lorentz transformation (the transformation of space and time coordinates between inertial frames of reference) as confirmation of his vitalistic epistemology.[31] The decisive factor is that it was difficult to avoid dealing with the theory of relativity at the beginning of the 1920s.

It was the ubiquity of the theory of relativity on the one hand, and the universality and fundamentality with which the foundations of physics had been rewritten on the other, that also almost forced the world riddle solvers to take a position as described in the following sections. For citizens interested in science who had not yet appeared on the scene as world riddle solvers, the theory of relativity sometimes represented just the right occasion to express their own views on time, space, or gravitation – something as false as the theory of relativity could not simply be left standing as the latest achievement of science. Thus, in a letter to Gehrcke, a Mr. Hauck stated: "But I was first led to write down my thoughts due to the fuss about the rel. theory."[32]

The question of whether the theory of relativity had any significance for politics, economics, religion, philosophy, or private life could be debated. But there was no

[28] The words of the Basel *Nationalzeitung* newspaper, reporting on a lecture by Rudolf Lämmel. Relativität und Weltanschauung [Relativity and ideology]. *Nationalzeitung* newspaper, November 30, 1921, GN 4, XII, 64r.

[29] In the daily press, however, many opinions were published that emphasized that the theory of relativity had *no* ideological significance; only a portion of these opinions can be interpreted purely as a defensive reaction.

[30] Cf. Henderson, Linda Dalrymple. (1985). Theo van Doesburg. Die 'vierte Dimension' und die Relativitätstheorie in den Zwanziger Jahren [Theo van Doesburg. The 'fourth dimension' and the theory of relativity in the Twenties]. In *Zeit. Die vierte Dimension in der Kunst [Time. The Fourth Dimension in Art]*, ed. Michel Baudson, [exhibit catalog], Weinheim, pp. 195–205; cf. also Note 30 in Chapter 1.

[31] Kupffer argued as follows: If the Lorentz transformation would mathematically allow insertion of a negative value for t, then it would be conceivable that, if time actually did run backwards, this would invert causality relationships, and finality (vitalism) would be directly plausible instead of causality (materialism). Cf. Kupffer, Karl Reinhold. (1925). 'Materialismus, Vitalismus und Relativitätstheorie'. Vortrag gehalten auf dem Jahresaktus der Herder-Gesellschaft zu Riga am 7. September 1924 ['Materialism, vitalism, and the theory of relativity.' Lecture held at the annual meeting of the Herder Society of Riga on September 7, 1924]. *Abhandlungen des Herder-Institutes zu Riga [Transactions of the Herder Society of Riga]* **1**(1), Riga, pp. 23–49.

[32] Hauck to Gehrcke, June 17, 1924, GN 29-E-1.

doubt about the fact that it appeared as a competing universal physical theory – in the eyes of the world riddle solvers.

2.2 The mental block to reception

The opponents of the theory of relativity were not concerned with the theory of relativity. This finding may seem paradoxical at first glance. That it is accurate nevertheless was first suggested in the presentation of world riddle solving in Chapter 1 and will become even more clearly apparent in the following discussion of the specifics of what the opponents of the theory of relativity actually knew about it and what motivation caused them to carry on the dispute. This section first presents the specific reception pattern of Einstein's opponents. The following section describes the threatening situation perceived by Einstein's opponents and their resulting defensive attitude.

Many of Einstein's opponents had not even understood the basics of the theory of relativity – and this applies not only to the non-academic opponents of the theory of relativity, but also to the academic physicists and philosophers among them. Thus, for instance, Kleinschrod, the hydropathic Kneipp physician, presented the theory of relativity in the introduction to his booklet against it, using the example of a train.[33] For reasons of clarity, however, he replaced the flashes of light from the original example with two thunderclaps that are heard from the train and from the embankment.

> He [the man in the train] will hear the thunder that he is travelling toward much earlier than the other thunder that he is hastening away from, although both thunderclaps occur simultaneously in nature. *Now Einstein labels this process as one of 'relativity of time' or the 'relativity of simultaneity,' and this is what his actual discovery consists of.*[34]

With his blithe replacement of the spread of light – lightning flashes – with the significantly slower spread of sound waves – thunderclaps – that are not spreading with constant speed independently from the state of motion of the observer, it is not just that Kleinschrod proves that he doesn't even have the ghost of an idea about the special theory of relativity; no, he also concludes from this that Einstein had discovered the classical addition of velocities. In fact, Kleinschrod goes on to speculate that the theory of relativity probably had its origin in the "personal equation" of the astronomers.[35] Many additional pieces of evidence can be cited

[33] Cf. p. 100 as well as Einstein, Albert. (1917a). *Über die spezielle und die allgemeine Relativitätstheorie. Gemeinverständlich [On the Special and General Theory of Relativity (A Popular Account)]*. Braunschweig (Sammlung Vieweg Vol. 38), pp. 16ff. English translation available in Kox *et al.* 1997, Vol. 6, Doc. 42.

[34] Kleinschrod 1920, p. 9.

[35] The "personal equation" includes the reaction times of the observing researchers; it played a role particularly in astronomy. On this topic, cf. e.g. Canales, Jimena. (2001). Exit the frog, enter the human: Physiology and experimental psychology in nineteenth-century astronomy. *The British Journal for the History of Science*, **34**, 173–97.

for the fact that Einstein's opponents' conception of the theory of relativity was characterized by more or less gross misunderstandings; this will become clear particularly in the content-based dispute with the theory of relativity presented in Chapter 3. However, it is more informative to ask what these misunderstandings resulted from. The lack of understanding of the theory of relativity is already a phenomenon in itself. According to the bibliographies in their works, many of Einstein's opponents had consulted literature that certainly would have been suitable to impart a solid basic knowledge of the significant content of the theory of relativity. This applies particularly to the physicists who had read Einstein's original works. In contrast, the academic philosophers and the non-academic critics predominantly referred to Einstein's popular writings and other mostly serious popular literature. Almost all of them knew Einstein's *Über die spezielle und die allgemeine Relativitätstheorie. Gemeinverständlich [On the Special and General Theory of Relativity (A Popular Account)]*. Why were Einstein's opponents so resistant to explanations of the theory of relativity?

This is where the mental block to reception comes into play. With regard to Einstein's opponents who argue from classical physics and philosophy, it is evident that in some way they had to deal with the challenge that the theory of relativity represented to their views and theoretical conceptions, such as those on space or time. They did so either by revising or immunizing their own convictions, or by opposing the theory of relativity, as shown in this book, or – in very few cases – by ignoring it.

The same holds true for the world riddle solvers. When Patschke, the engineer, read a description of how well the theory of relativity can explain the outcome of the Michelson–Morley experiment (cf. p. 138), he had a perspective that was limited by his firm belief in his universal law; he was not open to the arguments and did not want to be convinced. He already had a theory in mind: he already knew that his light ether theory was correct and that the theory of relativity must be false. He therefore searched the secondary literature on the theory of relativity, on the one hand for statements that in his opinion confirmed his light ether theory more than the theory of relativity, and on the other hand for statements that proved the theory of relativity false from the perspective of light ether theory. It was not his intention to understand the theory of relativity, but rather to show that an interpretation from an atomistic point of view likewise explains the result of the Michelson–Morley experiment – thus rendering the theory of relativity superfluous.[36] In criticizing the theory of relativity, the world riddle solvers wanted above all to present their own theories as the victor over the theory of relativity. When they read texts about the theory of relativity in preparation for this project,

[36] Cf. Patschke 1920, pp. 11ff.

this consequently did not involve an attitude that was open to new knowledge and instruction, but rather selective reception from a latent defensive attitude. This is a crucial difference between them and the reception behaviors of the public interested in science and the amateur scientists who are willing to learn. For a world riddle solver, everything that he read was considered from the point of view of his own theory as either supporting it – "forerunners" – or running counter to it – "opponents." Their own theory was the scale by which all previous and current science was measured. Thus the world riddle solvers were only able to categorize the theory of relativity in a very specific framework and therefore developed a specific version of the theory of relativity that seemed to be "limited" from a different point of view.

Another determining factor for the reception of the theory of relativity turns out to be the fact that, due to this selective reception and their conviction that they themselves possessed the true theory, a problem of understanding usually did not even exist for the critics of the theory of relativity. While the eagerly curious public wrestled with understanding the theory of relativity, the errors of the theory of relativity were evident from the perspective of the world riddle solvers. Usually, they found a "quite elementary mistake": "[T]he *error in Einstein's theory* lies in the complete neglect of the ether."[37] It is certainly possible to argue against this view of Fricke's based on academic physics, but it will hardly be convincing to Fricke, who considered academic physics to be completely wrong. Who will be able to dissuade a world riddle solver like Fricke, who had developed an ether vortex theory over a period of years, from considering the existence of the ether to be the central argument against the theory of relativity?

The lack of understanding of the theory of relativity among the world riddle solvers thus went back primarily to incommensurability effects. However, this was a reciprocal relationship – the lack of understanding occurred on both sides. Since the relationship between non-academic and academic research was highly asymmetrical, the situation was such that one side vehemently tried to convince the other, while the other side had no need whatsoever to deal with the critics. Because academic physics had good reasons for accepting the theory of relativity, it thus defined what constituted physics and what was "outdated" and "unscientific." Understandably, this was perceived differently from the underdogs' perspective. For example, Fricke held the view that there were two opposing camps in physics – "relativists" and "ether physicists" – with equal rights. Fricke made extensive observations about why "the physicists" didn't understand him, whereas conversely he could understand modern physics, after all. "People have [. . .] said to me: 'You

[37] Fricke 1920a, p. 18.

speak a different language.' I would like to answer: 'I speak German. What do you speak?'"[38] For his part at least, no communication problem existed. "I understand quite well what the relativists think."[39]

In fact, the world riddle solvers strongly opposed their disqualification as inadequate discussion partners. In a polemic against Theodor Lessing, who had made critical remarks about the polemics that completely missed the point of the content of the theory of relativity (cf. pp. 88f.), Georg August Grote, the founder of the popular science periodical *Der Forscher [The Researcher]*[40] remarks, "The only objection that the theoreticians of relativity raise against us [Einstein's opponents] is that we don't understand the theory of relativity. Personally, I only don't understand what would be so difficult to understand about the foundations of the theory of relativity?"[41]

The academic physicists among Einstein's opponents, who adhered to classical physics, found themselves confronted with similar disqualifications as discussion partners (cf. pp. 193ff. on the special features of the marginalization of the academic physicists) and showed the same reaction pattern as the non-academic researchers, such as the physicist Johannes Stark (1874–1954):

> [O]n the side of its [the theory of relativity's] defenders, today they are hiding from the criticism of experimental physicists behind the convenient excuse that they would not have the higher mathematical training that would be necessary to understand the general theory of relativity. Accordingly, those physicists who criticize it do not have the necessary mathematical talent to understand it anyway; they are relegated to a lower class compared to the theoreticians of relativity.[42]

Gehrcke also defended himself against alleged "knockout arguments" by the "relativists." Anyone who criticized Einstein "is either called an anti-Semite or someone who is too stupid to be able to grasp the theory of relativity, or both."[43] The "experimentally-oriented physicists" felt particularly forced to rebel "against what

[38] Fricke 1939a, p. 89. The fact that he had correctly understood the theory of relativity himself, "this is guaranteed to me by his [Einstein's] own presentation in the document 'On the Special and General Theory of Relativity'." In a 1936 letter to Gehrcke, he spoke about the difficulties of discussion with academic physicists, saying, "My view is, so to speak, a slap in the face of the generally prevailing opinion. Now you try and speak objectively and calmly with a man who believes that you have just slapped him." Fricke to Gehrcke, December 21, 1936, GN 33-H-3.

[39] Fricke 1939a, p. 132. Although Fricke himself was a physicist with a doctoral degree, he obviously did not count himself on the side of "the" physicists, which once again shows that boundaries drawn using formal criteria do not work with the world riddle solvers.

[40] This periodical belonged to the Bund deutscher Forscher [League of German Researchers], a popular science association founded in 1909; cf. Daum 2002, pp. 187f.

[41] Grote, August. (1922). [Reply to Lessing 1922]. *Der Forscher [The Researcher]* **8**–10, 14.

[42] Stark, Johannes. (1922). *Die gegenwärtige Krisis in der deutschen Physik [The Current Crisis in German Physics]*. Leipzig, pp. 10f.

[43] All quotes from: Gehrcke, Ernst. [(ca. 1921)]. Einsteins Relativitätstheorie – ein Bluff! [Einstein's theory of relativity – a bluff!]. [Unpublished typescript], GN 4-B-4. On the role of anti-Semitism among Einstein's opponents, cf. pp. 230ff.

was in their opinion the unphysical spirit of the theory of relativity"[44] that had propagated itself in their discipline. They transformed their non-reception into a reproach to modern physics, which made them, the competent classical physicists, out to be "stupid." This mental block to reception that some experimental physicists had, resulted from their deep rootedness in the mental approach of classical, "lucid" physics, which so decisively shaped their understanding of reality that it could not be given up (cf. particularly pp. 147ff.).

Investigations of where Einstein's opponents got their – often deficient – knowledge of the theory of relativity are not very informative. This is because the world riddle solvers were less concerned with understanding the theory of relativity or what they thought was the theory of relativity, because they were instead concerned above all with interpreting the theory of relativity within the framework of their own theory. It did happen that incorrect statements from the popular literature were taken to be "the theory of relativity," but that ultimately had hardly any influence on the motivation and structure of the criticism. The question of whether Einstein's opponents had used original sources or popular adaptations bears no relationship to their views about the theory of relativity and says nothing about *who* understood these works, and *how*. This becomes apparent in the popularity of Einstein's book *Über die spezielle und die allgemeine Relativitätstheorie. Gemeinverständlich [On the Special and General Theory of Relativity (A Popular Account)]*. It became widely used as a basis for discussion, not because it explained the theory of relativity particularly well, but rather because it legitimized the criticism in a special way. Einstein had stated that with this document, it would be possible to obtain a fundamental understanding of the theory of relativity even without extensive knowledge of mathematics.[45] If this were the case, the world riddle solvers then argued, taking Einstein literally, the theory of relativity could also be criticized on the basis of this generally comprehensible book. The jurist Eddo Thedinga, for example, justified his dispute with the theory of relativity in precisely this way:

> In Einstein's document 'On the Special and General Theory of Relativity,' which he described as generally comprehensible, he attempted to impart to laymen as exact an insight into the theory of relativity as possible. He assumes that readers have 'a high school education, quite a lot of patience and willpower.' I can provide documentary evidence of the first condition, and I have made a respectable effort to satisfy the other conditions as well. From this I derive the authorization to also make statements about the theory of relativity as a layman.[46]

[44] Stark 1922, p. 16. [45] Cf. Einstein 1917a, Foreword.
[46] Thedinga, Eddo. (1922). *Einstein und wir Laien: Wege zu einer neuen Licht-Theorie [Einstein and We Laypeople: Pathways to a New Light Theory]*. Leipzig. Foreword.

The fundamentally Janus-faced nature of the "dominant view"[47] of science popularization is particularly apparent in the popular science works on the theory of relativity. Certainly, simplified and processed knowledge is supposed to be provided to the public, but it is considered a lesser form of knowledge because it is merely summarized in a popular way, after all. Any challenge to this knowledge by the public is then blocked using the argument that the public does not have *real* knowledge of the material. The point where general comprehensibility stops and the exclusivity of scientific discourse begins – described by Holm Tetens as "The Boundary" in an article using the general theory of relativity as an example[48] – is remeasured constantly, not least with each attempt at popularization. In order to participate in the technical discussion about the general theory of relativity, one would doubtless need to have a more comprehensive foundation than a popular introduction can offer. However, Einstein's opponents were not at all interested in this technical discussion; instead, they categorically challenged the existence of such a "boundary" in the context of their different understanding of science (cf. pp. 41ff., pp. 147ff.).

2.3 The defensive attitude to the "attack" of modern physics

"New! A generally comprehensible defense of common sense *against* Einstein's attacks!" This phrase, in shocking pink, was used to advertise a book (Fig. 2.3), and it unambiguously expressed the attitude of the opponents of the theory of relativity – in this case, the physician Karl Vogtherr (born 1882). It was not that an individual person was attacking the theory of relativity; instead, it was an act of resistance or even self-defense against this new affront by academic science opposing "correct natural science," which had never occurred before on this scale. The fact that Einstein's opponents usually explicitly stated their motivation is informative; they wanted to justify their criticism, to be convincing about it using good reasons, and to ensure the legitimacy of an attack on as famous a scientist as Einstein. In the following, the motives of Einstein's opponents are presented using these justifications.

One frequently advanced justification for the opposing works was the ubiquity of the theory of relativity in the broad public and the major world-view significance that was attributed to it. They had to defend themselves against this one-sided reporting and get themselves heard. Hans Christiansen (1866–1945), known as an Art Nouveau artist and as a writer primarily from his treatises against equal rights, but also as a solver of world riddles,[49] stated as an introduction to his

[47] Hilgartner 1990. [48] Tetens, Holm. (September 9, 1999). Die Grenze [The Boundary]. *Die Zeit* newspaper.

[49] Cf. e.g. Christiansen, Hans. (1930). *Der Gleichberechtigungswahn [The Delusion of Equal Rights]*. Wiesbaden; ibid. (1931). *Offener Brief an Seine Heiligkeit Papst Pius XI. [An Open Letter to His Holiness Pope Pius XI.]* Wiesbaden; Christiansen 1915.

Fig. 2.3 Booklet opposing the theory of relativity by Karl Vogtherr, 1923.

criticism of the theory of relativity, "Einstein's 'relativity principle' is thus a gauntlet thrown down on the tournament field of universal knowledge, so to speak, which the author of the following discussion allows himself to formally take up."[50] Since the theory of relativity was perceived as a declaration of war, it was not only permitted, but almost imperative to react to it. Patschke, the engineer, became acquainted with the theory of relativity relatively late for a world riddle solver – that is, not until 1919. He fancied himself in the role of the riddle solver who liked to ask himself questions, his house was open to all interested parties in the evening, and he liked to take opportunities to defend his light ether theory and to obtain recognition for it. He had been asked by his friend, the astronomer Wilhelm Foerster, "to check the Einsteinian theory of relativity for correctness."[51] It is not surprising that this investigation turned out to be devastating for Einstein (cf. particularly pp. 133f.). At this point, it is important to note that Patschke could not leave it at the falsity of the theory of relativity that was obvious to him. Instead, he literally saw it as his duty to add a new twenty-first section to his comprehensive work *Electromechanics* (the mechanics of his postulated light ether atoms), which was dedicated exclusively to the theory of relativity and also appeared as an offprint:

> Since this theory [of relativity] has now already begun to occupy the scholars of the entire world, and since it has caused substantial confusion in that it has considerably shaken the foundations of mechanics, which have been thoroughly tested, which has even led many scholars down false paths of understanding nature, thus I consider myself obligated to publish the twenty-first section of my treatises on the uniform explanation and mechanics of the natural forces in the present form in order to rescue all friends of the natural sciences from the nightmare of the theory above by proving the correctness of the time-tested rules of mechanics, and to point out my developments as well as discoveries that make the eternal growing and passing, the uninterrupted merging of the natural forces understandable mechanically.[52]

Beyond this educational obligation that most world riddle solvers imposed on themselves, however, modern physics also often became a real bogeyman. They absolutely wanted to help their own theory to become acknowledged by academic science, and the presence of the theory of relativity in science and the public sphere created even more difficult conditions for doing so than before. In the eyes of a world riddle solver, modern physics was at best an annoyance, but more frequently an enemy attack. "The theses that Planck advanced against ether mechanics in 1910 [referring to the lecture 'On the limits to understanding nature through mechanics']

[50] Christiansen, Hans. (1920). *Absolut und relativ! Eine Ablehnung des 'Relativitäts-Prinzips' Einsteins auf Grunde einer reinen Begriffs-Mathematik [Absolute and Relative! A Rejection of Einstein's 'Relativity Principle' based on new Conceptual Mathematics]*. Wiesbaden, p. 1.

[51] Patschke 1925, p. 45. [52] Patschke 1920, p. 8. An expanded edition was published in 1922, cf. Patschke 1922.

can easily be proved to be errors. He is the one who attacked, fought, and supposedly 'overturned' physics. He was the *attacker*,"[53] said Fricke, who also shows with this opinion that the "false physics" was blamed not only on Einstein, but also on the other theoretical physicists.

This vehement defensive attitude towards modern physics can only be understood in the context that much was at stake here and that the world riddle solvers' own theories had to be protected. Leo Gilbert[54] (born 1862), the technical editor of the Viennese *Die Zeit* newspaper, complained to Gehrcke shortly after the publication of his work *Das Relativitätsprinzip, die neueste Modenarrheit der Wissenschaft [The Relativity Principle, the Latest Fashionable Folly of Science]*:

The physicists completely [. . .] misunderstand the intention of my satire against relativity. I would never have thought of wasting so much time on this stupidity. I also couldn't care less about the Fizeau problem.[55] I did write this little book, but it only happened because I wanted to show the gentlemen how their deficient understanding of the fundamental conditions of physics (and natural philosophy) necessarily had to lead them to absurdity and ridiculousness.[56]

In fact, the *Modenarrheit* is by no means just a satire, but rather – despite its satirical elements – a criticism of the theory of relativity based on Gilbert's own *Neue Energetik [New Energetics]*, which he had published two years earlier with a scope of 228 pages as Volume 1 of a planned multi-volume work, *Fundamente des exakten Wissens [Foundations of Exact Knowledge]*. *Neue Energetik* was a world riddle solver project, a monistic natural philosophy, within whose framework Gilbert defined a new axiom of physics. For him, the "principle of oppositeness as identity" is the fundamental method of research, and he created unconventional redefinitions of physical terms[57] – for instance, he defines "density" as "universal energy." *Neue Energetik* asserted comprehensive world-view

[53] Fricke, Hermann. [(ca. 1942)]. *Ein Protest gegen die Planck-Physik [A Protest Against Planck Physics]*. Berlin-Schmargendorf (Deutsche Gesellschaft für Weltätherforschung [German Society for Universal Ether Research], Publication No. 57), GN 18-C-11.

[54] Actually Leo Silberstein; very brief biographical information on Silberstein can be found in the anti-Semitic biography collection published by Philipp Stauff, *Semi-Kürschner*, Berlin-Lichterfelde 1913.

[55] In 1851, the physicist Fizeau wanted to measure how the propagation velocity of light changes with or against the flow of current in a moving medium. In hindsight, the result already indicated a deviation from the classical principle of addition of velocities for the speed of light and was therefore considered an important confirmation of the special theory of relativity. Cf. e.g. Einstein 1917a, pp. 26f.

[56] Gilbert to Gehrcke, May 18, 1914, GN 83-C–24.

[57] Assigning new meanings to established terms or symbols was a common practice of the world riddle solvers, often without explicitly labeling these new meanings. Cf. also Gehrcke's criticism addressed to Rudolf Orthner: "The difficulty of understanding [Orthner 1928] is considerably increased in my opinion by the fact that you use common mathematical symbols [. . .] but these symbols are not supposed to have the common mathematical meaning, as you emphasized repeatedly; you talk about the fact that the meaning should be a *similar one* to that of mathematics, but it is not entirely clear which one." Gehrcke to Orthner, August 21, 1928, GN 83-A-4.

claims in the area of society and esthetics. It stood outside the academic discourse. Gilbert's comment that "the physicists" did not understand him (made, interestingly, to the physicist Ernst Gehrcke, who obviously did not belong on the side of "the" physicists for Gilbert) shows that he was conscious of this. For Gilbert, who avowedly "couldn't care less" about the Fizeau problem, *Neue Energetik* simultaneously represented the motivation and the documentation of his refutation of the theory of relativity.

Even an academic philosopher such as Oskar Kraus perceived modern physics as a threat. "I am acting in self-defense. The attack was made by Einstein," he assured Gehrcke on the occasion of the publication of the presentation that he had made at the "As-If" conference in Halle, with which he was not entirely satisfied.[58] The theory of relativity caused Kraus to become extremely insecure. Starting in 1913, he expressed criticism of the special theory of relativity in the correspondence with Gehrcke; his first publication against the theory of relativity appeared only significantly later, in 1919. The fact that he had not published something any earlier was probably due to the fact that, although Kraus did reject the theory of relativity, he was at the same time wrestling with difficulties of comprehension, which he openly admitted (in contrast to the world riddle solvers) not only to Gehrcke, but also to his colleagues in Prague, the mathematician Kowalewski, and the physicist Frank.[59] Moreover, what his position on the general theory of relativity should be was not clear to him for a longer period, and he therefore felt too uncertain for a publication. Thus he stated to Gehrcke in 1913, "It is risky to publish something myself. Unfamiliar with the mathematical methods, I would not be able to leave the territory of pure philosophy and primitive mathematics. Also, what I say has not the least authority, either among physicists or among mathematicians, or among natural philosophers."[60]

[58] Kraus to Gehrcke, June 16, 1920, GN 76-A-18. The conference was organized by Hans Vaihinger, the founder of the "philosophy of as-if," and was dedicated to the relationship between the "philosophy of as-if" and the theory of relativity. Einstein himself had also been approached to make a presentation, but had ultimately withdrawn his participation. Cf. particularly Hentschel 1990, pp. 168ff.

[59] He had his article, which developed from the presentation at the "As-If" conference, checked by Kowalewski before printing with regard to the correctness of the mathematical content. Cf. Kraus to Gehrcke, June 19, 1920, GN 76-A-18. The presentation "Fiktion und Hypothese in der Einsteinschen Relativitätstheorie. Erkenntnistheoretische Betrachtungen" [Fiction and Hypothesis in the Einsteinian Theory of Relativity. Epistemological Considerations] was published as an article in: Kraus, Oskar, Lipsius, Friedrich, Lincke, Paul F., and Petzoldt, Joseph. (1921). *Zur Relativitätstheorie [On the Theory of Relativity]. Annalen der Philosophie [Annals of Philosophy]*, No. 3, Vol. 2, p. 335–96. Cf. also Kraus' statements to Gehrcke: "Unfortunately, I am lacking any physical-mathematical support. For even though colleague Frank is not infrequently so friendly as to talk with me about these things, he is, after all, just like Privatdozent Fürth, too entangled in the bonds of the philosophical absurdities not to consider his thinking habits to be axioms, and my axioms to be thinking habits." Kraus to Gehrcke, April 1, 1920, GN 82-I-10.

[60] Kraus to Gehrcke, October 17, 1913, GN 72-A-4.

Oskar Kraus (1872–1942)

Photograph courtesy of Franz Brentano Archive, Graz.

Oskar Kraus studied jurisprudence and philosophy starting in 1890, the latter with Anton Marty, a student of Brentano's. In 1895, he earned a doctorate of laws and joined the Financial Procurator's Office in 1896. In 1902, he completed his degree in philosophy with his habilitation thesis *Zur Theorie des Wertes. Eine Bentham-Studie [On the Theory of Value. A Bentham Study]*. In 1907, he took the bar examination. Kraus converted from Judaism to Protestantism. In 1909, he became an adjunct associate professor, and in 1911 an associate professor at the German University in Prague. In 1916, Kraus became Marty's successor and thus a full professor of philosophy. In 1931, the Brentano Society of Prague was founded and Kraus took on the leadership. In March 1939, Kraus was arrested when German troops marched into Prague. After six weeks he was released and fled to the United Kingdom. There, he held the Gifford Lectures, "New Meditations on Mind, God, and His Creation," at the University of Edinburgh in 1941. He died of cancer on September 26, 1942, in Oxford.

Source: Franz Brentano Archive, Graz

Kraus overcame this restraint in the following years, but he always remained conscious of his comprehension difficulties; this self-doubt made him an absolute exception among Einstein's opponents. Nevertheless, Kraus felt forced to take a position on the theory of relativity to defend his philosophical views, just as did the non-academic critics:

The unusual stir that Einstein's theory of relativity is causing even in the philosophical world is based on certain extraordinary philosophical assertions that are linked to it. [...] It is therefore necessary for philosophers to take a position on the new doctrine and certainly understandable if they ask themselves whether the theory of relativity, if it is correct, requires such propositions?[61]

Kraus' criticism also did not primarily relate to the physical theory of relativity. Kraus, who belonged to the Prague Brentano School, cites explicitly in his correspondence with Gehrcke competing philosophical directions as his "main enemies," above all Paul Natorp "and other Kantians[, who] believe they can obtain new confirmation for the infamous transcendental idealism in this 'caricature' of a scientific theory."[62] At the same time, this defensive attitude was not only destructive; Einstein's opponents often thought it important that their dismantling of the theory of relativity would finally create the prerequisites for a reorganization of physics (or alternatively, philosophy or all of science). In a letter to Reuterdahl, the philosopher Melchior Palagyi emphasized:

I want to attempt to solve the positive task of *rendering* Einsteinism and relativity *superfluous* and to smooth the way for a reorganization of physics and natural philosophy. It is not sufficient to prove the falsity of the destructive direction of relativity; it is further necessary to also show how it can be done better and how science can be helped to achieve real progress.[63]

Publications opposing the theory of relativity always also contain views on what the alternative to Einstein's theory is. Criticisms of the theory of relativity prioritized dealing with the question of *why* the theory of relativity was wrong and not merely *that* it was wrong. What does that mean for the criticism of the theory of relativity that will be investigated in more detail in the following chapters? On the

[61] Kraus 1921.

[62] Kraus to Gehrcke, October 12, 1913, GN 72-A-3. Cf. also Kraus to Gehrcke, September 8, 1920, GN 71-G-3. "[...] that I am primarily fighting Einstein's philosophical speculations as superfluous." However, Kraus did also repeatedly criticize the physical theory of relativity, which he justified to Gehrcke as follows: "I am following your advice of avoiding specifically physical issues as much as possible; but they cannot be entirely avoided. It is useless to say that this or that is philosophically and logically absurd when the other side declares all of that to be thinking habits! The source of error must be discovered and the formulae themselves discussed." Kraus to Gehrcke, June 6, 1920, GN 76-A-17.

[63] Palagyi to Reuterdahl, October 12, 1922, RP 4–40. Cf. also Fricke 1920a, p. 5: "It will not be possible to construct a criticism of Einstein merely on negation; instead, where Einstein's view must be denied, something better will have to be put in its place."

Melchior Palagyi (1859–1924)

Palagyi (in Hungarian, Menyhért Palágyi, actually Silberstein) was born in Paks, Hungary, in 1859. He studied mathematics and physics at Budapest Technical College in preparation for a teaching position, concluding with a professorial examination. He initially worked as an author for several periodicals and newspapers, then he passed through several positions in school teaching, becoming a high school teacher after several years; in 1899, he received his doctorate from the University of Budapest. During his studies, Palagyi had developed an interest in philosophy; from 1900, he lived in Germany for several years, attending lectures by Wundt, Volkelt, and Heinze in Leipzig and by Husserl in Halle. In 1905, Palagyi became a Privatdozent for natural philosophy in the mathematical-scientific department of the University of Klausenburg. After World War I, he became a director of studies at the Hungarian Ministry of Culture, but due to the political unrest, he left Hungary and went first to Frankfurt. In 1923, he moved into the Darmstadt artists' colony, where he intended to complete his "Universal Mechanics," a monistic natural philosophy. He died in 1924 of a stroke.

Source: Schneider 1942

one hand, it has become clear that the world riddle solvers almost had to attack and refute the theory of relativity if they did not want to give up their claim of offering an all-encompassing physical explanation of the world. Due to the public response to the theory of relativity – and not to the world riddle solver theories, which they would have actually deserved (in their opinion) – the problem had become so virulent that sweeping polemical blows against academic science were no longer sufficient. Experimental physicists like Stark or Gehrcke and philosophers like Kraus likewise took a defensive position. The theory of relativity was (unconsciously) constructed as the concept of the enemy appropriate to each case. It was the theory that negated the truth of the world riddle solver theories, it was the theory that tossed out "lucid" classical physics or, for Oskar Kraus, the theory whose philosophical consequences forced him to take a position as a philosopher ("self-defense").

This procedure, often used by Einstein's opponents, of first constructing a specific concept of the enemy with the name "theory of relativity" and then smashing it with great pleasure based on their own theories, was criticized by Theodor Lessing in an article in the popular science periodical *Der Forscher [The Researcher]*, in which he came out against a theory of relativity polemic that had appeared there:

First blaming researchers and thinkers for the exact opposite of their teachings and then afterward fanatically proving that the entirely banal precondition, which even every child can understand and on the basis of which their teaching was developed and made possible, is the actual and correct truth; that is, a degree of ambiguity or misunderstanding that probably cannot be outdone. But it is also not honest.[64]

Lessing did describe Einstein's opponents' procedure accurately, but it was not used, as he thought, due to dishonesty or a calculating intention. Instead, it was attributable to their extremely selective and tendentious perception of the theory of relativity, which in turn had its roots in the mental block to reception just described, and the resulting defensive attitude. This is shown not least in the reply by the editor-in-chief of the *Forscher*, Georg August Grote, to Lessing. Grote, a world riddle solver, certainly noticed that people were talking at cross-purposes. But instead of reflecting on this fact, Grote wanted to convert Lessing and once again explained to him his "Weltanschauung" ("world view," which he labeled as such himself) according to which, matter is the eternal and absolute thing. The dispute that did not occur was symptomatic. Ultimately, Grote was not at all disturbed by the fact that they were talking at cross-purposes. He did not want to discuss things; after all, he was speaking from the standpoint of the solved world riddle, in possession of the absolute truth, while Theodor Lessing was just a "specialist" with "blinders" who "does not yet [understand] my very simple world view due to all the scientific theories."[65]

A large portion of the intensified debate about the theory of relativity in the 1920s and the public campaigns against it can be explained by this mental block to reception and discussion, as well as by the defensive attitude on the part of Einstein's opponents. This attitude resulted not only in content-based criticism, which essentially bypassed the physical theory of relativity and which will be presented in the next chapter, but also and particularly in a political-strategic reaction that will come into focus in the chapter thereafter.

64 Lessing, Theodor. (1922). Zu 'Der Irrtum Einsteins' in Nr. 6/7 des Forschers« [On 'Einstein's error' in No. 6/7 of the Forscher]. *Der Forscher [The Researcher]* **8–10**, pp. 13–14.

65 All quotes in this paragraph from: Grote 1922, p. 14. But the academics also displayed a similar attitude. Cf. Kraus to Gehrcke: "It makes no sense to debate with people who have fundamentally forgotten how to think." October 24, 1929, GN 33-J-4.

3

The debate on the content of the theory of relativity

The opposition to modern physics that is explored in this and the following chapter had an epistemic dimension and a strategic dimension, which were expressed in content-based criticism and in strategic opposition. The epistemic dimension covers the bodies of knowledge that were cited against the theory of relativity and which, as will be shown below, had partly been crowded out of the academic discourse as "obsolete" (the classic physical explanations of phenomena that were now explained "modernly") and partly fell completely outside the limits of statements that could be called science at that time (the metaphysical explanations, for instance). This chapter examines the content-based criticism of the theory of relativity. Two different levels of criticism are identified. First, the criticism focused on fundamental physical concepts that had received a new meaning due to the theory of relativity. Einstein's opponents were convinced that they were able to refute the theory of relativity using their own concepts of time, light, space, gravitation, and ether, or to demonstrate its irrelevance. Since these attempts at refutation were conducted by the world riddle solvers on the basis of their own theories, and by the physicists and philosophers mostly based on their individual views of classical physics or philosophy, Einstein's opponents constituted a heterogeneous group in terms of the type of content refutation. Second, the status of the theory of relativity as a physical theory was attacked. Using a different understanding of what "correct" science had to be, modern physics was denied the status of a science. On this point, academic and non-academic opponents of the theory of relativity agreed. Moreover, the content-based criticism by the world riddle solvers showed a consequence of the mental block to reception discussed in Chapter 2. Convinced that they were able to encompass "everything that can be [. . .] an object of science"[1] with their theories, many made accusations that the theory of relativity was a plagiarism of their own theory. Their conviction that they had conclusively

[1] Ziegler 1931, p. 96.

refuted the theory of relativity and declared it superfluous resulted in a self-confident demand for recognition of these refutations, which led to the strategic opposition to modern physics examined in Chapter 4.

3.1 The criticism of the reorganization of fundamental physical concepts by the theory of relativity

3.1.1 The simultaneity of the non-simultaneous as a determining factor for the criticism

In the first chapter, the phenomenon of world riddle solving was examined and it was shown that "independent" research existed in the non-academic space of the early twentieth century with a claim to universal explanations, whose content and methods were incompatible with those of academic science. The second chapter showed that the omnipresent theory of relativity was perceived by non-academic world riddle solvers, but also by academic physicists and philosophers, as a threat to their own theories or their central scientific or philosophical postulates. Therefore, Einstein's opponents attacked the theory of relativity based on various principles and motivations. The world riddle solvers saw it as the latest development in academic science, which had a hostile attitude towards the world riddle solver theories. The academic physicists saw it as an unjustified replacement of classical physics with a mathematical theory lacking a connection to reality. The philosophers were convinced that in this case, a physical theory was illegitimately presuming to take on philosophical significance. The various emphases and the heterogeneity of the content-based criticism resulting from this situation will become clear in the individual sections. However, the content-based criticism of the theory of relativity had a common framework that was independent from the content advanced in each case against the theory of relativity. Both the world riddle solvers and Einstein's academic opponents had been left behind by the developments in modern physics, or could not or would not follow it. The areas of knowledge left behind will be described briefly here, both for academic physics and for the non-academic area.[2]

The people left behind by progress – academic science

In periods when systems of knowledge are being fundamentally reorganized, there are always conservative tendencies. These tendencies were apparent in the academic physicists among Einstein's opponents who would not or could not accept the

[2] The philosophers are not included at this point, since they had asserted claims of being able to review the philosophical consequences from within a different discipline, and they did not assert any claims (at least, not explicitly) of making a physics-based refutation of the theory of relativity.

movement from the comprehensibility of classical physics, with its mechanical models, to a mathematical investigation of nature. For the physicists, this transformation was in fact not merely about accepting a new theory, but rather about placing their discipline on completely new foundations. "The transition from classical physics to modern physics is [. . .] characterized by the fact that, for the first time, intuitive and practical models are replaced with theoretical models with similarly universal claims."[3] Their refusal to go along with this paradigm shift was thus not a sign of deficient intelligence, but was rather attributable to socialization by training and the orientation of their own research work within a physics that was largely focused on mechanics around 1900. In fact, the revolution in the foundations of physics in connection with relativity and quantum theory was not obviously in the air at this time. Even just a few years prior to the formulation of the special theory of relativity, it seemed that there were no longer any great mysteries in physics. True, work had to be done on the details in the individual specialty areas, but significant discoveries were not to be expected, much less any restructuring.[4] When "classical physics" was mentioned at this time, what was meant was not the contrast to "modern physics," that is, to relativity and quantum theory, instead expressing the fact that physics was a mature science with well-defined concepts and assured methods, which, moreover, held the unquestioned leadership role among the scientific disciplines.[5]

Certainly, thermodynamics and electrodynamics had established themselves as independent branches of physics alongside classical mechanics in the course of the nineteenth century. Their central theoretical assumptions were no longer straightforwardly compatible with the basic concepts of mechanics; it was only possible to produce a certain degree of compatibility by introducing supplemental mechanical models. In fact, one of the greatest challenges in theoretical physics at the end of the nineteenth century was to place the three branches, mechanics, thermodynamics, and electrodynamics, on a unified basis. However, this did not succeed. Instead, problems existed at the boundaries between these branches, such as the problem of the electrodynamics of moving bodies at the interface between mechanics and electrodynamics – whose solution ultimately led to the development of a modern physics that could not be reduced to the subdisciplines.

[3] Renn, Jürgen. (2006). *Auf den Schultern von Riesen und Zwergen: Einsteins unvollendete Revolution [On the Shoulders of Giants and Dwarves: Einstein's Unfinished Revolution]*. Weinheim, p. 99.

[4] Cf. Planck, Max. (1982 [1942]). *Selbstdarstellung [Self-Portrait]*. Berlin, pp. 6f.

[5] Cf. Stichweh 1984, p. 456. Even after 1905, "classical physics" was understood to mean various things, and not always the contrast to "modern physics" (as in this book). Cf. Jungnickel, Christa, and McCormmach, Russell. (1986). *Intellectual Mastery of Nature. Theoretical Physics from Ohm to Einstein*, 2 vols., Vol. 2: *The Now Mighty Theoretical Physics 1870–1925*. Chicago, Ill. [*et al.*], p. 313; Staley, Richard. (2008a). World views and physicists' experience of disciplinary change: On the uses of 'classical' physics. *Studies in the History and Philosophy of Science*, **39**, 298–311.

This initial situation in physics around 1900 was the reason why academic physicists, who mostly subscribed to the mechanical conception of the world but also subscribed in considerable numbers to the electromagnetic[6] or thermodynamic one (energetics was unable to establish itself in academic physics and was more often represented in the non-academic space), had difficulties, at least initially, detaching themselves from it and accepting the relativistic conception of the world. In particular, experimental physicists who advocated a specific view of physical research into natural phenomena due to their practical interactions with nature and who asserted special competence in this regard, found little to relate to in the new description of natural phenomena, which in their eyes was less clear (cf. particularly pp. 147ff.). However, only a very small number of physicists became professed opponents of Einstein, even though they sometimes had reservations about the conceptual restructuring of physics. The reason why some physicists such as Gehrcke and Lenard fundamentally rejected the theory of relativity will be presented later in this chapter.

The people left behind by progress – the non-academic area

In the non-academic area, identifying what had been left behind by the paradigm shift within physics is more complex. Since they were located outside of academic physics, the protagonists were less sharply confronted with paradigm shifts in academic physics and could therefore ignore their progress more easily and over longer periods of time. Electromagnetism had already been received in non-academic circles largely either as a purely mechanical phenomenon, or as a phenomenon to be interpreted from an occult perspective. Conceptions of a substantial ether as the medium filling space, and as the cause of all forces and of primordial atoms as the smallest building blocks of matter, were widespread. But conceptions of heat as substance also continued to be advocated.[7]

In contrast to the specialized research in academic physics, the important questions in non-academic research were of a general nature, but not identical to the basic issues that academic physics was faced with in light of the boundary problems of its subdisciplines. The special knowledge and overview necessary in order to ask such questions did not exist in the non-academic space. Instead, the "great questions" of science were in play here, those that had already appeared in textbooks 100 years earlier – What is the nature of heat? What is gravity? – or, questions that had rather been assigned in the meantime to the competence of

[6] On electromagnetism as a new conception of the world within academic physics, cf. McCormmach, Russell. (1970). H.A. Lorentz and the Electromagnetic View of Nature. *Isis* **61**(4), 459–97.

[7] "It must thus also again become possible to treat *heat* as *substance*." Fricke, Hermann. ([no date]). *Neue Weltdeutung [New Interpretation of the World]*, (Original Works No. 7), GN 86-B.

religion or philosophy – Will the universe come to an end? Does the world have a purpose? The answers to these questions allowed a far greater variety of content to be advocated than in the academic forums, and this content deviated both from the state of academic physics as well as from its claims to validity.

The content cited in the non-academic space against the theory of relativity was therefore not simply the result of a failure to connect with modern science. It is true that in his work *Weltätherforschung [Universal Ether Research]*, Fricke reported to his readers, "People have said to me: The way that you do physics is the way people did it 100 years ago!",[8] and in fact, the world riddle solvers did conduct anachronistic science in a certain sense. They demanded largely mechanical explanations in physics; only very few had followed the development of electromagnetics, and not one of them participated in the revolution in physics associated with the theory of relativity. Their scientific concepts were certainly "old," but still extremely widespread in the 1910s and 1920s, and they had been adapted and further developed in a variety of non-academic contexts. These non-academic bodies of knowledge were handed down in sub-academic networks and publication forums, as well as through personal, professional, and private meetings (cf. pp. 126ff. as an example). New developments in academic science such as the development of radioactive radiation, Planck's constant, or even the development of the theory of relativity were attentively followed by the world riddle solvers, who attempted either to refute them or to integrate them into their non-academic theories and thus to absorb them.

Numerous booklets by engineers, physicians, military officers, chemists, or authors, therefore did not often criticize the theory of relativity on the basis of classical physics around 1900, as was usually the case with the experimental physicists, but rather on the basis of non-academic bodies of knowledge. These will be demonstrated shortly in the discussion of the non-academic alternative views of the concepts of time, light, space, gravitation, and ether, which had been newly conceived of by modern physics (or were no longer used, in the case of ether). At the same time, not every one of Einstein's opponents had detailed individual ideas for each concept; instead, a concept such as light or ether usually played the leading role in the theory solving the mysteries of the universe, and the refutation project was then also focused on this concept. With their universal theories, on the other hand, the world riddle solvers had their own views to offer on every scientific topic. They often developed ad hoc concepts based on the background of their own theory for other terms that had only become virulent due to the theory of relativity, such as the concept of space and time.

[8] Fricke, Hermann. (1939a). *Weltätherforschung. Ein Aufbauprogramm nach dem Umsturz in der Physik [Universal Ether Research. A Development Program following the Revolution in Physics]*. Weimar, p. 89.

3.1.2 Time

First, the conception of time will be explained that dominated in everyday understanding, but also in classical physics, before it was called into question by the theory of relativity, whose concept of time will be outlined thereafter. This will be followed by a discussion of the criticism of this new conception of time. This criticism sometimes sweepingly rejected the relativity of simultaneity, but sometimes also called it into doubt based on other bodies of knowledge.

The everyday understanding of time and the understanding of time in classical physics

The relativization of the concept of time was primarily met with scarcely reflected but deeply anchored conceptions of time, both among academic physicists as well as among Einstein's non-academic opponents. The everyday understanding of time in particular, which was reinforced among some of Einstein's opponents by the classical physical concept of absolute time, determined the concepts of time that were thrown onto the scale in their sweeping rejection.

All individuals experience the fact that time cannot be turned back, that they age, and that they cannot make done things undone. Along with this individually experienced temporality of life, the everyday understanding of time depends to a great degree on culturally customary methods for determining time sequences. In early cultures, the determination of time was oriented to regularly recurring events, such as the motion of the Sun and the Moon, or harvest time. Time determinations based on constantly recurring or even singular events ("at the time of the great flood"), however, do not yet result in the time conception of the modern age. It seems only natural to us that we expect every person to know how old he or she is and what the date is. Our modern everyday understanding of time is characterized by the existence of standardized clocks, continuous calendars, and time scales spanning millennia (we live in the "year 2013 after the birth of Christ"). The anchoring of these time conceptions in everyday life and everyday life's temporal structuring, shape the modern understanding of time as something that persistently occurs in one direction.[9] The relativization of simultaneity was therefore the most-discussed consequence of the theory of relativity in the public sphere. The understanding of time particularly among physicists, but also among scientifically trained physicians and engineers, was additionally influenced by the Newtonian concept of "absolute time," which Newton distinguished from "relative time."[10]

[9] Cf. Elias, Norbert. (2004). *Gesammelte Schriften [Collected Works]*, Vol. 9: *Über die Zeit [On Time]*, ed. Michael Schröter. Frankfurt (Main), particularly pp. 54ff.; cf. also Galison, Peter. (2003). *Einstein's Clocks, Poincaré's Maps. Empires of Time*. New York.

[10] Cf. Newton, Isaac. (1687). *Philosophia naturalis principia mathematica*, English edn. (1729). *The Mathematical Principles of Natural Philosophy*, translated by Andrew Motte. London (see p. 9).

Newton introduced absolute time as an absolute quantity to which measurable occurrences in nature could be related. "Absolute, True, and Mathematical time, of itself, and from its own nature, flows equably without relation to anything external, and by another name is called Duration..."[11] Both the topological as well as the metric structure of absolute time are fixed in Newtonian mechanics. Because of this, in classical physics it was possible to determine the absolute simultaneity of two events independently of their individual frames of reference by reference to absolute time. Another influential reference with regard to the absolute character of time, along with everyday understanding and Newton, was Kant. Like Newton, Kant assumed that the simultaneity of two physically distant events was unequivocal.[12] Therefore, the experimental physicists in particular often referred to Kant in order to further justify their adherence to absolute time.[13]

The relativization of simultaneity

In his paper "On the electrodynamics of moving bodies," Einstein dealt at length with the problem of determining simultaneity. He begins his problematization of this concept, which is apparently so unequivocal, at a fundamental level with the statement "that all our propositions involving time are always propositions about *simultaneous events*. If, for example, I say that 'the train arrives here at 7 o'clock,' that means, more or less, 'the pointing of the small hand of my clock to 7 and the arrival of the train are simultaneous events.'"

However, this determination of simultaneity using a clock is only possible for the location of the clock. In order to know whether an event distant from a Clock A occurred at the same time as an event at the location of Clock A, a signal must be sent from a Clock B located at the site of the distant event. However, the transmission of a signal takes time. If one assumes the principle of a constant speed of light, then one can synchronize the two distant clocks using light signals (whose speed in a vacuum always amounts to about 300,000 km/s). This

11 Ibid.

12 Cf. Kant, Immanuel. (1787). *Kritik der reinen Vernunft [Critique of Pure Reason]*, particularly A 183–184/B 226–27. However, Kant's conception of time was not exhausted by the Newton's physical conception of time. According to Kant, space and time are forms of intuition that are different in character, which first make it possible to experience spatial and temporal events. Because of this, the popular argument that Einstein had confused (relative) time measurement with the (absolute) a priori temporal form of intuition could refer to Kant's authority. On the presentation and discussion of the Neo-Kantian immunization and revision strategies, cf. Hentschel 1990, pp. 199–239.

13 Gehrcke, for instance, appealed to Kant for reinforcement in order to rescue the concepts of Newtonian mechanics. Cf. Gehrcke, Ernst. (1912). Über den Sinn der absoluten Bewegung von Körpern [On the meaning of the absolute motion of bodies]. *Sitzungsberichte der Königlich Bayerischen Akademie der Wissenschaften [Transactions of the Royal Bavarian Academy of Sciences]*, Vol. 12, pp. 209–22; reprinted in Gehrcke 1924a, pp. 12–20 (see p. 16). Ever since physics developed as an academic discipline at the beginning of the nineteenth century, Kant was an important epistemological point of reference for physicists, as Jungnickel and McCormmach show using textbooks from this period. Cf. Jungnickel and McCormmach 1986, Vol. 1, pp. 23ff.

results in the following definition of time and simultaneity: "The 'time' of an event is the reading obtained simultaneously with the event from a clock at rest that is located at the place of the event and that for all time determinations is in synchrony with a specified clock at rest."[14]

The relativity of simultaneity is a consequence of this definition. Events that occur simultaneously for an observer in frame of reference A will not occur simultaneously for another observer in frame of reference B that is in motion relative to A. This is a result of the two postulates of the special theory of relativity: the relativity principle and the principle of the constancy of the speed of light. If simultaneity can be verified with regard to two light signals in a system A at rest, then system B moves toward the one light signal and away from the other. Therefore, simultaneity is not established in system B with regard to the two light signals. Furthermore, it can be shown in a similar way that lengths and time intervals change depending on the state of motion. Both length contraction and time dilation are physical effects which are certainly very small, but which can be observed at high speeds using very precise measurements. However, these effects are not among those that can be experienced directly in everyday life, which was sufficient proof of their supposed absurdity for many of Einstein's non-academic opponents.[15]

The sweeping rejection of the relativization of simultaneity

The criticism of the relativization of simultaneity was largely of a sweeping nature due to the deeply anchored conceptions of time outlined above, which were, however, not deeply reflected. Apart from a few exceptions, particularly in the context of vitalistic thinking, where detailed alternative concepts of time had already previously played a central role (cf. pp. 101ff.), it was assumed as self-evident that time was a temporal stream permanently flowing in one direction.

Most of the critics of the relativistic concept of time argued at an elementary level. The term "simultaneity" itself was said to already contain absolute simultaneity, just as the term stallion contains the idea of a male horse. In this respect, a discussion about relative simultaneity was just as absurd to them as a discussion about a female stallion. "The only new thought that Einstein has brought to light consists

[14] Einstein, Albert. (1905b). Zur Elektrodynamik bewegter Körper [On the electrodynamics of moving bodies]. *Annalen der Physik*, **322**(10), 891–921. English quotes are taken from Stachel, John *et al.* (eds.). (1989). *The Collected Papers of Albert Einstein*, Vol. 2: *The Swiss Years: Writings, 1900–1909*. (English translation supplement), translated by Anna Beck. Princeton, Doc. 23 (see pp. 141, 143).

[15] The situation was different in the academic space. The special theory of relativity had rapidly become established in physics and only a very few academic physicists such as Ernst Gehrcke had a fundamentally negative attitude from the outset. Lenard and Stark, on the other hand, only criticized the general theory of relativity. In the 1920s, however, Lenard also turned against the special theory of relativity. Cf. Lenard, Philipp. (1920a). Über Äther und Uräther [On ether and primordial ether]. *Jahrbuch der Radioaktivität und Elektronik*, **17**, 307–56 (see p. 309).

of the fact that in his opinion two events occurring at different times can be considered to be simultaneous. But this is a contradiction in terms and must be labeled utter nonsense."[16] And where would it ultimately lead if the passage of time would not unequivocally run uniformly forwards? For Erich Ruckhaber, it was clear that it was not the measurement, but rather the direction of the motion, which was always the same, that characterized time as absolute. "The direction of all events is an *unambiguous, absolute* one, and this expresses the concept of time. Lacking such an absolute basis and unambiguity, logical thinking and the exchange of ideas would be impossible; at best, a chaos of whirling perceptions and memories would be possible."[17]

Like Ruckhaber, many critics suspected that the relativization of the concept of time is linked to a relativization of the causality of world events.[18] Above all, it was the thought experiment of the twins paradox[19] that was used in different variations of spontaneous opposing reactions as evident proof of the absurdity of the theory of relativity, compared to which the clarity of one's own theory would particularly obviously stand out. In this thought experiment, one of two twin brothers travels at almost the speed of light into the universe and then returns after some time to his brother who remained on Earth. Since, due to the high speed, according to the theory of relativity, much less time has passed for the space traveler than for the twin on Earth, he would have to be younger than the Earth twin. The Earth twin, however, could argue for his part, due to the relativity of the motion, that he had moved relative to his traveling brother and would therefore be the younger one. However, this construct ignores the fact that one of the twins is subject to acceleration and in this way the physical equivalence of the motion is canceled. For the elementary

[16] Mewes 1922a, p. 417. The engineer Hans Wittig expressed the same thought, "There is only one physically real time." Wittig, Hans. (1921). *Die Geltung der Relativitätstheorie: Eine Untersuchung ihrer naturwissenschaftlichen Bedeutung [The Validity of the Theory of Relativity: An Investigation of its Scientific Significance]*. Berlin (see p. 38); cf. likewise Weinmann, Rudolf. (1926). *Versuch einer endgültigen Widerlegung der speziellen Relativitätstheorie [An Attempt at a Final Refutation of the Special Theory of Relativity]*. Leipzig, particularly p. 19.

[17] Ruckhaber, Erich. (1928). *Die Relativitätstheorie widerlegt durch das Widerspruchsprinzip und die natürliche Erklärung des Michelsonversuchs: das dreidimensionale Raum-Zeit-System [The Theory of Relativity Refuted by the Contradiction Principle and the Natural Explanation of the Michelson Experiment: The Three-Dimensional Space-Time-System]*. Leipzig, p. 24.

[18] Cf. likewise Gilbert 1914, pp. 14f.; Vogtherr, Karl. (1923). *Wohin führt die Relativitätstheorie? Kritische Betrachtungen vom physikalischen und erkenntnistheoretischen Standpunkt aus [Where does the Theory of Relativity Lead? Critical Reflections from the Standpoint of Physics and Epistemology]*. Leipzig, pp. 22f.

[19] The twins paradox is the popular variant of the clock paradox. In a popular science presentation on January 16, 1911, before the Naturforschende Gesellschaft Zürich [Zürich Natural Science Society], Einstein first sent two imaginary clocks on the trip and then illustrated the result using a "living organism" that is sent on the trip. Cf. Einstein, Albert. (1911a). Die Relativitäts-Theorie [The theory of relativity]. *Vierteljahresschrift der Naturforschenden Gesellschaft Zürich*, **56**, 1–14. English translation available in Stachel, John *et al.* (eds.). (1994). *The Collected Papers of Albert Einstein*, Vol. 3: *The Swiss Years: Writings, 1909–1911*. (English translation supplement), translated by Anna Beck. Princeton, Doc. 17. Fölsing points out that Langevin was the first to use the example of the twin brothers a few months later. Cf. Langevin, Paul. (1911). L'évolution de l'espace et du temps. *Scientia*, **10**, 31–54; cf. Fölsing 1997, p. 192.

criticism[20] of the twins paradox, however, that did not play a role: The fact *that* the moving person should age more slowly was already proof of the absurdity.

For some of Einstein's opponents, such as Kleinschrod, the hydropathic Kneipp physician, the twins example could only be understood as a joke. According to the theory of relativity:

Every locomotive engineer would have to live a much longer time than his colleague the mailman, who only walks. What bold prospects would open up for humanity in this way! One could easily cheat death. Think of an elegant airship. With 500 kilometers of speed [*sic*] it travels along without a sound. One notices nothing at all of the motion – but one would be able to extend life enormously in time, if, yes, if Einstein were right. Bismarck would perhaps still be living today, and our Fatherland would have been saved. This is not a fairy tale that I am telling here. The followers of Einstein believe something like this.[21]

For Patschke, the engineer, it was quite clear that this thought experiment made publication of his universal law absolutely necessary, since the twins paradox makes:

...every unbiased, clearly thinking reader clearly aware of the fact that time is falsely interpreted here, that it is treated falsely mathematically, and that it was necessary to publish my far-reaching scientific discoveries in opposition to such contrived notions in order to show the paths on which one can clearly follow the natural forces intellectually even with simple mathematical tools, and in this way can make the natural forces useful more quickly and better than before to the benefit of all mankind.[22]

Sweeping rejection of the relativity of simultaneity, or limiting criticism of the relativistic conception of time to its allegedly obvious ridiculousness, shows on the one hand that the world riddle solvers did not even attempt to understand the problems that Einstein was able to solve with the relativization of simultaneity. Firmly convinced that they themselves had formulated a theory that would solve all of the problems of science, they also saw no occasion whatsoever to do so. For them, the supposed absurdity of the relativization of simultaneity was a building block in

[20] When Gehrcke dealt with the clock paradox, on the other hand, he attacked the explanation that the equivalence of the motion should have been canceled due to the acceleration. Here he saw a contradiction to the equivalence of frames of reference in the special theory of relativity. However, this equivalence in the special theory of relativity is only assumed for non-accelerated frames of reference. Cf. Gehrcke, Ernst. Zur Diskussion der Einsteinschen Relativitätstheorie [On the discussion of the Einsteinian theory of relativity], in: Gehrcke 1924a, pp. 34f. On the criticism, cf. e.g. Einstein, Albert. (1918b). Dialog über Einwände gegen die Relativitätstheorie [Dialogue on objections to the theory of relativity]. *Die Naturwissenschaften*, **6**(48), 697–702, reprinted in English translation in Janssen, Michel *et al.* (eds.) (2002b). *The Collected Papers of Albert Einstein*, Vol. 7: *The Berlin Years: Writings, 1918–1921*. (English translation of selected texts), translated by Alfred Engel. Princeton, Doc. 13; Thirring, Hans. (1921a). Über das Uhrenparadoxon in der Relativitätstheorie [On the clock paradox in the theory of relativity]. *Die Naturwissenschaften*, **9**(13), 209–12.

[21] Kleinschrod 1920, p. 15. Gartelmann said the same thing, describing the twins paradox as an "intermezzo from magic." Cf. Gartelmann, Henri. (1920). Zur Relativitätstheorie [On the theory of relativity], Part 2. *Neue Weltanschauung*, **9**(7), 198–214 (see p. 202).

[22] Patschke 1920, p. 26.

the disqualification of the theory of relativity as a physical theory to be taken seriously.

Along with this widespread denunciation of the relativity of simultaneity as nonsensical, it was often assumed that it would have to be a mistake. The relativization of time was so inconceivable for many people that a misunderstanding was assumed: Einstein's relativization of simultaneity would not mean the relativization of time "in and of itself," but was instead merely a problem on the level of time measurement. According to this interpretation, one could get rid of the relativity of simultaneity by assuming that in reality the theory of relativity did not affect absolute time at all.

In his popular introduction, *Über die spezielle und die allgemeine Relativitätstheorie. Gemeinverständlich [On the Special and General Theory of Relativity (A Popular Account)]*, Einstein clarified the relativity of simultaneity (which says that the speed of two events depends on the state of motion of the observer) using what is called the train example: The assumption that the speed of light has the same constant value for an observer on a train as for an observer on an embankment leads to the fact that two rays of light that appear simultaneous to one observer appear unsimultaneous to the other (moving) observer, since the paths traveled by the light differ for the two observers due to the relative motion.[23] A widespread view among the opponents of the theory of relativity was that a clock on a tower standing in the middle of the embankment and visible to both observers would render the theory of relativity superfluous. The problem of determining the simultaneity of two events with regard to two frames of reference that are in motion relative to each other is not solved in this way, but this was not perceived. This also applied to trained physicists such as Fricke and Gehrcke. In his work opposing the theory of relativity, Fricke frankly admits:

> However, it is completely incomprehensible to me what the changes in the movement of the clocks as a consequence of motion should have at all to do with the measurement of the speed of light. The two observers are not dependent upon their pocket watches, after all; instead they can jointly use any clock on a tower standing at right angles to the direction of motion, or otherwise use any astronomical occurrence suitable for time measurement, such as the motion of Jupiter's moons.[24]

Gehrcke, who carried on a controversy with Hans Thirring about the clock paradox in the *Naturwissenschaften* periodical in 1921, which Thirring ultimately resignedly ended due to Gehrcke's inability to follow the argumentation of a physicist grounded in modern physics,[25] expresses his view on the nature of time most clearly in an unpublished typescript:

[23] Cf. Einstein 1917a, pp. 16ff. [24] Fricke 1920a, p. 13.

[25] Cf. Thirring 1921a; Gehrcke, Ernst. (1921a). Über das Uhrenparadoxon in der Relativitätstheorie [On the clock paradox in the theory of relativity]. *Die Naturwissenschaften*. **24**, 482; Thirring, Hans. (1921b). Erwiderung

The concept of the simultaneity of two events that take place at different locations is meaningless, says Einstein. This statement constitutes the central point of the 'definition of time' customary in the theory of relativity, and this point is ascribed the greatest of importance by various sides. I am of the opinion that there is nothing at all deep or difficult in Einstein's words. It is not in the concept of the simultaneity of events at spatially separate locations, but rather in *the technical* determination of this simultaneity that a certain problem lies. In my opinion, then, the difficulty is not at all a philosophical or epistemological or theoretical-physical one, but rather an experimental-technical one.[26]

This quote makes it clear why the debate with Thirring had to fail. First, Gehrcke did not accept the fact that a different concept of simultaneity is actually asserted in the theory of relativity. Second, he located the problem at the level of time measurement and thus claimed an advantage of competence as an experimental physicist over the theoretical physicists. Granted, the exact determination of simultaneity was "a technically difficult task." "But such an imperfection of technology [. . .] does not affect the concept of simultaneity," Gehrcke insisted, and continued, "it is certainly useful to develop the *concept* of ideational simultaneity, such as every reasonable person has developed. That is, the technical determination of simultaneity is always only approximately possible, but the thought of simultaneity is as sharp as any thought can be." In the same way, after all, one could also develop the geometric concept of a perfect circle without ever being able to draw one. Oskar Kraus also endorsed this view, writing to Gehrcke, "It is nothing other than the error of Protagoras. The measurement is being confused with the thing measured."[27]

Both the supposition that relativity only affected the level of time measurement and the insistence on the inconceivability of relative simultaneity were widespread. In addition, the world riddle solvers claimed that they had completely refuted the theory of relativity on the basis of alternative concepts of time.

Criticism based on alternative concepts of time

Particularly in vitalistic and occult thinking, criticism of the relativistic concept of time was centrally linked to the assumption that a genuine "biological time" or "life time" existed that was differentiated from physical time. In the area of academic life philosophy, the criticism by Henri Bergson (1859–1941) of the concept of time in physics is particularly well known. Bergson did not consider the special theory of relativity to be mathematically false, but rather to be a theory that only applies to a particular form of time. The time of physics and the exact sciences was called "temps

hierzu [Response thereto]. Ibid., 482–83; Gehrcke, Ernst (1921). Die Erörterung des Uhrenparadoxons in der Relativitätstheorie [The discussion of the clock paradox in the theory of relativity]. Ibid. **28**, 550–51; Thirring, Hans. (1921c). Erwiderung hierzu [Response thereto]. Ibid., 551.

26 This and the following quotes from: Gehrcke [about 1921], GN 4-B-4. On the great significance of the exact technical determination of simultaneity around 1900, cf. Galison 2003.

27 Kraus to Gehrcke, October 17, 1913, GN 72-A-4.

scientifique," which was to be differentiated from "durée réelle." *Durée réelle* was a time inherent in every living creature, because it was experienced, which differentiated the living creature from dead matter. This time, he said, was an absolute time.[28] Vitalistically motivated criticism of the physical concept of time was also expressed in the non-academic area. This was by no means a popular version of academic life philosophy, but rather an independent development in the non-academic space, which was characterized by vitalism and the occult movement.[29] An academic representative of vitalism such as Johannes Reinke even expressly distanced himself from such a radical vitalism as that of Kleinschrod, the hydropathic Kneipp physician.[30] Of Einstein's opponents considered in this book, Kleinschrod was the one who applied vitalistically motivated criticism to Einstein's concept of time. He had already developed his vitalistic view of time long before becoming aware of the theory of relativity. For Kleinschrod, the vitalistic concept of time caused the need to criticize the theory of relativity. As "life time," it had the same status in Kleinschrod's theory of "meta-mechanics" as life; it was thus superior to all mechanical-physical, non-vitalistic descriptions and was therefore not on the same plane as space.[31] Since the life forces could not be active at constant strength, for instance due to fatigue and sleep, Kleinschrod arrived at a definition of time that is relative in the sense that it is always subjective and linked to the individual creature:

> Time is a duration and an intensity [...] it thus must be a 'meta-material' force that stands 'above' matter, or a 'meta-mechanical' force that stands 'above' the mechanical forces of lifeless matter and controls them. [...] Then, however, considered from the material side of things, all life processes must be meta-mechanical processes of controlling the mechanics of the external world. [...] And thus the true nature of life force has been discerned for all time.[32]

[28] Cf. Bergson, Henri. (1922). *Durée et simultanéité à propos de la théorie d'Einstein*, Paris (Bibliothèque de philosophie contemporaine), Introduction, as well as pp. 54ff. in particular. As secondary literature on Bergson's criticism of the theory of relativity, cf. particularly Čapek, Milič. (1971). *Bergson and Modern Physics. A Re-interpretation and Re-evaluation*. Dordrecht (Boston Studies in the Philosophy of Science Vol. 7); ibid. (1991). *The New Aspects of Time. Its Continuities and Novelties*. Dordrecht (Boston Studies in the Philosophy of Science Vol. 125), particularly pp. 296ff.; cf. also Hentschel 1990, pp. 441–56.

[29] An alternative concept of time was cited by the anthroposophists, for instance, against modern physics. In his critical discussion of the theory of relativity, Rudolf Steiner likewise referred to a biological time that was immanent to organisms as lifespan. Cf. the questions answered by Rudolf Steiner following a lecture by Dr. Hermann von Baravalle on "The Theory of Relativity" on March 7, 1920, held during the "Heat Course" in Stuttgart, March 1920, in Steiner, Rudolf. (1982). *Texte zur Relativitätstheorie [Texts on the Theory of Relativity]*, ed. Wim Viersen. Dornach, pp. 30–39 (see pp. 37f).

[30] Cf. Reinke, Johannes. (1911). *Einleitung in die theoretische Biologie [Introduction to Theoretical Biology]*, 2nd revised edn. Berlin. Introduction to the 2nd edn., p. XV. In contrast to Hans Driesch or Henri Bergson, Reinke himself did not expressly take a position on the theory of relativity since he did not consider himself competent to truly comprehend it. Cf. Reinke, Johannes. (1926). *Das dynamische Weltbild: Physik und Biologie [The Dynamic Conception of the World: Physics and Biology]*. Leipzig. Foreword, p. III.

[31] "Professor Einstein himself has [...] fallen victim to a serious error, even as a mathematician, for the time coordinates do not play the same role in a four-dimensional space as the three spatial coordinates, but rather a higher one, namely that of meta-mechanical control." Kleinschrod 1920, p. 43.

[32] Ibid., pp. 26f.

And thus the task of the world riddle solvers would actually have come to an end, if Einstein had not put forward a theory that was highly threatening to their conception of life time. The primary motivation for this reaction by a very busy hydropathic Kneipp physician to the theory of relativity lay in rescuing meta-mechanical time from the attack of materialistic science, embodied in the theory of relativity. Kleinschrod had to take such drastic action against the theory of relativity because as a physicist, Einstein considered space and time together and thus subsumed time (and consequently life) under a "materialistic" natural law, in Kleinschrod's view. In contrast, Kleinschrod stated that as a fourth dimension, time, or life, respectively, could only be a scientific concept in "future biological mathematics."[33] Kleinschrod's criticism of Einstein's train example shows particularly clearly the specific epistemological framework in which arguments were advanced against the theory of relativity's concept of time:

One can probably ask, along with Einstein: At what speed (W) does the man move forward relative to the embankment while walking, and set up the related formula $W = v + w$ [v = train speed]. *But this is just disguising the facts. The issue is not at what speed the man moves forward, but rather whether his personal motion inside the traveling train car is physically equivalent to the mechanical train motion.* Only when the two are equivalent can the R[elativity] P[rinciple] be applied to them.

But this would first have to be established scientifically, which, however, has not yet been the case. Quite the opposite. Up to now, it has specifically not been possible to explain personal motion mechanically, and it was on that point that the entire mechanical theory of life has foundered until now. But for Einstein's theory of relativity, the whole thing is again of the highest significance. In his investigation of the speed of light relative to a travelling railway train, Einstein has the moving ray of light play the role of the man moving himself; that is, he sets the mechanical motion of light as physically equal to the self-motion of the man, and arrives at a formula that is in contradiction to the one above [of classical addition of velocities]. Again, Einstein builds up his theory of relativity on the resolution of this contradiction.

It is clear that we are dealing with a real *petitio principii* here. Einstein accepts a completely unproven statement of the physical equivalence of mechanical motion and self-motion as proven and draws further conclusions from that.[34]

For Kleinschrod, rescuing the vitalistic self-motion of the man in the train was of such central significance that he was only able to relate even elementary statements of the theory of relativity to that issue. Therefore, his criticism did not even deal with defending the classical law of addition of velocities, dealing instead with the fact that the motion of a living being (a man) was seen as equivalent to the motion of dead matter (light) in order to illustrate the addition of velocities in Einstein's train example. Kleinschrod ended his dispute with the theory of relativity with a

[33] Cf. Kleinschrod 1928, pp. 405f. [34] Kleinschrod 1920, p. 14.

declaration of war. "In conclusion, I oppose the Einsteinian R[elativity] P[rinciple] in space and time with the P[ositivity] P[rinciple] I have discovered, and defend its existence to the utmost."[35]

In contrast to Kleinschrod, other world riddle solvers, Gilbert and Fricke for example, developed a concept of time within the framework of their own theories only after their confrontation with relativity. In his satire of the relativity principle, Gilbert labeled the "falsification" of clock times and scales as "arbitrarinesses" that the "mathematicians" justified by decreeing:

> . . .some completely new 'natural laws'! They are turning everything we know about things on its head. [. . .] All objects are said to contract into themselves like hedgehogs [. . .] Velocity bewitches them, makes them internally contractile. This is how Professor Lorentz arrived at his famous Lorentz contractions. [. . .] Expressed in tangible terms: Motion eats time.[36]

Thus far, this is the usual spontaneously critical reaction to the relativistic concept of time, this time in satirical exaggeration, but Gilbert now also saw cause to anchor the concept of time, which had not yet played a role in his *Neue Energetik [New Energetics]* of 1912, into its framework using an ad hoc explanation. Instead of letting the matter rest with the absurdity of the relativity of time, he states in his satire against the theory of relativity, making reference to his *Neue Energetik*:

> How should time be erased by motion? They are two very different things, after all. Motion is something energetic: A body changes space during a time flow, that is, during the vanishing of the moments. Time, on the other hand, is something un-energetic; it can neither be shortened, nor lengthened, neither pushed, nor pinched, nor eaten.[37]

In his satire, Gilbert suggests – quite unsatirically – speaking of two types of time, on the one hand of simultaneity, in which infinitely many things happen next to each other, and on the other hand of the succession of the infinitely small moments (time as such). Space would be a dimension of time insofar as space could be considered as the events next to each other. Space and time would then be functions that together would enable spatial-temporal occurrence.

Fricke developed an unconventional notion of time in his ether vortex theory (cf. pp. 142f.); for him as well, the concept of time was only interpreted in his own theory due to the confrontation with the relativistic concept of time:

> The ether vortex theory leads [. . .] directly to a physically comprehensible explanation of the concept of time. Time is a physical process, just as is light. Space or ether flows, and its mysterious vortex-like flow is time. The absolute universal constant *c* that Einstein's formulae use as a basis is thus not, as he thought, the speed of light in a vacuum, but rather time itself, flowing absolutely evenly. Now Einstein's disastrous error of natural philosophy becomes

[35] Ibid., p. 63. [36] Gilbert 1914, pp. 22f. [37] Ibid., p. 51.

completely clear; he has confused two related concepts with each other. He thought that the speed of light was absolutely fixed, and time a relative concept. Now because he kept the speed of light in his calculations, he had to distort space and time in the most terrible way, but without arriving at a truly satisfying conception of the world.[38]

In this way, Fricke had killed two birds with one stone. He had both rescued the absolute character of time and declared the principle of the constancy of the speed of light to be null and void.

3.1.3 Light

Fricke's statement above already shows that the criticism of the principle of the constancy of the speed of light is so closely intertwined with the criticism of the concept of time that one cannot investigate the critical discussion about the speed of light without the context of the topic of time. On the other hand, the criticism due to different views of the nature of light is independent from the concept of time and will be presented thereafter.

Time and the principle of the constancy of the speed of light

The relativization of the concept of simultaneity in favor of the principle of the constancy of the speed of light practically provoked the opposite standpoint, because it is splendidly possible to disagree about this setting of priorities. Is it not more reasonable to vary the speed of light in favor of the absoluteness of simultaneity and thus to rescue the classical addition of velocities right along with it? Fricke, for instance, inverted the setting of postulates without further ado based on his ether vortex theory. "I thus oppose Einstein's principle of the constancy of the speed of light with the principle of the constancy of time."[39]

Rejection of the principle of a constant speed of light unified Einstein's academic and non-academic opponents. The fact that the velocity of light in a vacuum is always measured at about 300,000 km/s independent of the state of motion of the observer does not in fact correspond to any experience that can occur in everyday life. For instance, if one throws an object from a moving train, the velocities of travel

[38] Fricke 1920a, p. 22. In a letter to Gehrcke, Fricke stated in addition that time, as "movement of space [. . .]," would have to be "at least likewise three-dimensional," but without discussing this any further. Fricke to Gehrcke, August 13, 1921, GN 3-G-9.

[39] Fricke 1920a, pp. 22f. Guillaume, Einstein's former colleague at the Patent Office, likewise attempted to rescue absolute time by setting other axioms. He wrote to Einstein on October 3, 1917, "I examined whether it was possible *mathematically* to arrive at the *known final equations* when 'universal' time is simply *postulated*. On the other hand, Minkowski has emphasized that the great merit of your 1905 paper lies in the fact that it explains the Lorentz contraction by the relativity of simultaneity. The reciprocal to it has has been lacking until now, though: if absolute simultaneity is introduced, the contraction must disappear." Guillaume to Einstein, October 3, 1917, in Schulmann, Robert *et al.* (eds.) (1998). *The Collected Papers of Albert Einstein*, Vol. 8: *The Berlin Years: Correspondence, 1914–1918*. (English supplement translation), translated by Ann Hentschel. Princeton, Doc. 385.

and the throw are added. In a letter to Gehrcke dated November 26, 1913, Oskar Kraus freely admitted his amazement about Einstein's definitional daring in dealing with the speed of light:

I am amazed about the naiveté with which the 'arbitrariness' of the so-called time definition is expressly conceded here [what is meant is Einstein's lecture in Zurich in 1911]. He expressly says: 'We stipulate that the velocity of propagation of light in a vacuum from a Point A to a Point B is equally as large as the velocity of propagation of a light ray from B to A.'[40]

For Kraus, this process of definitionally establishing the speed of light was unthinkable due to the related consequences. He got to the heart of this criticism in another letter to Gehrcke. "If the difference of times results, then it merely follows that light rays are wholly unsuitable to determine the synchrony of clocks."[41] Added to this unequivocal setting of priorities to the detriment of the principle of the constancy of the speed of light and to the benefit of absolute simultaneity was the fact that for many, the principle of the constancy of the speed of light in relation to observers in any kind of motion was, as it was for Fricke, "logical nonsense."[42] He himself assumed – as did Lenard and other adherents of ether[43] – that an ether existed that was carried along by bodies in motion and that the speed of light would therefore be measured as the constant *c* relative to ether, but the law of addition of velocities would of course also have to apply to light. Fricke continued:

Now Einstein's train of thought was entirely different. [. . .] He thought that the speed of light would have the odd characteristic *in general* of always displaying the same value at every measurement relative to an observer *moving in any way at all* or at rest. Einstein's strange principle actually represents a constraint on experimental physicists; they are ordered to always measure the same velocity *c* whether they are moving along with the light ray or counter to it.[44]

[40] Kraus to Gehrcke, November 26, 1913, GN 72-A-5. Cf. Einstein, Albert (1911a). Die Relativitäts-Theorie [The Theory of Relativity]. Naturforschende Gesellschaft, Zürich, Vierteljahresschrift **56**, 1–14. Reprinted in English translation in Stachel *et al.* 1994, Doc. 17.

[41] Kraus to Gehrcke, October 11, 1913, GN 72-A-2.

[42] Fricke, Hermann. (1934). Die moderne theoretische (relativistische) Physik – eine Wissenschaft, die nicht stimmt (Der Justizmord am Weltäther) [Modern theoretical (relativistic) physics – a science that is not right (The judicial murder of universal ether)]. *Docentra*, **38**, September 23, 1934, reprinted in Fricke 1939, pp. 9–17 (see p. 13). Cf. also his remark in a letter to Gehrcke, "Why should the speed of light in particular be constant? There is no reason whatsoever for such an assumption." Fricke to Gehrcke, August 24, 1948, GN 2-E-2. The assumption of a constant speed of light was also "impossible" and "contrary to logic" for Gilbert as well. Cf. Gilbert 1914, p. 25.

[43] In his conception of an ether carried along by the heavenly bodies and an ether at rest, Lenard assumes that light always propagates at velocity *c* to the ether of the individual light source (the velocity of the sun's rays is equal to *c* to the solar ether, etc.). Cf. Lenard, Philipp. (July 1921). Fragen der Lichtgeschwindigkeit [Questions of the speed of light]. *Astronomische Nachrichten*, **213**, 303–08.

[44] Fricke 1920a, p. 12.

Karl Vogtherr's criticism aims in the same direction. He would allow either the light source (if light is emanated as particles) or the medium of propagation (if light were a wave in the ether) to be considered the reference body for measurement of the speed of light. But reference to the moving observer, following Einstein, was "untenable from the outset because the observer is not in any causal relationship to the progress of the light process."[45] This is where he saw the actual problem of the Einsteinian conception of light, which in his view did not bother with causality and was therefore subjective. The philosopher Joseph Petzoldt (1862–1929) also stated to Gehrcke that he "doubted that the propagation of light is independent from the motion of the light source."[46]

In the alternative mechanical theories of the world riddle solvers, the speed of light was also specified at different values than the approximation 300,000 km/s that had been very precisely determined by Armand Fizeau and Jean Foucault's experiments as early as the mid-nineteenth century.[47] Patschke's electromechanics, for instance, assumes that "free light ether flows" in space move at about 300,000 km/s, but can be braked by impact contacts (a light ether atom meets a large body) and also accelerated. This might occur, for instance, when two differently dense light ether flows collide and the lower density flow receives a thrust that accelerates it to 500,000 km/s or more.[48]

Reuterdahl criticized the constancy of the speed of light at an even more elementary level. In his view, the introduction of the speed of light as an absolute quantity already refuted the theory of relativity, writing "Postulating one Absolute against the relativities destroys Relativity as a true generalization." He also insisted that the speed of light could not be an exception to the classical theorem of addition of velocities, stating "Everything that moves, including light, has a relative and not an absolute velocity in reference to an observer."[49]

The criticism of the principle of a constant speed of light is unusual in that it mostly does not refer to an unconventional interpretation of the theory of relativity, but rather actually addresses a central pillar of the special theory of relativity. Most of Einstein's opponents certainly understood that Einstein had elevated the constant speed of light to a principle, but this procedure was fundamentally not

[45] Vogtherr 1923, p. 19.

[46] However, Petzoldt did not think anything of Gehrcke's criticism of the theory of relativity. Cf. Petzoldt to Gehrcke, November 13, 1912, GN 83-B-4.

[47] Astronomically, the speed of light had already been quite precisely determined by Ole Römer at the end of the seventeenth century from the shadow phase of Io, one of Jupiter's moons, and by James Bradley at the beginning of the eighteenth century from the aberration of light.

[48] Cf. Patschke 1921, p. 209.

[49] Both quotes from Reuterdahl 1931a. For additional positions analogous to Reuterdahl's first objection, cf. Hentschel 1990, pp. 94ff.

accepted. Einstein's opponents held to the approaches of classical physics or to their own theories and saw no reason to make an exception to the addition of velocities for the speed of light. Criticism of the principle of the constancy of the speed of light frequently appealed to specific views on the nature of light that were often metaphysically charged for the world riddle solvers. They thus introduced a level of argumentation into the content-based criticism that fell in principle outside of the sphere of authority of academic physics, but for which a claim to scientific validity could be asserted and which will also be found in subsequent topics.

Wave, particle, or something completely different – what is light?

A different critical examination of the principle of the constancy of the speed of light shifted the discussion of the speed of light onto the plane of ontology. How could Einstein elevate the constancy of the speed of light to a principle as long as no certain knowledge existed about the nature of light? With regard to the question of the nature of light, reference was made on the one hand to the question discussed in academic physics of "wave or particle," and on the other hand, world riddle solver views of light showed similarities to occult views of light. Both views on the nature of light play a role in the following presentation of the alternative conceptions of light. First, on the more narrowly physical question of the nature of light, Vogtherr, the physician, complained:

> Although the theory of relativity now makes this most peculiar conception of the process of light propagation, which certainly does not unquestionably result from the observations, into the foundation of its theoretical construct, it does not consider it necessary to form a precise conception of the nature of light.[50]

For him, the cardinal question of the theory of relativity consisted of whether or not light needed a medium of propagation, whether it was or was not a wave; he stated, "The problem that the theory of relativity unfurls is the question of the nature of light," and he found it improper that Einstein did not "make very specific physical assumptions,"[51] instead considering the development of a new light theory to be a desideratum of theoretical physics and nevertheless setting the speed of light as a constant.[52] Vogtherr made no secret of his own conviction regarding the nature of light. "Today we know quite certainly that light is based on

[50] Vogtherr 1923, p. 14. [51] Ibid., p. 17.

[52] In fact, Einstein called for a coherent light theory and by no means supported the contemporary state of affairs, which he made clear, among other things, in a lecture intended for the broad public. Cf. the report in the *Vossische Zeitung* newspaper about a lecture by Einstein on "The Crisis in Light Theory" at Berlin's Marmorsaal. *Vossische Zeitung*, February 8, 1922, GN 4, XVI, 31r.

wavelike disruptions in a medium; after all, we have these waves practically before our eyes in the interference fringes."[53] With his admonition to first determine the nature of light, Vogtherr was referring to the "wave or particle" question. Newton's conception of light as consisting of light particles had long been authoritative. Important explanations such as James Bradley's of the phenomenon of aberration were based on the particle model of light. This only changed in the course of developments in optics in the early nineteenth century. The experiments by Thomas Young and Augustin Fresnel on interference made the decision in favor of wave theory. The interference patterns hardly permitted any other conclusion – as mentioned in Vogtherr's quote above – than that light waves were reinforcing or canceling each other. With the formulation of electrodynamics, the conception of light as a wave – now an electromagnetic wave – in the ether received further reinforcement.

This was also the initial state of affairs when Einstein developed a "very revolutionary"[54] conception of the nature of light in his "annus mirabilis," 1905. However, this did not occur in conjunction with the special theory of relativity, which makes no statement about the nature of light, but rather in the paper "On a heuristic point of view concerning the production and transformation of light."[55] In 1900, Planck had already introduced the quantum of action h, a physical constant that states that electromagnetic radiation is not transmitted continuously, but rather in discrete portions. But only Einstein assumed that light actually possesses a quantum nature. To many contemporary physicists, this re-adoption of the concept of light particles initially seemed to be a return to antiquated conceptions and rejection of the electromagnetic conception of the world, which was considered certain at this point in time; however, it became widely recognized after some time.[56] Among Einstein's opponents, a difference can be clearly determined in this context between the criticism of Einstein's conception of the nature of light by academic physicists (academic philosophers

[53] Vogtherr 1923, p. 26. Cf. also Fricke's statement, "Before Einstein, however, people did not consider light to be an abstract computing formula, but rather an oscillational state of the ether." Fricke 1920a, p. 15.

[54] Einstein to Habicht, May 18 or 25, 1905, reprinted in Klein, Martin J., Kox, A. J., and Schulmann, Robert (eds.). (1995). *The Collected Papers of Albert Einstein*, Vol. 5: *The Swiss Years: Correspondence, 1902–1914*. (English translation supplement), translated by Anna Beck. Princeton, Doc. 27.

[55] Einstein, Albert. (1905a). Über einen die Erzeugung und Verwandlung des Lichtes betreffenden heuristischen Gesichtspunkt [On a heuristic point of view concerning the production and transformation of light]. *Annalen der Physik*, **322**(6), 132–48, reprinted in English translation in Stachel *et al.* 1989, Doc. 14.

[56] Following Einstein's lecture, the light quanta hypothesis was still being discussed skeptically as late as 1909 at the annual meeting of the GDNÄ in Salzburg. Max Planck in particular came out against it; only Johannes Stark, later an opponent of the general theory of relativity and quantum theory, came out clearly in favor of it. Cf. Einstein, Albert. (1909). Über die Entwickelung unserer Anschauungen über das Wesen und die Konstitution der Strahlung [On the development of our views concerning the nature and constitution of radiation]. *Physikalische Zeitschrift*, **10**, 817–25, reprinted in English translation in Stachel *et al.* 1989, Doc. 60; Discussion, ibid., 825–26, reprinted in English translation in Stachel *et al.* 1989, Doc. 61.

were scarcely interested in the nature of light) and by world riddle solvers. While the academic physicists among Einstein's opponents were more likely to accuse Einstein of returning to Newtonian light theory with the conception of light quanta[57] (when they even considered the introduction of the particle nature of light to be worthy of discussion at all in their criticism of modern physics),[58] many world riddle solvers sympathized with the light quanta, which fitted perfectly into their atomistic theories. They therefore thought that they had determined the nature of light long before Einstein. And in fact, the "wave or particle" question had indeed been explicitly taken up by some of them, and answered on the basis of their own theories.

Reuterdahl's light theory, which he also labeled "light quantum theory" and which he praised as the first substantiation in physics of Planck's constant *h*, is such an atomistic light theory. It is based on assuming the existence of the "monon," a postulated primordial atom with a radius one million times smaller than the hydrogen atom. According to Reuterdahl, monons were transmitted by radiation sources at absolutely regular time intervals. Einstein on the other hand, Reuterdahl notes in this context, had no correct conception of light whatsoever. "Einstein has failed utterly in providing a mechanical model for the physical changes involved in the production and transmission of light."[59] According to Reuterdahl, the regular transmission of the monon should explain that the speed of light is constant, but not – as with Einstein – absolute. Reuterdahl did not consider experimental detection of the existence of the monon to be relevant. Due to the necessary role that the monon performed in his system, its existence was certain in his eyes.[60] Characteristically, Ziegler, the primordial light theorist, was the greatest adherent of Reuterdahl's light theory, and extremely thankful for the introduction of the monon. He held the view that Reuterdahl's monon was identical with Ziegler's supernatural primordial light atom, which could not be detected per se due to its supernatural nature, but had now been determined mathematically by Reuterdahl.[61]

[57] The astronomer T. J. J. See, for instance, commented polemically that with his conception of light, Einstein was returning to Newton's emission theory. Cf. See, T. J. J. (May 13, 1923). Objections to relativity theory, Part 1, *The San Francisco Journal*, Part 2, ibid., May 20, 1923, Part 3, ibid., May 27, 1923.

[58] Lenard, for instance, accepted the light quanta hypothesis, which was, however, completely irrelevant to his criticism of the general theory of relativity.

[59] Reuterdahl, Arvid. (October 28, 1922). What the Einstein test can and cannot prove. Fact that he "forgot" to credit others before him can't be overlooked. *The Dearborn Independent.*

[60] Reuterdahl argues that the atom, the electron, and the "positon" – this is what he calls it – could not initially be detected experimentally, but that their existence was considered certain based on physical phenomena that made assuming them necessary. Cf. Reuterdahl, Arvid. (1923a). *A Synthesis of Number, Space, Time and Energy, and a Physical Basis for Planck's and Rydberg's Constants.* Brooklyn, NY, pp. 22ff.

[61] Cf. Ziegler to Reuterdahl, December 29, 1923, RP 4–66.

Thomas J. J. See (1866–1962)

Source: Reuterdahl papers; courtesy of the Department of Special Collections, University of St. Thomas, St. Paul, MN.

Thomas Jefferson Jackson See was born in Missouri on February 19, 1866. He completed his studies of science with a focus on astronomy at the State University of Missouri in 1889, and then earned a doctorate in astronomy at the University of Berlin in 1892. He worked first at the University of Chicago, then at the Lowell Observatory. In 1899, he became a professor of mathematics in the US Navy at the US Naval Observatory in Washington, DC, moving to the Naval Observatory at Mare Island in 1903. He came into conflict with his colleagues early on due to his presumptuous behavior as well as his publication of speculative results, but was very popular in the general public as the author of the periodical *Popular Astronomy.* Among other things, he developed an unconventional theory about the creation of the universe ("capture theory") and an extensive ether theory starting in the 1920s. See died in Oakland, California, in 1962.

Source: Sherrill 1999

The pattern that everything in science was related to one's own theory also applied to the world riddle solvers. The view was widespread among adherents of kinetic theories that the "true" physical explanation for Planck's quantum hypothesis had been provided by these primordial atoms such as the monon or the light ether atom, because the explanation was mechanical. Along with Reuterdahl and the astronomer T. J. J. See,[62] Patschke also thought that he had been the first to provide this explanation; he considered the quanta to be the light ether particles that were so important in his theory. Unfortunately, Patschke said, Planck had not been granted the privilege of setting up the all-encompassing light ether theory, as Patschke had, but it was typical for the ethos of the world riddle solver that Patschke, an engineer, praised Planck for the fact that he "probably foresaw" the goal, "but was not yet able to discern it sharply enough." Patschke emphasized the significance of his own works "for the works of Planck that must still be revised and supplemented accordingly."[63]

Among the world riddle solvers, the question of the nature of light could not be reduced to rejecting or accepting what Einstein's quantization of light consisted of in their view. Along with ether, light was the concept that was most likely to be amenable to a religious interpretation, and was understood as a connecting link between science and religion, as the "primordial phenomenon" of natural science. This level of the world riddle solvers' conception of light must be seen in a broader context. Ever since science began investigating light as a purely physical phenomenon, metaphysical and religious interpretations of light have continued to exist separately from academic physics. Light plays a prominent role in many religions and religious movements.[64] Around 1900, a specific conception of light was advocated, particularly in occult thinking, which in part was quite consciously set in opposition to conceptions of light in physics.[65]

[62] According to Sherrill, in See, T. J. J. (1940). *Wave-theory!* Fifth Book. Lynn, Mass. This book was not accessible to me. Cf. Sherrill, Thomas J. (1999). A Career of Controversy: The Anomaly of T. J. J. See. *Journal for the History of Astronomy*, **30**(98), Part 1, 25–50.

[63] Patschke 1925, p. 10. Patschke solved the "wave or particle" question in his *Elektromechanics*: If the electrons – this is how he labels his light ether atoms – were to wander unhindered through space, the classical situation of emission theory would result: the light ether atoms would continue to move ahead almost in a straight line, with only a slight longitudinal oscillation, undisturbed by other atoms. However, the closer they would come to heavenly bodies, which transmit atoms in Patschke's theory, the more probable it would be that they would encounter other atoms, and would thus begin oscillating; the atoms would continue moving in a zigzag line with transversal oscillations. In this way, the problem of unifying wave and emission theory was solved for Patschke. Cf. Patschke 1921, pp. 34f.

[64] Examples cover a wide range, from the Christian creation story, according to which the creation of light is God's first work (Genesis 1:3), to occult groups such as the "Empowered Brotherhood of Old Rites," a fundamentalist masonic group with masters who were designated "Radiances of the Primordial Light." Cf. Glowka, Hans-Jürgen. (1981). Deutsche Okkultgruppen 1875–1937 *[German Occult Groups, 1875–1937]*. Munich (Hiram Edn. Vol. 12), pp. 100ff.

[65] In fact, the physical and occult conceptions of light existed alongside each other in the non-academic space at the beginning of the twentieth century. Particularly in art, mostly occult interpretations of the scientific views of light were taken up and physical and occult concepts were combined. Cf. Hoormann 2003, p. 37.

Theosophy and anthroposophy in particular influenced these occult conceptions of light. Theosophy, which interpreted the electromagnetic world view ideologically according to its understanding, defined "astral light" as "anima mundi" – as the soul of the world, which was a force of the ether that worked "magnetically."[66] Here, light was understood as a symbol and means of representation of invisible, intangible phenomena.[67] Rudolf Steiner explicitly contrasted the anthroposophical concept of light with Einstein's concept of light, which he criticized as representative of the "materialistic" conception of light in science:

> I see [. . .] the one fundamental error in Einstein in that he applies the usual mechanical formulae – because that's what they are – to the propagation of light, and hypothetically assumes that the propagating light can be measured like any other body flying through space. He does not take into account the fact that the propagating light is not worldly particles flying away, but rather something where something happens in space, where a trace is left behind with the effect of shining [. . .], and I cannot speak of the fact that it propagates itself with a particular speed, but rather only the contour surface [propagates itself].[68]

The theistic and anthroposophical conceptions of light did claim to be scientific, but explicitly set themselves apart from being "merely" scientific/physical. The world riddle solvers, on the other hand, claimed that their conceptions of light were the actual scientific ones, as occult as they ultimately were. They mixed occult interpretations of light with physical ones, or appealed to the former to legitimize the latter, and postulated that they had discovered the true physics of light. Patschke was among those for whom the religionization of light made their own conception of light not more unscientific, but on the contrary made the light ether atomic theory "more true":

> As long as there have been logically-thinking people on Earth, these people will have placed the light of the stars, to which the sunlight, the moonlight, and the light of all the planets also belong, in the sphere of their detailed examination of nature. That is probably also why it happened that in earlier times, simpler natural people had clearer and more universal basic concepts of the structure of space than many of today's specialized scholars, who have allowed themselves to be misled by the Einsteinian theory of relativity and other falsely developed dogmas. That is, the ancient natural peoples primarily honored the light of the heavenly bodies as the giver of strength, as the

66 Here, magnetism was understood in the sense of "animal magnetism" (mesmerism). On ether, cf. Blavatsky 1877, pp. 125ff.

67 Cf. Hoormann 2003, p. 41.

68 Steiner 1982, pp. 32f. However, Steiner did not want to assert any new physics against academic science, but rather to establish an "intellectual science" that was to discover the laws of the life of the soul and the intellect, in contrast to science, which was oriented towards knowledge of external nature.

omnipotence of the world. And they were right! The omnipotence of light is derived in the simplest manner according to my universal light theory.[69]

The religious dimension that light possessed for Ziegler has already been emphasized in the first chapter in the sketch of primordial light as the cornerstone of his occult primordial light theory. For Ziegler, light, and more precisely the primordial light atoms of primordial light, was ultimately God. "The primordial light is thus not only the primordial material and positive substance in all things, it is at the same time also the incomprehensible and inviolable world spirit that eternally animates the world and keeps it in motion. It is what Saint Augustine called God in his Confessions, but was unable to grasp."[70]

Ziegler also cited religions of light to support his own theory, even writing a treatise on the "Sun God of Sippar" for this purpose. He assumed that all sun cults and religions of light (which for him included "ancient sun cults," but also the conceptions of light in Heraclitus and Goethe) would have an instinctive feeling, instinctive because still unspoiled, that the truth lay in light. However, these precursors would not have been in a position to develop a scientific formulation of a theory of light; instead, only he himself had succeeded in doing so, and had been plagiarized by Einstein – this was his accusation (cf. p. 163, pp. 168ff.).[71]

These conceptions of light were brought to bear as competition to the conception of light in academic physics and presented as opposing scientific concepts. They were based on a metaphysical elevation of the concept of light that did not play a role in academic physics, which certainly asked the question about the physical nature of light but did not assign it any religious dimensions. There was an ontological difference here between a metaphysical contemplation of the essence of light and "ordinary" physics. In these conceptual designs, just discussed, light is much more than light in physics – it is "omnipotence," it is "God," it is transcendent. In comparison to these conceptions of light, any physical, metaphysics-free concept of light seems unidimensional, limited, and false.

69 Patschke 1925, pp. 102f.

70 Ziegler 1923, pp. 14f. Cf. also Ziegler's statement, "He [Ziegler] begins his creation of the world, like Moses, again with light, the simplest working of the primordial light, and therefore calls his theory primordial light theory." Ziegler 1931, p. 19.

71 Ziegler, Johann Heinrich. (1904). Über den Sonnengott von Sippar [On the sun god of Sippar] in ibid., *Die wahre Einheit von Religion und Wissenschaft [The True Unity of Religion and Science]*. Zurich, pp. 175–93. Ziegler expressly presents the primordial light theory as a scientific expression of "ancient" truth. The "striking correspondence between the new primordial light theory and the tradition that we possess of the primordial light theory of the ancient Teutons, Romans, Greeks, Persians, Jews, Incas, and Aztecs and their esoteric doctrine, has probably already been able to guide you [the reader] to the supposition that my primordial light theory does not involve an absolutely new discovery, but rather only the rediscovery of age-old wisdom that had long been lost and forgotten." Ziegler 1932, p. 15.

3.1.4 Space

Relativistic space – a physics of curved nothingness?

For the majority of physicists and scientific laypeople who had learned Euclidean geometry in school and in their training, Euclidean space with its dimensions of height, width, and depth corresponded readily to experiences with space in everyday life, and was "real" space. In the course of the nineteenth century, the possibility of non-Euclidean geometries was discovered in mathematics, and spaces with higher dimensionality were discussed. For physics, this represented a challenge to ascertain the extent to which these developments in mathematics related to the physical concept of space.[72] While physicists were still reacting to developments in a neighboring discipline at this time, the situation was considerably aggravated when non-Euclidean geometry was brought in and applied extensively in their discipline with the general theory of relativity.[73]

The general theory of relativity generalized the relativity principle by including gravitation in the theory.[74] In this framework, gravitation is not described as a force, but rather as a geometrical distortion of four-dimensional spacetime. The first consequence that results for the concept of space is that physical space is not described by Euclidean geometry, but rather by non-Euclidean geometry. Moreover, space is not an uninvolved container in which physical events take place; instead, it co-determines these events. The structure of space is conditioned by the distribution of mass and energy in the universe and in turn conditions the motion of matter in space. The motion of light rays and heavenly bodies is not caused by an attractive force; instead, what appears to be an attractive force is the force-free motion of light rays and heavenly bodies in a curved space.

[72] Cf. Jammer, Max. (1993). *Concepts of Space: The History of Theories of Space in Physics*, 3rd enlarged edn. Mineola, New York, pp. 145ff.

[73] The special theory of relativity itself resulted in a challenge to the classical conception of space, since time was added to space as a fourth coordinate. Minkowski's much-quoted statement "From henceforth, space in and of itself and time in and of itself should descend entirely into shadow, and only a union of the two should maintain independence," was the impetus for many to emphatically take an opposing position. Cf. Minkowski, Hermann. (1909). Raum und Zeit [Space and Time], Lecture held at the annual meeting of the Society of German Natural Scientists and Physicians in Cologne in 1909. *Physikalische Zeitschrift*, **10**(3), 104–11 (see p. 104). The criticism of the conception of space in the special theory of relativity accordingly concentrated on showing that time and space were fundamentally different from each other. This was a concern of the Neo-Kantians above all, who saw this as a threat to a core element of their philosophy. Cf. Hentschel 1990, particularly pp. 208ff.

[74] Einstein, Albert (1915a). Zur allgemeinen Relativitätstheorie [On the general theory of relativity]. *Sitzungsberichte der Königlich Preußischen Akademie der Wissenschaften*, 778–86, reprinted in English translation in Kox *et al.* 1997, Doc. 21.

Stjepan Mohorovičić (1890–1980)

Source: Reuterdahl papers; courtesy of the Department of Special Collections, University of St. Thomas, St. Paul, MN.

Stjepan Mohorovičić was born in Bakar (Croatia) in 1890 as the son of the well-known geophysicist Andrija Mohorovičić. He studied in Zagreb and Göttingen from 1913 to 1915 and earned his doctoral degree in 1918. He did not succeed in attaining his desired career as a university physicist. He first worked as a senior primary school teacher at high schools focusing on language, math and science; in 1918 he was the head of a field weather station, and from 1917 to 1921 he was a professor at Zagreb Commercial Academy. He worked at Zagreb Academic High School from 1923 to 1939, and from 1948 to 1953 he was head of the scientific department of Industrie Zagreb, an optics company, where he was Scientific Counselor from 1951 to 1953. He died in 1980 in Zagreb.

Sources: Kertz 2002

The criticism of the Einsteinian concept of space referred particularly to Einstein's lecture on "Ether and relativity theory."[75] In that 1920 lecture, Einstein had stated that according to the general theory of relativity, space was equipped with physical qualities and one could also call this space ether. On the one hand, this reintroduction of ether by Einstein represented a free pass for the ether advocates among Einstein's opponents (cf. pp. 143f.), but on the other hand, a space equipped with physical qualities likewise provoked critical comments. How could Einstein claim that matter could have an effect on nothingness and that this nothingness could also even bend? How should a curved nothingness, which itself had an effect on matter, be imagined? Stjepan Mohorovičić objected:

By universal ether, the relativists only understand empty space with physical qualities, qualities that matter transfers to empty space in such a way that they impose various curvatures on it at various locations. How matter can have an effect on empty space, this is the *most unclear* point in the theory of relativity![76]

This "most unclear point" is resolved if one accepts the concept of the physical field, which is an established concept in physics at this time. But the possibility of thinking of space dynamically and of ether not substantially, but rather as a field, did not exist for those of Einstein's opponents who had remained closely attached to substance thinking. For the physician Karl Vogtherr, as for many others, it was entirely clear what a proper physical field had to be. "A physical 'field,' like any real part of space, can only exert effects if it is filled with matter."[77]

Similarly to the issue of the concept of time, many of the objections by Einstein's non-academic opponents to empty, curved space were of a general nature.[78] Reuterdahl, for instance, did not stoop to discussing the dubious nature or even just the debatability of the definitions of space, time, and matter. In his postulated "space-time potential," the intrinsic separability of space, time, and matter is a basic assumption. Accordingly, space and time are interdependent elements, but specifically not inseparable elements. Therefore, he merely noted:

Matter can only affect other matter. [. . .] Space and Time are not matter. Therefore, the presence of matter in Space-Time cannot change the nature of Space and Time. Space is not

[75] Einstein, Albert. ([1920a]). Äther und Relativitätstheorie, Rede gehalten am 5. Mai 1920 an der Reichs-Universität zu Leiden [Ether and relativity theory, a speech made on May 5, 1920, at the Imperial University of Leiden]. [Berlin], reprinted in English translation in Janssen *et al.* 2002b, Doc. 38.

[76] Mohorovičić 1923a, p. 48. [77] Vogtherr 1923, p. 15.

[78] The question of whether space is empty or filled with ether was already being discussed in popular science periodicals before the theory of relativity and thus the "abolition" of ether had reached them. Cf. e.g. Ist der Weltraum leer? [Is space empty?]. *Unsere Welt*, **5**(5), (1913), 335.

a material thing which can close like the walls of a box. It is not curved, because curvature pertains only to material objects existing in Space. Consequently Space is not limited and enveloping.[79]

Kurt Geissler, the senior primary school teacher, raised an additional general objection:

Just as 'bumps' are conceivable on the surface of a sphere, like the surface of the Earth, for instance, space is also supposedly bumpy, quasi-spherical. [. . .] [S]pace itself [should] be able to receive small changes from the effect of 'masses' – as if such bumps would be comprehensible at all, and would have any meaning at all if one does not assume that something non-bumpy exists that the bumpy part is differentiated from.[80]

This argument was popular and could be expanded as desired – the curved requires the straight, the relative the absolute, etc. But even a professor of theoretical physics like Ludwig Zehnder (1854–1949) supported very strict views about the nature of space; for him, the real space of a scientist had exactly three dimensions, no more, and no less. Zehnder insisted, "We cannot be permitted to attribute qualities to space itself, which is defined as absolute emptiness, after all, that can only be granted to matter, as Einstein and others have recently suggested in order to be able to deny the existence of ether."[81] Zehnder did not make the level of definition a topic of discussion. The idea that one could perhaps define space differently is inconceivable ("we cannot be permitted," "it is not feasible").[82] In the network of Einstein's opponents, those dealing more extensively with the concept of space were particularly the academic philosophers such as Oskar Kraus or Lenore Kühn, and the academic physicists.

"Thing-space" or "container space" – alternatives to relativistic space

While Gehrcke and Mohorovičić decisively favored Euclidean space as the physical space, they commented more cautiously than the objections just cited, and not categorically dismissively, on the possibility that a non-Euclidean space actually existed. In 1921, Gehrcke published a book on a topic that was rather unusual for an

[79] Reuterdahl, Arvid. ([1930]). *Einsteinism. Its Fallacies and Frauds*, RP 1–28. This is the English-language typescript of Reuterdahl's contribution to the German anthology *100 Autoren gegen Einstein [100 Authors against Einstein]*. It was translated into German for this publication.

[80] Geissler, Kurt. (1931). Schluss mit der Einstein-Irrung [An end to the Einstein error], in Israel, Ruckhaber, and Weinmann, pp. 10–12 (see p. 11).

[81] Zehnder, Ludwig. (1927). Ueber die Grundlagen der Naturwissenschaften [On the foundations of the sciences]. *Verhandlungen der Naturforschenden Gesellschaft in Basel*, **38**, 84–100 (see p. 88). T. J. J. See was significantly more polemical: "A fundamental postulate of Einsteinism is that the ether does not exist, and gravity is not a force but a property of space. These crazy vagaries scarcely require mention, beyond the remark that such discussion is a disgrace to our age." Quoted from Einstein Theory is 'Old Stuff.' US Astronomer Says German Scientist is Plagiarist. [Unknown newspaper], April 1923 [no specific date], GN 4, XX, 84r.

[82] Zehnder 1927, p. 88.

experimental physicist, namely *Physik und Erkenntnistheorie [Physics and Epistemology].*[83] It was exceedingly important to him to ascertain the philosophical foundations of his discipline.[84] Gehrcke had developed this interest independently from the challenge that the theory of relativity represented to those foundations, but this interest did represent the decisive cause for his fundamental opposition to modern physics. In *Physik und Erkenntnistheorie*, as well as in his correspondence with Oskar Kraus, who as a philosopher was particularly suited as an interlocutor for such discussions with Gehrcke, he presented his views on physical space more precisely.

Gehrcke made a distinction in *Physik und Erkenntnistheorie* between physiological space (subjective perceptions of reality), physical space (extracted as objective reality from the subjective perceptions of space), and mathematical space (lacking a connection to reality). While physiological space was real and perceivable, physical space was likewise real, but not necessarily perceivable. Mathematical space was neither real nor perceivable. Gehrcke particularly emphasized the objectivity of physical space, which he saw as follows: Since the spatial aspect was a characteristic that various subjects could experience using their perceptions such as sight or touch, one could infer the objective nature of these subjective experiences of space:

> [T]he spatial aspect is already contained in the perceptions, which as such cannot be considered unreal even by the greatest skeptic, in something real, in something existing; there is thus no reason to accept why the spatial aspect, which possesses empirical reality as the content of perception, should *not* be able to be objectively real in *physical* processes.[85]

Certainly, it was no more possible to prove the realistic position any more than an idealistic position, but the former possessed "the advantage of having a connection to the realism of the common man." Gehrcke understood his own position as "purified realism."[86] According to him, Euclidean space had been confirmed up to that point by experience and measurements. Gehrcke saw a rewarding undertaking in continuing to investigate space experimentally in order to come closer to "real" geometry in this way, which for Gehrcke was most likely Euclidean geometry. In a draft letter to Hugo Dingler, Gehrcke expressed this in a marginal note with the indication "not to Dingler": "The core question is this: Must experienced space be Euclidean, or is experienced space actually Euclidean within the precision of our previous measurements, that is, *approximately* Euclidean, but *can* it be non-Euclidean under some circumstances?"[87]

83 Gehrcke, Ernst. (1921). *Physik und Erkenntnistheorie [Physics and Epistemology].* Leipzig/Berlin (Wissenschaft und Hypothese Vol. 22).

84 Cf. also Schlick's remark in his review of the book, stating "that the author [Gehrcke], [. . .] obviously really feels the need to obtain clarity about the principles of his science." Schlick, Moritz. (1921). [Review of Gehrcke 1921c]. *Die Naturwissenschaften*, **9**, 779.

85 Gehrcke 1921, p. 62. 86 Both quotes ibid., p. 63.

87 Cf. Gehrcke to Dingler, November 8, 1921 [draft], GN 38-H-13.

Consequently, Gehrcke did not categorically exclude the possibility that physical space could be non-Euclidean, and Mohorovičić also remained open in principle to developments in the physical conception of space, although he, like Gehrcke, thought it to be probable that real space did not show any greater curvature, but was instead largely Euclidean. Mohorovičić held the view that empirical space could only be defined approximately at any rate.[88] Oskar Kraus had a significantly different conception of space. He laid out his conception of space very succinctly in his correspondence with Gehrcke, in which both men endeavored to understand each other's conception and its plausibility. These reciprocal attempts by a student of Brentano and an experimental physicist to convince each other are particularly informative. Oskar Kraus started the preserved correspondence[89] on this topic:

You ask whether I understand space as a *subject*, as a *modification*, as *both* or as *neither of the two*? The answer: Space is a fiction; there is no 'space,' but rather only spatial-temporal things, 'bodies.' [. . .] [H]ere, the word 'body' should be taken in a very broad sense, let us thus say [. . .] there is no 'space,' but 'real spatially extended things' or perhaps only a single continuous spatially extended thing – we can call this aether.[90]

Space freed of all weighable bodies and physical states was a fiction, he said; real space was the space filled with ether and thus a "thing."[91] In his reply of January 25, 1921, however, Gehrcke insisted that ether could not be "the ultimate real thing" and argued against it with his conception of an unperceivable, real physical space:

This theory [of space as a thing] is still incomprehensible to me; it appears to me to be improper: If ether should be able to be in a state of absolute motion or rest, then I think that space that is free from ether is also implicitly assumed as real. For how should *real* ether be able to *move* if space is *not real*? Can something *real* move in something *unreal*?[92]

Kraus answered four days later:

Only when space is not anything real can aether (and anything at all that is spatially extended) move. For if space were something real, then it would be something real and extended, that is, something spatial itself, i.e. something material in the broadest sense of the word. [. . .] After all, 'space' is only a *façon de parler*, a figure of speech, that corresponds to nothing or rather corresponds to something quite different than one, misled by the noun, normally thinks.[93]

As becomes clear in this discussion between Gehrcke and Kraus, Einstein's opponents did not agree with regard to the concept of space. There were advocates of

88 Cf. Mohorovičić 1923a, p. 17.
89 The correspondence preserved in Gehrcke's papers comprises twenty letters as well as a postcard from Kraus, and eight copies or drafts of answers by Gehrcke from the years 1913–1929.
90 Kraus to Gehrcke, November 22, 1920, GN 71-G-6. 91 Ibid.
92 Gehrcke to Kraus, January 25, 1921, GN 71-G-9. 93 Kraus to Gehrcke, January 29, 1921, GN 71-G-10.

empty space as well as advocates of a materialistic conception of space. Material space usually combined ether and space or set them as equivalent. In the context of academic physics, identifying space with electromagnetic ether was an obvious step. If the ether was conceived of as absolutely at rest, as formulated by Lorentz, then it could function as a frame of reference for absolute motion, just as absolute space could. In addition, this absolute ether-space was seen as a comprehensible conception of space. Although this conception did not become established in physics and was soon overtaken by the theory of relativity, the conception of a substantial ether-space continued to exist among both academic and non-academic researchers. Fricke's theory, for example, results in a material space. He simultaneously labeled his ether vortex theory (cf. pp. 142ff.) a new and comprehensible theory of space; he identified space with ether. "If the word 'space' is interpreted in the concrete-physical sense and not in the mathematical-abstract sense, then it can be said that the basic concepts of space, ether, and energy become identical."[94]

While the adherents of material space placed the comprehensibility of this concept in the foreground, the other direction of criticism specifically denied the materiality of space. The second direction referred back primarily to Kant and assumed space to be a form of intuition a priori. Thus Lenore Kühn also stated, "Did we not already have the great warner in Kant – who denoted space and time as particularly phenomenal intuitional regularities, described them, and separated them from the actual elements constituting experience?"[95] For her, it was certain that "thing" and "space" were mutually exclusive. Space could not be attributed any effects on things, otherwise "the strict chain of physical causes" would be undermined by introducing an "uncontrollable and unforeseeable factor of a different (geometrical) kind."[96] Kühn further thought she had discovered a fundamental contradiction in that space does not only appear as a cause but also as an effect in the general theory of relativity, in that it is thought of as being "generated" by matter. For Kühn, Einstein's physics ultimately had to be classified as "mere phenomenalism and a perspectivism that is only applicable as an illustration, and is subjective on top of that." Her conclusion, which was also attractive to Gehrcke (he had placed double underlines under this passage in Kühn's typescript), was that Einstein had transferred causality from its proper place (physics) to geometry

94 Fricke, Hermann. (1918). *Eine neue und einfache Deutung der Schwerkraft und eine anschauliche Erklärung der Physik des Raumes [A New and Simple Interpretation of Gravity and a Comprehensible Explanation of the Physics of Space]* [printed as a manuscript], Berlin, p. 25.

95 Kühn, Lenore. (August 17, 1921). Kant contra Einstein. *Hellweg: Wochenschrift für deutsche Kunst [Light Path: Weekly for German Art]*, GN 4, XII, 16r.

96 All of the following quotes in this paragraph from: Kühn, Lenore. Der Raum als angebliche Ursache oder Wirkung physikalischer Zustände (Frage des inhomogenen Raumes und Frage der Erzeugung des geometrischen Raumzustandes durch Zustand der Materie) [Space as the alleged cause or effect of physical states (the question of unhomogeneous space and the question of the generation of the geometrical state of space by the state of matter)], [unpublished, undated typescript], GN 38-I-9.

in order to somehow still ensure a causal connection in natural events despite his subjective theory (in their opinion), but causality simply had no business being in geometry at all.

Lenore Kühn (1879–1955)

Source: Detlef Kühn, Berlin.

Lenore Kühn studied at the College of Music (Berlin) from 1895 to 1897 and worked as a pianist and piano teacher after graduating. She studied philosophy in Berlin, Erlangen, and Freiburg starting in 1903, earning her doctorate in 1908 with Heinrich Rickert with a dissertation on the problem of esthetic autonomy. Kühn began working as a journalist in 1903 and was active in the German folk nationalist movement. Starting in 1916, she was editor of the national-liberal periodical *Der Panther* and from 1917 she was assigned to press relations in the national women's committee of the German National Folk Party and publisher of the periodical *Frau und Nation [Woman and Nation]*. From 1926, Kühn worked at the Nietzsche Archive; in later years, she traveled abroad as a reporter.

Sources: Degeners "Wer ist's" [Who's who] 1955; Kühn 2010

The differentiation between physical, three-dimensional, Euclidean space (singular "reality") and the multi-dimensional, non-Euclidean spaces of mathematics (pluralistic "possibilities") was repeatedly emphasized and was the main unifying topic of the criticism of the concept of space in the general theory of relativity, which Einstein's opponents assigned to the category of mathematical possibilities. Independent of whether a real "container space" was assumed in which bodies and ether could be located, or a "material space" was assumed that could not exist independently from real corporality, a single real space was vehemently demanded. It was asserted in opposition that the dynamic concept of space in the general theory of relativity did not treat real space, but rather only one of many possible mathematical spaces lacking a connection to reality.[97] The fact that the connection to reality was lacking was apparent, according to the conclusions of the circular argument, simply due to the fact that if this mathematical possibility were to apply to reality, such absurd results as the curvature of space would be the consequence.

3.1.5 Gravitation

The debate about gravitation

Gravity, like the passage of time and the perception of three-dimensionality, is a phenomenon that directly affects every person. Nevertheless, it is perhaps the most mysterious area of natural science. Why bodies always fall downward, but never upward, is a question that was answered historically in various ways. Aristotle, for instance, assumed that bodies move toward their natural location. For him, this was the center of the world, and in the age of the geocentric conception of the world, the fact that all bodies fall to the Earth was a convincing argument for this conception. Newton, on the other hand, assumed that two masses attract each other and formulated a law of gravity that he applied not only to earthly phenomena, but also to the motion of the heavenly bodies, and which is applied even today in many astronomical calculations. With Newton's law of gravity, however, the question of the nature of gravity became virulent in a particular way. Newton's law implies that gravity works instantaneously without requiring time for propagation. Newton himself was dissatisfied that he was unable to discover the mechanism of gravitation. Nevertheless, Newton's law of gravity remained without serious competition until the development of the general theory of relativity due to the enormous successes of its application over a long period of time.

[97] This was the assessment by Henri Gartelmann, for instance, "The question: is space present in reality, or not? They [the physicists] disregard this as less important." Gartelmann, Henri. (1920). Zur Philosophie Kants [On Kant's philosophy]. *Neue Weltanschauung*, **9**(2), 33–41 (see p. 40).

However, alternative approaches were repeatedly developed with the goal of providing Newton's law of gravity with a physical explanation, or to find an explanation whose results at least approximated the law. Newton himself thought about the possibility that a medium, ether, could be the cause of gravitation, as did his contemporary Huygens. However, there continued to be a need for an explanation as to how gravitation could spread as a long-distance effect in the ether in accordance with Newton's law. Many people also asserted the opinion that the speed of propagation of gravitation would have to be finite. In the second half of the nineteenth century, an attempt was made to establish ether as a medium of gravitation using a modified scientific basis. Advocates of the electromagnetic conception of the world such as Karl Friedrich Zöllner endeavored to interpret gravitation as an electromagnetic phenomenon, and the suggestion was made to assume the propagation of gravitation at the speed of light. Weber's electrodynamic law was generalized and applied to the motion of the planets.[98]

Along with the electromagnetic interpretation, the mechanical theory of gravitation[99] existed as another attempt at an explanation. Among the non-academic bodies of knowledge that were cited as alternative explanations for gravitation opposing the theory of relativity, this was the one most often mentioned. In the academic world, the gravitational theory of Georges-Louis Le Sage (1724–1803) became particularly well known.[100] Le Sage assumed that the supposed attraction of two bodies arose from the fact that the pushes of the smallest particles ("corpuscules ultramondains") moving very quickly in a straight line in space were pressing these bodies towards each other. The gravitational theories of Angelo Secchi (1818–1878) and William Thomson (Lord Kelvin, 1824–1907) were also based on mechanical models of gravitation. However, these approaches were beset by massive problems that had to be discussed when this mechanical theory was dealt with in the context of academic discussion. Due to the empirically confirmed proportionality of the force of gravity with mass, for instance, the mechanical theory of gravitation must thus assume a high degree of permeability in matter in order for the gravitational particles to be able to interact evenly with all parts of a body. However, the more permeable the body, the less it can be assigned a shielding effect, which is used in this model to

[98] Cf. Zöllner, Karl Friedrich. (1882). *Erklärung der universellen Gravitation aus den statischen Wirkungen der Elektricität und die allgemeine Bedeutung des Weber'schen Gesetzes [Explanation of Universal Gravitation from the Static Effects of Electricity and the General Meaning of Weber's Law]*. Leipzig; as secondary literature, cf. van Lunteren 1991, p. 246; on the criticism, cf. Whittaker, Edmund Taylor. (1951 [1910]). *A history of the theories of aether and electricity*. 2 volumes, Vol. 1: *The classical theories*, 2nd revised and expanded edn. London [*et al.*], pp. 201ff.; on the influence of Weber's work on theories of gravitation, cf. ibid., pp. 207f.

[99] German *Drucktheorie*. Sometimes also called *Stoßtheorie* ("impact theory"); often called "kinetic theory of gravity" in English, and sometimes also "pushing gravity."

[100] An earlier approach came from Nicolas Fatio de Duillier (born 1664). Cf. Zehe, Horst. (1980). *Die Gravitationstheorie des Nicolas Fatio de Duillier [The Gravitational Theory of Nicolas Fatio de Duillier]*. Hildesheim (Arbor scientiarum, Series A: Abhandlungen Vol. 7).

explain gravitation.[101] In academic science, the developments in electrodynamics and thermodynamics in the second half of the nineteenth century additionally confronted the mechanical gravitation models with the knowledge that many other physical phenomena cannot be explained purely mechanically – so why should it be possible for gravitation, of all things?

The non-academic gravitational researchers were not impressed by this objection. They held the view that thermodynamics and electrodynamics could very well be reduced to mechanics, and that gravitation was of course also to be explained as a mechanical phenomenon. In fact, the academic physicists were often explicitly accused of having moved away from mechanics. "These gentlemen [relativists] are still in the dark as to the fact that all knowledge of the things of nature must be traced back to 'matter' and mechanics: Explaining means mechanizing! (materializing, monizing)."[102]

As van Lunteren also remarks, the project of fathoming the physical cause of gravity in the last third of the nineteenth century was less a topic in academic physics[103] than in non-academic research.[104] Possible approaches to an explanation were a favorite topic in the popular science periodicals. Karl Müller (1818–1899), one of the founding fathers of *Natur*, the flagship of popular science journalism in Germany, stated there in 1880:

> [W]e make no secret of the fact that for us, 'attraction' has long been one of those spectres that only slowly depart from the world because it is so difficult to set a positive position in its place. The professional physicists and astronomers have little cause to occupy themselves with correcting this questionable superstition [. . .] In our opinion, the Physics Association of Breslau, headed by Mr. Aurel Anderssohn, whose many communications about the aforementioned object we have published particularly in these pages, has provided the significant impetus to pass this matter into the hands of laymen trained in physics, and thus it appears that a conception that is no longer tenable is gradually being undermined by the laity, until the scientific world will ultimately have to follow it.[105]

101 For a brief overview of the problems of mechanical theories of gravitation, cf. Renn, Jürgen and Schemmel, Matthias. (2007c). Introduction, in *The Genesis of General Relativity*, ed. Jürgen Renn, 4 volumes, Dordrecht (Boston Studies in the Philosophy of Science Vol. 250), Vol. 3: Renn, Jürgen and Schemmel, Matthias (eds.), *Gravitation in the Twilight of Classical Physics. Between Mechanics, Field Theory, and Astronomy*, pp. 4f.

102 Gilbert 1914, p. 113.

103 On the discussion of gravitation in academic physics, cf. Renn, Jürgen and Schemmel, Matthias. (2007c). Introduction, in Renn 2007c, pp. 1–18; as contemporary skeptical voices on the conceivability of the force of gravity, cf. e.g. Du Bois-Reymond, Paul. (1888). Über die Unbegreiflichkeit der Schwerkraft [On the inconceivability of gravity], *Naturwissenschaftliche Rundschau*, **3**(14), 170–78; Mach, Ernst. (1969 [1872]). *Die Geschichte und die Wurzel des Satzes von der Erhaltung der Arbeit [The History and the Root of the Theorem of Conservation of Work]*, ed. Joachim Thiele, photomechanical reprint of the Prague edn., Calve, 1872, Amsterdam, p. 32.

104 Cf. van Lunteren 1991, pp. 196f. and p. 246. A few of these non-academic studies were discussed in the academy, including one by von Isenkrahe, for instance, and even this one only on a limited basis.

105 Müller, Karl. (1880). Eine drastische Kritik der Anziehungskraft [A drastic criticism of attractive force]. *Die Natur*, New Series, **6**(15), 194. On Anderssohn's work, cf. this chapter, Notes 110–112.

Due to this strongly advocated position for mechanical explanations of gravitation in the non-academic space, Einstein's opponents defended classical physics in the area of gravitation – Newton vs. Einstein – less often than might have been expected. When it did occur, it was usually done by academic physicists.[106] The world riddle solvers considered Einstein's conception of gravity to be rather a mere continuation of Newton's "mathematical" theory of gravitation; indeed, it will become apparent below that the non-academic criticism of Einstein with regard to gravitation was in the main a criticism of Newton and thus completely missed the point of the theory of relativity.

First, an example will be used to examine how a non-academic gravitational theory was passed on. A second step will discuss the criticism of the theory of relativity based on non-academic gravitational theories.

The non-academic search for the nature of attractive force

As mentioned in Karl Müller's assessment of gravitational research just quoted, Aurel Anderssohn (1824–1896), an industrialist in lead and tin foil, played a major role in the non-academic dissemination of the mechanical theory of gravitation. Many opponents of the conception of gravitation in the theory of relativity – including Zacharias, Fricke, and Reuterdahl – referred to Anderssohn.[107] As an industrialist, Anderssohn was an independent scholar – his student Zacharias referred to Anderssohn's large library of scientific works[108] – and chairman of the Breslau Physics Association, which was dedicated entirely to the mechanical theory of gravitation under his leadership. The ophthalmologist and medical historian Hugo Magnus (1842–1907), a professor at the University of Breslau and likewise

[106] Cf. e.g. See, T. J. J. (October 12, 1924). New Discoveries showing the Complete Triumph of Newton over Einstein, [press release], RP 4–49. See understood his ether theory as a perfection of Newton's theory. See's theory is likewise a mechanical theory. "[. . .] the *Aether* is a corpuscular medium, the Aetherons moving with a velocity of 294,000 miles per second." See, T. J. J. (April 19, 1923). To the editor of the New York Times, RP 4–49. Cf. also Zehnder's statement, "Gravitation [. . .] [is the] perfect remote force, working directly through empty space." Idem (1938). *Die tiefsten Grundlagen der Physik und Chemie. Aus den Universitätsvorlesungen in den Jahren 1935 bis 1938 [The Deepest Foundations of Physics and Chemistry. From the University Lectures in the years 1935 to 1938]*. Zurich/Leipzig, p. 108.

[107] Cf. Zacharias, Johannes. (1922). *Verborgene Gewalten im Weltgeschehen: Eine neue Raum-Kraft-Lehre [Hidden Powers in the Events of the World: A New Space-Force Theory]*. Munich, p. 14; Fricke, Hermann. (1919). *Eine neue und einfache Deutung der Schwerkraft und eine anschauliche Erklärung der Physik des Raumes [A New and Simple Interpretation of Gravity and a Comprehensible Explanation of the Physics of Space]*. Wolfenbüttel, p. 93; Reuterdahl, Arvid. (1921a). Einstein and the New Science. *Bi-monthly Journal of the College of St. Thomas*, **9**(3), 3–26 (also as an offprint), p. 3. Anderssohn also was in contact with Secchi; in 1875, he had visited him in Rome, and Secchi was listed by Anderssohn's association as an honorary member until his death in 1878. On the meeting between Anderssohn and Secchi: cf. the reprint of the interpreter's minutes of the meeting in Zacharias, Johannes. (1912b). Die sechsteilige Kugel zur Erklärung der Gravitation [The six-part sphere for explaining gravitation]. *Weltwissen*, **1**(2), 33–34 (see p. 34). Secchi's honorary membership is mentioned in Erdmenger, Otto. (1881). Bericht über eine Sitzung des Breslauer physikalischen Vereines [Report on a meeting of the Breslau Physics Association]. *Die Natur*, New Series, **6**(52), 628.

[108] Zacharias, Johannes. (1921). Ein Pionier der Physik als Brückenbauer zur Metaphysik. Autobiographisches vom Ingenieur Johannes Zacharias [A pioneer of physics as a bridge builder to metaphysics. Autobiographical information on the engineer Johannes Zacharias]. *Der Forscher*, **2**, 4–6 (see p. 5).

a member of the association, indicated that the "absolute negation of attraction" was the motto of the association. Looking back, he reported in the 1884 annual report:

When the Association was founded by a few men, it was certainly not with the intention of founding a society that denied attraction. On the contrary, all of the gentlemen who assembled to found the Association were unconditional adherents of attraction and for them the issue was exclusively to understand the nature of attraction.[109]

These attempts to discover the nature of the force that causes attraction, however, led the association down unusual paths and ultimately on a confrontational course with academic physics. Magnus felt it necessary to justify this anti-academic turn of the association's members. Even though the association had been founded "completely legally on the territory of modern physics" (what is meant here is classical physics), it later felt obliged to leave this territory:

If we have gradually been driven from this territory of exact modern physics, and the supporters of modern physics have become opponents thereof, at least in some respect, then this is certainly not due to the members of the association, but rather due to the nature of attraction. Despite the fact that the six men who initially assembled to found the Association spared no effort and work, it was not possible to comprehend attraction.

The number of meetings was increased from monthly to weekly and they often lasted until after midnight. However, as the chronicler Magnus reports:

Attraction absolutely could not be found. Despite very diligent experiments, although all the authorities had been consulted and research conducted without respite, attraction vanished more and more. No lever could be found with which it would have been possible to more closely approach attraction. In short, the members of the Association were ultimately forced to an awareness that attraction did indeed exist in the handbooks of physics, but not in space.

The members of the Breslau Physics Association now became aware that they were thus moving away from academic physics, "that a breach with physics was unavoidable, and a difficult battle now began for many of us." From then on, the Breslau Physics Association called itself the Central Association for the Solution of the Problem of Attraction, and for a brief period also the Breslau Physics Association for Establishment of the Theory of the Pressure of Masses from a Distance. They consciously took the step into research that argued non-academically, precisely for the reason just cited by Magnus, that academic science was unable to grasp the nature of attraction; discovering this nature was of central importance to its members. The association members felt compelled not only to deny attractive force, but also to set something else against it, specifically a universal mechanical theory which had been developed particularly by Anderssohn in various articles in *Natur*

[109] Breslauer Physikalischer Verein. (1884). *Physikalisches Jahrbuch*, **1**, 35. Breslau. The following quotes are ibid., p. 35, p. 36, pp. 36f., p. 37.

since the mid 1870s[110] and set forth in two works, *Die Theorie vom Massendruck aus der Ferne [The Theory of Mass Pressure from a Distance]*[111] of 1880 and *Physikalischen Prinzipien der Naturlehre [Physical Principles of Science]*[112] of 1894. Anderssohn advocated a classical mechanical theory. "There is no direct distant effect, particularly no attraction and repulsion. The motor in the mechanism of the world is solely the pressure imparted by the ether." This pressure had "its mechanical explanation [. . .] in the ether pushes, which follow one after another with very great velocity and hit the weighable atoms of the relevant body."

The Breslau gravitation researchers wanted to document a "central motion" caused by mechanical forces experimentally. The motion of the heavenly bodies was also said to be of such a nature. In the experiment, a sphere was placed in a basin of water. The water represented the ether and the sphere the Earth. The pressure transmitted by the ether and affecting the heavenly bodies was to be simulated by jets of water coming from a "water sun" in the center of the basin and also from its edge.

This mechanical theory was expanded to cover all other physical forces. Thus Anderssohn interpreted magnetism and electromagnetism as phenomena caused by the pushes of ether particles.[113] The other members of the Central Association for the Solution of the Problem of Attraction developed the theory further. Magnus, the chronicler, for example, whose specialty area was ophthalmology, worked on a mechanical theory of optics, particularly of vision. A radiant body, he said, caused an oscillation of the ether molecules, which rapidly propagated to the eye and exerted pressure on the optic nerves, causing the optic nerves to send signals to the brain.[114]

The Breslau gravitation theorists were able to sever their ties to academic physics in the awareness that they were part of a larger movement. Magnus noted the almost monthly appearance of works that were attacking the concept of attraction.[115] This non-academic scientific community was extremely

[110] Cf. Anderssohn, Aurel (1876a). Zur Lösung des Problems der Anziehung [On the solution of the problem of attraction]. *Die Natur*, New Series, **2**(33), 357; ibid., (1876b). Ein Apparat zum praktischen Beweise der Naturnotwendigkeit des Gravitationsgesetzes [An apparatus for the practical proof of the natural necessity of the law of gravitation], ibid., **2**(35), 383–84; ibid. (1877a). Zur Lösung des Problems der Anziehung [On the solution of the problem of attraction], ibid. **3**(4), 53–54; Breslauer physikalischer Verein zur Begründung der Lehre vom Druck der Massen in der Ferne. (1877). Neues zur Lösung des Problems der Anziehung [New information on the solution of the problem of attraction]. Ibid. **10**, 136–37; Anderssohn, Aurel. (1877b). Ebbe und Flut nach der Lehre 'vom Druck der Massen aus der Ferne' [Low and high tide according to the theory 'of the pressure of masses from a distance']. Ibid. **3**(15), 207–08; ibid. (1878). Attraktion oder Massendruck? [Attraction or mass pressure?] Ibid. **4**(16), 581; ibid. (1879). Aufruf an die deutschen Naturforscher [Appeal to German natural scientists], ibid., **5**(47), 591.

[111] Ibid. (1880). *Die Theorie vom Massendruck aus der Ferne in ihren Umrissen [The Theory of Mass Pressure from a Distance in Outline Form]*. Breslau.

[112] Ibid. (1894). *Physikalische Prinzipien der Naturlehre [Physical Principles of Science]*. Halle (Saale). The following quotes are ibid., p. 16, p. 19.

[113] Ibid., p. 90. [114] Breslauer Physikalischer Verein 1884, p. 39. [115] Ibid., p. 45.

Fig. 3.1 Experiment on pressure radiation (central motion by means of centrifugal and centripetal pressure radiation). Source: Anderssohn 1880, Table 5.

important. It was only in this way that adherents of mechanical theories of gravitation, who had no doubts as to their status as outsiders with regard to academic physics, obtained confirmation that they were members of a group of pioneers and not isolated private researchers who had gone astray.

It was thanks above all to Anderssohn's student Johannes Zacharias, who had moved to Berlin, that the theory became accepted in monistic contexts after Anderssohn's death, for instance by the dissemination and development of the theory in Zacharias' own periodical, *Weltwissen*, which was used as a discussion forum by many opponents of modern physics in the 1910s and 1920s. Zacharias himself advocated a mechanical theory that was strongly colored by the occult, and his periodical *Weltwissen* was entirely focused on offering a forum to assorted varieties of mechanical theories. Articles such as "On the mechanics of gravitation"[116] or "On the value of theory and the replacement of attraction theory"[117] were published in almost every issue of the bi-weekly periodical. Anderssohn and his network were an example of non-academic research in the area of mechanical theories of gravitation, but were not an isolated case. At the end of the nineteenth century, mechanical theories of gravitation were widespread, partly within the scientific community, but particularly in contexts outside of academic physics;[118] however, a more comprehensive investigation extending beyond the German-speaking area is still lacking. This context of mechanical theories of gravitation, which could only be outlined here, played the central role in the criticism of Einstein's theory of gravitation.[119]

The criticism of the theory of relativity based on mechanical theories of gravitation

The world riddle solver criticism of academic science with regard to gravitation applied much more often and more extensively to Newtonian attractive force than to

[116] Zacharias, Johannes. (1912a). Zur Mechanik der Gravitation [On the mechanics of gravitation]. *Weltwissen*, **1**(1), 9–10.

[117] Beckenhaupt, Carl. (1912a). Über den Wert der Theorie und den Ersatz der Anziehungslehre [On the value of theory and the replacement of attraction theory], *Weltwissen*, **1**(1), 4–6.

[118] Cf. e.g. Dellinghausen, Nicolai. (1880). *Das Räthsel der Gravitation [The Riddle of Gravitation]*. Heidelberg; Isenkrahe, Caspar. (1879). *Das Räthsel von der Schwerkraft: Kritik der bisherigen Lösungen des Gravitationsproblems und Versuch einer neuen auf rein mechanischer Grundlage [The Riddle of Gravity: Criticism of Previous Solutions to the Gravitation Problem and Attempt at a New One on a Purely Mechanical Basis]*. Braunschweig; Sahulka, Johann. (1908). *Über die bisherigen Versuche, die Gravitation aus dem Stoße der Ätherteilchen zu erklären [On the Previous Attempts to Explain Gravitation from the Impact of the Ether Particles]*. Vienna; Staub, Julius Bernhard. (1898). *Die thatsächliche Widerlegung der Newtonschen Hypothese von der allgemeinen Anziehungskraft durch den naturgemäßen Ersatz derselben als Grundlage einer neuen monistischen Weltanschauung [The Actual Refutation of the Newtonian Hypothesis of the General Force of Attraction by Natural Replacement thereof as a Basis for a New Monistic World View]*. Leipzig-Lindenau. For the US, cf. Brush, Charles Francis. (1911). A Kinetic Theory of Gravitation. *Science*, **33**(845), 381–86; ibid. (1928). Some Experimental Evidence Supporting the Kinetic Theory of Gravitation. *Journal of the Franklin Institute*, 206(2), 143–50; ibid. (1928). Gravitation. *Proceedings of the American Philosophical Society*, pp. 55–68; Stevenson, Robert. (1914). Do Bodies Fall? *Harper's Weekly*, 1st part on August 22, then a continuation at weekly intervals until the 11th part on October 31, 1914 (published under the pseudonym Kinertia).

[119] However, not all world riddle solvers were adherents of mechanical theories. Mewes followed Zöllner and assumed an electrodynamic theory of gravitation. Cf. Mewes 1920a, p. 93. The electrodynamic explanation of gravitation was particularly widespread among Einstein's academic opponents, Lenard, for instance. Cf. also Note 422 in Chapter 4.

the Einsteinian interpretation of gravitation as curvature of space.[120] Here, an already existing conflict between non-academic and academic bodies of knowledge was transferred to the theory of relativity as the latest development of academic physics. The curvature of space was "absurd" and thus any further criticism was unnecessary; there was no reason to delve into the deeper swamps of details such as field equations. While excessive calculations were carried out in the dispute with the special theory of relativity (particularly in order to "correctly" calculate the result of the Michelson–Morley experiment and to demonstrate the needlessness of relativistic length contraction), hardly any of Einstein's opponents attempted to refute the general theory of relativity mathematically. Instead, the criticism of the general theory of relativity started particularly at its incomprehensibility, complexity, and lack of clarity, and turned them against the theory. For the world riddle solvers, Einstein's "un-physical" theory of gravitation only represented the pinnacle of a whole series of mistakes in physics that had already found a beginning with Newton – so why not tackle the root of the problem? Sometimes, the theory of relativity was used almost thankfully by the "independent researchers" as a peg on which to hang a criticism of Newton:

Since the foundations of this [relativity] theory are Newton's foundations of his gravitational mechanics, the law of inertia, the attractive force, and completely empty space, a technically educated layperson could thus easily arrive at the thought of subjecting Newton's gravitational mechanics of our solar system to an investigation. Is this mechanics, the way Newton thinks of it, really possible? Do the specialists believe it and can we, must we laypeople believe it?[121]

Alfred Brandhoff asks this question in the periodical *Neue Weltanschauung [New World View]* and then presents his criticism of Newton. Like *Weltwissen*, *Neue Weltanschauung*, the periodical of the Humboldt League, a splinter group from the Monist League, offered a forum in the 1910s and 1920s for mechanical theories that usually related to gravitation, but were also applied to other areas, from electromagnetism to human consciousness. In a footnote to an article on "The Driving Force in Space," the editors noted, "It is known to our readers that doubts have often been expressed in recent years as to the existence of an attractive force in the sense

120 Cf. e.g. Koller-Aeby, Hermann. (1931). *Der Grundirrtum Newton's als Ursache des Einstein'schen Grundirrtums [Newton's Fundamental Error as the Cause for Einstein's Fundamental Error]*. Leipzig. Six days after the opening event of the public series of lectures against Einstein at the Berlin Philharmonic, Julius Hermann Klemt wrote to Gehrcke, "I am just reading in the *Rostocker Anzeiger* newspaper of your critical picking to pieces of the Einsteinian theory of relativity. I just returned a year ago directly from the African bush, but I have always been ashamed of the fashion in which Einstein's ideas were gushed over. I am glad of your honesty and therefore allow myself to approach you with some scientific ideas that attack Newton's gravitation theory differently from Einstein's." Klemt sent Gehrcke some papers and requested experimental validation of his gravitational laws. Klemt to Gehrcke, August 30, 1920, GN 4-B-11. Any answer from Gehrcke is not preserved.

121 Brandhoff, Alfred. (1921). Newtons Gravitations-Mechanik. Eine Laien-Betrachtung [Newton's gravitational mechanics. A lay examination]. *Neue Weltanschauung*, **10**(10), 65–72.

of Newton; some want to replace it with radiation pressure."[122] In fact, the mechanical theory of gravitation was the most popular alternative to Newton's theory of gravitation in the monistic contexts of the 1910s and 1920s, particularly to the even more abstract general theory of relativity. At the moment when the general theory of relativity was acknowledged by academic physics, a community existed in the non-academic space that understood itself as scientific, and accepted and valued the contemporary non-academic mechanical theories that were largely ignored by the academic world. In *Neue Weltanschauung*, Emil Horst concluded a review of a book by Johannes Oltmanns on *Die Mechanik des Weltalls: Grundlagen einer einheitlichen mechanistischen Weltanschauung [The Mechanics of Space: Foundations of a Uniform Mechanistic World View]*[123] in 1921 with the following words:

> Oltmanns' book is a tremendously welcome enrichment to the mechanistic literature. It is apparent that researchers like Anderssohn, Zacharias, Hertz, Beckenhaupt, Fricke, Frölich, Newest, Horst, are no longer working in absolute isolation. The problem of pressure theory is in the air today, it is ready to be defined.[124]

Like the reviewer Horst himself, who had published a multi-part treatise on the "Epistemological Grounds for Pressure Theory"[125] in *Weltwissen* in 1913, among other things, the people Horst lists are contemporary non-academic researchers, including the world riddle solvers already mentioned in this book, Hermann Fricke, Cark Beckenhaupt, Johannes Zacharias, and Aurel Anderssohn. The mechanical theorists known in the academic world, Le Sage, Secchi, and in part also Isenkrahe, were sometimes cited as references and to reinforce individuals' own theories. However, their approaches were hardly discussed, since the world riddle solvers had developed their mechanical theories largely autonomously, that is, without any connection to the current state of research and – as in Patschke's case, for instance – had only been made aware of Isenkrahe, for example, by someone else.[126] Based on these non-academic bodies of knowledge, the criticism of the general theory of relativity was carried on from a particular perspective – "Einstein = Newton", and

122 Comments by the editors to Tölle, Oswald. (1913). Die Triebkraft im Weltall [The driving force in space]. *Neue Weltanschauung*, **6**(4), 142–44.

123 Oltmanns, Johannes. (1920). *Die Mechanik des Weltalls: Grundlagen einer einheitlichen mechanistischen Weltanschauung [The Mechanics of Space: Foundations of a Uniform Mechanistic World View]*. Hamburg.

124 Horst, Emil. (1921). [Review of Oltmanns 1920]. *Neue Weltanschauung*, **10**(1), 16–19 (see p. 19).

125 Horst, Emil. (1913). Erkenntnistheoretische Begründung der Drucktheorie [Epistemological justification of pressure theory], Part 1. *Weltwissen*, **1**(9), 121–25, Part 2. Ibid. **1**(10), 133–35, Part 3. Ibid. **1**(11), 141–43.

126 While an academically trained physicist like Fricke knew the relevant literature and claimed everything from Descartes to Maxwell for himself in various papers on his ether theory without dealing in any depth with them, the majority of the engineers' theories involved independent approaches. For Patschke, it was his experience with the forces of the steam turbine that inspired him to formulate his universal law. In this formulation, he referred to the foundations of mechanics, which had been imparted to him in the context of his training at Mittweida College of Technology. But he was grateful when other theories were pointed out to him. For instance, during a discussion about his universal law at the Düsseldorf Scientific Association, a member of the audience referred him to Isenkrahe. Cf. Patschke 1925, pp. 17f.

due to a specific motivation – "the mechanical theory explains gravitation much better than all other possible explanations". In his booklet against the theory of relativity with the title "The Primordial Forces are Pushing Forces not Pulling Forces," Patschke, the engineer, subjected the conception of gravitation in theoretical physics to a criticism of its principles. For him as well, there were more commonalities than differences between Newton's and Einstein's conceptions of gravitation:

> Einstein concludes from the Newtonian theory that the material world would probably have to be limited, and even calculates the size of the allegedly spherical space. His trains of thought [. . .] rest on the assumption that the gravitational force, the so-called Newtonian force, is an inexplicable attractive force.[127]

This equating of the Einsteinian and Newtonian conceptions of gravitation as theories of "attraction" misses the significant difference between Einstein's geometrical conception of gravitation and Newton's concept of attractive force, but allowed Patschke to place Newtonian attractive force in the focus of his criticism and, in his view, precisely also the Einsteinian one. Patschke's criticism of the general theory of relativity had three principal points, two of which related to Newton.

First, the stability of the planetary system would become questionable if the assumption of attractive force were correct.[128] According to Patschke's theory, in contrast, the universe was a perpetual motion machine whose eternal stability supposedly resulted from the equilibrium of three forces, namely two "distancing" forces and a "pushing" force. The distancing forces affecting the Earth were centrifugal force and the Sun's light ether rays, which had a repulsive effect; the approaching force likewise consisted of light ether rays, but from the average of all stellar rays in space, which were very diluted and caused gravity as actual "gravitation rays." Second, the attractive force was not a mechanical force. Patschke had already stated in his Universal Law of Force in 1905 that all forces were pressure forces, however, and he was not willing to move away from that position – neither for Newton nor Einstein. The third point, finally, related to Einstein, specifically to the area of cosmology. Contrary to Einstein's assumption of a bounded universe, Patschke held the view that the universe was infinitely large. It was possible without further ado "to conclude from the uniformity of the natural forces and from the

[127] Patschke 1920, p. 19.

[128] A contemporary academic discussion of the stability of the planetary system can be found in Karl Schwarzschild. Following Poincaré's investigations of the nature of the three-body problem in 1890, it was, according to Schwarzschild, "very well possible, one may almost say probable, that the planetary system is unstable to the highest degree over the long term." Schwarzschild believed that Poincaré was indeed correct, but that the stability of the planetary system was certainly guaranteed for a million, "probably also for another 1,000 million years." Schwarzschild, Karl. (1903). Über Himmelsmechanik [On heavenly mechanics]. *Verhandlungen der Gesellschaft Deutscher Naturforscher und Ärzte*. General part, pp. 1–14, reprinted in Schwarzschild, Karl. (1992). Gesammelte Werke *[Collected Works]*, Vol. 1, ed. Hans H. Voigt. Berlin [*et al.*], pp. 84–95 (see pp. 91f).

structure of our own world system [. . .] that infinitely many other solar systems must also exist in infinite invisible space."[129] These "light worlds" (Patschke) were supposedly held in eternal equilibrium by the stellar rays of the other light worlds.

Like the other critics of the conception of gravitation in academic physics, Fricke was also of the opinion that an explanation of gravity would first have to liberate it from the "mathematical paralysis and isolation" that it had suffered from both Newton and Einstein, "and [that one must] replace the abstract operands which have made the considerations of the gravity problem so bleak and unpleasant in the past with the living forces that are well known to us."[130] For him, Newton's theory "merely represented an imperfect mathematical skeleton of the real natural process."[131] Fricke assumed that the "gravitation rays" that were emanated by a body were "a kind of reversed light rays" that supplied the bodies affected by them with energy, which they then re-radiated in the form of light radiation.[132] Accordingly, gravity would be energy radiation of a type hitherto unknown, and he thought he had discovered a new natural law, that is, a connection between temperature and gravity.[133] He held the opinion that the heavenly bodies were a kind of transformer and that energy supply and energy drain would balance each other. Fricke considered his transformer theory to have a decisive advantage in that the heavenly bodies could thus continue to transform throughout all eternity and were not "dwindling condemned bodies,"[134] and he asked:

Is it not much more plausible and self-evident that despite their sometimes quite high temperatures, all of these worlds [of the radiating suns] are in the best of harmony with the universe as a whole, after all? Indeed, isn't this view the only one possible at all? Because if the universe really did want to go cold, it would have long since had time enough to do so.[135]

It would have to be expected that Fricke, for whom gravity was a substantial flow in the ether, considered the geometrization of gravitation in the general theory of relativity to be fundamentally false, but sometimes unconventional interpretations

[129] Patschke 1920, pp.19ff. (see p. 21). Here, Patschke refers to the paper by Einstein, Albert. (1917b). Kosmologische Betrachtungen zur allgemeinen Relativitätstheorie [Cosmological considerations in the general theory of relativity]. *Sitzungsberichte der Königlich Preußischen Akademie der Wissenschaften*. [Berlin], pp. 142–52, reprinted in English translation in Kox *et al.* 1997, Doc. 43.

[130] Fricke 1919, p. 7. [131] Ibid. p. 9 [132] Ibid. pp. 8ff.

[133] Fricke's gravitation theory is mentioned in Wiechert, Emil. (1921). Anmerkungen zur Theorie der Gravitation und über das Schicksal der Gestirne [Comments on the theory of gravitation and on the fate of the stars]. *Vierteljahresschrift der Astronomischen Gesellschaft*, **56**(3), 171–91; Riem, Johannes. (1927). Welten-Werden [Becoming worlds], in *Erdball und Weltall [Globe and Space]*, ed. Oskar Prochnow, Berlin-Lichterfelde, pp. 202–68; Mohorovičić, Stjepan. (1922). Eine elementare Theorie der Gravitation [An elementary theory of gravitation]. *Naturwissenschaftliche Wochenschrift*, New Series, **21**(11), 145–53.

[134] Many world riddle solvers endeavored to explain away the supposedly imminent end of the universe. Particularly the "heat death" – the end of the world derived from the second law of thermodynamics assuming a bounded universe – was repeatedly proven to be unfounded using their own theories. Cf. e.g. Patschke 1921, p. 29 and pp. 354f.; Zacharias, Johannes. (1914a). "Geniale" Offenbarungen der "exakten" Forschung ["Brilliant" revelations from "exact" research]. *Weltwissen*, **2**(35), 125–29 (see p. 126).

[135] Fricke 1919, p. 19.

of the theory of relativity did not lead to a "concept of the enemy theory of relativity," but rather to a benevolent conception of Einsteinian theory. This was also the case for Fricke, who assumed that Minkowskian spacetime was a "somewhat abstract and veiled form of the old unitarian ether vortex theory;" he derived quite a positive assessment of Einstein's general theory of relativity ("very remarkable") and hoped, "Perhaps in his gravity theory Einstein will still gradually cast off the adolescent attempt of his impossible relativity principle entirely and, after so much storm and stress, return peacefully to good old 'times,' to ether, and to Euclidean space."[136]

Occult and vitalist conceptions of gravitation

While the mechanical theories of gravitation often embodied "outdated knowledge," thus showing a certain interface with academic physics, the occult and vitalist conceptions of gravitation were concepts that lay beyond any possible points of contact with the discourse of academic physics from the start. They were not classified by the academic discourse as "outdated knowledge," but rather as incompatible with science. In occult thinking, Ziegler's, for instance, gravitation was explained by "electromagnetic" attraction and repulsion, where the concept of electromagnetism refers to Karl von Reichenbach's (1788–1869) Odic theory. Baron von Reichenbach, independent scholar, inventor, and unconventional theoretician, became famous above all due to his postulate of "Odic force," a "natural force spanning the universe"[137] that supposedly poured out of all things that had a polar nature, whether human, animal, plant, or matter. This Odic radiation supposedly did not reduce weight and passed through all materials, but produced concrete and visible effects. Sensitive people could see the escaping Odic force – a kind of colored aura – and reacted strongly to Odic forces. Reichenbach and others carried out many attempts to measure the Odic reactions and to capture them on photographic plates. All physical phenomena such as magnetism, radioactivity, or even gravity itself were derived from Odic force with its universal effect. For Ziegler, making the dualism of positive and negative forces absolute in Odic theory corresponded to his epistemological premise of the fundamental contrast as the basic structure of the universe, and the radiating nature of Odic force was consistent with his preference for the supernatural. For Ziegler, gravitation was therefore a phenomenon of universal contrast that could manifest itself in very different ways: as gravity in the narrower sense in free fall, as the mutual attraction of large bodies, but just as well as "attraction," as sympathy between persons. Placing "gravitation" in such a broad context was not unusual in

[136] Fricke 1920a, p. 23.

[137] Cf. Reichenbach, Karl. (1850). *Physikalisch-physiologische Untersuchungen über die Dynamide des Magnetismus, der Elektricität, der Wärme, des Lichtes, der Krystallisation, des Chemismus in ihren Beziehungen zur Lebenskraft [Physical-physiological Investigations of the Dynamides of Magnetism, Electricity, Heat, Light, Crystallization, Chemism in their Relationships to Life Force]*, 2 volumes, 2nd corrected edn. Braunschweig (see Vol. 1, p. 162).

occult thinking.[138] Reuterdahl also introduced a "gravitational law of the spiritual world" as an ethical principle in his work *The God of Science*.[139]

The vitalist Kleinschrod's conception was just as far removed from the types of concepts that could be discussed in academic physics. He got by entirely without any gravitational forces conceived of in any way whatsoever. As a hydropathic Kneipp physician, he saw the cause of motion as grounded in the masses themselves; he defined matter as "not something dead in and of itself" but rather as an "active space being that can make motions."[140] For him there were "no remote forces [. . .], just as little as attractive and repulsive forces";[141] instead, the stars moved themselves along their paths – each according to its individual law – without further physical influence of force. From the lofty standpoint of meta-mechanics, Kleinschrod was able to state self-confidently:

> We thus arrive at a very different definition of space than Einstein does in his theory of relativity. According to Einstein, the world is a 'non-Euclidean four-dimensional continuum in space and time.' – This definition must be a false one if the material world is based on a dynamic substantial principle of space.[142]

Here again, the restrictive interpretation pattern of Einstein's opponents becomes clear – according to which the theory of relativity must be false as long as one's own theory is correct.

Similarly, but even more clearly than on the topic of "space," the criticism of the Einsteinian conception of gravitation as well as that of classical physics focused on the demand for a simple, comprehensible explanation of gravity. As was apparent in the outline of the mechanical, occult, and vitalist conceptions of gravitation, differing conceptions of this "simple" essence of gravity were indeed advocated, but the motivation to criticize Newton and Einstein's incomprehensible mathematical conceptions of gravitation, which showed more commonalities than differences in the eyes of Einstein's non-academic opponents, was the same.

3.1.6 Ether

Ether in classical physics

Conceptions of ether have a long tradition in physics, having been introduced to explain various phenomena, particularly gravitation and the propagation of light. Huygens held the opinion that light was a wave phenomenon, and in *Traité de la lumière* he introduced ether as a carrier medium for the propagation of light,

[138] Theosophy declared gravitation to be a "magnetic" effect (in the sense of mesmerism). Cf. Blavatsky 1877, pp. 206ff.
[139] Reuterdahl, Arvid. (1928). *The God of Science*. Minneapolis, pp. 166ff. [140] Kleinschrod 1928, p. 352.
[141] Kleinschrod 1920, p. 21. [142] Ibid.

analogously to air as a medium of propagation for sound waves.[143] Newton, on the other hand, was able to interpret the straight-line propagation and reflection phenomena mechanically using his particle theory of light and did not require a light ether.[144] Newtonian light theory dominated for a long time, although the conception of ether as a medium for the propagation of light continued to be advocated and enhanced in the eighteenth century, by Leonhard Euler and Benjamin Franklin, among others. Light ether theory experienced a terrific boost at the beginning of the nineteenth century with Thomas Young and Augustin Fresnel's work in the field of optics. Developments in the area of research on heat, electricity, and magnetism ushered in the next stage in the development of the conception of ether. Michael Faraday assumed that there was a connection between these phenomena and light ether, but it was James Clerk Maxwell (1831–1879) who finally succeeded in providing proof in his electrodynamics that the fields of electricity, magnetism, and light, which had hitherto been seen as largely independent from each other, belonged closely together and could be described as electromagnetic waves in ether. Light ether had become electromagnetic ether.[145]

Great unity did prevail among physicists with regard to the existence of the electromagnetic ether, but there was a fundamental problem. It was certainly possible to describe the characteristics of the ether very precisely, but it was ultimately impossible to provide proof of the existence of this medium. Moreover, to do justice to all known natural phenomena, ether had to have very peculiar characteristics. Due to the fact that light propagates itself with transversal oscillations and not with longitudinal waves, as would be expected in compressible fluids or gases, it was decided that ether would not be permitted to be compressible, but on the other hand, it would have to be extremely elastic, since light waves propagated very well in it (as in a solid body).[146] Max Born summarized it in 1919: "The conclusion was thus reached: Ether is an incompressible, completely elastic, solid body. And nevertheless, the heavenly bodies move with enormous speeds completely free of friction and disruption."[147]

More closely determining the characteristics of ether, and proving its existence, had long been a prominent topic in experimental physics. For many physicists trained in

143 Cf. Huygens, Christiaan. (1920 [1678/1690]). *Traité de la lumière*, Paris.

144 Nevertheless, the concept of ether also played a role for Newton, particularly as a possible cause for explaining gravitation. Cf. Newton, Isaac. (1704). *Opticks or a treatise of the reflections, refractions, inflections and colours of light*. London, Queries 17–24, particularly Query 21 on gravitation. Cf. also van Lunteren 1991, pp. 54ff.

145 Cf. e.g. Maxwell's finding, "[W]e can scarcely avoid the inference that *light consists in the transverse undulations of the same medium which is the cause of electric and magnetic phenomena*." Maxwell, James Clerk. (1861–1862). On physical lines of force. *Philosophical Magazine*, **21**, reprinted in ibid. (1890). *The scientific papers*, Vol. 1, ed. W. D. Niven, Cambridge, pp. 451–513 (see p. 500).

146 On the concept of ether in the nineteenth century, cf. Whittaker 1951; Cantor, Geoffrey N. and Hodge, Michael J. S. (eds.). (1981). *Conceptions of Ether, Studies in the History of Ether Theories 1740–1900*. Cambridge [*et al.*]; Schaffner, Kenneth F. (1972). *Nineteenth-Century Aether Theories*. Oxford [*et al.*]. For a contemporary discussion, cf. Maxwell, James Clerk. (1879). Æther. *Encyclopedia Britannica*, Vol. 8, pp. 568–72.

147 Born, Max. (1919). Vom mechanischen Äther zur elektrischen Materie [From mechanical ether to electrical matter]. *Die Naturwissenschaften*, **7**(9), 136–41.

the last third of the nineteenth century, physics was primarily ether physics, even at the beginning of the twentieth century. The experiment by Michelson and Morley, which subsequently became so important in connection with the special theory of relativity, was probably the most famous attempt to determine the motion of the Earth compared to the ether, using the speed of propagation of a ray of light. It was expected that small variations in the speed of light would be observable, depending on the angle of the ray of light in relation to the Earth's motion to the assumed ether. However, no such motion could be proven.[148]

Mechanical conceptions of ether continued to exist in parallel with this concept of ether anchored in electromagnetism, particularly in the non-academic space. Ether was thought of as composed of the smallest particles, and ether movements were explained mechanically.[149] In place of an adaptation of the electrodynamic world view, electromagnetism had been integrated here into the mechanical world view using atomistic interpretations of the phenomena.[150] Even a university-trained physicist like Fricke (who was now conducting non-academic research) completely failed to see the difference between mechanics and electromagnetism. "Incidentally, I do not really understand why a difference should exist between 'mechanical' and 'electromagnetic' theory. It seems to me that there is a misunderstanding here."[151]

The defense of ether

While ether was accepted in classical physics, its undetectability was regretted, but its existence was not fundamentally called into question; the ether adherents felt increasingly under pressure following Einstein's statement that this light ether would be superfluous in his theory.[152] They now had to prove that the ether really did exist, and furthermore to present a conception of ether that was able to neutralize the successes of the theory of relativity – above all, to explain the result of the Michelson–Morley experiment that had turned out so embarrassingly for proving the existence of the ether. In fact, ether theories initially enjoyed a new upturn due to the provocation of the etherless theory of relativity, as one of the persons directly involved, the astronomer Johannes Riem, also noted. "Specifically *because* Einstein had announced that ether was abolished, practical physics pushed ether physics so strongly to the fore as never before. Think of the names Lenard, Wiechert, Gehrcke, Fricke, Michelson, Eddington, Sagnac, See."[153]

Criticism of the "abolition" of ether as a medium of propagation for light waves and as the cause of additional physical phenomena was widespread. This uneasiness

[148] Extensively on this point, cf. Swenson, Lloyd S. (1972). *The Ethereal Aether: A History of the Michelson–Morley–Miller Aether-drift Experiments. 1880–1930.* Austin.

[149] Cf. e.g. Anderssohn 1894, p. 10.

[150] Academic scientists such as Maxwell had also initially attempted to interpret electromagnetism mechanically. Cf. e.g. Siegel, Daniel M. (1981). Thomson, Maxwell and the universal ether in Victorian physics, in Cantor and Hodge 1981, pp. 239–68.

[151] Fricke to Gehrcke, January 6, 1927, GN 76-A-15.

[152] Cf. Einstein 1905b, p. 892.

[153] Riem, Johannes. (1923). Die Bestätigung des Soldner-Effekts [The confirmation of the Soldner effect]. *Rheinisch-Westfälische Zeitung* newspaper, April 23, evening edn., GN 4, XX, 85r.

was also shared by many physicists who accepted the special theory of relativity, such as Lorentz and Michelson. However, they were aware that it was no longer possible to turn the clock back to before the special theory of relativity, and although they assumed the existence of the ether, they did not become professed opponents of Einstein like Gehrcke and Lenard.

Johannes Riem (born 1868)

Source: Reuterdahl papers; courtesy of the Department of Special Collections, University of St. Thomas, St. Paul, MN.

Johannes Riem was born in Pomerania in 1868. He studied in Halle, Strasbourg, and Leipzig starting in 1889. In 1894, he earned the degree of Doctor of Natural Sciences (Dr. rer. nat.) at Strasbourg. Riem was a student of E. Becker, the director of the Imperial University Observatory in Strasbourg. He was first an assistant and then, from 1905, an observer at the astronomical mathematical institute of the University of Berlin, where he held a professorship. In 1932, he became emeritus. Riem was the local chairman of the Berlin Kepler League and wrote many contributions for its popular science media. In this context, he himself carried on deviant science. As a deeply religious Christian, he attempted to provide scientific proof for the existence of God. For a time, he also sympathized with Hörbinger's cosmic ice theory.

Source: Poggendorff, Johann C. (1938). Biographisch-literarisches Handwörterbuch [Concise Biographical-Literary Dictionary], 6, 1923–1931, Part 3

Philipp Lenard (1862–1947)

Source: Archive of the Max Planck Society, Berlin.

Philipp Lenard was born in Pressburg in 1862 and first studied chemistry in Vienna and Budapest. From 1883 to 1885, he studied physics in Heidelberg and Berlin, followed in 1886 by a doctoral degree with Quincke, whose assistant he was from 1886 to 1889. This was followed by a brief stay in England, after which he went to Breslau as an assistant and in 1891 to the University of Bonn to work with Heinrich Hertz. This was followed by several more positions in his career, until Lenard finally became Quincke's successor and director of the Physics Institute in Heidelberg in 1907. He became emeritus in 1932. He received the Nobel Prize in 1905 for his research on cathode rays. Lenard was a nationalist and anti-Semite, and was already a follower of Hitler in the 1920s. During the Nazi period, he was an important role model for "German physics."

Source: Hoffmann 2004a

Lenard did accept the special theory of relativity,[154] but wanted to retain a very specific conception of ether[155] that plunged him into a conflict of principles with modern physics. He thought that only the concept of ether would correspond to the "second-order pictures" which a natural scientist would think in. While

[154] On Lenard's later renunciation of the special theory of relativity, cf. this chapter, Note 15.

[155] Cf. particularly Lenard, Philipp. (1911). *Über Äther und Materie: Vortrag, gehalten in der Gesamtsitzung der Heidelberger Akademie der Wissenschaften am 4. Juni 1910 [On Ether and Matter: Lecture held at the Plenary Meeting of the Heidelberg Academy of Sciences on June 4, 1910]*. 2nd, more extensive edn. with supplements. Heidelberg; ibid. (1921). *Über Relativitätsprinzip, Äther, Gravitation [On the Relativity Principle, Ether, Gravitation]*, 3rd expanded edn. Leipzig, p. 12.

the "first-order pictures" existed in the mathematical formulation, the "second-order pictures" assumed that all natural processes were attributable to the motion of a substance, which could be described mechanically. He therefore rejected the general theory of relativity as a sufficient description of gravitation merely due to its incomprehensibility and abstractness (cf. also p. 148). Gehrcke did not deal in detail with the development of an ether theory, as Lenard did; and in contrast to Lenard, he extensively criticized the special theory of relativity.

But Gehrcke also assumed an ether. He followed the theory advocated by the physicist George Gabriel Stokes (1819–1903) of an ether carried along by the Earth, and the consequential interpretation of aberration. He attempted to counter criticism of Stokes' ether theory with the assumption that ether possessed extremely low internal friction.[156] Gehrcke argued against the objections that the experiments did not indicate an ether that was carried along, by saying that ether was only carried along by large masses at high velocity.[157] However, the assumption of ether being carried along was unable to point to positive experimental confirmation; instead, this assumption referred particularly to the result of the Michelson–Morley experiment, according to which no motion of the Earth against the ether could be determined.

For the world riddle solvers among Einstein's opponents, ether was anchored at a significantly more concrete level in their thinking, more or less as if anyone would doubt the existence of air today. This is shown, for instance, in Ruckhaber's objection to the "abolition" of ether. "The 'abolition' of ether, which caused so much argument some while ago, has been adequately refuted by radio."[158] Because didn't radio waves wander, even literally, "through the ether"? And he goes on,

All waves are a propagating motion of a substance whose rhythm is created by the fact that the shift in the substance encounters resistance, the pressure encounters counter-pressure, which in turn meets counter-pressure. This process is just as impossible in empty space as in an ether-lacking resistance, for in empty space nothing would be possible at all, and without resistance there would neither be any kind of formation (or wave form either) nor time nor measurable velocity.

[156] Gehrcke to Lenard, [about 1921], GN 71-H-1. On contemporary criticism of Stokes' theory, cf. Lorentz, Hendrik Antoon. ([1909]). *The theory of electrons and its applications to the phenomena of light and radiant heat. A course of lectures delivered in Columbia University, New York, in March and April, 1906.* (Publication Ernest Kempton Adams Fund for Physical Research Vol. 2), pp. 168ff.; cf. also Einstein, Albert. (1918a). Bemerkung zu Ernst Gehrckes Notiz "Über den Äther" [Comment on Ernst Gehrcke's Note "On the Ether"]. *Verhandlungen der Deutschen Physikalischen Gesellschaft*, **20**, 261, reprinted in English translation in Janssen *et al.* 2002b, Doc. 15.

[157] Cf. Gehrcke, Ernst. (1918). Über den Äther [On the Ether]. *Verhandlungen der Deutschen Physikalischen Gesellschaft*, **20**, 165–69, reprinted in Gehrcke 1924a, pp. 44–47, particularly p. 45.

[158] This and the following quote from: Ruckhaber, Erich. (1941). *Die Ätherdynamik des Sonnensystems [The Ether Dynamics of the Solar System]*. Berlin, pp. 6f. Fricke also argued using radio. Cf. Fricke, Hermann. ([Undated typescript]). *Rettet den Weltäther! Ein Aufruf an alle, besonders an Physiker und Astronomen [Rescue Universal Ether! An Appeal to All, Particularly Physicists and Astronomers]*. GN 86-C.

Ruckhaber assumed that all bodies were permeated by ether and that an interplay of forces occurred in turn between this “internal ether” and the “external ether.” The attempt to document motion of heavenly bodies through the ether, as the Michelson–Morley experiment had attempted for the Earth, was based on the completely false assumption – according to his view – that matter would be ether-free and would move against the ether.[159] In fact, most of the alternative explanations for the outcome of the Michelson–Morley experiment referred to the concept of the ether being carried along by the Earth. Physicists like Lenard and Gehrcke referred to theories of a carried and moving ether in the same way as ether proponents advocating non-academic theories, or the philosopher Kraus did.[160]

Like Ruckhaber, with whom he collaborated for a longer period (cf. pp. 293ff.), Fricke advocated an ether vortex theory established on a mechanical foundation, which he developed over several decades. He considered the conclusions of various approaches to formulating a theory of a vortex ether conceived of as mechanical. He was quite liberal when it came to listing his predecessors, including among them Descartes as well as Thomson and Maxwell. No serious discussion of these different ether theories[161] occurred in Fricke’s work.[162] In his various works ranging from meteorology to astrophysics, Fricke attributed all physical phenomena to the motion of the ether. The uniform flow of “energy-filled space,” as he also called ether, was for him the primordial phenomenon from which all other phenomena and physical forces resulted. “Our aether model encompasses the characteristics of empty space,

159 Ruckhaber 1941, p. 7.

160 Cf. the remark by Kraus in a letter to Gehrcke, “I now think that the ultimate substance must be gradual; but the so-called light ether could be discrete after all, it would just not be the ultimate substance, but rather a state of it. It would then also be ‘observable,’ while the ultimate substance is not observable. If this were true, then a possibility could also be devised to allow the light ether to participate in the motion of large masses and to explain the Michelson experiment using ‘carrying along’.” Kraus to Gehrcke, December 29, 1920, GN 71-G-8; cf. Gehrcke, Ernst. (1923). Die Gegensätze zwischen der Äthertheorie und Relativitätstheorie und ihre experimentelle Prüfung [The contrasts between the ether theory and the theory of relativity and their experimental testing]. *Zeitschrift für technische Physik*, **4**(9), 292–99; reprinted in Gehrcke 1924a, pp. 83–95 (see p. 92); Gehrcke 1918; Lenard, Philipp. (1922a). *Über Äther und Uräther [On ether and primordial ether]*, 2nd expanded edn. with an admonishment to German natural scientists. Leipzig, p. 31. Lenard already stated his conception in a lecture in 1910 that ultimately each atom would have its own ether, which was independent from the ether of the other atoms; cf. Lenard 1911, pp. 20f.; cf. Vogtherr, Karl. (1922b). Über Aberration und Michelsonversuch [On aberration and the Michelson experiment]. *Astronomische Nachrichten*, **217**(5203), 381–96. The Swedish physicist Sten Lothigius (1877–1959) made critical remarks about the conceptual design of the moving ether. He wanted to explain the outcome of the Michelson–Morley experiment based on his own “string” light theory. Cf. Lothigius to Gehrcke, April 25, 1925, GN 29-E-3.

161 In addition, Fricke’s assignment of Maxwell’s approach to the mechanical ether theories is questionable. Cf. Schaffner 1972, p. 82: “The theory which Maxwell gave in 1865 is non-mechanical, though it is clearly not anti-mechanical.” On the discussion and presentation of Maxwell’s conception of ether, cf. ibid. pp. 78ff. and the references there.

162 Fricke obtained his knowledge primarily from the secondary literature. In his presentation of the ether theories of the nineteenth century, he referred inter alia to the dissertation by Block, Heinrich. (1912). *Die erkenntnistheoretische Rolle des Äthers in der Entwicklung des Elektromagnetismus [The Epistemological Role of the Ether in the Development of Electromagnetism]*. Bonn. For a discussion of the mechanical ether theories of the nineteenth century, cf. Whittaker 1951; cf. Schaffner 1972, pp. 45ff.

corpuscular rays, gases, liquids, and solid bodies, which stand out alternately depending on the circumstances."[163]

Fricke's ether theory claimed to be all-encompassing; with its hydrodynamic analogy it was comprehensible, but it had one major problem: Its creator did make all kinds of untested assumptions, preconditions, and generalizations, but he did not even attempt what the mechanical ether theories of the nineteenth century had failed at – which was to do a mathematical formulation. For this reason alone, his ether theory, like the other non-academic ether theories, had no chance of success in academic physics. However, Fricke thought that he did not require a mathematical formulation; his theory would be convincing enough due to its comprehensibility and simple evidence (cf. pp. 150f.). However, such a conception of how a scientific theory proved its worth lay outside of what was acceptable in academic science, where pure comprehensibility was not a criterion for the validity of a theory (cf. also pp. 148ff.). For Fricke as an ether theoretician, Lorentz in particular became a target of criticism; after all, Einstein had ultimately only invented a mathematical compromise for Lorentz's problems of "abstract" ether at rest. For Fricke, the actual origin of the "error in the theory of relativity"[164] thus came from Lorentz.

> Lorentz separates the force field moving together with the body, and its energy, from the ether in a very arbitrary way. Ether is artificially stripped of all perceivable effects and fades away into a shadowy intangible creature that the theoreticians can then easily mock. The curious manner in which Lorentz treated the ether problem prepared the soil for Einstein's anti-ether direction.[165]

His assessment in the last sentence was not wrong. In fact, the concept of ether advocated in the context of Lorentz's electrodynamics was the most elaborated concept at the time the special theory of relativity was formulated. In order to formulate a consistent theory with an ether that was absolutely at rest, Lorentz used Fitzgerald's length contraction and introduced local time. The Lorentz transformation equations represented an important component of the theory of relativity. Therefore the opponents of the theory of relativity often did not criticize Einstein exclusively; instead, they also included other theoretical physicists such as Lorentz in their criticism.

Einstein's opponents felt reinforced in their defense of ether by the presumed reintroduction of ether by Einstein. On May 5, 1920, Einstein gave a lecture in Leiden with the title "Ether and the theory of relativity." Following a historical outline of the development of the concept of ether, Einstein raised considerations of a concept of ether in the context of the general theory of relativity which was clearly different

[163] Fricke 1948b. [164] This is the title of Fricke 1920a.

[165] Fricke 1920a, p. 9; cf. also Mewes 1920b, p. 7 on the "most consequential defects of Lorentz's theory and the investigations by A. Einstein (1905) and H. Minkowski (1907) that are based on it."

from the mechanical and electromagnetic concept of ether. "The ether of the general theory of relativity is a medium which is itself devoid of *all* mechanical and kinematical qualities, but helps to determine mechanical (and electromagnetic) events."[166] For Einstein's opponents, this was a free pass: Einstein was re-introducing ether! Here, the definition of Einsteinian ether was either:

- rejected as incorrect – "unphysical and empty of content" (for instance, Reuterdahl, Kraus, Mohorovičić, Gehrcke, Vogtherr);[167]
- or exposed as a "tactical reintroduction through the back door" – Einstein was said to have recognized that it was impossible to get along without ether, but after his vociferous abolition of ether, he only dared to introduce this pale imitation of "correct" ether (e.g., Riem);[168]
- or largely ignored (for example, See, Zehnder, in some ways Kraus).[169]

The way Einstein's reintroduction of the ether was dealt with varied depending on the orientation of the argumentation from "We won" to "That's not ether." The former was happily announced publicly. Thus Johannes Riem celebrated the success in the daily newspapers, writing, "Under the weight of the facts and reasons advanced by his opponents, Einstein in fact felt forced to acknowledge the ether in a roundabout way."[170] But in a letter to Kraus, Gehrcke stated simply and concisely: "Einstein's concept of ether = not ether, but rather non-Euclidean space."[171]

In this subsection, it has only been possible to touch on some aspects of the adherence to the concept of ether. The demand to adhere to ether was asserted from a

166 Einstein 1920a, p. 12, English quote taken from Janssen *et al.* 2002b, Doc. 38, p. 177.

167 Cf. Reuterdahl 1931b, p. 43. For Kraus, it was established that the "aether as the underlying substance" would have to possess a motion state, even if it could not be detected, and he stated, "But if Einstein only understands ether to be ether states – and leaves these states without a carrier, then that is absurd." Kraus to Gehrcke, November 22, 1920, GN 71-G-6; cf. Mohorovičić 1923a, pp. 47f.; on Gehrcke, cf. the information in Note 160 (this chapter); Einstein's introduction of an ether without mechanical characteristics was also not legitimate to Vogtherr. Thinking of ether as something other than a "thing" was out of the question for him. "Every 'thing' must have at least one mechanical characteristic, that is, a state of rest or motion, otherwise it is simply not a 'thing,' and if a thing has no mechanical characteristics at all, then it does not exist in space and is no longer an object of physics at all. What physics should do with absolutely nothing, with a 'thing' that not only does not have a state of movement, but does not even have a state of rest, is entirely incomprehensible." Vogtherr 1923, pp. 16f.

168 Cf. Riem 1923.

169 "I am also astonished about Einstein's article 'Ether and the Theory of Relativity.' Have you seen it? On p. 11, he adopts your theory almost verbatim, 'the significant thing is, after all, only that another, imperceptible thing must be considered as real along with the observable objects in order to be able to consider acceleration or rotation as something real.' For him, this real thing is now – the ether!!!! The retreat is complete: 'according to genl. RT [general theory of relativity] space without aether is unthinkable'!!! – granted, *this* ether is supposedly something completely new; however, that is an outrageous assertion; for it is nothing else, after all, than that concept of aether that we are just now discussing, for ex., as I understand it following Brentano (cf. my Brentano book)," and how it would also have been conceived of by others. Kraus to Gehrcke, November 22, 1920, GN 71-G-6; cf. See 1923a; cf. Zehnder, Ludwig. (1922). *Der Aufbau der Atome aus Uratomen: Vortrag, gehalten in der Kantgesellschaft, Ortsgruppe Basel am 12. Dezember 1921 [The Construction of the Atoms from the Primordial Atoms: Lecture held at the Kant Society, Basel Group, on December 12, 1921]*. Tübingen, pp. 2f.

170 Riem 1923. 171 Gehrcke to Kraus, December 7, 1920, GN 71-G-7.

defensive position, and ether theories often continued to be advocated despite their shortcomings. This is explained by the fact that the most important issue for the ether adherents was not setting up a plausible, flawlessly formulated ether theory or devising experiments to verify the ether, but rather that by adhering to ether, they took an ontological position that demanded comprehensible and "true" physics. Gehrcke expressed his opinion very clearly in a treatise on ether that appeared in the proceedings of the German Physics Society. The following statement appears there, which is astonishing considering the target group for this publication: "[T]he issue of the mathematical form of the equations for the ether is entirely eliminated here, and the issue is rather the existential issue of whether it is *true* that there is an ether and what its true characteristics are."[172] This preeminence of the "being" of the ether was particularly apparent in the religionized and metaphysical conceptions of ether of Einstein's opponents, who introduced a plane of argumentation into the discussion, as with the topic of light, that was beyond what was considered by academic physics to be compatible with science.

The ether as a metaphysical anchor

Ascribing a religious dimension to ether and seeing it as a link between God and the world was not new.[173] In the second half of the nineteenth century, the religious dimension of ether found an advocate in the Scottish physicist Peter Guthrie Tait, among others. In his book, co-written with Balfour Stewart, *The Unseen Universe*, he pleaded for a harmonious relationship between Christian belief and science.[174] With regard to the "visible" universe, the authors argued that its atoms consisted of ether vortices, and these in turn from finer ether, and so on. The finer and more immaterial this matter, the more permanent and full of energy it would be, "until it rises into existences absolutely immaterial and spiritual."[175] At the beginning of the twentieth century, Oliver Lodge advocated the view that ether would be a substance from which the spiritual bodies of humans were also constituted and in which the soul has its seat.[176] These views do not reflect the mainstream of scientific

[172] Gehrcke 1918, p. 45. Cf. also Gehrcke's remark, "But ether, if it exists, cannot be killed by any thinking, no matter how hard." Gehrcke, Ernst. (1919). Zur Diskussion über den Äther [On the discussion about the ether]. *Verhandlungen der Deutschen Physikalischen Gesellschaft*, **21**, 67–68, reprinted in Gehrcke 1924a, pp. 47–48 (see p. 48). The irony of these immunizing arguments – *if* ether exists, then it can in fact not be "killed" – was certainly not intended by the author.

[173] Cf. Cantor, G. N. (1981). The theological significance of ethers, in Cantor and Hodge 1981, pp. 135–55, and the references there.

[174] Cf. Stewart, Balfour, and Tait, Peter Guthrie. (1882). *The Unseen Universe or Physical Speculations on a Future State*. London, foreword to the 1st edn.

[175] Ibid., p. 20. In *Isis unveiled*, Blavatsky referred approvingly to that book and hoped that its theories would become accepted by science. Cf. Blavatsky 1877, p. 159. Cf. also Blavatsky's statement, "*modern* ether; not such as is recognized by our scientists, but such as it was known to the ancient philosophers, long before the time of Moses; ether, with all its mysterious and occult properties" ibid., p. 134.

[176] Cf. Lodge 1933, particularly chapters 20 and 21 on A Psychical Function Suggested for the Ether of Space, and Ether and the Soul, pp. 221ff.

conceptions of ether at the end of the nineteenth and beginning of the twentieth century, but they do reflect the efforts to unite science with spirituality in a new way, and can be seen in many defenders of ether in the 1920s in one form or another.

In fact, ether often had to perform the function of a metaphysical anchor for physical phenomena. Kleinschrod, the hydropathic Kneipp physician, for instance, asked himself where matter came from. "Matter also cannot have developed from nothingness, it must have emerged from some kind of substantial principle. After all, new stars are still coming into being in space! Where do they come from?"[177] In a later work, he defined the ether as "the very first substance, the primordial seed, so to speak, from which the material world existing in our experience first emerged."[178] In his mechanical theory, Patschke, the engineer, ultimately dissolved human beings into the ether as creatures composed of ether atoms, and embedded them in the eternal cycle of the atoms. In his view, life was a force and thus was motion of matter like all other forces. Since forces could be transformed, but not newly created, he concluded that life also would have to be eternal. "Accordingly, the light ether must also be considered the soul of life, which is immortal due to its material, atomistic quality; for the light ether with its atoms is indestructible and its motion must also have already been present from the beginning of eternity."[179] This knowledge of the "divine universal soul of the light ether" would then lead humans to the "true universal religion" and "this natural religion, which preaches to us the omnipotence of the Sun and the other stars in the universal language of nature that is comprehensible to all peoples, also show us the way to world peace that is dreamed of by all benevolent people."[180]

Fricke ended his *Neue und einfache Deutung der Schwerkraft [New and Simple Interpretation of Gravity]* with a section on "New Principles of Natural Philosophy." There, he justified his far-reaching revision of classical physics with the fact that "laboratory physics, which is founded merely on the laws of mechanics and matter" failed completely in the face of "all phenomena of life and above all the great problems represented by the human spirit."[181] With his "more comprehensive" and "more complete" physics, Fricke wanted to prepare the way for a unified science – "'metaphysics,' so to speak" – in which "ether" and "spirit" would be forms of expression of the one universal substance, and which understood the world according to the models of Leibniz, Goethe, and Fechner as a prestabilized harmonic organism. Therefore:

> The space or ether problem [was . . .] not a special question of theoretical physics [. . .] but rather [. . .] the [matter] here is the basic questions of our culture, questions of logic and

[177] Kleinschrod 1920, p. 19. [178] Kleinschrod 1928, p. 347. [179] Patschke 1925, p. 122. [180] Ibid.
[181] Fricke 1919, p. 114. All of the following quotes ibid., p. 114, p. 137, p. 115, p. 118.

common sense, which theology, law, and medicine are just as interested in as are philosophy and the sciences.

And on the significance of ether, he dramatically stated, "Universal ether is space and time, it is light and gravity, electricity and magnetism. It is the empty and the full. It is heaven and earth, life and death, spirit and matter. It is nature, is the world. It is the 'interrelation' of all things."

These homages to divine light ether and the metaphysical overburdening of the ether concept with the task of representing the fundamental substance of the world clearly show that Einstein's "abolition" of the ether extended beyond a revolution in the narrower circles of physics for many of his opponents. It removed the ontological and metaphysical anchor from physical phenomena and was therefore perceived as a threat to their world view.

3.2 Two conceptions of science

3.2.1 The rift between the modern and the popular conceptions of science

It has already become clear that there was no unity among Einstein's opponents as to what the correct understanding of central concepts of physics should be, in contrast to the relativistic ones. In the following, a common basis among Einstein's opponents will be brought out that has often been apparent in the individual criticisms, namely their shared conception of science, which only indirectly involves the specific content of their opinions.

Einstein's opponents' conception of science differed in fundamental ways from the conception of science in modern physics. It goes without saying that I am working with a gross simplification here, and that in practice there were not two conceptions of science existing in pure form. The conception of science hereinafter called "popular" that was widespread in the non-academic area, but also in the academic one, was internally differentiated, as is the conception of science used by representatives of modern physics. This section does not attempt to identify possible commonalities in the theoretical physicists' conception of science; instead, the "modern" conception of science is illustrated in the following using reactions to Einstein's opponents' criticism of the conception of science in modern physics by leading theoretical physicists.

Nevertheless, it is justifiable to speak of a fundamental difference between two conceptions of science, since Einstein's opponents fought against a specific conception of science that they found (or thought they found) represented in modern physics and set a different one against it. What was fought in the modern conception of science, and what was set against it, show great commonalities. Places where significant differences are apparent, particularly between the world riddle solvers and Einstein's academic opponents, will be pointed out.

The popular conception of science demanded comprehensibility and simplicity as mandatory requirements that a proper physical theory would have to fulfill. Here, the conflict with the modern conception of science became apparent with the rejection of the increasing significance of mathematics in natural science. Furthermore, criticism of the conception of science in modern physics showed that the popular conception of science was strongly shaped by metaphysics. This was expressed particularly in the demand that science should comprehend the ultimate substance of reality.

3.2.2 Comprehensibility

By rejecting incomprehensibility and complexity in physics, which were particularly embodied in the general theory of relativity for Einstein's opponents, it was not only the world riddle solvers, but also many experimental physicists, some (mostly older) theoretical physicists, and the philosophers, who were all in agreement.

These differing conceptions of the significance of comprehensibility in physics were brought to a head in the controversy between Einstein and Lenard at the annual meeting of the GDNÄ in Bad Nauheim on September 23, 1920.[182] In his discussion with Einstein, Lenard emphatically demanded comprehensibility, which he did not consider ensured by the concept of gravitation in the general theory of relativity. He assumed the existence of first- and second-order pictures in physics. While the former expressed "all processes by equations," the latter interpreted "the equations as processes in space." Lenard continued, "In the second-order pictures, ether is indispensable. It was always one of the most important tools in making progress in natural science, and its abolition means abolishing the thinking of all natural scientists using second-order pictures."[183] Lenard therefore did not want to accept Einstein's objection that the physical reality of the gravitation field resulted "because it satisfied the general differential equation and because it could be reduced down to the effect of all remote masses."[184] Einstein, on the other hand, was simply unable to do anything with this demand for comprehensible second-order pictures and referred Lenard to the historicity of "comprehensibility," saying, "I would like to say that what man considers comprehensible and what he does not has changed. The view of comprehensibility is to some extent a function of time. I think that physics is conceptual, not comprehensible."[185]

[182] For the secondary literature on this event, see Note 11 in the Introduction.
[183] (1920). Discussion. *Physikalische Zeitschrift*, **21**, 650–51, 662, 666–68, reprinted in Janssen *et al.* 2002a, Doc. 46.
[184] Ibid., p. 666. [185] Ibid.

Planck argued in the same way, stating in his lecture on "The position of more recent physics on the mechanical view of nature" at the annual meeting of the GDNÄ in Königsberg in 1910:

The standard for evaluating a new physical hypothesis is not found in its comprehensibility, but rather in its effectiveness. If the hypothesis has once proven itself fruitful, then one gets used to it, and then a certain comprehensibility gradually develops completely on its own.[186]

This relative and historicizing consideration of comprehensibility contradicted Einstein's opponents' conception of science, which conceived of comprehensibility as historically invariant. This open confession to the conceptual and symbolic interpretation of reality, not just by the new physics elite,[187] the theoreticians, was inconsistent with the concept of a science tracking down the essence of nature, of intuitive, direct access to nature by the researcher. Experimental physicists such as Ernst Gehrcke, Johannes Stark, and Philipp Lenard positioned themselves against the increasing importance of theoretical physics. For them, the important thing was preventing the "overgrowth"[188] of natural science research by mathematical theory, and they advocated the view that it was not possible to investigate the essence of reality with mathematical descriptions, but only with insight into the essence – "eye to eye" with nature, so to speak. Johannes Stark complained in his pamphlet *Die gegenwärtige Krisis in der deutschen Physik [The Current Crisis in German Physics]* of 1922:

If only Einstein had gone to the mathematicians with his theory from the start! Then German physics would perhaps have been spared the paralyzing poison of the thought that from imaginative fictions ('thought experiments') with the aid of mathematical operations, it would be possible to obtain physical knowledge or, as it is generally called, the 'conception of the world'.[189]

Stark's criticism of the theory of relativity peaked in his accusation of "physical meaninglessness."[190] Shortly after the debate in Bad Nauheim, Lenard made a brief statement to Gehrcke on the relationship between mathematics and physics:

The mathematicians, such as Mr. Weyl, naturally take great pleasure in the four-dimensional space-time construct, but they forget, or do not comprehend at all, that observational

[186] Planck, Max. (1958). Die Stellung der neueren Physik zur mechanischen Naturanschauung, Vortrag gehalten am 23. September 1910 auf der 82. Versammlung der Gesellschaft Deutscher Naturforscher und Ärzte in Königsberg [The position of more recent physics on the mechanical view of nature, Lecture held on September 23, 1910, at the 82nd meeting of the Society of German Natural Scientists and Physicians in Königsberg], in: Planck, Max. (1958). *Vorträge und Reden [Lectures and Speeches]*, ed. Max-Planck-Gesellschaft zur Förderung der Wissenschaften/Verband Deutscher Physikalischer Gesellschaften. Braunschweig, pp. 30–46 (see p. 41).

[187] Philosophers like Ernst Cassirer were also counted among them, cf. particularly Cassirer (1994[1910]) and Cassirer (1994[1929]).

[188] Stark 1922, p. 7. [189] Ibid., p. 9. [190] Ibid.

physics, which should check their results afterward, only deals with very ordinary, real space and with very ordinary time.[191]

Gehrcke stated the difference between the procedure of modern physics and "correct" natural science less polemically as follows:

A general rule, the principle of relativity is derived from the mosaic of existing experiences and formulae by logical analysis. This latter method [of modern physics] is without a doubt the most objective and logically irreproachable one, but it does not make use of logical sensitivity; for it sees things in principle only from the outside and makes no attempt to sympathize with the physical processes and to grasp them from the inside.[192]

For Einstein's academic and non-academic opponents, "sympathy" and "comprehensibility" clearly took priority over the mathematical formulation of a theory. As a world riddle solver, Fricke countered objections that his ether hydrodynamics would result in a highly complicated mathematical formulation – due to the many additional assumptions required – by stating that it was not necessary to depend on mathematical formulations at all. "A simple observation of nature in flowing water can also help us. We need visualization."[193] And this was clearly of overriding importance for Fricke. He placed comprehensible science above incomprehensible science, whose beginnings extended very far back, in his opinion:

The crisis in theoretical physics does not come first from Einstein, it is as old as science itself, it is based on a congenital defect in science. Theoretical physics is not right. The physicists have not recognized that their science is built up on two very different systems, that it possesses two souls that are in conflict with each other.[194]

The first system was the "incomprehensible" one from Galileo and Newton: the conception of masses moving in empty space according to the law of inertia; the second was based on the "comprehensible" ether vortex theories of Descartes and Maxwell, in which ether was the fundamental substance and empty space did not exist. Fricke positioned himself as an "experimental physicist" in opposition to the "theoretician" Einstein and denied that Einstein had any competence to interpret physical reality. He considered himself to be in a "battle against the 'reign of terror' of the mathematicians"[195], stating that:

... the theoretician Einstein has completely misunderstood the results of experimental physics. This is the point that no mathematical wizardry, no juggling with formulae will

[191] Lenard to Gehrcke, November 6, 1920, GN 3-F-3.

[192] Gehrcke, Ernst. (1911). Nochmals über die Grenzen des Relativitätsprinzips [Once again on the limits of the relativity principle]. *Verhandlungen der Deutschen Physikalische Gesellschaft*, **13**(21), 990–1000, reprinted in Gehrcke 1924a, pp. 4–11 (see p. 11).

[193] Fricke [ca. 1942]. Cf. also Fricke to Gehrcke: "A hydrodynamic ether theory can only be permitted to be developed based on the comparison with water, not on imagined 'fundamental equations' of 'frictionless' fluids." August 13, 1921, GN 3-G-9.

[194] Fricke 1937. [195] Fricke 1920a, Preface.

help Einstein get over. For mathematicians or philosophers do not have the floor here, only practical experimental physicists. Here the battle can be fought independently of all dialectics and philosophy, with pure facts.[196]

This polemic by Fricke reflects the direction of the majority of Einstein's opponents' criticism of mathematics. It involved a different view of "correct" physics and claimed competence in regard to who must decide what "correct" physics is.[197] Experimental physicists – which Fricke himself was not, incidentally, but which he positioned himself as – and engineers thought that they were closer to nature due to their work and therefore possessed a better understanding of nature than theoretical physicists.

For Einstein's opponents, comprehensibility was not a value in and of itself; instead, it stood for closeness to reality and compatibility with truth. Conversely, the accusation of incomprehensibility was always linked to the view that it was an indication of remoteness from reality, of falsity, and of an unscientific nature. Ziegler assessed it this way: "Nothing is more typical for the agnostic operations of current physics than the fact that it falsely imagines that it is able to develop a conception of things without taking into account their form. For without a conception of form, there can be no comprehensibility, no clarity, no understanding."[198] Einstein's opponents were completely immune to the arguments of someone like Planck, who frequently took positions on the criticism of the new, strongly mathematical orientation of modern physics. For instance, Planck stated in a lecture in 1937:

Therefore, whoever denies the value of careful cultivation of modern atomic theory, and asserts that its abstractness and mathematical complexity is a defect that could be remedied by turning back to the more primitive, more easily handled conceptions of the classical epoch, only proves that a deeper understanding of the essence of the problems that physics currently must struggle with has remained alien to him. It is true that current theory is difficult and uncomfortable, but this unpleasant characteristic was by no means devised by the theoreticians in order to be able to display their mathematical acrobatic arts better; instead, after many attempts, it has proven itself to be a last, more or less desperate way out of the thicket of mysteries and apparent contradictions that experimental research itself gave to the theoreticians to solve.[199]

Although Planck is correct with this assessment of the problems of physics at the beginning of the twentieth century from the points of view of the academic physics of the time and of current historical research, it is a fact that this assessment was not accepted by many people. Fricke, who quotes Planck in his essay "Protest against

[196] Ibid., p. 6. [197] On the issue of the legitimacy of the speaker, cf. Bourdieu 1991.

[198] Ziegler 1914, p. 80.

[199] Planck, Max. (1937). Zum 25jährigen Jubiläum der von W. Friedrich, P. Knipping und Max von Laue gemachten Entdeckung [On the 25th anniversary of the discovery made by W. Friedrich, P. Knipping, and Max von Laue]. *Verhandlungen der Deutschen Physikalischen Gesellschaft*, 3rd series, Vol. 18, pp. 77–80.

Planck physics" with this stance, remarks indignantly, "That is simply false!"[200] Einstein's opponents lacked the "deeper understanding" (Planck) of the problems of the physics of their time, neither did they intend to acquire any such understanding because they, being closely attached to a different conception of science, fundamentally rejected contemporary physics as a misguided development. At the same time, the modern conception of science was seen by Einstein's opponents as more counterintuitive than it actually was,[201] since in the popular conception of science, intuitive access always described the opposite of abstract, rational, mathematical access, which was conceived of as the access used by modern science. In a review of Ziegler's *Die Umwälzung in den Grundanschauungen der Naturwissenschaft [The Upheaval in the Fundamental Views of Science]*, Ruckhaber emphasized:

Unified world views [such as Ziegler's] are something rare because they require not only erudition, but also the gift of intuition. The rational, analytically working school must have an aversion to all intuition from the outset, and will never have anything to spare for inspired syntheses. It boycotts the natural philosophy of Spencer and Haeckel and instead feels enthusiastic about Einstein's knowledge-corroding work.[202]

A further accusation was that the mathematical formulation asserted a claim of exclusivity or that obfuscation was being used. While the experimental physicists (who likewise did not possess the mathematical tools to penetrate the general theory of relativity[203]) did not widely announce this circumstance, the non-academic researchers used it more offensively and often turned it into an accusation against modern science. A scientifically competent citizen of the nineteenth century could no longer join in natural science in the twentieth century with the application of higher mathematics, or, as Gilbert pointedly stated in his satire, "For most people, a formula is a divinely obscure symbol before which they pay obeisance in their feeling of utter nothingness."[204]

Ziegler differentiates between the "mathematical physicists à la Planck and his gang," the "mathematical tightrope walkers," who deserved "only abhorrence," and

[200] Fricke [ca. 1942].

[201] In his article on the "Motives for research," Einstein expressly emphasized the role of intuition: "The supreme task of the physicist is to arrive at those universal elementary laws from which the cosmos can be built up by pure deduction. There is no logical path to these laws; only intuition, resting on sympathetic understanding of experience, can reach them." Einstein, Albert. (1918c). Motive des Forschens [Motives for research], in *Zu Max Plancks sechzigstem Geburtstag. Ansprachen, gehalten am 26. April 1918 in der Deutschen Physikalischen Gesellschaft von E. Warburg, M. v. Laue, A. Sommerfeld und A. Einstein [On Max Planck's Sixtieth Birthday. Speeches held on April 26, 1918, at the German Physics Society by E. Warburg, M. v. Laue, A. Sommerfeld and A. Einstein]*. Karlsruhe, pp. 29–32, reprinted in English translation in Janssen *et al.* 2002b Doc. 7 (see p. 44).

[202] Ruckhaber, Erich. (1921b). Die Umwälzung in den Grundanschauungen der Naturwissenschaft [The upheaval in the fundamental views of science] [Review of Ziegler 1914]. *Neue Weltanschauung*, **10**(7), 215–18.

[203] The specialization of a physicist in the area of theoretical physics demanded extensive training in higher mathematics. The experimental physicists were lacking it. Cf. Jungnickel and McCormmach 1986, Vol. 2, p. 346. While some regretted this (cf. ibid.), others, like Einstein's opponents among the experimental physicists, saw it as a "takeover" of physics by mathematics.

[204] Gilbert 1914, p. 21.

the experimental physicists (Edison, Marconi), who "belong among the greatest promoters of civilization and culture due to their practical inventions" and therefore earned admiration.[205] The fact that the theory of relativity was contradictory and absurd was said to be hidden behind incomprehensible formulae that were only accessible to mathematicians:

Until now, no one has truly understood it [the theory of relativity] because it is incomprehensible in the first place due to its logical contradictions. But until now, it was also not possible to finally refute it because its purely mathematical explanations could only be understood by mathematicians, and even they are still trapped in the delusion that mathematics would be the sole correct evidence for true understanding, so that they were still incapable of any correct refutation.[206]

In his universal light theory, Patschke, the engineer, generalized:

People have often clothed unclear, incorrect theories in vector or tensor analysis equations in order to give them an apparently greater mathematical weight. Clear theories, on the other hand, do not in the least require such types of calculation for their foundations. The general, clear differential and integral calculations are entirely sufficient for them.[207]

Fricke stated, "The effort of the mathematics leads to a type of infallibility arrogance. If someone states a different opinion, then he must be stupid or ignorant, consequently one need not deal with him."[208]

Now, neither the world riddle solvers nor Einstein's academic opponents were avowed opponents of mathematics; however, they wanted it to recommit itself to its traditional role as a supplemental science. It must be emphasized with regard to this rejection of the mathematization[209] (which exists across the board for all of Einstein's opponents) that in this way the theory of relativity was disqualified as

[205] Ziegler 1931, p. 124.

[206] Ziegler 1923, p. 20. Cf. also Ziegler to Gehrcke: "[. . .] while according to my actual profession I am an organic chemist, and one from the old school, where mathematics, at least higher mathematics, did not play any role at all." August 19, 1922, GN 82-D-6. Kühn was also aware of her limited knowledge in physics and mathematics. In 1929, on the occasion of a press report about a lecture by Einstein about unified field theory, she asked Gehrcke to take an opposite point of view: "Don't you want [. . .] to – state your opinion? I cannot betake myself onto the *special* physical field. The general, *philosophical* theory remains just as nonsensical as it was. *Don't you want to speak out?*" Kühn to Gehrcke, December 12, 1929, GN 12-J-18.

[207] Patschke 1925, p. 43.

[208] Fricke, Hermann. (1936). *Luftfahrt und Wetterkunde – gehemmt durch die "moderne" Physik [Air Travel and Meteorology – Inhibited by "Modern" Physics]*. Berlin-Schmargendorf (Deutsche Gesellschaft für Weltätherforschung, Publication No. 33), GN 86-A.

[209] Cf. e.g. Mewes 1920a, p. 88; Kleinschrod 1920, p. 11; Reuterdahl, Arvid. (1920). *Scientific theism versus materialism. The space-time potential*. New York, p. 45; Driesch, Hans. (1924). *Relativitätstheorie und Philosophie [Relativity Theory and Philosophy]*. Karlsruhe (Wissen und Wirken: Einzelschriften zu den Grundfragen des Erkennens und Schaffens, Vol. 14), p. 49. Cf. also Mohorovičić: "I am convinced that the general theory of relativity as a mathematical system is neither true nor false." Mohorovičić to Gehrcke, December 20, 1924, GN 29-F-1. Several weeks later, he stated "that the entire Einsteinian theory is pure mathematical fiction." Mohorovičić to Gehrcke, January 28, 1925, GN 29-F-4.

"only a mathematical theory" and the fact that the theory of relativity is a physical theory was not accepted in any way at all.

3.2.3 The metaphysical level of the popular conception of science

The demands for comprehensibility arose from a particular understanding of what science is, what reality is, and how the two stand in relationship to each other. This can be characterized as a strongly metaphysically shaped conception of science, in contrast to the modern conception of science. In his article "On the 'comprehensibility' of physical theories,"[210] Philipp Frank describes the demand for comprehensibility on the part of the opponents of modern physics as "metaphysical materialism." Behind the demand for comprehensibility would be "partly a certain need for comfort, and partly unconscious dependence on some philosophical systems that have been handed down." Frank demonstrates this "metaphysical materialism" using the example of Lenard's first- and second-order pictures, which Lenard had used in the discussion with Einstein in Bad Nauheim – for instance, in Lenard's view "that all processes in nature [. . .] are mere movement processes, that is, only consist of changes in location of a material that exists once and for all." It is this search for a "material that exists once and for all," for a substance of nature, that Frank criticizes as metaphysical materialism.

However, this is where the dispute really starts. A criticism such as Frank's, convincing as it might be, would not have been accepted by Einstein's opponents. What Frank considered to be holding fast to handed-down philosophical systems was seen by the opponents of modern physics as representing the preservation of correct science. In the world riddle solver theories in particular, religion, world view, and science were collapsed (cf. pp. 41ff.), and they specifically emphasized the fact that their theories contained the ultimate grounds for all scientific questions. For the world riddle solvers, arguing metaphysically was thus generally not a fault that would justify an accusation; on the contrary, it was asserted offensively. Einstein's academic opponents, on the other hand, were aware that a scientific theory in the twentieth century would be disqualified and not distinguished if it could be labeled metaphysical. They therefore applied the accusation of metaphysics against modern physics first and foremost. This occurred by setting mathematics and metaphysics as equivalent without further ado as symbols for speculation removed from reality.

"Physics is slowly but ever-increasingly turning into mathematics and metaphysics,"[211] complained Mohorovičić, and Henri Bouasse, a physics professor in

[210] Frank, Philipp. (1928). Über die 'Anschaulichkeit' physikalischer Theorien [On the 'comprehensibility' of physical theories]. *Die Naturwissenschaften*, **16**(8), 121–28. All of the following quotes ibid., p. 126, p. 121.
[211] Mohorovičić 1923a, p. 32.

Toulouse, stated, "[L]a théorie d'Einstein ne rentre pas dans le cadre des théories physiques: c'est une hypothèse métaphysique."[212] Current physics was sinking into a "morass of unbalanced licentiousness," Heidenreich warned, but simultaneously clarified, "But this is not science. On the contrary, it is a new brand of metaphysics,"[213] and Oskar Kraus expressed his opinion: "Modern theoretical physics is speculative; it is sinking into similar confusion as did the natural philosophy of Schelling and his gang."[214] In 1922, Lenard stated that he did not need to take seriously the criticism of the "specialists" (whom he consistently placed in quotes – what he meant was the theoretical physicists); they were "predominantly mathematicians who, in a sort of relapse into ancient scholasticism, are seeking new knowledge in their own heads instead of outside in nature."[215]

On the other hand, there was no awareness whatsoever that the demands that science would have to be comprehensible and recognize the "substance" of reality, which were also advocated by physicists such as Gehrcke and Lenard, were ultimately metaphysical and were no longer being asserted in this form in modern physics. It is true that they formulated their criticism of modern physics against the background of classical physics and are differentiated in this way from the world riddle solvers, who integrated metaphysical justifications into their argumentation. However, their motivation for criticizing modern physics originated in particular conceptions of nature that were shaped metaphysically in a manner similar to the world riddle solvers.

If nothing else, the tendency of Einstein's academic and non-academic opponents to "primordialize" the concepts of physics clearly shows this metaphysical dimension. It was not sufficient to speak of atoms, ether, space, and time; wherever these concepts were suddenly defined in a way that made them unreal to many people (atoms could be subdivided, ether was a field, space was a function of matter, and time depended on the state of motion), they were simply shifted a level "backwards."[216] Science had already been speaking for some time of primordial atoms, primordial forces, or primordial substances;[217] primordial vortices or simply the

[212] "Einstein's theory does not fit into the framework of physical theories: it is a metaphysical hypothesis." Bouasse, Henri. (1923). *La question préalable contre la théorie d'Einstein*. Paris, p. 8.

[213] Heidenreich, Eyvind Lee. (1929). Reuterdahl's light quantum theory [Prologue to Reuterdahl 1929], in Reuterdahl 1929, pp. 1–6 (see p. 2).

[214] Kraus to Gehrcke, November 26, 1913, GN 72-A-5. [215] Lenard 1922a, p. 7.

[216] Carl Christian Bry has aptly called this desire to detect the things behind the phenomena "Hinterweltlertum" ("backworldliness"). Cf. Bry 1964.

[217] Cf. Zehnder 1938, p. 1 and pp. 25ff., who saw the hydrogen atom as the primordial atom of weighable matter and the ether atom as the primordial atom of matter that was as yet not weighable; cf. also Mohorovičić to Gehrcke: "[A]ll bodies [are] constructed from the same primordial atoms." March 3, 1921, GN 3-G-1. "Primordial substances" and "primordial forces" are found in Zehnder 1922, pp. 8f.; cf. also Fricke's statement: "We no longer need the many small physical forces, electricity, magnetism, mechanics with inertia; everywhere we see the uniform flowing of space as primordial force. In this way, the physical world view is not only simplified, it is also significantly completed." Fricke 1918, p. 45.

Primordial[218] occurred primarily in world riddle solver writings, but primordial ether, primordial space, and primordial time developed as a reaction to modern physics.[219] The German prefix "Ur-" ("primordial") is a symptom of the effort to ontologically anchor the reality described by scientists by postulating a level of reality lying "beyond," a different, absolute, and "more real" level of reality. Why didn't Lenard name his two ethers "Ether 1 and Ether 2" or "ether in motion and ether at rest" instead of "ether and primordial ether"? As a philosopher, Oskar Kraus found it easier than the physicists to openly admit the metaphysical function of the ether. Along with light ether, a "subaether" supposedly existed "which is unobservable as such (i.e. as a substance), while light ether is tangible. [. . .] I require the latter ether due to *actio in distans* and for other metaphysical epistemological reasons."[220] The demand for a metaphysical, substantial substructure of physics was often a decisive motivation for attacks on the theory of relativity. Mohorovičić, for example, saw the theory of relativity as untenable:

> . . .for the reason that it is not only encumbered with several fundamental errors, *but also because it is too narrow.* Force is an inherent characteristic of substance, and it is only *partially* 'explained' by the *Einsteinian* theory, and unnaturally to boot. But I am convinced that it is possible to set up a theory which will truly encompass all natural phenomena and which will interpret force in the sense indicated here. I myself am working on it, but I have not yet gotten very far.[221]

Ultimately, Einstein's opponents had the same goal as the modern physicists: setting up a unifying theory of natural phenomena. However, they loaded this goal metaphysically, and due to the perspective of their substance thinking, the theory of relativity appeared as an incomprehensible caricature of a physical theory deprived of any comprehensibility and ontological anchoring. The content-based criticism of the theory of relativity therefore did not only apply to individual scientific concepts (time, space, ether, etc.), but was instead strongly shaped by the topicalization of what was, in the eyes of Einstein's opponents, the highly questionable status of the theory of relativity as a physical theory. This was simultaneously linked with an argumentative strategy opposing the theory of relativity. If what Einstein had set up was not a physical theory at all, then these "speculations" could be rejected on principle. Additional argumentative strategies opposing the theory of relativity, which were less motivated by content than by politics, are examined in Chapter 4.

[218] "Primordial vortices" e.g. in Fricke, Hermann. ([ca. 1938]). *Eine Bitte an alle Goethe-Freunde [A Request to all Friends of Goethe]*. Berlin-Schmargendorf (Deutsche Gesellschaft für Weltätherforschung, Publication No. 30), GN 86-A; the "primordial" is found in Zacharias 1922, p. 12; "primordial motions," "primordial forces," "primordial pressures," and "primordial impacts" are found in Patschke 1925, pp. 35f.

[219] On "primordial ether," cf. Lenard 1922a; on "primordial space" and "primordial time," cf. Ziegler 1923, pp. 26f.

[220] Kraus to Gehrcke, December 29, 1920, GN 71-G-8.

[221] Mohorovičić to Gehrcke, March 3, 1921, GN 3-G-2.

3.3 The content-based accusations of plagiarism

The accusations of plagiarism raised against the theory of relativity are a special feature and highly informative for understanding the background of the content-based criticism of the theory of relativity. These accusations are therefore discussed in the following in a separate section as a transition to the strategic dispute with the theory of relativity in Chapter 4.

3.3.1 The standard of originality in science

In his lecture "Science as a Vocation," Max Weber describes science more as a calling than merely as a way to earn a living. In modern specialized science, he said, the issue was not about knowledge of a universal truth; instead, one would have to be in a position to dedicate oneself passionately to a highly specialized area. "And anyone who lacks the ability to don blinkers for once and to convince himself that the destiny of his soul depends upon whether he is right to make precisely this conjecture, and no other at this point in his manuscript, should keep well away from science."[222] In his answer to the question of what drives scientists to science and what they receive as a reward, Weber emphasized that scientists' rewards are primarily of an ideational nature. Along with the personal satisfaction of actually having made the right conjecture at the right point, which should not be underestimated, it consists of reputation and recognition. Scientists receive this recognition from their scientific communities for original contributions that are seen as relevant.

The demand for originality and relevance is established in the structure of the scientific production of knowledge. Merton in particular has emphasized its role as the cause of disputes about priority.[223] The question of priority is the social flip side of professional originality and therefore a fundamental and important question which, however, is often not easy to answer. Given a state of knowledge that is shared by many scientists, it is thus not unusual, for instance, for discoveries to be made in close temporal proximity and for theories to develop contemporaneously, so that clarifying who can claim priority is only possible with difficulty or even not at all.

So what is the measure of priority? The first publication? The first manuscript? What if a new idea is only expressed and explained orally in a discussion? Does priority already exist? Couldn't perhaps the insight of the scientist count, his or her first thought, which he or she perhaps does not dare utter or write down due to external circumstances? If that is the case, does the correct thought also count as

[222] Weber, Max. (2004 [1917]) Science as a Vocation. In *The Vocation Lectures: "Science as a Vocation", "Politics as a Vocation"*, ed. David S. Owen, Tracy B. Strong, and Rodney Livingstone. Indianapolis (see p. 8).

[223] Cf. Merton, Robert. (1957). Priority disputes in scientific discovery: A chapter in the sociology of science. *American Sociological Review*, **22**(6), 635–59.

having priority if it was stated on the basis of a false or outdated theory?[224] Priority disputes are usually complex processes of a social nature. Initially, they involve demanding payment of the scientist's ideational reward, of recognition, which in this case is claimed by more than one scientist for themselves. But recognition is something that cannot be directly demanded, even though it is possible to solicit it or fight over it. In fact, the harmless question "Who was first?" often degenerates into full-blown priority campaigns in which the rivalries of individual scientists or entire scientific communities, and their pride and injured vanity, play a role – sometimes more among the adherents of a person involved in a priority dispute than among the actual protagonists.

The accusations of plagiarism against Einstein that are the focus of the following discussion are of a different nature than normal priority disputes in two respects. First, they do not relate to the actual content of the theory of relativity, as was the case in the Einstein–Hilbert priority dispute, for instance, where the question was: "Who had first set up the final mathematical formulation of the general theory of relativity?" Instead, these accusations were raised with regard to statements that usually only had anything to do with the Einsteinian theory of relativity in the eyes of the person who was charging Einstein with plagiarism. Second, in this case it was Einstein's opponents who claimed priority for the theory of relativity or portions thereof; insofar as they did it for themselves, they fell into an obvious self-contradiction, and insofar as they did it for others, their behavior nevertheless requires an explanation. The following section is focused on tracing this phenomenon.

First, however, a differentiation should still be made between a dispute over priority and an accusation of plagiarism. While a dispute over priority asks about the timing of the individual discovery – for example, whether it was Einstein or Hilbert who first set up the field equations of the general theory of relativity – the accusation of plagiarism involves dishonest behavior and scientific fraud. While the question in the priority dispute is: "Who was first?" the accusation of plagiarism charges "He stole it!" The fact that the transition from one to the other can be gradual should not hinder us from paying attention to who used plagiarism terminology and who did not among Einstein's opponents.[225]

[224] As early as 1887, for instance, Woldemar Voigt had specified a transformation of the space and time coordinates that corresponded to the Lorentz transformation except for one factor. On the occasion of a reprint of his paper, he stated that this had occurred on a completely different basis than in Lorentz's electrodynamics and that he had therefore not pursued it further. Cf. Voigt, Woldemar. (1915). Über das Dopplersche Prinzip [On the Doppler principle]. *Physikalische Zeitschrift*, **16**, 381–86 (preface to the reprint of his paper); cf. also Hentschel 1990, pp. 151f.

[225] However, avoiding the term "plagiarism" also occurred for tactical reasons. This was true for Ruckhaber, for instance, who advocated against the use of the word "plagiarism" in the project of translating *100 Autoren gegen Einstein [100 Authors Against Einstein]* (cf. pp. 288ff.) with the justification that his co-editors "Dr. Weinmann and Dr. Israel were very afraid of getting into trouble with regard to the expression 'plagiarism'; it would be difficult to prove or decide what plagiarism is, and a possible lawsuit might cost a lot of money and

I have placed particular importance on the priority disputes and accusations of plagiarism in my analysis of the content-based criticism and have examined them from a specific point of view that differs particularly from Hentschel's approach, but which has likewise paid great attention to the "attempts at re-excavation of the history of science"[226] and has above all demonstrated the content of these accusations and their unfoundedness. Since it has already been shown in this book that the content-based criticism was carried out to protect people's own theories from a threatening situation that was ascribed to the theory of relativity, what is interesting here is the motivation for the priority disputes and accusations of plagiarism. A differentiation can be made between content-based motivations on the one hand (this chapter) and strategic motivations on the other (Chapter 4). An investigation of the motivation for the content-based accusations of plagiarism must go beyond confirming the groundlessness of the asserted claims. It must start with the fundamental question of what Einstein had actually plagiarized, from the point of view of Einstein's opponents.

3.3.2 The theory of relativity as plagiarism of a world riddle solver theory

An accusation of plagiarism or an assertion of priority claims were popular argumentation patterns in the struggle against the theory of relativity because they were effective. If it could be shown that Einstein was not a great scientist and original thinker at all, if it could even be proven that he was a copier and a fraud, then the acceptance of the theory of relativity would dissolve of its own accord. This argumentation was used strategically, but was rooted in the deepest of convictions, as becomes clear in the example of Reuterdahl. Reuterdahl wrote an extensive article, "Einstein and the New Science," just on the "forerunners" of the theory of relativity. One would have expected discussion about the contributions by Poincaré, Lorentz, or Voigt, who are also named in the booklet. But according to Reuterdahl, the actual and unrecognized forerunners of "New Science," which was only linked publicly with Einstein, were instead: Karl von Reichenbach, Aurel Anderssohn, Johannes Zacharias, "Kinertia" Robert Stevenson, Johann Heinrich Ziegler, Robert T. Browne, Charles Brush, Melchior Palagyi, and Rudolf Mewes. Philipp Lenard, Ernst Gehrcke, Arthur Patschke, Paul Weyland, and Edouard Guillaume were also mentioned in connection with "New Science." This "New Science" had nothing to do with modern physics, and the forerunners listed by Reuterdahl are in large part individuals who have been discussed earlier in this book. How did this connection come about?

would be of doubtful result, as the Jews have the best advocates and many judges are Jews. The German jurisdiction is today corrupted by political passion; it has become relative." Ruckhaber to Reuterdahl, February 11, 1931, RP 4–48.

226 Cf. Hentschel 1990, pp. 150ff.

The typical accusations of plagiarism or priority claim were rooted in a specific understanding of what the theory of relativity was, an understanding that often had nothing to do with the physical theory of relativity, referring instead to specific content that was identified as "the theory of relativity" (such as the "abolition" of the ether or the unification of space and time). The conflict about priority frequently resulted simply from the fact that when a world riddle solver heard of the theory of relativity for the first time, he had already developed an all-encompassing theory. He therefore knew what space, time, and the cause of gravitation were, and accordingly everything in the theory of relativity that did not agree with that was false, and everything that could be made to coincide with it had already been discovered by the world riddle solver.

The fact that one's own theory and Einstein's had completely different goals played no role, even if it was recognized. Reuterdahl, who had asserted a priority claim to the link between space and time (cf. p. 167), had to justify himself to his publisher, Devin-Adair Company, stating "The purpose of my work is radically different from that of Einstein." The company had anxiously inquired whether Reuterdahl could rather be accused of plagiarizing Einstein's theory. In an eight-page memorandum to his publisher, Reuterdahl stated in December 1919 that he had not had any knowledge of Einstein's work due to the war, and had only found out after completion of his paper in February 1915 that Einstein and Lorentz were interested in the Michelson–Morley experiment. Reuterdahl did not take into account the fact that priority was being asserted in this context involving the special theory of relativity, which had already been known about for ten years in 1915. Instead, he stated that he knew nothing about the theory of relativity except for a newspaper clipping from the *New York Times* that his publisher had sent to him, "in which it appears that Dr. Albert Einstein's theory, in its final form, was presented to the Royal Society at a meeting in London, Nov. 6, 1919."[227] In fact, this article in the *New York Times* involves the meeting that has become famous in which the experimental confirmation of the general theory of relativity was announced, but the final version of the general theory of relativity had been completed four years earlier. Along with the confession that he had had no knowledge of the theory of relativity, Reuterdahl further assured his publisher:

> The primary purpose of my work is to correlate science, philosophy, and religion. This involves a proof of the existence of God [. . .] My entire position will therefore be different *in toto* from the position of a physico-mathematician [. . .] Einstein naturally will proceed from the facts of physical science to certain conclusions within the limited realm of physical science. [. . .] Einstein's relativity is merely physical and leaves the world problem unanswered. [. . .] His interest probably ends with the physical universe.[228]

[227] Reuterdahl to Garrity, December 14, 1919, RP 3–23. [228] Ibid.

However, this insight that a religious or metaphysical theory was being advocated in contrast to Einstein's by no means prevented accusations of plagiarism. Instead, a perception pattern came into play here that is extremely important for understanding the accusations of plagiarism. The fact that the theory of relativity had a different form from the world riddle solver theory showed that not just plagiarism was involved, but also falsification. In a letter to Mohorovičić, Reuterdahl clarified, "When I first struggled with these ideas I had not heard of Einstein. In fact, I am convinced that Einstein's 'mollusks' were suggested to him by my Space-Time Potential. He did not quite dare to include the genuine energetic element in his scheme which is at the basis of my system."[229]

Reuterdahl was convinced that the Swedish physicist Gösta Mittag-Leffler (1846–1927), to whom he had sent his theory in 1915 for the purpose of publication, had not only failed to publish it, but had also secretly forwarded central insights to Einstein, who had then published them in his own form and under his own name.[230] "I have no direct proof that professor Einstein gained access to my manuscript," Reuterdahl had to admit, "but the many peculiar circumstances lead me to believe that such was the case. The remarkable similarities between his later works and the theories I advanced at that time constitute an astounding coincidence."[231] At the time when the first suspicion of plagiarism was expressed, Reuterdahl said that his only knowledge of the theory of relativity came from the *New York Times* report already mentioned. His accusation of plagiarism against Einstein was based on a finding of "similarities" between his and Einstein's theory, which referred exclusively to correlations in the choice of words. Specifically, Reuterdahl claimed to have been the first scientist to coin the terms "Space-Time," "Space-Time-Potential," and "gravitational mass." In fact, he held copyright to a lecture with the title: "The Space-Time Potential. A New Concept of Gravitation and Electricity," but not to the terms listed.[232] His accusation of plagiarism completely ignores the fact that these terms had a totally different meaning in his system. Instead, his conviction that Einstein had plagiarized him led to several vehement polemics.

229 Reuterdahl to Mohorovičić, May 24, 1923, RP 4–35.

230 Cf. e.g. the presentation in "Authorship of the famous theory of light bending contested by St. Paul scientist; lost paper is key." *St. Paul Sunday Pioneer Press*, February 15, 1920. Reuterdahl's accusation of plagiarism was supported by Erik Sjöstrand, the secretary of the Swedish Journalists Association in America (Svenska Journalistförbundet i Amerika), who in a letter also referred Reuterdahl to an article in the *Literary Digest* about "Einstein's cosmic religion" in which Einstein is quoted as saying that mankind had a "cosmic religious sense." Sjöstrand suspected that this involved plagiarism of Reuterdahl's concept of "theocosmism." "As you have shown how he [Einstein] has 'appropriated' the idea about 'relativity' [. . .] so it seems that he also wants to appropriate your 'theocosmism'." Sjöstrand to Reuterdahl, November 30, 1930, RP 4–51.

231 Reuterdahl quoted from "Prof. Reuterdahl to write book to show Einstein fallacies." *Evening Tribune*, April 18, 1921.

232 The "Certificate of Copyright Registration" for the lecture is reproduced in Reuterdahl, Arvid. (1923i). First attempt of the unification of electricity and gravitation. *Univerzum*, **2**(9–12), 292–94 (see p. 293). The article is a section from a letter from Reuterdahl to Mohorovičić.

On the occasion of Einstein's trip to Norway in the summer of 1920, Heidenreich and Reuterdahl succeeded in raising interest in the plagiarism story not only in the US press, but also in the Norwegian and Swedish press. The fact that both of them were Scandinavian emigrants – Reuterdahl originally came from Sweden, and Heidenreich from Norway – was presumably to their advantage. In an interview with the Norwegian daily newspaper *Aftenposten* on June 18, 1920, Einstein was confronted with Reuterdahl's accusations of plagiarism – in fact, the interview is only about Reuterdahl's accusations of plagiarism, because Heidenreich, who had been in Norway for several months, had used his stay to personally familiarize *Aftenposten*'s editor with Reuterdahl's theory and the accusation of plagiarism – apparently successfully. Einstein provided a longer explanation stating that he was not interested in the accusations of plagiarism, since neither Reuterdahl nor his theory was known to him, and that Mittag-Leffler, whom he did know, had not pointed out Reuterdahl's work to him.[233]

For Einstein's opponents, however, the fact that Einstein did not deal any further with these accusations could only be interpreted as a conscious strategy of avoidance. The fact that Mittag-Leffler would not have seen any occasion to familiarize Einstein with Reuterdahl's esoteric theory, that he himself would probably hardly have wasted a glance on Reuterdahl's writings – was unimaginable to Reuterdahl and Heidenreich, who were convinced of the significance of "Space-Time-Potential." The fact that Einstein did not actually know Reuterdahl, head of a modest institute of technology in the Midwest dealing primarily with industry-oriented training and lacking relevant contributions in physics, was unthinkable to them; in their eyes, after all, Reuterdahl's "Space-Time-Potential" was the most important scientific discovery of all time, which even Mittag-Leffler and Einstein would have to have recognized. So if Mittag-Leffler and Einstein nevertheless acted as if they knew nothing of "Space-Time-Potential," it would only be in order to deny that they had plagiarized Reuterdahl's theory.

Over the years, Reuterdahl repeatedly raised claims of priority, whether for individual aspects such as the abolition of the ether, or for the complete theory of relativity, which he considered plagiarism of his "Space-Time-Potential," both for himself and for others. As late as 1929, when Einstein announced that he had finally come to a conclusion of his work with the Unified Field Theory, which caused some stir in the media, Reuterdahl immediately wrote a press release in order to register

[233] "Diskussionen om relativitetstheorien." *Aftenposten* newspaper, morning edn., June 18, 1920. Similarly on the occasion of the renewed accusations during a trip by Einstein to the US in 1921: "Professor Einstein said he never heard of Professor Reuterdahl, and that he was not in the least interested in the latter's challenge to a written debate on the subject of relativity. He intimated that he might read an article written by Professor Reuterdahl, if he happened to come across it, but as for entering a controversy, he could not waste his time." Heidenreich, Eyvind Lee. (May 16, 1921). Calls Einstein's statements irreconcilable. *The Minneapolis Morning Tribune*.

priority to the Unified Field Theory as well.[234] For Reuterdahl, the issue was never merely the question of priority; instead, he always immediately and clearly expressed accusations of plagiarism. His preferred forum was the daily newspapers; articles about Reuterdahl's accusations appeared repeatedly, particularly in Minnesota's local papers.[235]

For Ziegler as well, the theory of relativity was just a garbled version of his own primordial light theory. His accusations of plagiarism referred partly to specific content, and partly to the entire theory. According to Ziegler's fundamental assumption of the complementarity of opposites, only both could exist: the absolute and the relative. A "relativity principle" alone would thus make no sense. The fact that the theory of relativity was a theory of *relativity* already showed Ziegler that it was incomplete and would be dependent on the related "absolutity" theory.[236] Reuterdahl spoke of the Einsteinian theory of relativity as "partial relativity" for the same reason, while his own "Space-Time-Potential" was the complete theory, including God as the absolute.[237] Ziegler was of the opinion that the Bernese philosophy professor Ludwig Stein (1859–1930), to whom he had sent his early work, had passed it onto Einstein. After all, wasn't Einstein living in Bern at this time and associating with Stein?[238]

This consciousness of possessing the correct theory is the key to the content-based accusations of plagiarism. Permeated by the firm conviction of having discovered something extraordinarily important, the world riddle solvers believed that professors such as Mittag-Leffler or Stein, to whom they forwarded their theories, would of course have recognized their significance, then secretly allowed them to reach their favorite, Einstein, and were now spending a great deal of time trying to hush up the whole affair. For Mewes, the conviction of having invented the

[234] "Communication by Dr. Arvid Reuterdahl of St. Paul, in reference to the new theory of Dr. Albert Einstein." [typescript], [1929b], RP. See reacted in the same way. In response to a brief Associated Press report stating that Einstein had announced a new discovery relating to "the connection between the earth's power of attraction and terrestrial magnetism," See asserted priority to *The Times* of London: "Will you be good enough to point out through the columns of your valuable paper that my own Discovery of the Law of Nature connecting terrestrial Magnetism with terrestrial Gravitation was finished a year ago this month [. . .] it would appear that five months after my Discovery is published in the greatest Scientific journal of the world [*International Journal of Astronomy*], Professor Einstein has seen the error of his way regarding the discredited Doctrine of relativity, and now is trying to follow the path which I opened up." See, T. J. J. (March 21, 1923). To the editor of *The Times* of London, RP 4–49.

[235] Cf. e.g. Heidenreich 1921: Reuterdahl asserts he is the real originator of Einstein's new theory. *The St. Paul Daily News*, January 27, 1929, RP 3–26; Authorship of the famous theory of light bending contested by St. Paul scientist; Lost paper is key. *St. Paul Sunday Pioneer Press*, February 15, 1920; St. Paul physicist concludes that relativity booster and his admirers are suffering from 'delirium mathematicum', scoffs at idea of curved space. *St. Paul Dispatch*, January 28, 1929.

[236] Ziegler 1923, p. 26. Gehrcke and Lenard were favorably disposed towards Ziegler's accusations of plagiarism against Einstein, although they did not actively support them. Cf. Lenard to Gehrcke, July 5, 1920, GN 3-F-14: "Mr. Ziegler from Switzerland wrote to me and sent me an older paper from which many things can be read with a certain amount of goodwill which are being attributed to Einstein. Mr. Ziegler set great store by this. Do you know him?" Cf. also Gehrcke to Ziegler, March 19, 1931, GN 83-L-5.

[237] Reuterdahl 1920, p. 273. [238] Cf. Ziegler to Gehrcke, February 11, 1931, GN 83-L-4.

correct theory of relativity himself went so far that he quite simply turned the tables and, as the inventor of the "real" theory of relativity, entitled one chapter of his *Raumzeitlehre [Spacetime Theory]* (which was predated to 1884): "Lorentz, Einstein, and Minkowski (rejection of their claims to priority)."[239] This context of the accusations of plagiarism explains why it was certainly not contradictory for Einstein's opponents to deny the theory of relativity and to simultaneously accuse him of plagiarism. The theory of relativity was perceived as a falsification of their own theories.

3.3.3 The content of the accusations of plagiarism in detail – four examples

The unification of space and time – copyright © Einstein?

The reason that the most frequently advanced accusations of plagiarism relate to the "unification of space and time" can be found in the generality of the topic. If a connection of any kind whatsoever – philosophical, psychological, etc. – of space and time is considered the core of the theory of relativity (and for many people that was precisely what the theory of relativity consisted of), then naturally the group of possible priority claims is large. Palagyi stated:

> The situation is actually such that Minkowski and Einstein drew their main thought, the unified theory of space and time, from my paper 'New Theory of Space and Time' [...] that already appeared in 1901. Of course, at the same time they misinterpreted and distorted my basic thought, and in this way they have brought about true anarchy of thinking in physics.[240]

Here again, one can see the interpretation pattern that the theory of relativity would be a falsification of one's own correct basic thought. Palagyi emphasized that his finding, namely that one could conceive of the flow of time in such a way "as if it would flow in a fourth, imaginary (or symbolic) dimension of space" would "certainly [have] nothing to do" with the theory of relativity, since he did not relativize space and time and was actually stating "only something self-evident."[241] Palagyi's assignment of the "time stream" to space was instead meant to express the idea that every point in time could be assigned to a particular point in space, that is, that all physical processes therefore occurred in space and time and thus were determined absolutely. Palagyi assumed that space was not something static, but

[239] Mewes 1920a, p. 62.

[240] Palagyi to Reuterdahl, October 16, 1921, RP 4–40. Here, Palagyi referred to his book published in 1901, *Neue Theorie des Raumes und der Zeit. Die Grundbegriffe einer Metageometrie [A New Theory of Space and Time. The Basic Concepts of a Meta-Geometry]*. Darmstadt. On Palagyi, cf. also Hentschel 1990, pp. 153ff.

[241] All of the following quotes from: Palagyi, Melchior. ([1922]). *Kurze und populäre Darstellung der Einheits- oder Zusammengehörigkeitslehre von Raum und Zeit [A Short and Popular Presentation of the Uniform or Connectedness Theory of Space and Time]*. [Unpublished manuscript], [1922], RP 4–40.

rather renewed itself again and again in the time stream; even apparently static physical phenomena – such as a mountain in the landscape – would permanently renew themselves, and the fact that this was not perceived was merely attributable to the narrowness of human possibilities of knowledge. Space "flows" in this way as a "phenomenon" in the flowing time stream:

The essence of this superior conception consists of the fact that the entire physical world of phenomena, consequently also space, is thought of as included in the time stream. In this way, space appears as a momentary cross-section through the time stream. Or conversely, the flow of time occurs in that fourth dimension in which space unceasingly renews itself.

According to Palagyi, his theory ultimately involves "nothing more than that each physical event must occur both at some place and at some time, so it would be including an absurdity to claim that a physical process could occur merely in time or merely in space."

What is remarkable here is neither Palagyi's philosophical conception of space and time as such, nor the fact that it in no way anticipates relativistic space-time; instead, what is remarkable here is that many of Einstein's opponents claimed priority for Palagyi's concept of a space-time union vis-à-vis Einstein's approach, and the justification with which it occurred. In fact, the justification was twofold: in content, "true," unfalsified space-time unification was set against its "distortion" by Einstein and Minkowski, while strategically Einstein was accused of the theft of intellectual property (cf. pp. 273ff. for additional, mostly strategically motivated accusations of plagiarism).

In his listing of space-time unifications prior to Einstein and Minkowski, Gilbert referred to the "well-known philosopher Melchior Palagyi," but assumed that the "relativists" had rather been inspired by Wells' time machine, "for the relativists, who possess no conception of the great mental work of humanity, of the philosophers, have not read Palagyi."[242] He was undoubtedly right, although for different reasons than he thought. Mohorovičić also emphasized that Palagyi had assumed four-dimensionality prior to Minkowski, specifically "without forsaking the very disparate characters of time and of space," which Mohorovičić considered a decisive advantage.[243]

Palagyi did not first assert his priority claim himself; instead, it was asserted and supported by Gehrcke. Gehrcke was very interested in possible claims to priority vis-à-vis Einstein and often marked relevant passages in written work opposing the theory of relativity in his possession. In his essay of 1911, "Once again on the limits of the relativity principle," he emphatically referred to Palagyi's work and additionally expressed a suspicion of plagiarism. "The Minkowskian form of the theory of

242 Gilbert 1914, p. 51 (Note 1). 243 Mohorovičić 1923a, p. 32; cf. also Reuterdahl 1921a, p. 9.

relativity likewise uses such spacetime diversity [like Palagyi], whereby it is an open question whether Palagyi's paper was or was not known to the architects of the theory of relativity."[244] A draft letter from Gehrcke to Palagyi also shows Gehrcke's interest in this accusation of plagiarism:

> If, as you appear to believe, these gentlemen [mathematicians and physicists who accept the theory of relativity] knew of the content of your old paper, then the accomplishment of these gentlemen consists of the fact that they have draped a tawdry mathematical garment over an original epistemological core.[245]

While Palagyi was firmly convinced of the fact that his space-time union was a predecessor of Minkowskian spacetime, and thus that the priority claim was motivated purely by content, things were different for Gehrcke. He supported assigning priority to Palagyi, although he was not convinced of Palagyi's space-time conception. That is, Gehrcke stated to Palagyi that he could not follow his introduction of the factor $\sqrt{-1}$, which Palagyi multiplied by the time axis and on which the accusation of plagiarism against Minkowski was substantially based:

> Now I have never previously understood, and even now still do not understand why you multiply the time axis by $\sqrt{-1}$ in contrast to the space axis. [. . .] Why don't you take any other factor, e.g. $-\sqrt{-1}$, or π or e, as far as I'm concerned? Why do you take any special factor at all? That is the point where I am no longer capable of following your further conclusions.[246]

This fundamental doubt as to Palagyi's argumentation, however, did not keep Gehrcke from publicly supporting Palagyi's priority with regard to the connection of space and time, as already mentioned, and expressly emphasizing that the time coordinate "occurs multiplied with the imaginary unit $\sqrt{-1}$" was written by Palagyi in 1901.[247] Despite these reservations as to the content, the great advantage of Palagyi's approach in Gehrcke's view as a physicist was that he combined space and time in a natural-philosophical manner that was absolutely meaningless in physics and which did not touch on the fundamental difference that existed for Gehrcke between the physical concepts of space and time.

Mewes also claimed the unification of space and time for himself and accused Einstein of having plagiarized his *Raumzeitlehre*, which allegedly dated from 1884. It was only after he had become aware of the theory of relativity, however, that Mewes published a volume with this name in 1920, which was a collection of older and newer papers. His claim to priority referred back primarily to a paper from 1889

[244] Gehrcke 1911, p. 9; cf. also ibid. (1924a). Die Relativitätstheorie, eine wissenschaftliche Massensuggestion [The theory of relativity, a scientific mass suggestion], reprinted in Gehrcke 1924a, pp. 54–68 (see p. 66).
[245] Gehrcke to Palagyi, [undated draft], GN 83-B-18. [246] Cf. ibid. [247] Gehrcke 1920, p. 66.

on "The essence of matter and of discerning nature," in which he stated, following Schopenhauer, "The entire essence of matter consists of its effect," and concluded:

> Matter is entirely relative, just as are space and time, since it only exists with respect to other matter on which it has an effect. We are unable to grasp matter in and of itself, that is, lacking a connection to its effect, just as little as we grasp space and time in and of themselves. [...] However, because matter has its essence in its effects and thus in causality, it thus assumes space and time, not merely each for itself, but rather both in association; for no causality would exist for all the juxtaposition in space and all the change in time as long as each of these two forms would have its continued existence and course for itself and without a connection with the other.[248]

For Mewes, the unification of space and time thus ultimately occurred with the same intention as for Palagyi, that is, in order to establish strict causality in that everything in space and time can be assigned to its specific place and its specific point in time.

For completely different reasons, Reuterdahl claimed the unification of space and time for himself and thought that he had anticipated the theory of relativity in his concept of "Space-Time-Potential." Reuterdahl's "scientific theism" constituted the theoretical framework for "Space-Time-Potential." It was based on the assumption that the universe was a "unitary, interacting, rational, purposive, and teleological system." Reuterdahl identified various levels of reality in this system (physical, psychic, etc.) that could ultimately be traced to the existence and effect of God. "Space-Time-Potential" related to the level of physical reality and therefore dealt with the interpretation "of God as the ultimate source of the Potential charted in Space and Time."[249] Reuterdahl claimed priority with regard to the theory of relativity with this form of merging space and time. And whoever should not immediately recognize the Einsteinian theory of relativity in this approach – according to the logic of an opponent of Einstein like Reuterdahl – had not discovered the groundlessness of the accusation of plagiarism, but had rather shown that Einstein was successful in his undertaking of concealing his theft of intellectual property by heavily falsifying "correct" space-time theory.

The "abolition" of ether

Due to the large number of ether theoreticians among Einstein's opponents, a claim of priority for the abolition of ether is rare. However, if a world riddle solver actually did work without ether in his theory and had begun the work prior to the statement of the special theory of relativity in 1905, then for him that was a convincing argument that Einstein had copied something. In fact, the abolition of ether in academic physics before 1905 did continue to be a conceivable possibility in the context of classical mechanical theories, but was not being discussed as a serious option in the

[248] Mewes 1920a, pp. 26f. [249] All quotes in this paragraph from Reuterdahl 1920, pp. 36f.

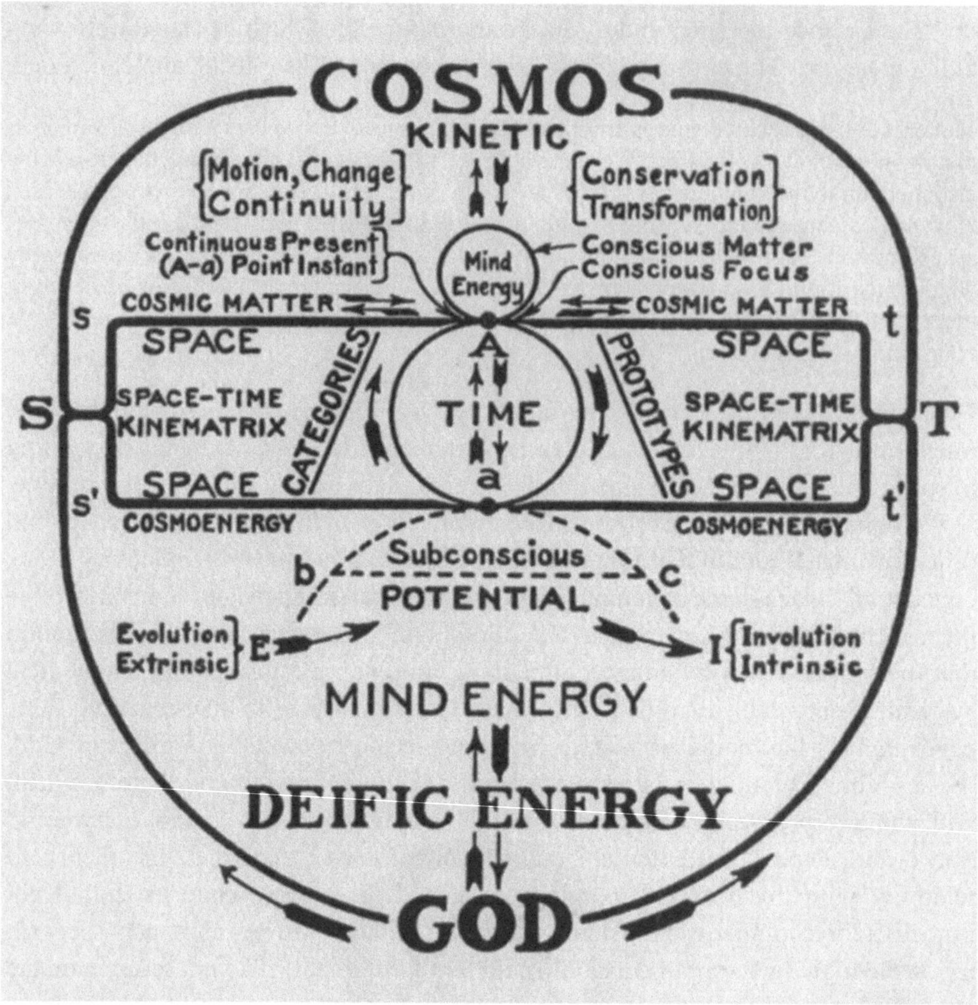

Fig. 3.2 Reuterdahl considered Einstein's theory of relativity to be a plagiarism of his own unification of space and time, represented here in the "Theocosmic Diagram." Source: Reuterdahl 1928, Plate 1.

age of electromagnetism. Things were quite different among the world riddle solvers, the majority of whom continued to work with mechanical theories.

Ziegler had already expressly labeled ether as abolished in the unified theory lectures of 1901/02. The fact that, as a chemist, he had a less close relationship to the concept of ether than a physicist may have played a role. In connection with the considerations of how a truly meaningful color theory would have to look, which would of course presume a precise knowledge of light and matter, Ziegler labeled the assumption of an ether "tangible nonsense" in his unified theory essay of 1902:

The immaterial matter filling space or materiallessness everywhere, which is called universal ether, is an inconceivable concept, and all theories that are based on it are precisely as imperfect and deceptive as their basis. None of the wave motions that are imputed to that insubstantial thing in order to explain the propagation of light actually exist. These are merely mathematical fictions that exist exclusively in the imagination of the physicists exactly in the same way as that elusive matter itself, which should at one moment perform oscillations like moving water, and at another like a struck taut cable.[250]

The fact that Ziegler, of all people, who deduces his entire theory from the metaphysical primordial light that he postulates, denounces the "elusiveness" of the unprovability of the ether does not lack a certain charm. But since he was unshakably convinced of the existence of primordial light, it did not even enter his mind to think that primordial light could be just as elusive in the eyes of the physicists as ether could be to his. As in the priority claim for establishing the constancy of the speed of light (cf. pp. 170f.), Ziegler assumed that no one before him had cast any doubt on the justification for the existence of ether. The only one who had "abolished" ether besides himself, that Ziegler was aware of, was Einstein. Ziegler wrote that Einstein didn't say anything about it:

. . .as if he knew something about the fact that Ziegler had already replaced the meaningless ether hypothesis with the meaningful theory of primordial light in empty space 23 years ago. Now if that 'assumption' of the constancy of the speed of light in Einstein's theory was already suspicious, then the new Einsteinian discovery of the substitutability of meaningless ether with full primordial light atoms and empty space must now appear to us as an unquestionable plagiarism, but as one that is still poorly understood, however.[251]

Ziegler's claim to priority was supported particularly by Reuterdahl, who saw strong similarities between his own theory and that of Ziegler. Reuterdahl linked highlighting Ziegler as the actual discoverer of the abolition of ether with expressing a suspicion of plagiarism against Einstein. "The fact that Dr. Einstein lived in Bern, Switzerland, at the time when Ziegler's theory of light was a topic of general discussion, leads one to justly question the extraordinary claims to originality of the founder of the Theory of Relativity."[252] The idea that Ziegler's theory was "a topic of general discussion" was an exaggeration, but the world riddle solvers were not stingy when assigning reciprocal honors – after all, they received little encouragement from elsewhere. Even Reuterdahl himself managed without the concept of ether. For him, ether was a supplemental mathematical construct that was not physically demonstrable, which was used to fill in the emptiness between the particles of matter, and which could be used to construct a vast number of ethers

[250] Ziegler 1902, p. 9. [251] Ziegler 1923, p. 31.
[252] Reuterdahl 1921a, pp. 4f., as well as in Reuterdahl, Arvid. (1921c). Science's 'baby-guy' was simple child till Einstein adopted it. *The Minneapolis Sunday Tribune*, May 22.

due to their purely theoretical character. "There are almost as many ethers as there are physicists,"[253] Reuterdahl remarks smugly, and continues:

> We are told by some physicists that the ether is structureless, incompressible, motionless, but capable of being set into motion, non-elastic, capable of indefinite subdivisions, and that the resulting parts can move over each other without friction. No known physical reality in the universe satisfies these specifications. Empty space (vacuum) qualifies better than any form of matter.

As with Ziegler, however, it was not enough to express these very modern-sounding doubts about ether; instead, Reuterdahl's main objection to the "materialistic" ether theories consisted of the accusation that the "materialists" (for him, all physicists not arguing religiously) were using the ether concept in order to remove God from the universe. But according to Reuterdahl's theistic science, God had to be the fundamental concept of every scientific theory:

> [I]n every form of the [ether] hypothesis one basic element is missing, without which the ether model becomes inoperative. This basic element is selective intelligence. [...] Any possible future substitute for the ether hypothesis will involve similar difficulties and inconsistencies unless the concept of God becomes the basic element of its hypothetical content. The concept of God solves the riddle of the physical universe.[254]

Reuterdahl did repeatedly point out the fact that he had already abolished ether in 1902, that is, three years before Einstein,[255] but in contrast to his other accusations of plagiarism, particularly with regard to the "Space-Time-Potential," this priority claim was not a central issue for him, although – like Ziegler's priority claim – it was supported by Mohorovičić, among others.[256]

Establishing the constancy of the speed of light

The criticism that Einstein considered the constancy of the speed of light to be a general principle has been presented in connection with the criticism of the relativity of simultaneity (cf. pp. 105ff.). In contrast, the priority claim for the first physical establishment of the constancy of the speed of light assumed that Einstein was claiming precisely this establishment for himself, which in reality he had never done. Ziegler assumed that the constancy of the speed of light had in fact first been logically proven by his deductive reasoning. He accused Einstein of plagiarism by identifying his concept of "primordial light" with the statement "the speed of light is constant," stating:

[253] Reuterdahl 1920, p. 29. [254] Ibid., pp. 30f. [255] Cf. e.g. in Reuterdahl 1921a, p. 5.
[256] "Reuterdahl in 1902 already (and at the same time J. H. Ziegler too) had rejected the hypothesis of cosmic ether." Mohorovičić 1923b, p. 2.

For actually it [the theory of relativity] should be called the theory of absolutity, since he [Einstein] used the absolute speed of primordial light as a basis. But this would have made it all too easily recognizable as a covert borrowing from primordial light theory, and so it seemed advisable to spread it among the people in a mask.[257]

The constant speed of light in empty space is a cornerstone of Ziegler's metaphysical primordial light theory – and Einstein's theory of relativity. This fundamental correspondence was already sufficient for Ziegler to devise an extensive plagiarism scenario:

For ever since I made my views public, I was boycotted by all of the representatives of that relative ignorance that is seen as true science. They no longer allow me to give lectures at their assemblies and hush up my papers. On the other hand, a theory from a pure mathematician, which hardly earns the name due to its internal contradictions, was immediately praised by them to high heavens and publicized around the entire globe with all means of advertising. This theory is based on the mere assumption of the constancy of the speed of light in a vacuum and, since it appeared a few years after the first announcements of my universal light theory, seems like a failed *contrefaçon* thereof.[258]

Ziegler's priority claim was supported by Reuterdahl,[259] who for his part also asserted claims with regard to the discovery of the physical basis for the constancy of the speed of light.[260]

"Kinertia" and the equivalence principle

The equivalence principle, that is, the equivalence of gravitational and inertial mass, was formulated in Newtonian mechanics, and its empirical validity had already been determined with great precision around 1900. In 1907, Einstein elevated the equivalence principle to a central fundamental assumption on whose basis he developed his theory of gravitation.[261]

Both in the criticism of the theory of relativity and in connection with accusations of plagiarism, the equivalence principle attracted much less attention to itself than the concept of time, the "mathematization" of gravitation, or the principle of the constancy of the speed of light. The accusation of plagiarism with regard to the equivalence principle originated with Reuterdahl in the US and was also adopted in

[257] Ziegler 1932, p. 37. Cf. also Ziegler 1914.

[258] Ziegler 1923, p. 20. Gehrcke reinforced him in the suspicion of plagiarism, but without agreeing in content with primordial light theory. Cf. Gehrcke to Ziegler, March 19, 1931, GN 83-L-5. Reuterdahl, in turn, was an avowed adherent of Ziegler's "divine science." "Ziegler [. . .], in the year 1902, laid the foundation of a cosmic theory in a lecture entitled, 'Die Universelle Weltformel und ihre Bedeutung für die wahre Erkenntnis aller Dinge [The universal unified theory and its significance for true insight into all things].' This theory is of basic and far reaching significance to the New Science." Reuterdahl 1921a, p. 4.

[259] Ibid. pp. 4f.

[260] Cf. Reuterdahl 1923a, pp. 35ff.; ibid. (1929a). *The Physical Basis of the Constant Velocity of Light*, Minneapolis.

[261] Cf. Renn 2006, particularly pp. 207ff.

the German-speaking area.[262] Reuterdahl did not assert the accusation of plagiarism for himself, but rather for the non-academic gravitation researcher "Kinertia." Kinertia was the pseudonym of Robert Stevenson, who had developed a mechanical gravitation theory under this name at the end of the nineteenth century. This time, the accusation of plagiarism even initially related to actual content. Stevenson, a student of William Thomson and Peter Guthrie Tait, whose turbulent biography also included a period as a mining engineer, had carried out experiments with free fall on himself in a gold mine. He let himself fall halfway down the length of the shaft in the elevator, and then the elevator braked. "That taught me that acceleration was the proximate cause of weight,"[263] Stevenson reported in retrospect.

Reuterdahl was not satisfied with the fact that Stevenson, like many before him, had inferred the equivalence of gravitational and inertial mass and then wanted to establish it experimentally; instead, he wanted to further show that Einstein's entire general theory of relativity represented a plagiarism of Stevenson's theory. The fact that Einstein in no way claimed discovery of the equivalence principle for himself, but rather that his contribution consisted of analysis and new interpretation of this well-known principle, did not play a role in the plagiarism discussion. Reuterdahl argued that some of Kinertia's papers had been sent to the Prussian Academy of Sciences in 1903 and thus it was obvious that Einstein, a member of the Academy, would have made use of Kinertia's insights in the archive. "Did Dr. Einstein avail himself of these easily accessible records?"[264] Reuterdahl asked in the *Dearborn Independent*, and immediately set about proving that the answer to this question could only be "yes."

In a long article in the *Dearborn Independent*, Reuterdahl compared quotes from a series of articles that Stevenson had published in the widely distributed periodical *Harper's Weekly* with quotes from popular works on the theory of relativity, particularly from the first English translation of Einstein's *On the Special and General Theory of Relativity (A Popular Account)*,[265] as well as from Eddington's *Space, Time, and Gravitation*.[266] In stringing together snippets of quotes out of context, Reuterdahl focused on the equivalence principle and particularly on the thought experiment of the elevator in space, which he made identical to the general theory of relativity without further ado. The thought experiment states that an observer in a closed elevator cannot determine by physics experiments whether the elevator is being accelerated in space or is located at rest on the Earth. He also cannot differentiate whether the elevator is moving uniformly in a straight line in space or is falling

262 The *Hamburger Woche* newspaper of June 9, 1921, includes a portrait of Reuterdahl and the comment, "Reuterdahl doesn't just call Einstein a 'crazy scientist with a hysterical theory,' but also a plagiarist and claims that the Einsteinian theory had already been discovered in 1866 by a scholar under the name Inertia [sic]."
263 Stevenson 1914, p. 210. 264 Reuterdahl 1921b.
265 Einstein, Albert. (1920b). *Relativity: The Special and the General Theory*. New York.
266 Eddington, Arthur Stanley. (1920). *Space, Time and Gravitation: An Outline of the General Relativity Theory*. Cambridge.

freely to the Earth. In both cases, he does not detect any attractive force. This permits the conclusion of the equivalence of gravitational and inertial mass.

Reuterdahl knew that Stevenson was above all interested in refuting the idea that gravitation was to be conceived of as an attractive force in the context of a kinetic theory of gravitation; but this fact did not play a role due to the reduction of the general theory of relativity to the thought experiment. Instead, Reuterdahl was certain of the following: "Kinertia's theory of gravitation is identical with Einstein's. Both men find the crux of the problem in acceleration, and the development of both theories is based upon the very same experiment."[267] But while Einstein only carried out the experiment in his thoughts, according to Reuterdahl, Stevenson carried out the experiment himself, years before Einstein set up the general theory of relativity. Reuterdahl concluded, "From the preceding is evident that 'Kinertia' derived his norm of gravitation before Einstein was born. The question of priority is therefore definitely and irrefutably established in favor of 'Kinertia' in the case of the General Theory of Relativity."[268] Reuterdahl tactically played off the accusation of plagiarism and the claim to priority against each other extremely skillfully:

> If Einstein was aware of 'Kinertia's' discovery then the appellation 'plagiarist,' bestowed upon him by his German professional colleagues, is eminently fitting. If, on the contrary, Einstein was unaware of this work, then he is, nevertheless, antedated by the work of 'Kinertia.' Einstein is at liberty to choose either horn of the dilemma.[269]

3.3.4 Summary of the content-based accusations of plagiarism

Several aspects of the accusations of plagiarism against Einstein investigated here are remarkable. The accusations of plagiarism referred to content that had nothing to do with the physical theory of relativity, or to isolated and generalized statements such as "space and time must be considered together" or "the speed of light is constant," which were interpreted as the core of the theory of relativity or simply as "the" theory of relativity. One special feature that was brought out in the discussion of the accusations of plagiarism is that the plagiarism referred to the accuser's own theory, and in fact almost every one of Einstein's opponents expressed suspicion at one point or another that he himself had already said what was correct about the

267 Reuterdahl 1921b.

268 Ibid. After the article "'Kinertia' versus Einstein" appeared, Reuterdahl received a letter from John Heslip, a pastor at Tabernacle United Presbyterian Church in Youngstown, Ohio, who had read both the article in the *Dearborn Independent* and Stevenson's original work in *Harper's Weekly* when it came out, and had been deeply impressed by his conception of gravitation since that time. Heslip suggested publishing the essays as a book in order to appeal to a larger audience. In addition, he suggested that it be made clear in book form "that 'Kinertia' was in the field long before Einstein." Heslip to Reuterdahl, April 30, 1921, RP 4–67.

269 Reuterdahl 1921b.

theory of relativity, just in simpler words.[270] Purely tactical accusations of plagiarism (such as Gehrcke's support of Palagyi) were rare. Einstein's opponents were deeply convinced of the correctness of their accusations. Reuterdahl truly believed that the general theory of relativity should be equated to the "man in the elevator." The fact that the theory of relativity was so sharply attacked despite such a lack of knowledge of it confirms the defensive attitude discussed in Chapter 2 against a "concept of the enemy theory of relativity" that was largely independent from Einstein's theory. This defensive attitude was traced back primarily to the vulnerability and sensitivity of the world riddle solvers with regard to their own theories, and resulted from a specific reception of the theory of relativity as a threat. Another remarkable point is that some of Einstein's opponents did not exclusively assert priority claims for themselves, instead attempting to give more weight to priority claims by naming other people. This sometimes caused paradoxical situations to arise – if someone conceded, as Mohorovičić did, for example, that Einstein's opponents Reuterdahl and Ziegler had abolished ether independently from each other, why shouldn't they then concede that Einstein had also developed a theory that got by without ether just as independently? This is often where the line of argument was brought up again, pitting the "true, sincere natural scientists" against the "dishonest relativists." If another natural scientist had arrived at the same or a similar result independently, this supported their own claim.[271] Einstein, on the other hand – according to this line of argument – was not a natural scientist at all; instead, he was said to be integrating the truths discovered by the other researchers into a fictitious mathematical system that had no connection to reality.

What is additionally noteworthy is the picture of Einstein and the scientists who were positively disposed towards the theory of relativity that prevailed among Einstein's opponents. Most of them insinuated that Einstein had behaved unscientifically and morally questionably and assumed that he would again make use of other people's theories at every opportunity. And the scientists who defended Einstein

[270] Alongside the positions presented in this section, cf. e.g. the statement by Geissler about the "relativity paper that I had already published in 1900." Geissler to Gehrcke, December 9, 1920, GN 76-A-7. Cf. also Kleinschrod 1920, p. 28. Gehrcke also thought he had been robbed by Einstein, but he did not refer directly to the theory of relativity, but rather to the areas of physics and epistemology. According to Gehrcke, Einstein had determined in his paper "The Foundation of the General Theory of Relativity" that classical mechanics had an epistemological defect and that Mach had been the first to clearly emphasize this fact. Cf. Einstein, Albert. (1916). Die Grundlage der allgemeinen Relativitätstheorie [The foundation of the general theory of relativity]. *Annalen der Physik*, **354**(7), 769–822, reprinted in English translation in Kox *et al.* 1997, Doc. 30, on Mach, p. 771. Now Gehrcke thought "that significant points of Einstein's considerations, insofar as they are not already found in Mach, are already contained in papers by me, and, as I believe, clearly contained." Gehrcke, Ernst. (1916). Zur Kritik und Geschichte der neueren Gravitationstheorien [On the criticism and history of more recent gravitation theories]. *Annalen der Physik*, **356**(17), 119–24, reprinted in Gehrcke 1924a, pp. 40–44 (see p. 40).

[271] Ziegler thought differently. He even asserted a claim to priority for the plagiarism that was allegedly committed against him. He wrote to Gehrcke "that I am, without a doubt, the first person who Einstein plagiarized." Ziegler to Gehrcke, February 11, 1931, GN 83-L-4.

against accusations of plagiarism or simply did not take them seriously were generally implied to be intentionally hushing up Einstein's plagiarism and thus acting equally questionably. These views resulted from a strong feeling of distrust towards academic science, which acknowledged the theory of relativity but not the solution of the world riddles, and from the related anti-academic attitude of the world riddle solvers. Both these factors codetermined to a great degree the strategic dispute with the theory of relativity that is presented in the next chapter.

4

Marginalization and protest: the strategic dispute with the theory of relativity

This chapter analyzes the vehement public controversy about the *theory of relativity* in the 1920s based on the results of the first three chapters. The first chapter demonstrated the existence of a non-academic science whose knowledge was incompatible with that of modern academic physics, and whose protagonists acted anti-academically on the offensive. The second chapter showed that the omnipresent theory of relativity was perceived as a threat by the non-academic researchers, but also by Einstein's academic opponents, against which they had to defend themselves, in their opinion. The third chapter discussed the fact that Einstein's opponents were convinced that they had completely refuted the theory of relativity based on their own theories or as physicists or philosophers, and had proven, furthermore, that the theory of relativity was not any kind of "correct" theory of physics at all, or that it was a falsification of their own correct theory. In terms of content, then, the matter was thus settled for them. But what was the use of correcting the falsity of the theory of relativity assumed by Einstein's opponents if it could not gain a hearing?

The content-based attempts at refuting the theory of relativity represented only one part, namely the content-based dimension, of the discourse opposing academic physics, whose strategic dimension demanded recognition of these refutations. In order to assert these claims of truth, an attempt was made to enter into the prevailing scientific discourse and thus to appropriate for themselves the "centralizing power-effects"[1] of the scientific discourse, which, in its function as a scientific discourse, passes judgment on the legitimacy of claims of truth and represents the framework of what is capable of being truth at all. The religious claims of the world riddle solver theories, for instance, did not possess this capacity of being truth from the outset in the context of academic physics in the 1920s.

Einstein's opponents had to commandeer the scientific discourse somehow – whether by a friendly takeover or by a putsch. In the face of such a powerful discourse

[1] Foucault 2003, p. 9.

as institutionalized academic science represented, this was not seen as a task for an individual; instead, liaisons and alliances with like-minded people were sought. This chapter is devoted to the subversion attempts and struggles, and the tactics and strategies against the theory of relativity. It will be shown that non-recognition of the individual refutation of the theory of relativity using their own theories, and non-acceptance of the objections raised by the academic physicists and philosophers, led to a situation where Einstein's opponents increasingly did not conceive of the question as one of correctness or falsity, but rather as a question of access to the system of recognition of claims to truth in the scientific system. The fact, and above all the permanent confrontation with the fact, that the "refutations" of the theory of relativity were not recognized led to a shift in the level of argumentation from content-based to political-strategic criticism, which was a primary cause for the sharpness of the criticism of the theory of relativity in the 1920s.

4.1 Marginalization

4.1.1 Dimensions of the marginalization process

Specialization and professionalization of science went hand in hand with the development of the scientific disciplines at the end of the eighteenth and the beginning of the nineteenth centuries.[2] The modern scientist was no longer a gentleman scientist or a man of independent means, but rather a professional employee of a university or a non-university research institution, trained according to specific standards and equipped with corresponding titles.[3] These developments represent the framework for the marginalization process to be analyzed in the following, multiple respects:

(1) the development of specializations in science goes hand in hand with intra-disciplinary differentiations, revaluations, and new developments of sub-areas in the discipline;
(2) its extensive impermeability to the non-academic sphere is structurally anchored in the institutionalization of scientific training and research;
(3) in addition – and this must be situated at the level of knowledge – a consequence of the organization of knowledge in a scientific discipline results in boundaries being drawn between scientifically compatible knowledge and scientifically incompatible knowledge.

[2] Cf. Stichweh 1984.
[3] These institutions and the titles and degrees awarded by them in turn strongly influence the power hierarchies within science. Pierre Bourdieu has analyzed this in detail using the French science system, which, compared to international standards, is strongly characterized by hierarchies in this respect. Cf. Bourdieu, Pierre. (1988). *Homo academicus*. Translated by Peter Collier. Stanford.

I will now briefly sketch out this framework,[4] which is stretched over the three points.

Intra-disciplinary marginalization

The specialization within scientific disciplines led to a situation in which a course of study, for instance physics, was unable to equip someone completing it with all the special knowledge of his discipline. Decisions were necessary at an early stage: did one wish to dedicate one's scientific career to optics or to cryogenics research? The answer would determine which teacher to seek out or which university to attend. Similar processes of specialization and exclusion often come into play between intra-disciplinary scientific communities in the same way as between scientists from different disciplines or between scientists and the public. Therefore, sociologists of science apply the expert–layman distinction within science and within disciplines.[5] In this chapter, this differentiation within disciplines will play a central role, primarily with regard to the experimental physicists opposing the theory of relativity, who were not "experts" in the theoretical physics that was now dominating the discipline.

Non-academic marginalization

The non-academic sphere has a special relationship with the academic, institutionalized scientific disciplines, since it represented the complement to the development of a largely impermeable academic sphere. Identifying a group of people who do not belong to academic science with the non-academic area and situating non-professional research on the side of popular and amateur science was not accepted by all non-academic researchers, however, as was presented in the first chapter. This exclusion of laypeople and the creation of the modern amateur was a process that took decades and encountered (and still encounters) resistance from some of those excluded.

Epistemological marginalization

The drawing of social boundaries by the internal differentiation and institutionalization of science is one level on which exclusionary mechanisms come into play. The other level is the level of knowledge where it is initially immaterial whether an academically trained physicist or an engineer is speaking. On this level, Einstein's academic opponents were likewise subject to marginalization. What mattered was the position that each one of them advocated and the extent to which it was compatible with science in the context of the internal organization of scientific knowledge, or marginalized as false, outdated, or as non-knowledge. Now systems

[4] This conceptual approach is further developed in Wazeck, Milena. (2013). Marginalization processes in science: The controversy about the theory of relativity in the 1920s. *Social Studies of Science*, **43**(2), 163–90.

[5] Cf. Note 5 in Chapter 1.

of knowledge are not static. Their internal organization develops and changes. Particularly in periods when knowledge systems are being restructured, differing views exist as to what falls within the scope of scientific compatibility and what does not. In the 1920s, however, the opponents of modern physics were confronted with a state of research from whose perspective their claims to knowledge did not represent any significant different conceptions, but rather only marginal ones.

Drawing social and epistemological boundaries

Due to these framework conditions, the boundaries drawn between what fell "within" and "outside" the scientific discourse took various forms, and they affected the various groups of Einstein's opponents in different ways. Einstein's non-academic opponents were already structurally outsiders because they had not passed through the academic institutions, while the academic physicists and philosophers were at least not excluded structurally due to their institutional connections and their academic training. However Einstein's academic opponents were subject to a marginalization process just like the non-academic critics due to their adherence to scientific positions deviating from the consensus state of research. Such mechanisms of marginalization, which came into play among all the groups of Einstein's opponents, particularly included processes of scientific quality assurance such as the peer review process by the professional journals and the decision to permit lectures in scientific associations. Additional factors were decisions on employment issues and distributions of positions on scientific boards and committees, which exclusively affected Einstein's academic opponents, while Einstein's non-academic opponents were for the most part subjected to the simplest and most effective form of marginalization: being ignored.

4.1.2 Early acquaintance with the theory of relativity

Persons interested in science, such as the world riddle solvers, had often become acquainted with the theory of relativity in the natural science associations and at lectures of the scientific associations years before its spectacular confirmation in 1919. No less than Einstein himself gave several lectures in forums that were not purely for specialists, particularly at the natural science societies in Switzerland.[6]

[6] Über die Theorie der elektromagnetischen Wellen [On the theory of electromagnetic waves] in 1903 before the Natural Science Society of Bern, listed in the minutes of the 992nd meeting, 1903, Burger Bibliothek Bern, GA BNG – Minutes of the Meetings, Vol. 10; Ueber die Natur der Bewegungen mikroskopisch kleiner, in Flüssigkeiten suspendierter Teilchen [On the nature of the movements of microscopically small particles suspended in liquids] in 1907 before the Natural Science Society of Bern, printed as an abstract in English translation in Stachel *et al.* 1989, Doc. 43; Die Relativitäts-Theory [The theory of relativity] in 1911 before the Natural Science Society of Zurich, cf. Einstein 1911a. The members of the Natural Science Societies were for the most part scientists, but the lectures were kept semi-popular since different disciplines met there.

The early critical reception shows two special features compared to the criticism of the theory of relativity in the 1920s. First, it is characteristic that the theory of relativity, or relativity principle (as it was called before the general theory of relativity was formulated) was considered merely to be one among several "false" developments in modern science and did not yet have the prominent significance that it had in the 1920s. Zacharias, for instance, described the situation in 1914 as follows:

> Attraction and entropy are impossible, relativity and Ostwald's energetics are nonsensical, the theory of magnetism is completely wrong and lacks a natural conception, meteorology and mathematics are inadequate, medicine is often going astray. These terrible circumstances, this unfounded confusion, these faulty theories and accomplishments are still being called 'science' today.[7]

Leo Gilbert's book Das Relativitätsprinzip, die neueste Modenarrheit der Wissenschaft [The Relativity Principle, the Latest Fashionable Folly of Science] appeared in 1914, and the title itself makes no secret of the fact that, for Gilbert, the relativity principle was just another one of the follies that were very widespread in science, which he had undertaken to expose based on his Neue Energetik.[8]

Second, this early criticism was noticeably friendlier for two reasons. Many of the critics in the 1910s still had direct contact with Einstein and they generally expressed themselves with more restraint in personal discussions. However, an even more decisive factor than these external circumstances was each individual's "career as a critic." Each person's first attempt at criticism – whether in 1913 or 1923 – was generally still kindly disposed. It was only due to disregard and rejection (regardless of how justified it was) that their annoyance grew.

4.1.3 "Prof. Einstein, Berlin. Re: Space"[9] – personal contacts with Einstein

Many opponents of the theory of relativity initially had nothing against Einstein as a person and were not determined opponents of the theory of relativity. In fact, they had first made contact with Einstein – before and after 1919 – in order to point out errors in the theory of relativity to him, scientist to scientist. The world riddle solvers followed the ethos of the scientists, and their comments were unpolemical and often stated in a respectful tone, for it was certainly conceivable that their colleague Einstein had not yet even noticed his errors. Perhaps he was simply not yet aware of the primordial light

[7] Zacharias 1914a, p. 128.

[8] Other "fashionable follies," according to him, included Lorentz's electron theory, for instance: "Professor H. A. Lorentz's electricity theory is just as phenomenal." Gilbert 1914, p. 15.

[9] This is how an open letter from Robert Büchler to Einstein begins. Cf. Büchler, Robert. (1921). *Lehrsätze über das Weltall mit Beweis in Form e. offenen Briefes an Prof. Einstein [Propositions About Space with Proof in the Form of an Open Letter to Prof. Einstein]*. Aachen.

theory or the universal mechanical theory, which made the theory of relativity unnecessary?

This attitude is clearly shown in Ziegler's attempt to make contact with Einstein. In the 1900s and 1910s, the two Swiss citizens were in some of the same groups, the natural science societies and scientific associations. Ziegler was a member of the Swiss Natural Science Society (SNG) and later also of the Zurich Natural Science Society, while Einstein was a member of the Berne Natural Science Society. In the years after completion of the special theory of relativity in 1905, although Einstein was not yet the public media star that he was after 1919, he was well known among scientists and those interested in science as an outstanding scientist.[10] Ziegler, who as a man of independent means spent a great deal of time travelling to scientific conferences, met Einstein for the first time when Einstein held a lecture "On the development of our views on the nature of the constitution of radiation" at the annual meeting of the GDNÄ in Salzburg on September 21, 1909.[11] This conference is considered to be the event where Einstein first spoke to the contemporary scientific luminaries at a large forum.[12] Among the roughly 100 participants, which included leading physicists of the time such as Planck, Born, Sommerfeld, and Stark, was also Ziegler, the world riddle solver. In the discussion following Einstein's presentation, Planck raised the objection that Einstein's consequence of even seeing light waves themselves as quantized went too far for him.[13] Ziegler got to speak after Planck and immediately had the solution at hand:

> If the *uratoms* of matter are conceived as invisible tiny spheres that possess unchanging speed of light, then it is possible to describe all interactions of corpuscular states and electromagnetic phenomena, and this would also establish the bridge between the material and non-material entities still missed by Mr. Planck.[14]

While the others present at this discussion did not discuss Ziegler's suggested solution, any authority on primordial light theory knew that Planck's differentiation between electromagnetic and mechanical processes could only be meant as a differentiation between intangible, spiritual primordial light and

[10] Cf. e.g. Fölsing 1997; cf. also Kraus' statement to Gehrcke in 1913: "[...] since great interest in the theory of relativity exists in large circles." October 17, 1913, GN 72-A-4, as well as the letter from Ziegler to Einstein in 1910 quoted on pp. 182f.

[11] Einstein, Albert. (1909).

[12] Cf. Benz, Ulrich, and Hermann, Armin. (1972). Quanten- und Relativitätstheorie im Spiegel der Naturforscherversammlungen 1906–1920 [Quantum and relativity theory as reflected in the annual meetings of the German Society of Natural Scientists and Physicians, 1906–1920], in *Wege der Naturforschung 1822–1972 im Spiegel der Versammlungen Deutscher Naturforscher und Ärzte [Paths of Science Research 1822–1972 as Reflected in the Meetings of German Natural Scientists and Physicians]*, ed. Hans Querner and Heinrich Schipperges. Berlin [*et al.*], pp. 125–37, particularly pp. 130f.

[13] (1909). Diskussion. *Physikalische Zeitschrift*, **10**(22), 825–26, reprinted in English translation in Stachel *et al.* 1989, Doc. 61.

[14] Ibid., p. 826.

matter. Ziegler was just as unsuccessful a year later, when he listened to Einstein's lecture on Boltzmann's principle before a smaller group at the Zurich Physics Society on November 22, 1910.[15] On this occasion, Ziegler stated, based on his unified theory, that the Relative would only be one half of the whole, which required the Absolute as a complement. Ziegler describes Einstein's reaction to this objection in this way:

At the end of a debate about a lecture by Prof. Einstein on his incomprehensible principle, in order to prove the internal unfoundedness of this theory of relativity in a convincing way, [I] intentionally [allowed] myself to address the simple question to him of whether he couldn't make it somewhat more comprehensible to me using a brief characterization of the absolutity principle.

The seeming naiveté of my question and Einstein's somewhat scornful counter-question of what that would be, triggered unrestrained laughter in the agnostic society, which ceased directly when I shortly proved that it would be inconceivable to have any thought at all without its counter-thought, and thus that every thing could also only be qualified by a comparison with its 'opposite.'

In this way, the Einsteinian relativity principle, which lacks a counter-principle, was proven to be unqualifiable in the truest meaning of the word and thus to be unfounded.[16]

Ziegler sees himself as having conquered Einstein here, although neither Einstein nor the others present took Ziegler's objection seriously. And why should they – as this report by Ziegler once again clearly shows, the relativity principle was something very different for him than the relativity principle in modern physics. Ziegler never found out what the Einsteinian relativity principle was really about. For him, the relativity principle could be nothing other than (only) one half of his own fundamental contrast between "absolute" and "relative." It was only 1910, and Ziegler was still favorably inclined to his Swiss colleague. He suspected that Einstein, with his "half theory" of relativity, would be close to his own "full theory" of primordial light and could use some help, and so he wrote a letter to Einstein on December 12, 1910:

Dear Sir,

A year ago, I had the pleasure of hearing your lecture in Salzburg, and recently another lecture on derivations from Boltzmann's principle as a guest at the Physics Society in Zurich. These mathematical derivations are too high for such a simple chemist as me and [therefore] I will not discuss them. But since you have brought a so-called principle of relativity *en vogue* – everyone is talking and writing about it now, Planck in Berlin and Gruner in Berne etc., etc. – and since I have likewise established a universal system of physics based on something very similar, I am taking the liberty of most humbly sending

[15] Either the relativity principle was also discussed at Einstein's cited lecture, or Ziegler has confused this lecture with another one by Einstein. Cf. Einstein, Albert. ([end of 1910]). Über das Boltzmann'sche Prinzip und einige unmittelbar aus demselben fliessende Forderungen [On Boltzmann's principle and some requirements directly derived from it], in Renn 2005c, pp. 76–89.

[16] Ziegler 1914, pp. 18f.

you one of my recent publications about these my views. Perhaps you can endorse them and take on the mathematical additions to my formula, if you do not wish to hold on to the view of a unilateral entropy.
Sincerely yours, Dr. H. Ziegler[17]

The letter is in a polite tone and is thus in contrast to the later polemics against Einstein, but also partially explains them. Ziegler was by no means an opponent of the theory of relativity from the start; after all, he did see it as "very similar" to the primordial light theory. But Einstein didn't answer. For a well-meaning colleague, as Ziegler understood himself to be, this was a snub and a decisive source of justification for his sharp attacks. An amazingly large number of Einstein's opponents, both academic and non-academic, not only sought out contact with Einstein, but also carried on more or less extensive personal discussions in the years before broad public reception of the theory of relativity. By and large, nothing is known of their precise content; however, the reports of Einstein's opponents do show that they went very unsatisfactorily for them.

Oskar Kraus, who was a tenured associate professor of philosophy at the German University in Prague, had been confronted with the theory of relativity for the first time in 1911 or 1912, when Einstein was likewise teaching in Prague.[18] The group of German-speaking academics in Prague was moderately sized; perhaps they had met at the university, or maybe at Berta Fanta's[19] salon, which both frequented – at any rate, Kraus absolutely wanted to find out more about the theory of relativity. He visited Einstein at the Physics Institute and got the theory of relativity explained by him quite extensively. This personal explanation later became an important source for Kraus, because it was particularly legitimate. He refers to it in the *Offene Briefe [Open Letters]* to Einstein and Laue.[20]

[17] Enclosed with the letter was Ziegler's booklet *Die Struktur der Materie und das Welträtsel [The Structure of Matter and the Riddle of the Universe]*. Bern 1908. A copy of the letter and the delivery note with Einstein's signature are in Reuterdahl's papers, RP 4–66.

[18] Cf. Kraus to Gehrcke, April 19, 1927, GN 33-J-11.

[19] On Berta Fanta, cf. Gimpl, Georg. (2001). *Weil der Boden selbst hier brennt ... : aus dem Prager Salon der Berta Fanta (1865–1918) [Because the Ground Itself Burns Here ... : From the Prague Salon of Berta Fanta]*. Furth im Wald [*et al.*].

[20] Kraus, Oskar. (1925). *Offene Briefe an Albert Einstein und Max von Laue über die gedanklichen Grundlagen der speziellen und allgemeinen Relativitätstheorie [Open Letters to Albert Einstein and Max von Laue on the Intellectual Foundations of the Special and General Theory of Relativity]*. Vienna [*et al.*], pp. 39ff. Kraus met Einstein one more time in January 1921, when Einstein was invited to Prague by the Lotos Scientific Association to give a lecture. Since the possibility of speaking directly with Einstein as an opponent had become rare in 1921, Kraus immediately made an offer to Gehrcke: "So Einstein is coming on the 7th. Today I was asked by telephone whether I would participate in a debate – to be held at a special location. I said yes. In response, they telegraphed him from the Lotos Association. If you [Gehrcke] would have any wishes regarding any questions to be asked of Einstein, then you should inform me of that without delay." Kraus to Gehrcke, December 29, 1920, GN 71-G-8. Kraus then spoke unpolemically about the development of the idea of relativity in philosophy; no discussion with Einstein came about. According to Kraus's recollection, the discussion after Einstein's lecture went very unfavorably for him. Cf. the retrospective assessment "like back then in the Urania Hall as an object of public pleasure in shaming." Kraus to Gehrcke, October 24, 1929, GN 33-J-4.

The philosopher Lenore Kühn, who polemicized in the 1920s that the theory of relativity would "destroy the concept of theoretical reality by a relativism that must bury reality itself under its rubble,"[21] had attended one of Einstein's lectures as early as 1915 and written him a letter expressed in respectful tones:

> If I allow myself to write these lines to you [Einstein], even though I must assume that you frequently receive communications, and certainly also from unqualified people, and that I will thus perhaps be placed in the group of rash know-it-alls for you, still, this does not occur due to the fleeting impression of an hour, but rather from a burning interest in the theory of relativity, with which I have been dealing particularly in the past few months.[22]

Fricke noted a longer personal discussion with Einstein incidentally in a letter to Gehrcke dated September 4, 1919. "Perhaps it interests you that I recently debated with Mr. Einstein for about two hours about these questions [the ether vortex theory]. I saw that Einstein did not even know about the ether theories, only the degenerated Lorentz form."[23] This disappointment about Einstein's "lack of understanding" was shared by other discussants such as Mohorovičić, who had had an opportunity to correspond with Einstein when he wanted to publish a paper on the possibility of multiple special theories of relativity in *Physikalische Zeitschrift* in 1916.[24] But he too was not very impressed by Einstein's objections. "Mr. A. Einstein still denied the physical value of my work, but he emphasized *that it is not illogical.* In this way, I got into a correspondence with Einstein, but his proofs were never compelling for me."[25]

In the 1920s, Einstein was inundated with mail, so he did not answer many letters, and probably in many cases never even got to see them. Also, the intensification of the controversy about the theory of relativity in the 1920s probably caused him to avoid such discussions. However, he did have to get involved in discussions with old acquaintances such as Edouard Guillaume, even though they were unfruitful. Guillaume was one of Einstein's former colleagues at the Patent Office in Bern. Like many other opponents of Einstein, he considered the relativization of time to be a problem on the level of time measurement and attempted to introduce an additional parameter t of an absolute time into the Lorentz transformation. In a letter dated

21 Kühn 1921.

22 Kühn to Einstein, June 26, 1915, GN 3-G-11. Kühn assumed that the relativization of the space and time concept by Einstein was similar to her own conception of space and time, which she had developed as part of her work on "fundamental forms of the Aesthetic" (ibid.). She attributed to Einstein that, with the theory of relativity, he had explained the differences between time in and of itself and time determination, as well as between space in and of itself and space determination. But Einstein specifically does not do this. It was only years later that she provided this unsent letter to Gehrcke.

23 Fricke to Gehrcke, September 4, 1919, GN 76-A-8.

24 Mohorovičić to Gehrcke, March 25, 1921, GN 3-G-3. Mohorovičić explained this thought in several papers, cf. e.g. Mohorovičić, Stjepan. (1916–17). Über die räumliche und zeitliche Translation [On spatial and temporal translation], Part 1 in *Académie des Sciences et des Arts des Slaves du Sud de Zagreb, Croatie, Bulletin des Traveaux de la Classe des Sciences Mathématiques et Naturelles*, 6–7, 46–72, Part 2 in ibid., 9–10 (1918), 21–33.

25 Mohorovičić to Gehrcke, March 25, 1921, GN 3-G-3.

October 9, 1917, Einstein said to him, briefly and kindly, but clearly: *"that parameter t just does not exist."*[26] A few letters still went back and forth before the correspondence broke off. In 1920, however, Guillaume took up the correspondence again,[27] and Einstein attempted once more to point out Guillaume's fundamental misunderstandings, but all the objections made no impression on Guillaume, who in turn repeatedly attempted to convince Einstein after all, until Einstein ultimately stopped responding. In the meantime, one of Einstein's closest friends, the mathematician Marcel Grossmann (1878–1936) spread the word in Switzerland that Einstein considered Guillaume's ideas nonsensical, and Guillaume pressed for a discussion in a letter to Einstein dated June 30, 1920:

As various persons have explained to me, however, it would be very fruitful if you could substantiate your position sometime; and for me personally it has become absolutely necessary that you speak up. For in a controversy in the press (in the N[eue] S[chweizer] Z[eitung]), your friend Grossmann wrote that my 'views about relativity had been completely repudiated (?) by you.'

When I thereupon asked Grossmann when and where you had expressed yourself in this way, he replied to me that in a letter, you had written that my 'views on relativity were nonsense.' This groundless judgment, which is apt to hurt me to the highest degree, forces me to insist that either it be given a basis or retracted.

That is why I must ask you, dear Einstein, for a reply that I can publicize. This you really can't refuse your old Office colleague![28]

Einstein extricated himself from the affair diplomatically:

Dear Guillaume,

My behavior must appear unkind and unloyal to you; it may seem that I wanted to condemn you by my silence. In reality, though, it is entirely otherwise. I often tried to understand your statements but always utterly failed. So if I wrote Grossmann that it was nonsense, this must be understood with reference to me, or better yet, to the present state of my brain; nonsense is what one calls whatever one cannot grasp; there is no other criterion.[29]

[26] Einstein to Guillaume, October 9, 1917, English translation taken from Schulmann *et al.* 1998, Doc. 387.

[27] "Dear Einstein, I do not need to tell you what pleasure the experimental confirmation of your fine theory gave me. That I value your theory very highly already arises from the fact that I have been working intensively on it for two years. Now, however, I have arrived at results that are so important for the development of the theory that I take it as my duty to inform you of them without delay. (…) For, my studies of space and time have led me to the conclusion that a *basic error* regarding time is hidden in the theory." Guillaume to Einstein, January 25, 1920, in Kormos Buchwald, Diana *et al.* (eds.). (2004). *The Collected Papers of Albert Einstein*, Vol. 9: *The Berlin Years: Correspondence, January 1919-April 1920.* (English translation of selected texts), translated by Ann Hentschel. Princeton, Doc. 280.

[28] Guillaume to Einstein, June 30, 1920, in Kormos Buchwald, Diana *et al.* (eds.). (2006b). *The Collected Papers of Albert Einstein*, Vol. 10: *The Berlin Years: Correspondence, May–December 1920, and Supplementary Correspondence, 1909–1920.* (English translation supplement), translated by Ann Hentschel, Princeton, Doc. 68; cf. also Grossmann to Einstein: "As you see, a cult is forming around Guilleaume that thinks it must correct essential points of your concepts." September 9, 1920, in Kormos Buchwald *et al.* 2006b, Doc. 142.

[29] Einstein to Guillaume, July 4, 1920, in Kormos Buchwald *et al.* 2006b, Doc. 71; cf. also Einstein's (unpublished) opinion on Guillaume's work in a letter to Grossmann that was intended for publication in the Geneva *Archives*

Einstein insisted that he could not discover any physical sense in Guillaume's equations and suspected that Guillaume had abandoned himself to "forbidden mysticism."[30] For Guillaume, on the other hand, it was clear that Einstein was the one who had gotten carried away with an absurd idea, and so he wrote to him compassionately, "Dear Einstein, No! I am not angry, but certainly sad, for it is always sad to see an outstanding person like you becoming confused in such an obvious way."[31]

This controversy, which up to that time had only been carried out between Guillaume and Einstein (Einstein attempted to dissuade Guillaume from his intention of publishing the correspondence, since Einstein feared that this would only lead to disgrace for his former colleague) took an unpleasant turn for Guillaume when he could not be prevented from publicly refuting Einstein.

In April 1922, Einstein traveled to Paris to give four lectures with discussions at the Collège de France; he also spoke before the Philosophical Society.[32] Guillaume had made the journey from Geneva specifically for this event, and had been widely advertised internationally in the press beforehand as the refuter of the theory of relativity.[33] He had made preparations to report extensively to the Collège de France about the results of his research on absolute time in a prominent appearance. To his disappointment, he was only given a few minutes and, what's more, he had to put up with Paul Langevin interrupting his lecture in order to allow Einstein to speak. The daily newspapers reported devastatingly about Guillaume's appearance. First, he was not described particularly flatteringly as a person. "M. Guillaume est petit, trapu, chauve, avec une barbiche noire et un sourire avantageux qui se changera tout à l'heure en un déplorable sourire d`écolier grondé qui se retient de pleurer."[34] *L'Œuvre* provided the most extensive report on Guillaume's objections to the concept of time in the theory of relativity and the reaction of the other scientists:

des sciences physiques et naturelles. Einstein to Grossmann, September 12, 1920, in Kormos Buchwald *et al.* 2006b, Doc. 148; cf. also Grossmann, Marcel. (1920). Mise au point mathématique. *Archives des sciences physiques et naturelles*, **2**, 497–99.

30 Einstein to Guillaume, July 4, 1920, in Kormos Buchwald *et al.* 2006b, Doc. 71.

31 Guillaume to Einstein, December 23, 1920, in Kormos Buchwald, Diana, Sauer, Tilman, Rosenkranz, Ze'ev, Illy, Josef, and Holmes, Virginia Iris (eds.). (2006a). *The Collected Papers of Albert Einstein*, Vol. 10: *The Berlin Years: Correspondence, May–December 1920, and Supplementary Correspondence, 1909–1920*. Princeton, Doc. 241.

32 On the trip, cf. particularly Biezunski 1991. Einstein cancelled a visit to the French Academy of Sciences after it had become known that a large number of the members of the Academy had decided to boycott a lecture by the "boche" ("Hun") Einstein in their halls. Cf. also various contemporary press reports, GN 4, XVII 124r to 128r.

33 Cf. e.g. Un adversaire des Théories d'Einstein. *Les Messin*, April 1, 1922; Les Théories d'Einstein? *Intransigeant*, April 3, 1922; Einstein Wrong, Swiss Declares. *Chicago Tribune*, March 31, 1922, GN 4, XVII, 129r.

34 "Guillaume is small, stocky, bald, with a black beard and an attractive smile that soon changes into the pitiful smile of a scolded schoolboy who is trying not to cry." Au Collège de France. M. Guillaume, de Genève, contre M. Einstein. *L'Œuvre*, June 6, 1922, GN 4, XVII, 130r.

AU COLLÈGE DE FRANCE

M. Guillaume, de Genève, contre M. Einstein

La troisième séance du débat contradictoire entre M. Einstein et les savants français s'est déroulée hier, mercredi, dans ce même petit amphithéâtre que le public, malgré la parcimonie avec laquelle les cartes ont été distribuées, emplit jusqu'aux couloirs.

Fig. 4.1 Extract from the Paris newspaper *L'Œuvre*, June 6, 1922.

Il avait cependant apporté son matériel de démonstration: une belle planche où sont dessinées des dents de scie bleues et d'autres rouges, et une feuille de papier avec une circonférence et une ellipse. Nous n'avons pas eu le loisir de savoir à quoi il les destinait. M. Guillaume fut mis hors de combat en deux rounds. Il s'aventura à dire que certaine sphère lui paraissait bien être une éllipsoïde. Peut-être bien que, dans un certain système, c'est une éllipsoïde, déclara M. Einstein avec un bon sourire; mais, quand c'est une éllipsoïde, ça ne nous intéresse pas du tout.[35]

[35] Ibid. "However, he had brought his demonstration equipment: a beautiful board with saw teeth drawn on it, some blue and others red, and a sheet of paper with a circle and an ellipse. We did not have the opportunity to know what he intended with them. Mr. Guillaume was knocked out in two rounds. He ventured to say that a certain sphere could well be an ellipsoid. It could perhaps be an ellipsoid in a certain system, said Mr. Einstein with a nice smile, but if it is an ellipsoid, it does not interest us at all." Guillaume's objection referred to the example of the transformation of the coordinates of a spherical wave in the special theory of relativity. Cf. Einstein 1905b, p. 901.

Following this experience, Guillaume was finished with his former colleague Einstein and the rest of the academic physicists. He complained bitterly to Reuterdahl, "Einstein reconnut l'existence de mon éllipsoïde mais déclaraqu'il n'etait pas 'intéressant', et les nombreux auditeurs présents se mirent à applaudir!"[36] It was only among opponents of Einstein like Reuterdahl that both Guillaume's refutation of the theory of relativity and his theory of absolute time met with any response.

The arguments that Einstein's opponents advanced in personal contacts with Einstein quite obviously did not cause Einstein to distance himself from the theory of relativity. They therefore turned to a larger audience in academic science.

4.1.4 Attempts to convince academic scientists

Among Einstein's opponents, it was the world riddle solvers in particular who also sought a hearing in academic science through other paths than via the creator of the theory of relativity. As always, they were firmly convinced that comprehensive enlightenment of the "ignorant" academic physicists would do the trick. If only they would listen! The world riddle solvers' reasonable and convincing theories would then be clear to them and allow them to recognize their "blindness." "The solution to the problem [of gravity] is easy and could be stated in a few words. But one problem remained insoluble for me: the physicists. One would think that the solution to a centuries-old problem would have to interest them somehow. But they have behaved entirely passively hitherto,"[37] complained Fricke, asking for attention. "I am prepared to publicly defend my views to any specialist." But the specialists were interested neither in his refutation of the theory of relativity, nor in his ether vortices, nor in his solar theory. The engineer Lamberty's attempts to teach Max Planck were also polite:

I recently again sent you a few lines relating to the two small academic errors of Fizeau and Olaf Römer. In the interest of our German science, I ask you most humbly to inform me whether you have recognized these two errors. And if not, then I will once again make the effort to describe them to you further, in more detail.

[36] "Einstein acknowledged the existence of my ellipsoid but said it was not 'interesting', and the large audience present began to applaud!" Guillaume to Reuterdahl, April 26, 1922, RP 4–19.

[37] Fricke 1937. The conviction that the solution to all problems of science were so evident in the world riddle solver theory that everyone would *have to* accept it was apparent both in the consistent repetitions as well as in the design of the written works. Some world riddle solvers worked excessively with the method of emphasis, using the following pattern, more or less: IT *FINALLY* **HAS TO** BE UNDERSTOOD AFTER ALL!!!!!! A kindly reader pointed out to Reuterdahl that the overemphasis could have the opposite effect. Cf. Myron Weiss, Associate Editor of *Time Magazine*, to Reuterdahl, February 28, 1929, RP 4–17: "May I offer a criticism on the mechanics of the book's [Reuterdahl 1928] printing? There are so many words beginning with capital letters in it that the emphasis of capitalization becomes lost. Further, I have found that excessive capitalization disturbs a reader and confuses him, makes him lose the continuity of thought."

Why, sneered world riddle solver Lamberty, didn't Planck use this assistance that was offered to him? After all, Lamberty would have been willing to explain everything to him personally "as long as it would be possible to listen to me for several days. For I cannot explain everything in a few hours."[38] Most scientists will have experience with non-academic theoreticians who want to express their views to them – astrophysicists presumably more than historians, although even they sometimes receive emails with longer discussions about the analogies between ages and the course of history. In most cases, a fruitful discussion does not occur because a common basis of shared scientific standards is lacking. Academic researchers usually recognize this at first glance, and they do not want to spend their scarce time proving the indefensibility of any particular ideas, as Einstein attempted to do with Guillaume.

This is not to say that insights that are worth taking up cannot come "from the outside,"[39] but experience shows that such cases are rather rare. And this is also not to say that divergent science is not conducted "within" science; in fact, such science can be of an innovative nature. Nevertheless, arbitrariness does not prevail in science. This results from scientists' epistemological socialization, which causes them to become participants in a body of knowledge in their time. Furthermore, quality assurance also exists that is structurally anchored in science, using self-control.

Despite the weaknesses that it has often been shown to have and the fact that it sometimes acts as a brake on innovation, in many cases this quality assurance mechanism functions satisfactorily in cases where the method of argumentation differs widely from the current state of research. Blondlot's "N-rays," for example, which caused a sensation around 1900, were quickly proven not to exist by other scientists.[40] And when an astronomer such as T. J. J. See shifted from observing to theorizing and claimed to have solved the problems of science ranging from the causes of earthquakes to gravitation, he was very quickly criticized by the respective specialists for the fact that the assumptions of his theories were outdated, or metaphysical, or both. Even a popular astronomer like See had experienced one thing above all: being disregarded. The professional world only reacted when he claimed that this disregard would be a tacit acknowledgement of his work, or when

[38] Both quotes from: Lamberty to Planck, May 4, 1925, [copy], GN 38-D-4.

[39] This is shown not least by the role of the amateur researcher Mandl in the story of the development of Einstein's paper on gravitational lenses. Cf. Renn, Jürgen and Sauer, Tilman. (2003). Eclipses of the stars – Mandl, Einstein, and the early history of gravitational lensing. In *Revisiting the Foundations of Relativistic Physics. Festschrift in Honor of John Stachel*, ed. Abhay Ashtekar, Dordrecht [*et al.*] (Boston Studies in the Philosophy of Science Vol. 234), pp. 69–92.

[40] The physicist René Blondlot thought that he had discovered "N-rays" (N for the city of Nancy, where Blondlot did his research) that were similar to X-rays. Cf. Ashmore, Malcolm. (1993). The Theatre of the Blind: Starring a Promethean Prankster, a Phoney Phenomenon, a Prism, a Pocket, and a Piece of Wood. *Social Studies of Science*, **23**(1), 67–106.

he began to publicly attack Einstein and attracted broader attention in this way.[41] It is therefore not surprising that academic science usually simply ignored the refutations of the theory of relativity based on an engineer's or physician's own theory.[42] And when the academic colleagues did occasionally answer, that meant a humiliating experience for most of Einstein's opponents. They received very clear rejections, which ranged, depending on the temperament of the individual, from friendly and restrained to brusque.

Fricke left nothing unattempted in trying to enter into the academic discourse, and as a world riddle solving but academically trained physicist, he at least had a foot in the door of academic physics (due to his membership in the German Physical Society (DPG), for instance). "I have even approached Mr. Planck and Mr. Laue with a request to make a clear discussion possible, and possibly even a personal consultation. My request was very firmly rejected. A sufficiently common basis for a discussion is supposedly lacking."[43] For him, this was entirely incomprehensible, although with Fricke's fundamental differences not only with modern physics, but even with classical physics, even a favorably disposed reader such as Ernst Gehrcke could sometimes only put question marks in the margin of particular discussions.[44] While Fricke addressed his earlier works on ether vortex theory primarily to theoretical physicists whom he wished to dissuade from accepting the theory of relativity based on Lorentz's ether theory (which he considered false), from the 1930s he clashed vehemently with astronomers who would not be convinced by his adherence to Herschel's solar theory. Along with an acquaintance who advocated for him, he failed particularly with the astronomers of the Astrophysical Observatory in Potsdam:

Mr. Bueren, the patent attorney, [...] who wanted to mediate due to his interest in science, was referred by Prof. Kienle[45] to Prof. Grotrian[46] and had to suffer rough treatment from him. They were not willing to have any objective discussion. Because I have given the views

[41] Cf. Sherrill 1999 on See's conflicts with academic science. A lack of reactions was often turned against academic science. The argumentation strategy was: Whoever does not react has not refuted the individual article, and whoever is not expressly against it, is for it. Cf. e.g. Ruckhaber's statement, "Not a single physician with whom I came in contact personally or in writing has contradicted it [biomechanics] thus far." Ruckhaber, Erich. (1952). *Biomechanik: das gelöste Lebensproblem, die Ursachen des Alterns und ihre alleinige Bekämpfung [Biomechanics: The Solved Problem of Life, the Causes of Ageing and their Sole Remedy]*. Berlin, Foreword to the 3rd edn.

[42] Cf. inter alia Lamberty's statement: "You know that, delighted about the results of my research and in the belief that this would be the right way, I have approached universities and scholars *since the beginning of 1924*. But all is silent. They are treating me like a zero." Lamberty to Gervai, May 11, 1925, [copy], GN 38-D-6.

[43] Fricke [ca. 1942].

[44] Cf. e.g. the marginal notes in Fricke, Hermann. (no date). *Sonne, Wind und Aether [Sun, Wind, and Aether]*, [typescript], GN 18-C-9, and in Fricke [ca. 1942]. Cf. also Lenard's statement about Fricke reproduced on p. 238.

[45] Johann Georg Kienle (1895–1975) was the director of the Astrophysical Observatory in Potsdam starting in 1939, and a professor of astrophysics and astronomy at the University of Berlin.

[46] Walter Grotrian (1890–1954) was appointed to the Astrophysical Observatory in Potsdam in 1922, and a professor of astrophysics at the University of Berlin from 1927.

of Herschel, Bode, Humboldt, and Arago, which were generally prevailing among astronomers until 1861, a new physical foundation, I am supposedly 'a person who belongs in a lunatic asylum' and whose views are only to be considered 'pathological.'[47]

But the boundaries between deprecatory academic science and world riddle solver science cannot be drawn quite so precisely. Interestingly, it was a friend of Einstein's, the French astronomer Charles Nordmann, who advocated a discussion of Fricke's solar theory in an article in *Le Matin* in 1933. Overall, this article shows openness and receptiveness to Fricke's project of taking up Herschel's solar theory once again. Nordmann even accepted Fricke's assertion that temperature and gravity were proportional (which was vehemently attacked by other astronomers).[48] Nordmann saw the spiral organization of nebulae as an additional corroboration of the vortex theory of ether. Nordmann concludes his article as follows:

Il sera curieux de voir la réaction des 'bien pensants' astronomiques en présence des audacieuses propositions que le docteur Fricke a étayées de technicité fort sérieuse. Gageons que ce soleil, devenu soudain, sous sa carapace de feu, tempéré selon le mode [...] naguère démodé [...] d'Herschel et d'Arago leur donnera froid dans le dos. Et pourtant elle tourne [...] la théorie.[49]

Such a favorable discussion on the part of a member of academic science represented a great triumph and Fricke immediately photocopied and disseminated a German translation. But that was absolutely an exception. Much more typical was the reaction of Paul Drossbach, a lecturer in Physical Chemistry and Electrochemistry at the Berlin College of Technology:

Dear Dr. Fricke, I have received your pamphlet in the field of astrophysics. I would like to point out a fundamental error right away. Specifically, you dispute the applicability of the

[47] Fricke, Hermann. [undated manuscript, no title], GN 19-G-5.

[48] Fricke's *Eine neue und einfache Deutung der Schwerkraft und eine anschauliche Erklärung der Physik des Raumes [A New and Simple Interpretation of Gravity and a Comprehensible Explanation of the Physics of Space]* (Fricke 1919) received a very favorable review from Johannes Riem in *Naturwissenschaftliche Wochenschrift* in 1920 (cf. Riem 1920). In response, Professor Becker of Heidelberg sent "an appraisal, differing significantly from the earlier review" of Fricke's work to *Naturwissenschaftliche Wochenschrift*. Becker stated: "The present book contains a series of essays in which the author believes he is able to provide an interpretation of gravitation that goes beyond our previous knowledge. The fundamental thought is that the force lines of the gravity field are to be understood comparably to light rays as the path of an energy ray (towards the attracting body). Consequently, energy would have to develop continuously in heavy masses, which would be expressed in the heavenly bodies by their temperature. In this context, the author believes he is able to make comprehensible a certain parallelism between the mass and the temperature of these bodies, which naturally follows without further ado from the proportionality of mass and heat content given a recorded temperature, but which was apparently not recognized. Unfortunately, these and numerous related reflections often cannot be reconciled either with our previous firmly grounded knowledge or with each other." Becker. (1920). [Review of Fricke 1919]. *Naturwissenschaftliche Wochenschrift*, New Series, **19**(29), 463–64.

[49] "It will be interesting to see the reaction of the astronomical 'do-gooders' in the face of the audacious proposals that Dr. Fricke has supported very seriously technically. We suppose that the sun suddenly got the shivers under its shell of fire, tempered in the manner [...] of Herschel and Arago [...] that was outdated only a short while ago. And yet it moves [...] the theory." Nordmann, Charles. (1933). Aux dernières nouvelles le soleil serait froid et entouré d'eau [The latest news is, the sun is cold and surrounded by water]. *Le Matin*, July 7.

Newtonian law of gravity for determining the masses of the stars. But the mass is only unequivocally *defined* by this law, so that any experimental determination – even in the laboratory – is thus also ultimately based on this law, on this *definition*.

Consequently, the determination of the masses of the stars is precisely only possible by applying the law of gravity, which first defines the mass. Your theories deviating from this principle are therefore simply false, precisely because they contradict the definition of mass. What you calculate is anything but mass. [...]

Your theories are not being rejected because they contradict, for instance, atomic physics, but rather because they are simply exactly as false as if someone mistook a cow for a tree and then told all sorts of fairy tales about the leaves and branches of the cow. If you want to practice science and carry out experiments at all, you simply never can contradict the unequivocal definitions of the basic concepts.[50]

Fricke's struggle for recognition in academic science began as a struggle against the theory of relativity and lasted for decades. Even after World War II, he was still trying to receive support; he approached Friedrich Schiller University in Jena and the Saxon Academy of Sciences with a request for support, but received refusals from both. While the university in Jena expressed its refusal diplomatically, the secretary of the mathematical-physical section of the Saxon Academy stated simply and concisely, "I cannot submit your request for sponsorship to the Academy since I cannot discern any scientific value in your paper."[51]

Ziegler was also initially convinced that the luminaries of physics would certainly be convinced by a precise presentation of the primordial light theory. He therefore visited Max Planck, whom he considered the most influential theoretical physicist, twice in Berlin; the first time in 1907 after the annual meeting of the GDNÄ in Dresden, but without success. "He [Planck] only said that nothing simpler could be thought of than my view of the primordial atoms."[52] In contrast to Planck, Ziegler specifically considered this simplicity to be convincing proof of the truth of his views. Some time later, he therefore asked Planck for another meeting, which Planck granted "with some reluctance." Ziegler failed once again to win the defiant Planck over to his side. "It [the meeting] took place and then extended for a full two hours, but due to Planck's incessant objections, nothing fruitful came of it."

In fact, the scientific luminaries were the worst conceivable people for Einstein's opponents to address. They were for the most part theoretical physicists who were

[50] Drossbach to Fricke, October 15, 1942, GN-18-7. Fricke sent the letter to Gehrcke with an accompanying letter on December 26. On Fricke's modification of the Newtonian law of attraction for masses in motion, cf. Fricke, Hermann. (1920). *Über eine neue Form des Wirkungsgesetzes für bewegte Massen [On a New Form of the Law for Moving Masses]*, May 1, [manuscript], GN 76-A-10.

[51] Hund, F. (secretary of the mathematical-physical section of the Saxon Academy of Sciences) to Fricke, August 28, 1946, [copy], GN 19-G-10. This paper was Fricke's Original Work No. 2, *Die Meerwassertheorie der Sonne [The Seawater Theory of the Sun]*. Bense (Registrar of Friedrich Schiller University) to Fricke, August 1, 1946, [copy], GN 19-G-10.

[52] All of the following quotes from: Ziegler, Johann Heinrich. (1935). *Das Geheimnis des Selbstverständlichen: der Schlüssel zum Welträtsel [The Secret of the Self-Evident: The Key to the World Riddle]*, Zürich, pp. 33f.

helping to develop modern physics. A different method of accessing academic science, other than through the great names and institutions, consisted of linking to existing networks from university studies and training. Sometimes they were successful, but often not, and their defeat was all the more bitter when they had known the former teachers and fellow students well.[53]

4.1.5 Special features of the marginalization of Einstein's academic opponents

Due to their fundamental rejection of the theory of relativity, Einstein's academic opponents were likewise subjected to a marginalization process. It made a decisive difference whether the academic critic came from their own discipline of physics, or from philosophy. Criticism by a professor of physics carried significantly more weight than that by philosophers such as Kraus or Palagyi. Moreover, the recognition hierarchy played a role. Einstein could hardly ignore someone like Lenard, while the dispute with Gehrcke was often carried on by other physicists such as von Laue or Thirring.[54] In the following, this intra-disciplinary marginalization will be placed in a broader context, since the motivation of academic physicists such as Gehrcke or (to some extent) Lenard to ally themselves with the non-academic opposition had its source substantially in this special context.

The experimental physicists among Einstein's opponents moved to the fringe area of the academic discussion due to their fundamental content-based opposition, by which they disconnected themselves from the current state of research and – in some areas of their discipline – carried out deviant science. This marginal position, which resulted from content divergences, was reinforced by the consequences of the new significance of theoretical physics to the organization of science. In the 1920s, who was at the head of scientific associations like the German Physical Society? Who sat on the executive boards of the institutions? Who was appointed to full professorships that opened up? Whose articles were published in the professional journals? And who held the editorships of these

[53] As a graduate of the Mittweida College of Technology, Patschke, the engineer, requested a review of his work *Lösung der Welträtsel durch das einheitliche Weltgesetz der Kraft [Solution of the Universal Mysteries by the Uniform Natural Law of Force]* by three lecturers at the College; the review was to be provided to the publisher. The administration notified him, however, that no one had been found willing to write the review, and excused this with the lecturers' workload. Cf. Patschke to Technikum Mittweida, November 21, 1905; Administration of Technikum Mittweida to Patschke, November 24, 1905, Technikum Mittweida, Student File No. 4393. Stevenson, the gravitation researcher, attempted to convince his former teacher, William Thomson (Lord Kelvin) of his mechanical pressure theory, also unsuccessfully: "I am afraid however we are no nearer for a physical explanation of gravitation than we were when you were in my class." Thomson to Stevenson, May 11, 1896, RP 4–56.

[54] On this "defensive belt" around Einstein, cf. Hentschel 1990, pp. 163ff.

journals and decided who published in them? Who gave the major lectures at scientific conferences? About whom and what did the newspapers write to a degree never seen before? Johannes Stark noted this development and judged it extremely critically:

> The increased significance of theoretical physics is also expressed in the fact that the number and scope of theoretical treatises in the physics journals had increased very much more than the number of treatises in which new phenomena, procedures, and devices are described or than improved measurements of physics factors are announced. It is also a sign of the times that at the scientists' conferences over the past few years, it is primarily theory that has been discussed in the general lectures from the physics side, for instance the quantum theory at the annual meeting of the GDNÄ in Nauheim, and the theory of relativity at this year's meeting in Leipzig. Indeed, the feeling of inferiority vis-à-vis theory seems to be gaining ground even among experimental physicists; [...] Given these circumstances, it is probably justifiable to speak of an overgrowth of theory in current physics.[55]

This situation was new and resulted from an intra-disciplinary transformation of significance at the beginning of the twentieth century away from experimental physics and towards theoretical physics, which was linked to a shift in intra-disciplinary recognition of experimental physicists and theoreticians, but also to a transformation at the level of the institutional organization of research and in social recognition. It had long been customary in physics to differentiate between applied or experimental physics on the one hand and theoretical physics on the other. Up until the end of the nineteenth century, physicists were easily able to switch from one area to the other. This was necessary not least because theoretical physics did not provide any promising career paths. In teaching, it was often covered by adjunct professors, and it only began to strengthen institutionally in 1870 by the creation of adjunct associate professorships. However they were primarily conceived of for supporting experimental physics and to relieve the burden on the professors who were working in it. Anyone who desired a scientific career had to work experimentally, even if he held a position for theoretical physics, since the full professors and physics institutes were almost exclusively found in the area of experimental physics.[56] Even Philipp Lenard, who became well known as an outstanding experimental physicist, had held an adjunct associate professorship in theoretical physics as a young scientist in Heidelberg in 1896.

The upturn in theoretical physics was caused by new, spectacular results in experimental physics and by developments within theoretical physics. It faced the challenge of integrating the new discoveries of X-rays, radioactivity, and the electron into a theoretical context. Now that thermodynamics and electrodynamics

[55] Stark 1922, pp. 1f. [56] Cf. Jungnickel and McCormmach 1986, Vol. 2, pp. 33ff.

had developed as new fields, it was also the task of theoretical physics to clarify the extent to which mechanics could be applied to them or whether alternatives would have to be developed.[57] Around the turn of the century, Berlin (Max Planck), Göttingen (Woldemar Voigt), and Munich (Arnold Sommerfeld) began to be established as centers of theoretical physics, and the numbers of students there increased.[58] Max Planck in particular had made a name for himself as the executive officer of German physics by taking on many offices and functions in physics.[59] Before World War I, three outstanding younger theoretical physicists, Max von Laue (Frankfurt), Albert Einstein (Berlin), and Peter Debye (Göttingen) had been installed as professors, who, moreover, had made their careers exclusively in theoretical physics in contrast to their predecessor generation.

Theoretical physicists increasingly moved into the influential positions; the dominance of theoretical physicists in the DPG[60] even led to the founding of a separate society for application-oriented physicists, the German Society for Technical Physics (DGTP) in 1919.[61] One of those involved, Johannes Stark, describes the motivation for founding it this way in his pamphlet *Krisis*:

> The physicists who undertook and carried out the foundation [of the DGTP] felt so little satisfied by the theoretical direction prevailing in Berlin, they had so little interest in the applications of physics in this direction, that they resolved to form a scientific group themselves and to take on the journalistic work for technical physics themselves.[62]

An attempt was made to give experimental physics more weight within the DPG. Gehrcke noted in the (unofficial) minutes of a meeting for reorganizing the journal operations of the DPG, "I then emphasized that something seriously had to be done and that the out-of-town gentlemen in particular would have to be satisfied; a

57 Cf. ibid., pp. 212ff.; Renn 2006, pp. 87ff. 58 Cf. Jungnickel and McCormmach 1986, Vol. 2, pp. 254ff.

59 Cf. Heilbron, John L. (1986). *The Dilemmas of an Upright Man: Max Planck and the fortunes of German science*. Berkeley; cf. Hoffmann, Dieter. (2004). Max Planck, in *Lexikon der bedeutenden Naturwissenschaftler [Lexicon of Important Scientists]*, eds. Dieter Hoffmann, Hubert Laitko, and Staffan Müller-Wille, 3 volumes. Heidelberg [*et al.*], Vol. 3, pp. 150–55; cf. ibid. (2008). *Max Planck. Die Entstehung der modernen Physik [Max Planck. The Emergence of Modern Physics]*. Munich, pp. 67ff.

60 In 1919/20, the executive board of the DPG consisted of Sommerfeld (chair), Rubens, Warburg, Planck, Haber, Nernst, Einstein, Boas, Goldstein (observer). Cf. *Deutsche Physikalische Gesellschaft, Uebersicht der Sitzungen f.d. Vorstandsjahr 1919/20 [German Physical Society, Overview of Meetings for the Executive Board Year 1919/20]*, GN 71-E-2.

61 The *Deutsche Gesellschaft für technische Physik* was founded by Georg Gehlhoff in 1919. Cf. Hoffmann, Dieter, and Swinne, Edgar. (1994). *Über die Geschichte der 'technischen Physik' in Deutschland und den Begründer ihrer wissenschaftlichen Gesellschaft Georg Gehlhoff [On the History of 'Technical Physics' in Germany and the Founder of its Scientific Society, Georg Gehlhoff]*. 2nd, revised and supplemented edn. Berlin (Berliner Beiträge zur Geschichte der Naturwissenschaften und der Technik, Vol. 1).

62 Stark 1922, p. 29. On Stark's tensions with the theoretical physicists, cf. Beyerchen 1977, pp. 106ff. On the motivation for founding the DGTP, cf. also Fricke's statement: "It [the DGTP] was founded with the eager participation of several gentlemen who did not approve of the manner of working and thinking of the German Physical Society under Planck's influence, and by technical physicists who hoped to cultivate a physics that was closer to reality and more tangible compared to the mathematical theories." Fricke [ca. 1942].

situation like the current one, where Lenard and Wien are dissatisfied with the Society, had not ever happened before."[63] The journal commission was later expanded primarily by including experimental physicists. However, these tensions at the beginning of the 1920s between theoretical and applied physics did not lead to a breach, let alone to a non-academic direction of applied physics. Neither the theoretical physicists nor the experimental physicists were interested in such a thing. Starting in 1923, the joint physicists' conferences no longer made a distinction between pure and technical physics.[64]

Along with the question of the representation of experimental physicists in the professional associations, issues of appointments and career paths were the topics where Einstein's academic opponents repeatedly made their marginalization and lack of influence the central theme:

> You can see from this course of affairs [the rejection of all candidates recommended by Lenard as possible successors for his chair] how well the power of the Einstein–Planck spirit has flourished. At the same time, any work to enlighten the university people is completely hopeless in my experience.[65]

This marginalization was often blamed not only on the dominance of theoretical physics, but also on an intentional conspiracy by the theoretical physicists, which will be examined below in the analysis of the conspiratorial patterns of interpreting marginalization (cf. pp. 227ff.). Another significant aspect of this marginalization, which particularly affected Einstein's opponents who argued based on their own theories, consisted of the rejection of papers by Einstein's opponents by the professional press, since publication in professional journals, after all, represents a central means of communication in academic discussions.

4.1.6 Publication problems and journalistic evasive actions

The professional journals

Einstein's opponents, who argued non-academically, initially attempted to place their articles in the relevant professional journals such as *Annalen der Physik*, *Physikalische Zeitschrift*, or *Zeitschrift für Physik*. This corresponded with their self-concept as serious scientists who wished to publish in a suitable forum. Publication in *Annalen der Physik* in particular was considered a must for anyone

63 "Aussprache über das Zeitschriftenwesen" [Discussion of journal operations], private minutes by Ernst Gehrcke, May 8, 1920, GN 71-E-1. On the division between the "Berliners" and the "Out-of-Towners," cf. inter alia Beyerchen 1977, pp. 107ff.

64 Cf. Hoffmann and Swinne 1994, p. 18.

65 Lenard to Gehrcke, May 26, 1927, GN 33-F-7. Lenard had recommended Stark and Gehrcke, among others. Cf. ibid. as well as Beyerchen 1977, p. 236, note 111.

who wanted to be recognized in academic physics.[66] The world riddle solvers' objections to modern physics almost always foundered in the professional press on the requirement of a positive expert review. Their content-based objections to the theory of relativity were unpublishable for the professional press because, as was shown in Chapter 3, they largely missed the point of the physical theory of relativity and were presented on the basis of Einstein's opponents' own theories – which were classified by the professional press as non-scientific. How could Ziegler's primordial light theory ever be published in *Annalen der Physik*? Or Patschke's Universal Law of Force? Mewes' periodicity? Gilbert's energetics?

For the professional press, quality assurance, either by an editorial decision or the review process, is indispensable. These checking procedures prior to publication are designed to ensure that the specific minimum standards for scientific publication are met. After publication, it is up to the scientific community to judge the work,[67] and Einstein's academic opponents in particular were affected by this, as they were less likely to be rejected by the professional press due to their membership in the group of academic physicists.

By 1920, some of the world riddle solvers had already had so many bad experiences in their attempts at placing their own theories with the flagship of the physics journals, *Annalen der Physik*, that the rejection of an article against the theory of relativity simply no longer surprised them at all. An article by Rudolf Mewes, in which he raises an accusation of plagiarism against Einstein, was "simply rejected" by the editors of *Annalen der Physik*, as Mewes reports. And he goes on, "From the experience of the past two decades, I know that I cannot expect accommodation from the large journals."[68] A popular option was to publish in the proceedings of academies and physics associations, but it was often unsuccessful. What particularly upset Reuterdahl was that even membership in the corresponding associations did not guarantee publication of polemical articles opposing the theory of relativity in their official publications. "I have had articles refused by Scientific Societies of which I am a member because they clearly exposed the Einstein Sham."[69]

None of Einstein's avowed opponents were to be found in influential positions in the professional press; instead these positions were occupied (not exclusively, but

[66] Cf. Stichweh 1984, pp. 394ff.; Jungnickel and McCormmach 1986, Vol. 1, pp. 114ff. Around 1906, *Annalen der Physik* only had a rejection rate of 5–10%. On the rejection practice of *Annalen der Physik* cf. ibid., pp. 310ff.

[67] Cf. also Stichweh 1984, pp. 394ff., particularly pp. 395f. The emergence of the professional journals was accompanied by a process of inclusion. In contrast to the general scientific periodicals, whose readers were often passive consumers, the readers of a professional journal often felt motivated to contribute to their professional specialty themselves. Cf. ibid.

[68] Both quotes from Mewes 1920a, p. 73.

[69] Reuterdahl to Fischer, October 11, 1931, RP 4–14. Reuterdahl was a member of the Mathematical Association of America, the History of Science Society, the Swedish Physical Society, the American Society of Mechanical Engineers, and the German Kepler League, among others.

mostly) by physicists who represented modern physics. It is true that *Annalen der Physik* was published by Wilhelm Wien, who viewed modern physics with extreme reservations, but Max Planck had been appointed as a member of the editorial board of the Physical Society in 1895 as successor to Helmholtz and acted as the expert reader for papers on theoretical physics.[70] *Zeitschrift für Physik* was run by Karl Scheel and, from 1921, *Physikalische Zeitschrift* was run by Max Born and Peter Debye. Arnold Berliner, who published the influential journal *Die Naturwissenschaften* that addressed broad areas of science, was also positively disposed towards the theory of relativity. Expert opinions on submitted articles and reviews of literature on the theory of relativity were likewise written almost exclusively by representatives of modern physics. Von Laue assessed the articles on the theory of relativity for *Zeitschrift für Physik*,[71] while most of the reviews of the literature on the theory of relativity in *Physikalische Berichte* came from Reichenbach. This dominance by the theoretical physicists was not interpreted by Einstein's opponents as merely reflecting further development of physics. For Einstein's opponents, this dominance instead resulted from a "takeover" of physics by the theoretical physicists, accomplished using questionable means.

In principle, the academic physicists among Einstein's opponents had access to the professional press, but there they discovered that they occupied a marginal position in academic physics, particularly in the form of sometimes devastating critiques. There were many reactions to Gehrcke's article against the theory of relativity. Max Born, Hans Thirring, and Hermann Weyl, among others, but also Einstein himself, took positions on Gehrcke's objections.[72] The same holds true for the book publications. Gehrcke, for instance, had to suffer the following review by Gustav Mie after publication of his collected works against the theory of relativity in 1924:

> Our author [Gehrcke] [...] considers his insufficient understanding of the Lorentz equations to be important enough to confront the professional world with it. Even after more than 10 years, he has still not succeeded in finding the solution out of the difficulty [that the relativization of simultaneity causes for him], although several truly capable professional colleagues have made an effort to help him in the meantime.[73]

In the 1920s, when every remotely conceivable fundamental objection to the theory of relativity had already been asserted in countless variations, it became noticeably more

[70] Cf. Jungnickel and McCormmach 1986, Vol. 2, p. 248 and pp. 254f.
[71] Cf. Scheel to Kraus, reprinted in Kraus 1925, pp. 75–77, particularly pp. 75f.
[72] Cf. Born, Max. (1913). Zum Relativitätsprinzip: Entgegnung auf Herrn Gehrckes Artikel 'Die gegen die Relativitätstheorie erhobenen Einwände' [On the relativity principle: Reply to Mr. Gehrcke's article 'The objections raised against the theory of relativity']. *Die Naturwissenschaften*, **1**(4), 92–94; Thirring 1921a; Thirring 1921b; Thirring 1921c; Einstein 1918a; Weyl, Hermann. (1921). Antwort an Herrn Prof. Gehrcke [Answer to Prof. Gehrcke]. *Die Umschau*, **25**(10), 123–24.
[73] Mie, Gustav. (1926). [Review of Gehrcke 1924a and Gehrcke 1924b]. *Deutsche Literaturzeitung*, **14**, 678–82 (see p. 680), GN 4, VII.

difficult for academic physicists to publish anything against the theory of relativity.[74] At the beginning of 1925, when Mohorovičić wanted to publish something quickly in the professional press in order to enhance his list of publications (due to the prospect of a chair at the University of Zagreb), Gehrcke advised him to write a short manuscript that did not name Einstein or the theory of relativity, for instance using the title "The general linear transformation of the measured values of space and time." He would then place it in *Annalen der Physik* after "personal consultation with Prof. W. Wien." He explained to Mohorovičić, "The matter is somewhat difficult due to the fact that our physics journals have recently gotten into the habit of rejecting manuscripts that refer [critically] to the theory of relativity."[75] It was never assumed that purely objective reasons could be the cause for rejecting a critical article on the theory of relativity; instead, they always saw themselves as victims of a conspiracy (cf. pp. 226ff.). Although they had access to the professional press, Gehrcke and Lenard commiserated with each other about the "domination of the physics 'professional press'" by the theoretical physicists, and disputed the right of this "professional press" (that they put in quotes) to represent "correct" physics.[76]

The situation was different for an academic philosopher like Kraus, who had been rejected by the physics journals in a similar way to Einstein's non-academic opponents. A manuscript that he submitted to *Zeitschrift für die Physik* was rejected by the editor, Karl Scheel, in spring 1924, making reference to Max von Laue's negative judgment. Von Laue wrote to Scheel that Kraus was one of the opponents of the theory of relativity "who are certainly trying to deal honestly with the issue, but who nevertheless do not get into the issue because they do not deal with the mathematical development of the theory."[77] But Kraus was a professor of philosophy, after all; he

74 Cf. also von Gleich to Gehrcke, February 3, 1923: "I therefore hope that the new version that I started working on right away complies not only with your wishes but also with those of the 'censorship bureau.' I had no idea that even the scientific literature was in such a straitjacket. [...] I now hear that even scholars such as you and Prof. Lenard are having difficulties when you want to express views in a professional journal that contradict the 'greatest scientist of modern Germany,' Mr. Einstein." GN 78-C-1. It is highly probable that this involved the article "Die allgemeine Relativitätstheorie und das Merkurperihel [The general theory of relativity and the perihelion of Mercury]" that was sent in to *Annalen der Physik* a few weeks later and published in Vol. 377, No. 19 (1923), pp. 221–35.

75 All quotes from: Gehrcke to Mohorovičić, February 4, 1925, GN 29-F-6.

76 Cf. inter alia the statements in a letter from Gehrcke to Lenard dated December 3, 1917, on "advertising" and the "domination of the physics 'professional press'," [draft], GN 83-E-4. "Science" was also often put in quotes. Cf. e.g. Lenard to Gehrcke, May 26, 1927: "[...] what a freak of 'science' this Jew E.[instein] is." GN 33-F-7.

77 Scheel to Kraus, April 14, 1924, reprinted in Kraus 1925, pp. 75–77. Scheel quotes this judgment by von Laue in his letter. Critical comments on Kraus are also in Frank, Philipp. (1979). *Einstein: sein Leben und seine Zeit [Einstein: His Life and His Time]*. Braunschweig [*et al.*], pp. 286f. Scheel also did not print a paper submitted by Gimmerthal to *Zeitschrift für Physik* because it "appeared to belong more in a philosophical than in a physics journal." Gimmerthal to Gehrcke, April 29, 1927, GN 76-A-4. In March 1921, Gehrcke attempted to convince the editor of *Umschau* to take what was in Gehrcke's eyes an "objective position on disputed scientific issues." Gehrcke wanted to get a reply printed to Hermann Weyl's reply to Gehrcke's reply to Weyl's report on the course of the discussion of the theory of relativity in Bad Nauheim. Cf. Gehrcke to Bechold, March 21, 1921, GN 4-G-10, and Gehrcke's statement, GN 4-G-15. This statement was published in *Die Umschau*, 25(17), 227. Cf. Weyl's report in ibid., 24(42), 609–11; Gehrcke's reply in ibid., 25(8), 99; Weyl's reply in ibid., 10, 123–24.

was deeply offended and therefore struck back. "Mr. Einstein himself is a philosophical layperson."[78] Kraus turned to an avenue that, until then, had been used primarily by Einstein's non-academic opponents. He wrote open letters to von Laue and Einstein in order to be noticed. In his open letter to Max von Laue, he emphasized that the area of his criticism was that of philosophy, and that he was only criticizing Einstein in this area, where Einstein had no expertise. This was not true, however; instead, Kraus repeatedly advanced into the area of physics criticism and conceded openly to Gehrcke that this was unavoidable for him as a philosopher.[79]

Despite the difficulties described, an astonishing amount of critical material was published on the theory of relativity (I refer here exclusively to fundamental criticism). Einstein's opponents, particularly when they came from academic physics and/or argued within the framework of classical physics (Gerold von Gleich, for example) and did not want to convert physics to solve the mystery of the universe, could certainly publish in the professional press and definitely in the journals addressing a wider audience, such as *Naturwissenschaften*, *Naturwissenschaftliche Wochenschrift* and *Umschau*, although not always without problems, as their correspondence with each other and with the editors shows.[80]

For example, Gehrcke clashed several times with the editor of *Naturwissenschaften*, Arnold Berliner. In 1914, he had written a notice about a statement Einstein had made on the clocks paradox in Ruben's Physical Colloquium in May 1914. He said that Einstein had admitted that accelerated motions would be absolute motions in the theory of relativity. The notice is unpolemical; in Gehrcke's view, Einstein had tangled up the theory of relativity in internal contradictions simply by characterizing accelerated motion as absolute motion. He now wanted to announce this to the professional world. However, Berliner stated that he would not print this notice at Einstein's request, as Einstein did not consider Gehrcke's very brief presentation to be adequate. Einstein stated to Berliner:

[78] Kraus 1925, p. 94.

[79] Ibid., pp. 94f. Cf. e.g. the comment made to Gehrcke about his lecture given at the "As-If" conference: "[…] it could not be done entirely without discussing the physics. If I make errors, one thing is certain: absurdities such as Einstein is committing […] I do not commit." Kraus to Gehrcke, June 6, 1920, GN 76-A-17. Cf. also Note 62 in Chapter 2.

[80] Gehrcke asked an engineer from Bremen, Max Drescher, to attempt to order the booklet *Massensuggestion [Mass Suggestion]* in a Bremen bookshop, first to see whether difficulties also existed in obtaining the work outside of Berlin, and second, to support Gehrcke by purchasing it. He said that the work could only still be obtained from a commission publisher "and the efforts to suppress this work extend up to this point." Gehrcke to Drescher, March 2, 1921, GN 71-G-11. In 1913, Gehrcke asked Kraus to recommend journals for the publication of an essay on the theory of relativity. Cf. Kraus to Gehrcke, October 17, 1913, GN 72-A-4; cf. also Palagyi to Gehrcke, February 20, 1923, GN 78-C-2. Vogtherr was referred by Gehrcke to the Heckner publishing house, which was, however, publishing one of Fricke's works at the time. Vogtherr therefore approached the W. Gente publishing house. Vogtherr to Gehrcke, July 27, 1922, GN 83-A-8. In 1932, Vogtherr was still having difficulties finding a publisher; several publishers had refused to publish *Das Problem der Gleichzeitigkeit [The Problem of Simultaneity]* "*without* having looked at the manuscript" even though Vogtherr offered a subsidy. Vogtherr to Gehrcke, August 10, 1932, GN 38-H-1.

I am entirely against this notice being published because – taken out of context – it can only cause confusion despite the fact that it is correct to a certain degree. The topic can only be made clear with a rather extensive explanation, as I recently gave in the colloquium.[81]

The fact that publishers forwarded such opinions directly to Einstein's opponents as justification for rejecting articles, as Berliner did to Gehrcke in this case, reinforced their view that the "relativists" were suppressing critical opinions. A longer dispute between Gehrcke and Berliner occurred at the beginning of 1919, when Berliner did not want to print Gehrcke's "Berichtigung über den Dialog über die Relativitätstheorien" [Correction of the dialogue about the theories of relativity], a reply to Einstein's "Dialog über Einwände gegen die Relativitätstheorie" [Dialogue on objections to the theory of relativity][82] in Gehrcke's desired prominent location as an opposing view. Gehrcke's article then appeared under the heading "Comments."[83] Gehrcke complained to Friedrich Poske, "Incidentally, the impression that Berliner must arouse in an impartial reader of his letter [which letter is meant is unclear], as if I would stand alone with my view on the Rel.th. like an obstinate eccentric, is entirely inaccurate!"[84] In 1927 and 1928, Gehrcke went so far as to file a lawsuit against Berliner in order to force publication of an opposing view on Gerlach's review of the *Handbuch der physikalischen Optik [Handbook of Physical Optics]* using legal means. Gerlach had found fault with the quality of the content of several contributions.[85]

Einstein's academic opponents also had difficulties with book publications. Thus Lenard remarked to Gehrcke about possible publishing houses for his work *Aether und Uraether*, "There is, [...] [besides Hirzel and Weicher] otherwise no sufficiently respectable publishing house for this case."[86] Although Lenard was dissatisfied with Hirzel, he ultimately saw no other possibility than to turn to this publisher. "I have concluded the contract with Hirzel; nothing better could be found."[87]

Mohorovičić was unsuccessful in placing his article "Die Folgerungen der allgemeinen Relativitätstheorie und die Newtonsche Physik [The implications of the general theory of relativity and Newtonian physics]" in the professional press; he had chosen *Physikalische Zeitschrift*, edited by Born and Debye. *Physikalische Zeitschrift* did not respond, and when Mohorovičić personally inquired of Debye

81 Quoted in: Berliner to Gehrcke, June 4, 1914, MPIWG Library, Papers No. 753/05, 4c. 82 Einstein 1918b.

83 Cf. Gehrcke, Ernst. (1919). Berichtigung zum Dialog über die Relativitätstheorien [Correction of the Dialogue about the Theories of Relativity]. *Die Naturwissenschaften*, **7**(9), 147–48. Berliner referred to Section 11 of the Press Act, which Gehrcke had been the first to refer to, and stated that he only had to correct facts, while the theory of relativity was a theory, not a fact. Berliner to Gehrcke, January 17, 1919, GN 83-B-8. Gehrcke also attempted to force publication of the response to Weyl in *Umschau* (cf. this chapter, Note 77) by referring to the Press Act.

84 Gehrcke to Poske, January 1, 1920, GN 83-E-9.

85 A file on this process is in Gehrcke's papers, GN 70-E. Cf. Gehrcke 1927; cf. Gerlach, Walther. (1927). [Review of Gehrcke 1927]. *Die Naturwissenschaften*, **15**(3), 410–12.

86 Lenard to Gehrcke, May 7, 1921, GN 3-F-10.

87 Lenard to Gehrcke, August 15, 1921, GN 3-F-15; cf. also Lenard to Gehrcke, July 24, 1921, GN 83-F-2.

whether he had received the paper, Debye stated that the paper was unsuitable for *Physikalische Zeitschrift*. Mohorovičić suspected that his clear attacks on the theory of relativity in that article were the cause for rejection.[88] He was unwilling to accept the fact that the article was in truth not rejected because Debye wanted to prevent justified criticism of the theory of relativity, but rather because it did not reflect the state of the professional discussion in physics. It was the same for Guillaume, who likewise wanted to publish an article with comments on his discussion with Einstein in Paris (cf. pp. 186ff.) in *Physikalische Zeitschrift*. This caused Born to inquire of Einstein what he thought of Guillaume's article, whether he should print it, and whether Einstein would have any additions. The answer reveals an Einstein disillusioned about whether disputes with the fundamental critics of his theory were worth the trouble:

Dear Born,
Guillaume is crazy and was also completely dotty in Paris. I have taken great pains to convince him, likewise Langevin. Everything is for naught. And then Mohorovičić is also so unhappy. I wouldn't print it. But if you find it more comfortable than rejecting it, that's also fine with me.[89]

The dispute with the professional press did not only relate to the German-speaking area. Einstein's American opponents even got their debate about "Einstein plagiarized Soldner" (cf. pp. 277ff.) into *Science*, but it was not of their own doing. After T. J. J. See had published articles in the American daily newspapers accusing Einstein of having plagiarized Soldner, William Wallace Campbell, the director of Lick Observatory, asked his colleague Robert Trumpler to write a correction, which he was happy to do – and his reply to See was then published in *Science*.[90] Because of this, See in turn had an opportunity to print a brief response in *Science*.[91] Reuterdahl had of course read See's publication in *Science* with great interest; after all, he was the one who had publicized the accusation of plagiarism and Soldner's work in the US (cf. p. 280). At any rate, Reuterdahl hoped that an "open" debate could now take place in *Science* (an article by Luther P. Eisenhart on "Einstein and Soldner"[92] had appeared on December 21, 1923) and wrote an article

88 Mohorovičić to Gehrcke, July 30, 1921, GN 3-G-4. The article then appeared in *Naturwissenschaftliche Wochenschrift*, New Series, **20**(52), 737–39 (also as an offprint).

89 Einstein to Born, [June 1922], the copy of the letter from Born to Einstein dated June 16, 1922, and Einstein's reply can be found in Gehrcke's papers, GN 4-B-6. Neither letter is included in Born, Max and Einstein, Albert. (1991). *Briefwechsel [Correspondence] 1916–1955*. Munich. Handwritten comment by Gehrcke on the back of the copy: "Transcript (reviewed) of the 'copy' of the original letter signed by Born and the handwritten Einsteinian answer that is added at the bottom on the same sheet on the original. Original owned by Prof. Guillaume in Berne."

90 Trumpler, Robert. (1923). Historical note on the problem of light deflection in the sun's gravitational field. *Science*, **58**(1496), 161–63.

91 See, T. J. J. (1923c). Soldner, Foucault, and Einstein. *Science*, **58**(1506), 372.

92 Eisenhart, Luther P. (1923). Einstein and Soldner. *Science*, **58**(1512), 516–17.

on "Einstein's Predecessors"[93] containing replies to Trumpler and Eisenhart for publication in *Science*. He told Mohorovičić about the hopes that he had initially set on *Science*: "I was gratified to learn that one scientific journal in the United States, at last, had come to the conclusion that further suppression of the historical facts concerning the development of Einsteinism was directly contrary to the principles of plain honesty and the high ideals that modest science claims to represent."[94]

The reply by Cattell, the editor, turned out to be different from what Reuterdahl had initially expected; Reuterdahl quoted from Cattell's reply, "it is undesirable to print the enclosed manuscript in *Science*" – the reviewers had rejected it, he said. The reviewers at *Science* knew that the content of Reuterdahl's priority claims in favor of Soldner and Gerber, which he asserted in his manuscript intended for publication in *Science*, were untenable for academic physics; but to Reuterdahl, this was inconceivable as a possible explanation. He was only able to explain the refusal to print his text by assuming that pressure had been exerted on Cattell. "I am led to infer that his action, which he seems to regret, is due to the pressure of extraneous advice," and he protested vehemently against *Science*, "whose editorial policies seem to be controlled not by the editor but by certain 'advisors' whose intentions [...] seem to be the exclusion of matter detrimental to Einstein and Einsteinism." How high the stakes were for Reuterdahl is apparent in his urgent plea that *Science* should not keep out of this controversy, because all of science would be in danger:

> The paramount purpose of science is the discovery of truth both historical and factual. [...] This incident reveals the presence of a malignant disease which threatens to destroy the very existence of science, intimidate the seekers for truth, and throttle all incentives to honest investigation. It is high time that sincere and honest scientists arise and protest, in no uncertain terms, against this corrupt condition before intellectual freedom and scientific honesty is completely crushed.

For Einstein's opponents, marginalization and exclusion legitimized their protests, as will be clearly shown in the presentation of the protest campaigns (cf. pp. 266ff.). Ziegler, who liked to attribute his failures to the direct influence of individuals, laid the blame for the fact that he was unsuccessful in publishing in the professional journals on the influence of his former teacher, Adolf von Baeyer, "who is currently reigning in Munich like Plutarch once did in Delphi."[95] It was improbable that von Baeyer, the 1905 Nobel Prize winner in chemistry and one of the leading lights of organic chemistry, actively tried to suppress the primordial light theory as Ziegler supposed. Instead, a brief look at Ziegler's papers, which use occult arguments, clearly shows

93 Reuterdahl, Arvid. ([1923d]). Einstein's Predecessors. [unpublished typescript], RP 3–88. A different article by Reuterdahl appeared under the same title in the *New York Times* (Reuterdahl 1923c).
94 All of the following quotes from: Reuterdahl to Mohorovičić [draft], [beginning of 1924], RP 3–88.
95 Ziegler 1908a, p. 17.

that no serious editorial office would have published any such article by Ziegler. While more unusual conceptions about "The true cause of the bright light radiation of radium"[96] could be published in the popular science and world view press, for instance in the occult *Psychische Studien* periodical, the established scientific journals opened themselves up to attacks to an extreme degree by including world riddle solver theories – as is shown by a letter from Paul Kirchberger, a mathematician in Berlin, to the editorial staff of *Forschungen und Fortschritte*:

> Some time ago you published a paper by Dr. Fricke, Senior Government Councilor at the Reich Patent Office, about an astrophysical law he had discovered [the connection between temperature and gravity]. As far as I can assess it, the work on which the paper is based can in no way be counted as scientific literature. Since I have always considered your worthy journal to be absolutely dependable, I was very surprised to find the said paper in it. It would very much interest me to hear whether I am mistaken in my appraisal of this paper, or whether you committed an error in including it.[97]

How *Forschungen und Fortschritte* reacted is unknown. But Einstein's opponents, confronted with rejection by the professional press, sought new forums for publication.

Publication niches

In their booklets, Einstein's opponents often described in minute detail when their papers had been rejected: by which journal and with what "ridiculous" justification – even though they had approached a broad spectrum of professional and popular science publication venues. The works of the retired senior military medical officer Johannes Haedicke, for instance, were rejected by *Zeitschrift für Physik*, *Unsere Welt* (the journal of the Kepler League), *Natur und Museum* (the journal of the Senckenberg Natural Science Society), *Geologische Rundschau*, *Koralle – Wochenzeitung für Unterhaltung, Wissen, Lebensfreude [Weekly Newspaper for Entertainment, Knowledge, Joie de Vivre]*, *Naturwissenschaften*, *Physikalische Zeitschrift*, and *Wissen und Fortschritt – Populäre Monatsschrift für Technik und Wissenschaft [Popular Monthly Magazine for Technology and Science]*, among others. This last magazine was the only one that issued a rejection with a justification that was positive from Einstein's opponents' point of view, namely that they were already publishing a critical piece on the theory of relativity in the upcoming issue. Haedicke's example also shows that Einstein's opponents did have a preference for professional journals, but ultimately attempted to publish in a broad spectrum of scientific, popular science, and world view publications.

[96] Cf. Ziegler, Johann Heinrich. (1905). Die wahre Ursache der hellen Lichtstrahlung des Radiums [The true cause of the bright light radiation of radium]. *Psychische Studien*, **32**, 449–55.

[97] Kirchberger to the editorial office of *Forschungen und Fortschritte*, sent to Gehrcke for his information, April 28, 1932, GN 38-I-7.

The best thing, of course, was when they had a journal of their own. Mewes was the editor of two engineering periodicals, *Turbine* and *Zeitschrift für Sauerstoff- und Stickstoff-Industrie [Journal of the Oxygen and Nitrogen Industry]*, which he also used for publishing his own and others' articles against the theory of relativity.[98] Ludwig Glaser, as the publisher of *Glasers Annalen für Gewerbe und Bauwesen [Glaser's Annals for Trade and Construction]*, could likewise accommodate his own articles and those of others.[99]

Gehrcke was friends with the publisher of *Kant-Studien*, Max Frischeisen-Köhler, since they had attended school together; they were very close, and when publications on the theory of relativity were forthcoming in *Kant-Studien* – including some by Einstein himself – Gehrcke was informed in advance. Gehrcke in turn was able to recommend his own articles or articles by other opponents of Einstein directly to "Dear Max" for *Kant-Studien*.[100] Due to his earlier work as a journalist,

98 Cf. e.g. Mewes, Rudolf (1919). Unterschiede zwischen der Relativitätstheorie von Mewes (1892/1893) und Lorentz (1895) [Differences between the theory of relativity of Mewes (1892/1893) and Lorentz (1895)], Part 1. *Zeitschrift für Sauerstoff- und Stickstoff-Industrie [Journal for the Oxygen and Nitrogen Industry]*, **11**(17–18), 70, Part 2. Ibid. **11**(19–20), 75–76; Idem. (1920c). Bemerkung zur Ableitung der Grundformeln der Relativitätstheorie [Comment on the derivation of the basic formulae of the theory of relativity]. Ibid. **12**(1–2), 6; Idem. (1920d). Bestimmung der Fortpflanzungsgeschwindigkeit der Schwerkraft nach Gerber und Einstein [Determination of the speed of propagation of gravity according to Gerber and Einstein]. Ibid. **12**(3–4), 13–14; Idem. (1920e). Die Versuchsformel von Fizeau und die Relativitätsgleichung von Einstein [Fizeau's experimental formula and Einstein's relativity equation]. Ibid. **12**(9–10), 38; Idem. (1920f). Elementarer Weg zur Ableitung der Einsteinschen Grundgleichungen [Elementary path to derivation of the Einsteinian basic equations]. Ibid. **12**(13–14), 65; Idem. (1921a). Lichtablenkung in bewegten Mitteln [Light deviation in moving media]. Ibid. **13**(1–2), 1–4; Idem. (1921b). Zur Beurteilung der abendländischen Naturerkenntnis nach Spengler (Entropie, Gas- und Relativitätstheorie) [On assessing occidental knowledge of nature after Spengler (entropy, gas theory, and relativity theory]. Ibid. **13**(11–12), 42–44; Idem. (1921c). Lenards und Reuterdahls Stellungnahmen zur Relativitätstheorie [Lenard's and Reuterdahl's opinions on the theory of relativity]. Ibid. **13**(17–18), 77–78; Idem. (1922b). Reuterdahl vs. Einstein. Festnagelung einer Täuschung [Reuterdahl vs. Einstein. Nailing a fallacy]. Ibid. **14**(5), 55; Idem. (1922c). Vergleichende Gegenüberstellung der Exponentialformeln von Mewes, Planck, Einstein und Nernst für die spezifische Wärme [Contrastive comparison of the exponential formulae of Mewes, Planck, Einstein, and Nernst for specific heat]. Ibid. **14**(10), 121–23; Idem. (1922d) Stimmen gegen die Relativitätstheorie [Voices against the theory of relativity]. Ibid. **14**(11), 134–35; cf. Fricke, Hermann. (1919). Eine neue Erklärung für die Ursache der Barometerschwankungen und des Windes [A new explanation for the cause of barometer fluctuations and wind]. Ibid. **11**(13–14), 49–53; Idem. (1922). [Review of Karollus 1921]. Ibid. **14**(1), 18. Cf. also Mewes' inquiry to Gehrcke as to whether he would be allowed to publish Gehrcke's treatise "Die gegen die Relativitätstheorie erhobenen Einwände [The objections raised against the theory of relativity]," which Gehrcke had sent to him, in *Zeitschrift für Sauerstoff- und Stickstoff-Industrie*. Mewes to Gehrcke, October 10, 1919, GN 38-H-9. The Association for the Oxygen and Nitrogen Industry also asked Fricke for a lecture. Cf. Fricke to Gehrcke, October 24, 1920, GN 76-A-13.

99 Glaser, Ludwig. (1920). Ueber Versuche zur Bestätigung der Relativitätstheorie an der Beobachtung [On attempts to confirm the theory of relativity using observation], Part 1. *Glasers Annalen*, 87(4) (entire series No. 1036), 29–33, Part 2. (1921). Ibid. 88(2) (entire series No. 1048), 30–32, and (3) (entire series No. 1049), 42–43; Fricke, Hermann. (1920). Eine neue und anschauliche Physik des Aethers [A new and comprehensible physics of the aether]. Ibid. 86(12) (entire series No. 1032), 95–96; Gehrcke, Ernst. (1921). [Review of Kraus, Lipsius, Linke, and Petzoldt 1921]. Ibid. 88 (entire series No. 1053), 80.

100 Frischeisen-Köhler wanted to send questions that *Kant-Studien* wanted to submit to Einstein, to Gehrcke, beforehand "for review." Frischeisen-Köhler to Gehrcke, March 7, 1919, GN 68-D-6. In 1914, Gehrcke himself published "Die erkenntnistheoretischen Grundlagen der verschiedenen physikalischen Relativitätstheorien [The epistemological foundations of the various physical theories of relativity]" in *Kant-Studien*, 19(4), 481–487, reprinted in Gehrcke 1924a, pp. 35–40 (also as an offprint). Cf. also Riedinger to Gehrcke: "The fact that I can count on your support in the Kantstudien, if necessary, makes me very glad." December 9, 1920, GN 71-K-13.

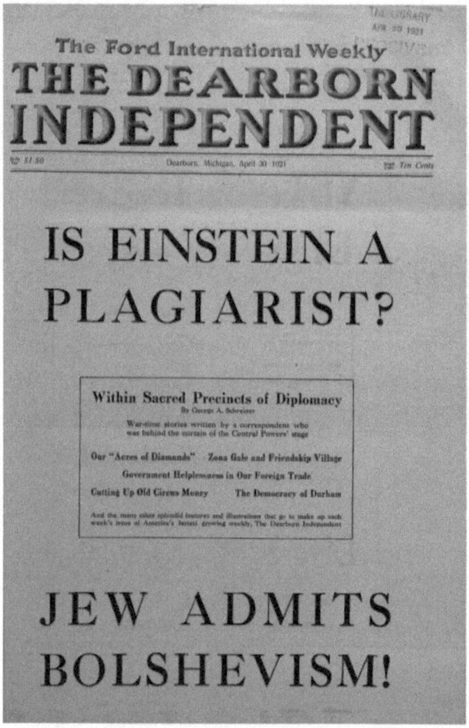

The Ford International Weekly

THE DEARBORN INDEPENDENT

Dearborn, Michigan, April 30 1921 — Ten Cents

IS EINSTEIN A PLAGIARIST?

Within Sacred Precincts of Diplomacy
By George A. Schreiner
War-time stories written by a correspondent who was behind the curtain of the Central Powers' stage

Our "Acres of Diamonds" — Zona Gale and Friendship Village
Government Helplessness in Our Foreign Trade
Cutting Up Old Circus Money — The Democracy of Durham

And the many other splendid features and illustrations that go to make up each week's issue of America's fastest growing weekly, The Dearborn Independent

JEW ADMITS BOLSHEVISM!

55. JAHRGANG

Deutsche Optische Wochenschrift
und
CENTRAL-ZEITUNG für OPTIK und MECHANIK

Nr. 29 — DOCENTRA — 22. Juli 1934

DRUCKEREI UND VERLAGSANSTALT R. BORKMANN • WEIMAR · JAKOBSTRASSE 2

Die Lehre Einsteins stürzt

Hermann Fricke's Kampf:

Die im Innern erdähnliche Sonne

Eine neue Anschauung von Aether, Schwe... aft und Sonne

VON DR. PHIL. HERMANN FRICKE
Oberregierungsrat und Mitglied des Reichspatentamts

Brosch. 73 Seiten. — Mit 10 Abbildungen. — Preis RM. 1.25

VERLAG R. BORKMANN, WEIMAR, JAKOBSTRASSE 2
(Vergl. Leitartikel in diesem Heft)

Fig. 4.2 Alternatives to the professional physics press: criticism of Einstein in *The Dearborn Independent*, Michigan, USA, April 30, 1921, and in *Docentra*, Weimar, Germany, July 22, 1934.

Palagyi had the best of relationships with the Frankfurt newspapers, in which he was able to place extensive articles about, for instance, Gehrcke's or Lenard's work.[101] These examples show that personal relationships played a prominent role in locating such – sometimes odd – niches for the criticism of Einstein. After he had personally visited the editors of *Deutsche Optische Wochenschrift (Docentra)*, Fricke, for example, published his papers in the 1930s and 1940s primarily in that venue – a reputable and established journal for the optical profession.[102]

Reuterdahl's alternative forum was less odd than it was politically questionable. Through the agency of Robert Stevenson, aka Kinertia, Reuterdahl received the

[101] Cf. Palagyi, Melchior. (1921a). Physik und Erkenntnistheorie [Physics and epistemology]. *Frankfurter Nachrichten* newspaper, August 7, morning edn., GN 4, XII, 87r; Idem. (1921b). Lenard und Einstein. *Frankfurter Nachrichten* newspaper, November 12, evening edn., GN 4, XII, 71r.

[102] On the occasion of Fricke's 60th birthday on December 16, 1936, an article entitled "Hermann Fricke's 30jähriger Krieg [Hermann Fricke's 30-year war]" appeared in *Docentra*, in which the journal not only acknowledged Fricke's struggle for recognition for his ether theory against the "prevailing Planck–Einstein movement," but also recapitulated its acquaintance with Fricke. Cf. Hermann Fricke's 30jähriger Krieg [Hermann Fricke's 30-year war]. *Docentra*, 57(25), 421–22 (see p. 421).

opportunity to design a science page for the *Dearborn Independent.*[103] The *Dearborn Independent* had been owned by Henry Ford since 1919, and Stevenson knew Ford's son Edsel, who had been running the Ford automotive empire since 1919. Even though a broad spectrum of topics was covered on the whole, the *Dearborn Independent* was an anti-Semitic paper – Ford's anti-Semitic pamphlet *The International Jew*[104] was a compilation of articles that had appeared in the *Dearborn Independent* since 1920. Publishing there was not necessarily the best testimonial for a scientist. Henry Garrity, president of the Devin-Adair Company, which published Reuterdahl's books, worriedly approached Reuterdahl's employer, the Catholic University of St. Thomas in St. Paul, Minnesota, stating, "Mr. Reuterdahl's attack on Einstein in that Anti-Jewish paper will work serious harm to him and us."[105] However, Garrity was more worried about the thought of vociferous protests from the Jewish side than about the fact that his author was publishing in the periodical of a known anti-Semite. From 1921 to 1926, Reuterdahl dealt with a broad field of scientific and semi-scientific topics on his science page, "International Science Briefs" in the *Dearborn Independent.* Along with popular science articles on the latest developments in technology ("The Fischer tide predicting machine", January 20, 1923), biology ("Protoplasm and heredity", November 10, 1923), or astronomy ("The origin and motion of planetary systems", September 15, 1923), multiple contributions against the theory of relativity were published.

The starting point was the article "Kinertia vs. Einstein" with the accusation of plagiarism that has already been discussed (cf. pp. 171ff.) relating to the general theory of relativity (April 30, 1921). In the article "What the Einstein test can and cannot prove" (October 28, 1922), objections were raised against the meaningfulness of the confirmation of light deviation as proof of the correctness of the general theory of relativity, and accusations of plagiarism were made once again that related particularly to the theories of Soldner and of Reuterdahl himself. The article "Reuterdahl vs. Einstein: Nailing a Fallacy" (March 25, 1922) by Reuterdahl's friend and follower Heidenreich was dedicated entirely to the plagiarism of Reuterdahl's own theory. "Einstein and the Nobel Premium" (January 6, 1923) opposed the award of the Nobel Prize to Einstein and made a big deal of the accusations of plagiarism asserted in advance by Oskar Westin. In "The Einstein film and the debacle of Einsteinism" (March 22, 1924), Reuterdahl took the "Einstein film"[106] as an occasion for an all-encompassing criticism of the theory of relativity.

[103] Cf. Stevenson to Reuterdahl, December 30, 1921, RP 4–56.
[104] Ford, Henry. (1922 [1920–1922]). *The International Jew. The World's Foremost Problem*. Dearborn.
[105] Garrity to Reverend Bush, May 28, 1921, RP 3–23. [106] Cf. Note 23 in Chapter 2.

The readers of the *Dearborn Independent* were not only informed by Reuterdahl about the theory of relativity and "Einsteinism"; Robert Morgan published an emphatic article on the "International Science Briefs" page about T. J. J. See, "Einstein theory declared colossal humbug by U.S. naval astronomer" (July 21, 1923), and See himself published a contribution about his view of light and ether in which he claimed to have proven that the theory of relativity was completely unfounded.[107]

With a circulation of about 750,000, the *Dearborn Independent* was the only channel of publication, along with the daily newspapers, for Reuterdahl's critical views of the theory of relativity that had a large range of coverage. At the same time, he was conscious of the fact that the *Dearborn Independent* was not a professional journal and only represented a makeshift solution. Following a bitter complaint by Mohorovičić about the lack of opportunities for publishing a paper about the solution of Euler's hydrodynamic equations, Reuterdahl told him, "If this development dealt with a subject which could be presented in a semi-popular manner I could promise you a proper place in The Dearborn Independent. Unfortunately that is not the nature of the subject." In response, Mohorovičić wrote a substitute article, "Worthlessness of Einstein's theory of relativity for the physics," his article with the sharpest tone against Einstein, for publication in the *Dearborn Independent.*[108]

Along with the popular science and world view press, which will be discussed in the next section, it was the daily newspapers that offered Einstein's opponents a forum for a variety of reasons. Political orientation was one factor; criticism of Einstein is usually found in nationalistic newspapers. However, the goal of printing a sensational headline ("Theory of relativity refuted!") or the pride of a local newspaper in its local opponent of Einstein ("Minnesota man calls relativity 'bunk'"), or both, also played a role.[109] The editors of nationalistic newspapers took sides for Einstein's opponents, particularly when they were likewise

[107] See, T. J. J. (1925). Travelling at speed of 294.375 miles a second. Scientist believes etherons provide pathway for light and radio waves. *The Dearborn Independent*, January 10. The first – and least polemical – article on the theory of relativity in the *Dearborn Independent*, however, was written before Reuterdahl's connection to the *Dearborn Independent* and discussed the lectures opposing the theory of relativity organized by Weyland, given at the Berlin Philharmonic. Cf. Einstein theory causes scientific war. *The Dearborn Independent*, October 30, 1920.

[108] Reuterdahl to Mohorovičić, May 24, 1923, RP 4–35; Mohorovičić to Reuterdahl, June 27, 1923, RP 4–35 (with the article as an attachment). The article was not published.

[109] "Einstein branded Barnum of science, Minnesota man calls relativity 'bunk.'" *The Minneapolis Sunday Tribune*, April 10, 1921. Pride in the local opponent of Einstein came into play, particularly for Ziegler of Switzerland and for Reuterdahl, living in Minnesota. Thus a portrait of Ziegler appeared in the *Zürcher Chronik*: "The 'Zürcher Chronik' can […] not have the task of introducing its readers to the details of Dr. Ziegler's articles; it also does not want to be involved in any polemic against Einstein. All the same, it does not want the work of a Zurich native and Swiss citizen to pale entirely into insignificance beside the somewhat suspiciously chauvinistically exploited theory of the German scholar […] The modest and yet justifiably self-confident Swiss scholar wants nothing more than to be heard. It is our task to hear him, to test him, to criticize him, and not to shy away from setting him against Einstein. Whether they will bless or curse us when the time comes to make the final judgment is unimportant; our duty is to keep Zurich and Switzerland from suffering!" "Dr. J. H. Ziegler." *Zürcher Chronik* newspaper, July 6, 1921.

nationalistically minded. The accusation of plagiarism by Mewes, a German nationalist, for instance, was defended in the *Deutsche Zeitung* in 1919 in a long article entitled "Scientific theft and bluff."[110] But an editor in the middle-class press could also be inclined against the theory of relativity and prevent it from coming across all that well in his newspaper. For instance, the *Leipziger Tageblatt* stated, "The position of the 'Frankfurter Zeitung' was always against Einstein because the editorial philosopher there rejected him."[111]

4.1.7 The scientific associations and their publications

The first chapter discussed the fact that the world riddle solvers can be placed in the context of the popularization of science, and they did in fact seek out the forums of the large and small scientific associations in order to present their solutions to the universal mysteries and to refute Einstein's "competing theory."[112] In this context, the diversity of the scientific and popular science associations is very significant, ranging from experiencing nature in everyday life up to representation of a professional group.[113] It made a difference whether someone who understood himself as an amateur reported about his most recent hike in the woods at the botany club, or whether a self-proclaimed world riddle solver applied to give a lecture in the physics section at a large natural science association like the SNG or the GDNÄ on eternal primordial light or gravitational radiation as the solution for all problems being discussed in physics: the self-proclaimed world riddle solver was in all probability rejected. In fact, the legitimacy of the founding of separate organizations[114] (cf. pp. 250ff., pp. 293ff.) was often derived from the fact that the traditional and established organizations did not offer a suitable platform either for solving the world riddles or for refuting the theory of relativity. While it was largely a new experience for Einstein's academic opponents to encounter resistance to their views in the scientific associations, the world riddle solvers were used to it.

[110] "Prof. Einstein was still attending school when the general theory of the relativity of forces or the generally applicable spacetime theory was developed, and thus cannot be considered the great intellectual hero with original creative power that he is being celebrated as everywhere today. Einstein is not a self-creator, but rather a parrot and a knowledge leech." Wissenschaftsraub und Bluff [Scientific theft and bluff]. *Deutsche Zeitung* newspaper, December 19, 1919, quoted from Mewes 1920a, pp. 74–76 (see p. 75).

[111] R[udolf] L[ämmel]. (1924). [Review of Gehrcke 1924b]. *Leipziger Tageblatt* newspaper, September 14, GN VII. The "editorial philosopher" could have been Melchior Palagyi or Walther Rauschenberger, the director of the Senckenberg Library, who published several critical articles on the theory of relativity in the *Frankfurter Zeitung* newspaper. Cf. e.g. Rauschenberger, Walther. (1921). Die philosophische Bedeutung der Relativitätstheorie [The philosophical significance of the theory of relativity]. *Frankfurter Zeitung* newspaper, June 21, 1st morning edn., GN 4, XI, 36r–37r1.

[112] Ziegler 1931, p. 20. [113] Cf. Daum 2002, pp. 89ff.

[114] Cf. also the assessment in Daum 2002, p. 110: "Where the amateur scientists once again encountered such exclusive tendencies [in the nature associations], the way remained open for them to become passive, to become readers, or to take the opportunity to use alternative options from the spectrum of associations or to create such options themselves."

The supra-regional scientific associations

Ziegler was the most persistent in seeking a connection to academic science through the associations. During his career as a chemist, Ziegler was not a member of the SNG; he only joined in 1901 as a world riddle solver with the express goal of finding a forum for his work on the "universal unified theory," and with the expectation of being acknowledged by the luminaries as the solver of all of the problems existing in science. Ziegler's first application to give a lecture, in 1901, was the only successful one in the next 35 years of his career as a world riddle solver. The lecture was assigned to the physics section and ended in catastrophe for Ziegler, who was rooted in organic chemistry and understood little of either physical chemistry or physics. There were no kindly disposed acquaintances from his university days among the physicists, as there were among the chemists, who leniently overlooked his world riddle solving. He was unknown there and not acknowledged as a professional scientist. It is not surprising that his first lecture had significantly different consequences than he had hoped for, as Ziegler describes in retrospect, "All of that [what he presented] was so unheard of at the time that it was not clear to the gentlemen whether I was speaking in earnest or in jest, or whether someone suffering from an idée fixe was making the section unsafe."[115]

Ziegler talked for around 15 minutes about the relationship of light and matter, color and chemical constitution, which were solved in his opinion by his unified theory, but when he commenced subdividing the colors into "masculine" and "feminine" according to his formula of absolute complementarity, "a real commotion broke out. The portly and generally admired chairman himself, Prof. Hagenbach-Bischoff of Basel, brayed with laughter so much that it shook his whole body."[116] On the whole, the result was, according to Ziegler's own assessment, "shattering."[117] Ziegler attempted to convince the physicists of his unified theory at a total of three annual meetings of the SNG in 1901, 1902, and 1903. In 1903, at the SNG annual meeting in Locarno, there was an altercation. Ziegler had registered two lectures and was again confronted by the physicist Eduard Hagenbach-Bischoff (1833–1910), who led the section. Hagenbach-Bischoff placed the lectures at the end of the session with the justification that they dealt less with physical issues than with philosophical issues. According to Ziegler's report, when Ziegler began to speak as the last speaker in the session, Hagenbach-Bischoff began to disrupt the lecture by chatting loudly with his assistant Veillon. In response, Ziegler immediately resigned from the SNG.[118]

Ziegler's brief career in the SNG had foundered after three years. In the following years, he repeatedly attended the meetings of the GDNÄ and had discussions with van t'Hoff and Ostwald, among others – but was unable to convince them to introduce the

[115] Ziegler 1914, p. 143. [116] Ziegler 1935, p. 31. [117] Ibid. [118] Cf. ibid., p. 32.

metaphysical primordial atom into chemistry. From around 1912, Ziegler resignedly no longer attended any scientific conferences and meetings; it was only in 1922 on the occasion of the protest by Einstein's opponents at the centennial celebration of the GDNÄ (cf. pp. 266ff.) that he again participated in such an event.[119]

Fricke, who had studied physics, was a member of the DPG, but soon found himself at odds with it. In 1906 and 1909, he gave lectures at the Berlin Physics Society in which he discussed his theory of the ether as a fluid with internal friction, but while the 1906 lecture was published in the Proceedings of the Deutsche Physikalische Gesellschaft,[120] the 1909 lecture was rejected by the chairman, Planck, with the justification that the significant material had already been contained in the first lecture. "I see this rejection of different opinions as suppression by Planck,"[121] said the snubbed Fricke. In fact, Fricke saw his chief opponent more in Planck, the theoretician and science executive, than in Einstein; he thus primarily polemicized against "Planck physics" and "Planck physicists" (among whom he of course included Einstein).[122]

In 1919, he brought another dispute with the DPG upon himself. He protested in a letter to the executive committee against the fact that the printing of a paper submitted in June 1919 had "been illegally prevented."[123] This protest, made by Fricke as a member of the DPG, was dealt with on June 24, 1919, in a meeting with Fricke, but although he was now able to defend his views personally before leading physicists, as he had so often demanded, in this case he did not agree with the process, since it was not the chairman (Arnold Sommerfeld at this time) who ran the meeting, but rather Karl Scheel, who had been managing director and secretary of the DPG since 1900. Fricke gave Scheel a brief lecture on the connection between temperature and gravity, but was then informed that his manuscript was not suitable for printing in the proceedings of the DPG.[124]

Following this first rejection, Fricke tried a second time the same year to publish works in conjunction with the DPG, but the DPG rejected his complaint in a notice dated December 19, 1919, that was based on an expert opinion by Arnold Sommerfeld and Wilhelm Westphal. For Fricke, this notice, just like "the behavior of the Executive Board" was "an insult to a member of the Society that was justified

119 Cf. the extensive report on his failure that Ziegler provided in the form of a letter to Gehrcke. Ziegler to Gehrcke, February 11, 1931, GN 83-L-4.

120 Cf. Fricke, Hermann. (1906). Über einen Versuch der Gebrüder Weber, transversale Luftschwingungen betr. [On an experiment by the Weber brothers relating to transverse air vibrations]. *Verhandlungen der Deutschen Physikalischen Gesellschaft*, **8**(12), 248–50.

121 Fricke [ca. 1942]. 122 Cf. ibid.

123 *Protest des Mitgliedes der Deutschen Physikalischen Gesellschaft Reg. Rat Dr. Fricke gegen die ihm vom Vorstande der Gesellschaft zu Teil gewordene Behandlung [Protest by the Member of the Deutsche Physikalische Gesellschaft, Gov't Councillor Dr. Fricke, Against his Treatment by the Executive Board of the Society]*, Fricke to the executive committee of the DPG, [September 1920], [carbon copy], GN 71-B-5.

124 Cf. ibid.

by nothing."[125] Here, he took the position of the "true scientist" – he had developed a scientifically formulated theory and did not understand why it was not being discussed. The DPG, however, did not get involved in any debate about Fricke's unconventional new definitions of physics concepts and his claims of being scientific. The Executive Board's notice read, "As you yourself will admit, the assumptions underlying your observations, as well as numerous comments in your paper, entirely contradict what the majority of physicists consider to be correct and proven today."[126] Fricke's immunity to such rejections is apparent in his scandalized statement, "I am unable to find any connection between this comment and the paper I submitted."[127]

Oskar Kraus likewise wished to move beyond the narrower circles of philosophy with his criticism of Einstein and to register for a lecture at the annual meeting of the GDNÄ in Innsbruck in 1924, but he was told that no lectures on the theory of relativity would be accepted for this conference:

> I refrained from pursuing the matter any further after I found out from other colleagues what difficulties are encountered at conferences by the opponents of the theory of relativity who wish to speak there, even if they are professors of the mathematical-scientific disciplines at German universities! What could an opposing philosopher hope for![128]

Publication niches in the popular science press

Opportunities for publication were available primarily through the popular science and world view associations, many of which had their own journals. Many world riddle solvers felt at home in monism, the most influential world view movement asserting a scientific nature at the beginning of the twentieth century (cf. pp. 35ff.). However, monism was predominantly concerned with disseminating the theory of evolution and thus dealt more with the areas of biology and religion, and less with the area of physics.

Moreover, the monistic movement did not have a unified world view and did not consist only of the Monist League; instead, many monistic splinter groups existed with various focus areas. Their adherents could have various conceptions of the theory of relativity or might not be particularly interested in modern physics at all – which was often the case. There is little on the theory of relativity in the *Monistische Monatshefte*, the official publication of the Monist League, in the years 1920–1922; the characteristic topic is the relationship between monist world view and religion. A popular introduction to the theory of relativity appeared in 1920 (as in almost all other periodicals).[129]

[125] Ibid [126] Executive Board of the DPG to Fricke, December 19, 1919, [copy], GN 76-A-13.
[127] Fricke 1920e. [128] Kraus 1925, pp. 5f.
[129] Cf. Schmidt, Harry. (1920). Wege zum Verständnis der Einsteinschen Relativitätstheorie [Paths to understanding the Einsteinian theory of relativity], Part 1. *Monistische Monatshefte*, **5**(6), 254–60; Part 2. Ibid. **5**(7–8), 305–11; Part 3. Ibid. **5**(9–10), 359–62; Part 4. Ibid. **5**(11–12), 404–8.

The reporting was very different in *Neue Weltanschauung*, the periodical of the Humboldt League, which was the most significant monistic organization alongside the Monist League. In *Neue Weltanschauung*, the reporting on the theory of relativity was done almost exclusively by two well-known opponents of Einstein, Henri Gartelmann and Erich Ruckhaber. The first critical discussion of the theory of relativity was published by Gartelmann in the June 1920 issue. In it, he argues primarily against Einstein's conception that the question about the truth of the axioms of geometry would not lead anywhere. For Gartelmann, on the other hand, geometry was based on experience and logic and was certainly capable of being truth because it could be verified experimentally. After all, it could be shown that several straight lines could represent the shortest connection between two points; in that case, Euclidean geometry would no longer be true, but rather false.[130] Just like Kraus, Gartelmann shifted the ability to assess the "logic" of the theory of relativity, with which it would stand or fall, into the field of philosophy. To him, length contraction and time dilation were only theoretical constructs and mere assertions.[131]

Starting about 1912, Erich Ruckhaber was a member of the editorial staff of *Neue Weltanschauung*, covering a broad spectrum in his articles. The topics ranged from "The conclusive logic of monism" through "Money, socialism, and intelligence" to "New production of energy."[132] Ruckhaber only started publishing booklets against the theory of relativity in 1928, but he took a position on the theory of relativity in *Neue Weltanschauung* as early as 1920, assigning Einstein to conceptual-abstract thinking (which was to be rejected), while asserting a sensualist position himself. He had spent several years in Africa as a businessman and was interested in the natives' conception of nature. He frequently referred to these impressions in order to legitimize his appeals to "common sense," which he saw embodied in the natives, whom he conceived of as original, unspoiled "children of nature."[133]

From 1920, almost every issue of *Neue Weltanschauung* includes one or more critical opinions on the theory of relativity, either in the form of articles by Gartelmann and Ruckhaber in the main section, discussions of lectures such as those given by Fricke, or reports about the discussion in Bad Nauheim, for example, in the column "Review."[134] In addition, *Neue Weltanschauung* offered a forum for

130 Cf. Gartelmann 1920b, Part 1, p. 165. He referred to Einstein 1917a, p. 1.
131 Cf. ibid. and Gartelmann 1920b, Part 2.
132 In *Neue Weltanschauung*, **6**(4), (1913), 125–28; ibid., **9**(5), (1920), 141–45; ibid., **9**(10) (1920), 303–4.
133 Ruckhaber 1920a, p. 2.
134 Cf. Gartelmann, Henri. (1920c). Zur Relativitätstheorie [On the theory of relativity]. *Neue Weltanschauung*, **9**(11), 341–42; cf. Anon. (1920). Die Relativitätstheorie auf der Naturforscher-Versammlung in Nauheim [The theory of relativity at the annual meeting of the German Society of Natural Scientists and Physicians in Nauheim]. *Neue Weltanschauung*, **9**(11), 342–44.

world riddle solvers' theories. Thus Fricke, Beckenhaupt, and Vogt[135] published there, and reports were also printed about Ziegler's and Fricke's work.[136]

The fact that a critical attitude towards the theory of relativity did not necessarily go hand-in-hand with a monist or anti-monist world view is also apparent, in that at least two of Einstein's opponents, Johannes Riem and Arvid Reuterdahl, were members of the anti-monist Christian Kepler League, and Riem, as the local Berlin chairman, was extremely active and in a prominent position.[137] The reporting on the theory of relativity in *Unsere Welt*, the journal of the Kepler League, was less extensive, in contrast to *Neue Weltanschauung*. Bernhard Bavink, a senior primary school teacher who had been trained by Woldemar Voigt in theoretical physics, was an editor at *Unsere Welt* with a positive attitude towards both the theory of relativity and quantum theory, and who popularized both theories himself.[138] The theory of relativity was certainly discussed in *Unsere Welt*,[139] but there were no polemics or firm opposing positions as in *Neue Weltanschauung*. Johannes Riem, who published numerous articles in *Unsere Welt*, did not publish any against the theory of relativity in that journal. This could be attributable to Bavink's influence as editor, but more probably related to the fact that the devout Riem was primarily interested in the reconciliation of religion and science in his capacity as a member of the Kepler League and was therefore pursuing a different goal with his publications in *Unsere Welt* than with his criticism of the theory of relativity. Riem argued in multiple popular science articles that the composition of the world, particularly the position of the Earth and the solar system in the cosmos, required the existence of God.[140] In contrast to the world riddle solvers, he did not criticize the theory of relativity using his own non-academic theory, instead concentrating first on the experimental confirmations of

[135] Cf. e.g. Fricke, Hermann. (1920b). Eine neue Erklärung der Schwerkraft [A new explanation of gravity]. *Neue Weltanschauung*, **9**(1), 9–18; Beckenhaupt, Carl. (1912b). Über die objektiven Grundlagen der Physik [On the objective foundations of physics]. Ibid., **5**(7), 250–61; Vogt, Johann Gustav. (1912b). Der neueste Stern [The newest star]. Ibid. **5**(6), 209–12.

[136] Cf. Ruckhaber, Erich. (1920b). Schwerkraft und Temperatur [Gravity and temperature]. [Comment on Fricke 1920b]. *Neue Weltanschauung*, **9**(4), 118–19; Idem. (1921). [Review of Fricke 1920a]. Ibid., **10**(5), 159–60; Ruckhaber 1921b.

[137] Here, monist world view is meant in the sense of the Monist League. In the broader sense, it would be possible to describe someone like Riem, who did not think much of the dualistic conception of the relationship between religion and science (which corresponded to the official line of the Kepler League), and who wished to produce a fusion of religion and science, as an adherent of a monistic world view.

[138] Cf. e.g. Bavink, Bernhard. (1920). Philosophische Folgerungen der Einsteinschen Relativitätstheorie [Philosophical conclusions from the Einsteinian theory of relativity]. *Unsere Welt*, **12**(4), 131–38; idem. (1925). Die Quantentheorie [The quantum theory], Part 1. Ibid. **17**(10), 253–61; Part 2. Ibid. **17**(11), 286–89.

[139] Cf. e.g. Sellien, Ewald. (1924a). Kant und Einstein. *Unsere Welt*, 16(5), 107–12; idem. (1924). Die Stellung des Aethers im Weltbild der Physik [The position of aether in the world view of physics]. Ibid. **16**(10), 217–20.

[140] Cf. e.g. Riem, Johannes. (1913). Hat die Welt einen Zweck? [Does the world have a purpose?] *Unsere Welt*, **5**(11), 774–78; idem. (1924). Astronomie und Religion. Ibid. **15**(1), 1–3. The Kepler League and particularly Riem were criticized from the perspective of a monist by Baege, Max. (1911). *Der Keplerbund und seine Gelehrten [The Kepler League and its Scholars]*. Frankfurt (Main), particularly on Riem, pp. 20ff.

the theory of relativity (which he did not consider a given) – which was natural for him as an astronomer – and second on attacking Einstein politically.[141] He carried out this nationalistic and anti-Semitic political polemic in the daily newspapers, usually in the *Deutsche Zeitung*; in addition, he also placed his content-based criticism in popular science journals such as *Umschau* and *Naturwissenschaftliche Wochenschrift*. The latter, a respected and widely circulated popular science journal, offered several of Einstein's opponents a forum. Mohorovičić, Vogtherr, Riem, and Fricke published their objections to the theory of relativity here, or expounded their own theories.[142] Riem and Fricke even acted as reviewers of literature on the theory of relativity, so they had the opportunity to discuss the books by Schlick, von Laue, and others on the theory of relativity, instead of the other way around: being reviewed negatively by the "relativists" or not even acknowledged at all. For example, Fricke judged Wilhelm Wien's *Die Relativitätstheorie vom Standpunkte der Physik und Erkenntnislehre [The Theory of Relativity from the Standpoint of Physics and Epistemology]*[143] as follows:

The book is a gratifying sign that the criticism of the theory of relativity is continuing to make progress even in the leading circles of the professional physicists. Thus the author objects to the premature abolition of the ether. [...] However, the author's reservations are initially directed primarily against the general theory of relativity, while the special theory of relativity is still discussed favorably. Lenard also originally held a similar point of view; however, he has abandoned it recently in favor of a complete rejection of the theory of relativity. In fact, it seems to me that Einstein's fundamental thought is incompatible with the conception of the ether. Perhaps the author will take a somewhat stronger position on this core issue in a later edition of his book.[144]

141 Thus he commented, for instance, with regard to Einstein's trip to France in spring 1922: "He [the Figaro reporter] considers it his [Einstein's] particular contribution that at the beginning of the war he excluded himself from the appeal of the 93 German scholars, but did sign an appeal with the opposite content along with Nicolai. That fits with the statement in Motzkowski's book that world history would have lost its meaning for Einstein when Wilhelm II rode through the Brandenburg Gate as the victor. Another adornment for the Royal Prussian Academy of Sciences!" Riem, Johannes. (1922a). Um die Relativitätstheorie [About the theory of relativity]. *Deutsche Zeitung*, January 7, GN 4, XVI, 9r.

142 Cf. e.g. Mohorovičić 1921; Riem, Johannes. (1921d). Keine Bestätigung der Relativitätstheorie [No confirmation of the theory of relativity]. *Naturwissenschaftliche Wochenschrift*, New Series, **20**(28), 420; idem. (1922d). Neues zur Relativitätstheorie [New information on the theory of relativity]. Ibid., New Series, **21**(1), 13–14; idem. (1922e). Rotverschiebung und Michelsonscher Versuch [Red shift and Michelson experiment]. Ibid. **21**(52), 717; Fricke, Hermann. (1921a). Wind und Wetter als Feldwirkung der Schwerkraft [Wind and weather as a field effect of gravity]. Ibid. **20**(7), 97–102; Idem. (1922d). Zur Klärung des Ätherproblems [On clarifying the ether problem]. Ibid., New Series, **21**(13), 169–73; Vogtherr, Karl. (1922a). Über Fragen der Aberration und Lichtausbreitung [On questions of aberration and propagation of light]. Ibid. **21**(2), 20–25.

143 Wien, Wilhelm. (1921). *Die Relativitätstheorie vom Standpunkte der Physik und Erkenntnislehre. Vortrag gehalten im Verwaltungsgebäude der Firma Siemens & Halske in Siemensstadt bei Berlin am 18. März 1921 [The Theory of Relativity from the Standpoint of Physics and Epistemology. Lecture given in the Administration Building of the Siemens & Halske Company in Siemensstadt near Berlin on March 18, 1921]*. Leipzig 1921.

144 Fricke, Hermann. (1922c). [Review of Wien 1921]. *Naturwissenschaftliche Wochenschrift*, New Series, **21**(12), 168.

Fricke's review of Max von Laue's *Die Relativitätstheorie [The Theory of Relativity]*[145] turned out to be significantly more negative, characterized by his fundamentally different conception of physics:

Von Laue thinks very little of the older mechanical interpretations of weight. 'Since the theory of relativity has of course taught us that empty space is completely free of anything substantial, also free of 'ether,' all these mechanical theories must be considered outdated.' How a physicist can write such a thing is not really comprehensible to me. [...] In v. Laue's discussion [...] I am unable to detect anything physical.[146]

4.2 Argumentative strategies against marginalization

Both Einstein's academic and non-academic opponents reacted to their rejection by the academic physicists relatively unreasonably, initiating a counteroffensive that differed from the content-based criticism and was essentially of a strategic nature in order to relativize the significance of the theory of relativity for their part. The theory of relativity was declared dead multiple times, or it was ascribed a subordinate role in physics (cf. pp. 216ff.). Therefore, the victory of the anti-relativists was repeatedly announced simultaneously with complaints about their rejection or disregard by the academic physicists, which was perceived as suppression. This initially baffling circumstance can be understood as a strategy for turning something into reality by declaring it to be a fact vehemently and continuously. The assertion that the theory of relativity had been refuted or was insignificant was thus supposed to work as a self-fulfilling prophecy and, not least, improve the morale of Einstein's opponents.

The most extensively used strategy against marginalization, however, was conspiratorial argumentation, which I will discuss separately in the same way as the anti-Semitic interpretation patterns that came into play against the theory of relativity (cf. pp. 227ff.).

4.2.1 Negation and marginalization of the theory of relativity

The Basel physics professor Ludwig Zehnder (1854–1949) did not mention the theory of relativity in his physics treatises in the 1920s and 1930s, or only marginally, and continued to develop his primordial atomistics. Such a strategy of consistently ignoring or even some form of accommodating the theory of relativity was out of the question for most of Einstein's opponents, however.[147] They sought ways

145 Laue, Max von. (1921c). *Die Relativitätstheorie [The Theory of Relativity]*. Braunschweig.
146 Fricke, Hermann. (1921c). [Review article on von Laue 1921c *et al.*]. *Naturwissenschaftliche Wochenschrift*, New Series, **20**(25), 373–76, on Laue, pp. 374f.
147 Cf. e.g. Zehnder, Ludwig. (1933). *Der Aether im Lichte der klassischen Zeit und der Neuzeit [The Aether in the Light of the Classical Period and the Modern Period]*. Tübingen. The theory of relativity was ignored primarily

to be able to declare the theory of relativity dead in order to elegantly solve the problem and to finally be able to dedicate themselves to their really important work.

Reuterdahl floated multiple reports in Minnesota's local papers according to which the groundlessness of the theory of relativity had been determined once and for all. The *Minneapolis Sunday Tribune* published the following headline in November 1921, that is, during the peak of the general enthusiasm for Einstein: "Einstein foes prove theory false, claim. Twin cities' mathematical association hears talk on relativity. Former exponents are now sorry, says St. Thomas engineering dean."[148] Reuterdahl is quoted in the article with declarations such as, "It is *literally* true that Europe is *seething* in revolt against the *yoke* of Einsteinism" and "Undoubtedly the great majority of American scientists are today solidly opposed to the theories of Einstein." Johannes Riem likewise wrote during the period of the greatest Einstein euphoria at the beginning of the 1920s, in the *Deutsche Zeitung*:

> Since the lectures in the Philharmonic a year ago carried the struggle against the theory of relativity into the broadest of public view, things have no longer become quiet, ever new material has been contributed by the opponents, philosophers as well as natural scientists, to whose value as evidence no one can close their eyes, and many who at that time were among the loudest voices in favor would be happy if they could make what was written then, unwritten.[149]

This publicly expressed assessment was partly in sharp contrast to the assessment of their own position in the correspondence among Einstein's opponents (cf. for example pp. 236f. and p. 238), but Einstein's opponents also partly reciprocally confirmed to each other that they were standing on what was now the victorious side.[150] Vogtherr assured Gehrcke in a letter in 1922, "There can probably be no doubt about the fact that the splendor and magnificence of Einsteinism is now approaching its end."[151] When Ruckhaber wrote his pamphlet against the theory of relativity in 1928, he justified it as necessary educational work in order to make the "former" adherents of the theory of relativity fully conscious of their error. "The

by adherents to those world views asserting a scientific nature that were very distant from academic science or whose specific interests had few intersections with modern physics. The adherents of the cosmic ice theory ignored the theory of relativity almost entirely, but embroiled themselves in fierce controversies with astronomers about their cosmogonic model. In occult circles, the theory of relativity played a far less important role than radioactivity.

148 *The Minneapolis Sunday Tribune*, November 20, 1921. Ten years later, Reuterdahl was still tirelessly stating in the press that opinion had now finally turned; for example, the following headline appeared on the occasion of the publication of the book *100 Autoren gegen Einstein [100 Authors Against Einstein]* in *The Kansas City Star* of March 24, 1931: "Scientific opinion turning against Einsteinism, says noted scientist [Reuterdahl]."

149 Riem, Johannes. (1922a). Um Einsteins Relativitätstheorie [About Einstein's theory of relativity]. *Deutsche Zeitung*, November 18, GN 4, XII, 70r.

150 Cf. e.g. See to Reuterdahl: "Einstein's sun has set – he is abandoned in Germany, Switzerland, Holland, and England, as well as France, Spain and Italy, so I hear." October 11, 1921, RP 4–49.

151 Vogtherr to Gehrcke, September 25, 1922, GN 83-A-10.

relativity fanatics have been healed from their delusion that the great messiah of philosophy had come, by the great opposition of the logically thinking people. They are finding it difficult to come to terms with the thought that it was merely a dream."[152] Ziegler was firmly convinced that it was not only the theory of relativity that had received a death blow from his primordial light theory. His immunity to any form of doubt about himself or the persuasiveness of his primordial light theory was expressed, among other ways, in that he placed the suicides of the well-known physicists Drude and Boltzmann in 1906 into a direct causal relationship with their reading his work *Die wahre Einheit von Wissenschaft und Religion [The True Unity of Science and Religion]*. According to Ziegler, this work had shown Drude and Boltzmann that they had carried out completely false science, and this was the unequivocal reason for their suicides. "[I]t was weariness with life and despair about the fact that they had wasted their best vital force on a misconceived matter."[153]

How independent of the times the argumentation strategy of "The theory of relativity is dead" was, becomes clear: as late as 1951, Gehrcke emphasized in a letter to Sten Lothigius[154] that "in the relativity movement" he expected "above all symptoms of fatigue in the public" – "One must only have the stamina and be able to wait." In light of the fact that at this point Gehrcke had already been waiting 40 years for the public to wake up from the "mass suggestion" of the theory of relativity, the question most emphatically arises of what period of time he was thinking of regarding the sustainability of an interest in the "unself-reliant, weak-thinking, sensation-hungry, and forgetful" public.

Oskar Kraus was the only person in the network of Einstein's opponents who pondered the acceptance of the theory of relativity in physics. "On the other hand, I had to say to myself that the special and general R[elativity] T[heory] had to possess physical-mathematical meaning, since the greatest portion of the physicists, or at least the most influential ones, are working on it."[155] However, Gehrcke attempted to dissuade him from this insight by return mail. "The fact that a large portion of the physicists are dealing with the R[elativity] T[heory] doesn't prove anything to me, as I do not think much at all of the *consensus gentium*."[156]

An additional defensive strategy consisted of marginalizing the theory of relativity. Here, the argument was, "The theory of relativity is not important." If the theory of relativity was not any kind of comprehensive new theory, but rather just an interesting speculative consideration for specialists, then one would not have to dwell on it any further. The weakness of this argument becomes apparent in that they felt the need to produce extensive refutations, instead of letting the allegedly so unimportant theory of

[152] Ruckhaber 1928, p. 10. [153] Ziegler 1935, p. 33; cf. also Ziegler 1914, p. 146.
[154] All quotes in this paragraph are from Gehrcke to Lothigius, April 15, 1951, GN 50-B-5.
[155] Kraus to Gehrcke, April 1, 1920, GN 82-I-10. [156] Gehrcke to Kraus, April 15, 1920, GN 82-I-11.

relativity rest. The "not-so-important" argument was usually advanced as a supplement to the criticism, as a sort of relativizing protection. "But one also should not overestimate the importance of the matter. This [the theory of relativity] involves very minor corrections to previous measurements and not a turnabout in our world view,"[157] Ruckhaber said in *Neue Weltanschauung*. Rudolf Weinmann's statement went in the same direction. "There can be no discussion of a new world view or even just a view of nature, of a 'Copernican deed' or even a deed of philosophical (epistemological) significance. Kepler and Copernicus remain untouched."[158]

Before launching into a scathing critique of the theory of relativity, occasionally, in accordance with the scientific ethos, a nod was made in the direction of their colleagues and the fundamental criticism was introduced by a vague statement such as, "As unusually interesting and noteworthy the Einsteinian observations and calculations doubtless are in and of themselves..."[159] followed by the loud "but," without any further discussion of what the "Einsteinian observations and calculations" actually stated. While the strategy of declaring the theory of relativity dead or unimportant was a defensive procedure with little power considering the broad acceptance of the theory of relativity, the political arguments were more promising.

4.2.2 Political arguments against the theory of relativity

If the theory of relativity was either insignificant or long since refuted, as Einstein's opponents claimed, an explanation had to be found for why it nevertheless continued to be accepted by the public and by science. The explanations for this almost always included conspiratorial interpretational patterns. This was particularly apparent in the interpretations of the acceptance of the theory of relativity based on conspiracy theories or anti-Semitism. Their significance to the struggle of Einstein's opponents will be discussed later in this section.[160] However, elements of a conspiratorial interpretation can also be found in the advertising argument discussed first, according to which the popularity of the theory of relativity was attributable to mass suggestion.

"The theory of relativity – a mass suggestion": the advertising argument

The "masses" are a phenomenon of the more recent modern age. The concept arose in the nineteenth century from the radical economic and social changes that went hand in hand with industrialization and was critically scrutinized, particularly by the

[157] Ruckhaber, Erich. (1920g). Noch einmal die Einsteintheorie [The Einstein theory once again]. *Neue Weltanschauung*, **9**(12), 356–60.

[158] Weinmann 1926, p. 15. [159] Christiansen 1920, p. 1.

[160] Other political arguments, particularly those motivated by nationalism, did also occur (in Riem and Mewes, for instance), but did not have the same strategic relevance as the conspiracy theories and are therefore not discussed separately.

intellectuals. For Nietzsche, the masses – the mediocre people, the rabble – took on a culturally threatening dimension.[161] Simmel investigated the "mass character of life"[162] from the viewpoint of a sociologist, and in the Weimar Republic, examinations of the "masses" were extremely popular, including examples such as Siegfried Kracauer's series of articles on the newly arisen social class of salaried employees,[163] or Ernst Toller's drama *Masse Mensch [Masses and Man]* written under the influence of the German Revolution of 1918–1919.[164] Along with cultural philosophers and sociologists, it was above all the psychologists who began to be interested in the masses, or more precisely, the "soul of the masses,"[165] around the turn of the century. In the 1920s, it was not just mass culture itself – the press, leisure-time activities, and film – that was popular, but likewise mass psychology as an approach to dealing scientifically with the new social phenomenon of the masses. Mass suggestion is a concept from mass psychology that had its beginnings particularly in the works of Gustave LeBon in 1895, and was further developed by Freud and others.[166] In this context, the interpretation of the theory of relativity as mass suggestion must be situated in the context of the popularity of the contemporary occupation with "the masses" – that undefinable, undifferentiable, influenceable, and sometimes also frightening thing.

It was Ernst Gehrcke who ensured dissemination of the view that the theory of relativity was a mass suggestion with his lecture given on August 24, 1920, at the Berlin Philharmonic, "Kritik der Einsteinschen Relativitätstheorie [Criticism of the Einsteinian theory of relativity]," which was simultaneously published under the title *Die Relativitätstheorie, eine wissenschaftliche Massensuggestion [The Theory of Relativity, a Scientific Mass Suggestion]*. However, the label "mass suggestion" or just "suggestion" had already been used in connection with the theory of relativity;[167] moreover, it was also a current term in the political

[161] Cf. Reschke, Renate. (1992). 'Pöbel-Mischmasch' oder vom notwendigen Niedergang aller Kultur. Friedrich Nietzsches Ansätze zu einer Kulturkritik der Masse ['Rabble-hodgepodge' or the necessary decline of all culture. Friedrich Nietzsche's approaches to a cultural criticism of the masses], in *Zwischen Angstmetapher und Terminus: Theorien der Massenkultur seit Nietzsche [Between Fear Metaphor and Term: Theories of Mass Culture since Nietzsche]*, ed. Norbert Krenzlin. Berlin, pp. 14–42.

[162] Cf. Müller, Gisela. (1992) . Der Massencharakter des Lebens und das rastlose Ich. Eine Lesart zu Georg Simmels Moderne-Bild [The mass character of life and the restless ego. A reading of Georg Simmel's picture of the modern], in Krenzlin 1992, pp. 43–72.

[163] Kracauer, Siegfried. (1971 [1930]). *Die Angestellten [The Employees]*. Frankfurt (Main) [first publication in 1929 in the *Frankfurter Zeitung* newspaper].

[164] Cf. Toller, Ernst. (1921). *Masse – Mensch [Masses and Man]*. Potsdam. In the first edition, the title "Masse Mensch" is written without a dash.

[165] LeBon, Gustave. (1895). *Psychologie des foules*, Paris (Bibliothèque de philosophie contemporaine Vol. 145) (see p. 11).

[166] Cf. LeBon 1895; Freud, Sigmund. (1922 [1921]). *Group Psychology and the Analysis of the Ego*, translated by James Strachey. New York.

[167] Drude, Paul. (1912). *Lehrbuch der Optik [Textbook of Optics]*, ed. Ernst Gehrcke, Leipzig, p. 470. Gehrcke had partially expanded this edition, including the section where the term "mass suggestion" is found. Cf. also Gehrcke, Ernst. (1913). Die gegen die Relativitätstheorie erhobenen Einwände [The objections raised against the theory of relativity]. *Die Naturwissenschaften*, **1**(3), 62–66, reprinted in Gehrcke 1924a, pp. 20–28 (see p. 28).

discussion.[168] Gehrcke's work *Die Massensuggestion der Relativitätstheorie [The Mass Suggestion of the Theory of Relativity]*, in which he expressly refers to LeBon and Freud, appeared in 1924. Gehrcke's strategic use of scientific arguments is clearly apparent. He was able to do without a cultural philosophy or social psychology analysis[169] because for him the phenomenon itself already served as proof, and the psychological terminology merely served to ennoble the argument. The fact that the theory of relativity had been recognized, that it was so very present in the public sphere and in science, was evidence to Gehrcke that something odd must have been going on and that it had to be a mass suggestion triggered by advertising.

The advertising argument was widespread not least because the observable overwhelming presence of the theory of relativity in the press (cf. pp. 69ff.) made it seem immediately evident. The advertising argument must be differentiated from media-bashing, however, which was expressed by many contemporaries, not least by Einstein himself[170] with a twinkle in his eye, whose reactions ranged from scandalized to helpless in the face of the "newspaper shrieking" (Einstein). Einstein's opponents, on the other hand, used media-bashing to explain why their refutations of the theory of relativity had been unsuccessful. They linked the advertising argument with Gehrcke's assertion that this public presence would have triggered something in the minds of people that could only be explained as mass suggestion, and it was only for this reason that the theory of relativity continued to be accepted. Many of Einstein's opponents supported Gehrcke's argumentation or argued analogously, like Mohorovičić: "Impressed with this religion [here, Mohorovičić refers to a statement made earlier that the theory of relativity was a "modern and dangerous religion"] the whole world almost had been hypnotized!"[171] Fricke also agreed with Gehrcke's hypothesis. "The matter reminds one of commands by hypnotists, and Einstein has in fact exercised strong suggestive

168 Thus the Monist League stated with regard to the first elections to the National Assembly on January 19, 1919, that the electorate "was dependent to an extraordinary degree on what the locally published press wrote. As the press, so the election. The masses succumbed to the suggestion of the press to a great degree." Massensuggestion [Mass suggestion]. *Mitteilungen des deutschen Monistenbundes [Newsletter of the German Monist League]*, **4**(3), 33–34.

169 Thus Lämmel also noted in 1924: "This is certainly not a psychological study." Hermann Neubert likewise in *Grundwissenschaft*, **VI**(3–4). But in the *Psychiatrisch-neurologischen Wochenschrift [Psychiatric-Neurological Weekly]* of all places, it was stated that the book was "of lasting value for mass psychology." **27**(19). Gehrcke himself understood his book as "documentary substantiation" of the hypothesis of mass suggestion. Cf. Gehrcke 1924b, preface; likewise his close colleague Ernst Lau in a discussion in *Annalen der Philosophie*, **4**(8). Lau's review was the second review of Gehrcke 1924a and Gehrcke 1924b in *Annalen der Philosophie*; Gehrcke had not agreed with the first discussion by Paul Volkmann in Vol. 4, No. 6 (1924).

170 Cf. e.g. Einstein's statement to Max and Hedi Born: "Just as everything he touched turned into gold for the man in the fairy tale, for me everything turns into newspaper shrieking: suum cuique." Einstein to Max and Hedi Born, September 9, 1920, in Born and Einstein 1991, p. 59.

171 Mohorovičić 1923b, p. 6. Cf. also Kraus: "The 'mass suggestion' is seizing physicists [...], seizing the philosophers themselves." Kraus to Gehrcke, October 12, 1913, GN 72-A-3; likewise Ziegler: "Scholars who think calmly and soundly [...], such as Prof. E. Gehrcke, considered the whole sorry effort [the theory of relativity] to simply be a bluff or a brazen 'mass suggestion.'" Ziegler 1931, p. 80.

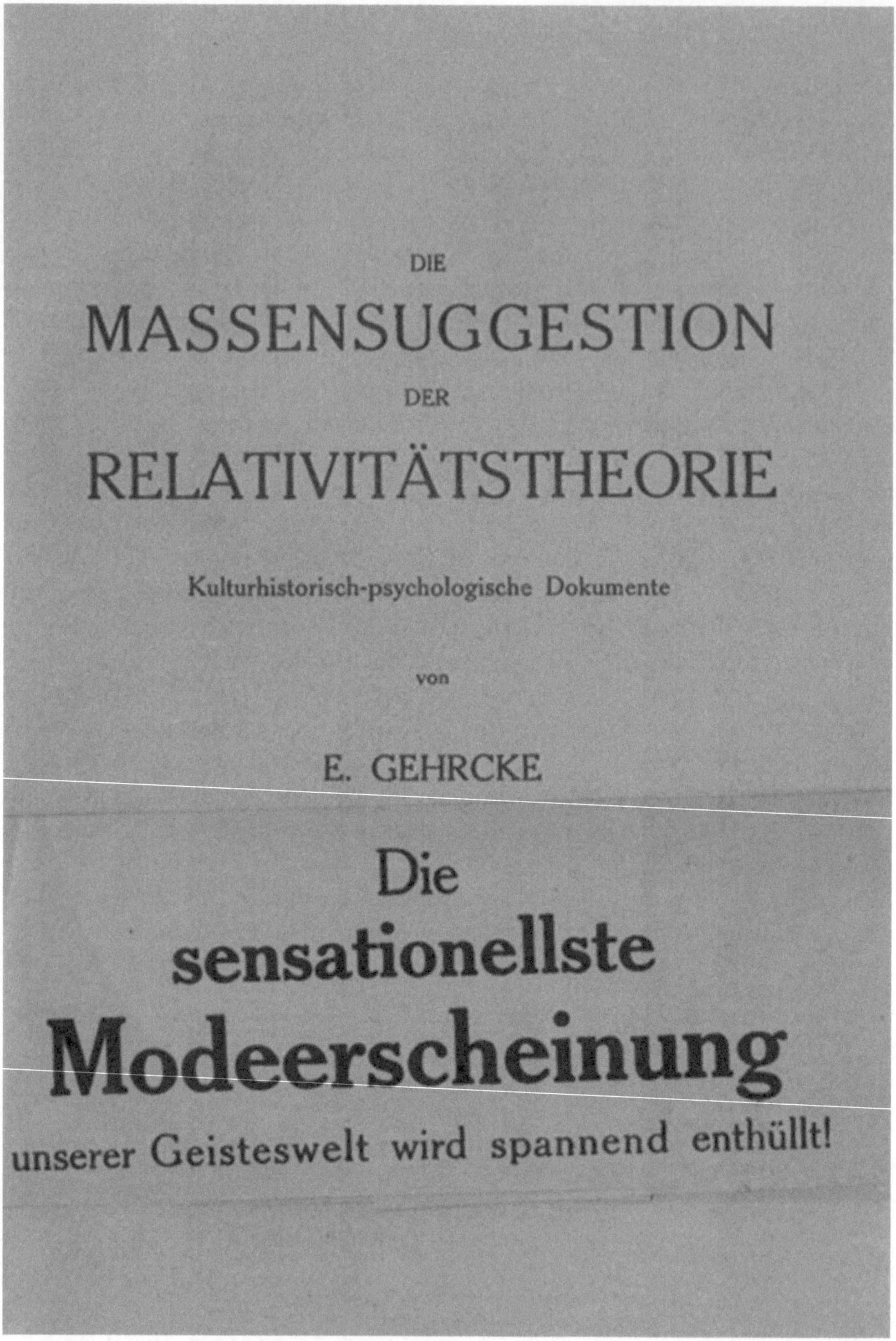

Fig. 4.3 This booklet criticizes the "advertising" for Einstein – but bears an advertising wrapper itself. Source: Gehrcke 1924b.

power – but more on the spectators than on the experimenters. Gehrcke justifiably describes the R[elativity] P[rinciple] as a great mass suggestion."[172] Vogtherr, the physician, likewise stated, "I believe that, along with the clock in the paradox, the structure of the theory of relativity, which has long been unsound, has been blown up. But it will still take some time and effort before the captivating spell of mass suggestion will be broken once and for all."[173]

Two different directions can be identified in the use of the advertising argument. One was that the press was responsible for the mass suggestion, while the other saw Einstein or the theoretical physicists or the Jews or all of them together (the more powerful the conspirators, the more heroic the struggle by the victim of the conspiracy) as the originators of the advertising. The latter is already a conspiratorial argument (cf. pp. 227ff.).

Kurt Geissler expressed moderate criticism to the extent that, although he did accuse Laue and Einstein of not fighting against the advertising, he did not imply that they had initiated it themselves. But for him as well, the advertising argument served as an explanation for the success of the theory of relativity. "Naturally newspaper advertising has brought about such a thing [the general enthusiasm for the theory of relativity]."[174] And in the preface to Ziegler's treatise *Das Ende der sogenannten Relativitätstheorie [The End of the So-Called Theory of Relativity]*, which he self-published in 1923, Ziegler pointed out that this article was intended to "fight the newspaper advertising that has recently been behaving obtrusively to the benefit of the prophet of relativity, Albert Einstein, and for the better understanding of his [Einstein's] incomprehensible theories."[175] According to the logic of an opponent of Einstein, it went without saying that the fact that Ziegler originally did not want to self-publish his article, but rather to publish it in a newspaper, was not to be seen as advertising, but rather as educational work. Lenard likewise justified an article in the *Deutsche Zeitung* newspaper to Gehrcke (who did not approve of scientists publishing in daily newspapers) using the duty to educate. "As for the 'D[eutsche] Zeitung', I completely understand your aversion to newspapers in general. But for my part, the thought was by no means just about R[elativity] Th[eory], but about sound scient[ific] education."[176] On January 3, 1920, a few

[172] Fricke 1920a, p. 12. [173] Vogtherr to Gehrcke, September 25, 1922, GN 83-A-10.

[174] Geissler to Gehrcke, December 9, 1920, GN 76-A-7. About his book *Gemeinverständliche Widerlegung des formalen Relativismus [Generally Comprehensible Refutation of Formal Relativity]*, Geissler stated that he had "only stayed with the topic; however, the strong advertising that had been done for Einstein also caused me to take a position against it." Geissler, Friedrich Kurt. (1921). *Gemeinverständliche Widerlegung des formalen Relativismus (von Einstein und verwandten) und zusammenhängende Darstellung einer grundwissenschaftlichen Relativität [Generally Comprehensible Refutation of Formal Relativity (Einstein's and Related Approaches) and a Coherent Presentation of a Basic Scientific Relativity]*. Leipzig, p. 7. Johannes Stark was even sharper: "Einstein cannot be spared the accusation that he has not opposed the dragging out of his theory at the fair, allowed the propaganda by his friends and supporters, and even encouraged writings by dilettantes for the glory of his theory." Stark 1922, p. 15.

[175] Ziegler 1923, Preface. [176] Lenard to Gehrcke, 1921, GN 83-F-7. Only the year is noted on this letter.

weeks after the press coverage of Einstein snowballed, Fricke asked, in a letter to Gehrcke, "The Einstein propaganda has continued to blossom merrily. [...] Can nothing be done against it?"[177]

These opinions make two things clear. First, Einstein's opponents felt challenged to become active. Their interpretation of the success of the theory of relativity as an effect of advertising made them feel legitimized in attacking it sharply. Second, and related to the first point, the advertising argument made it possible to explain why Einstein's opponents had not yet been able to succeed in their refutations of the theory of relativity. If the public and science were of unsound mind or at least suffered from impaired judgment, then wasn't it almost logical that the "correct" science of Einstein's opponents would not get a hearing? The primarily strategic significance of blaming the advertising is apparent not least in the fact that Gehrcke had no misgivings about sharply attacking the "advertising" for the theory of relativity, while effectively advertising criticism of Einstein to the public. On the occasion of the lecture that Kraus was to give in Berlin in 1920 (cf. pp. 241ff.), he said to Kraus that "the public" should be "made aware of your lecture in the correct way" and that he would arrange for "the necessary advertising."[178]

For Einstein's opponents, the advertising argument primarily functioned as an exoneration for them, because they attempted to explain their failure in this way. However, nationalistic and anti-Semitic circles thankfully took up this argument in reviews of Gehrcke's booklet *Die Massensuggestion der Relativitätstheorie [The Mass Suggestion of the Theory of Relativity]*, among other places. In this booklet, Gehrcke had apparently strung together bare facts – articles from his extensive collection of newspaper clippings – without any value judgments. However, the organization of the articles into particular clusters of topics, and the sequence and selection of the articles, reflect Gehrcke's own version of the public reception of the theory of relativity. Articles by Riem or Einstein's other opponents are often found among them and they give the strung-together reports a critical conclusion. The booklet was highly praised primarily by the nationalistic press. What is particularly noteworthy in these reviews is their vehement anti-Semitism. Among the roughly 2,700 preserved articles in Gehrcke's collection of newspaper clippings from 1921 to 1923, vehement political attacks do occur, particularly about Einstein's international travel after WW1, but there are no such concentrated anti-Semitically motivated reactions to an event or to an anti-relativistic book.[179]

At the same time, *Massensuggestion* was not an anti-Semitic work; Gehrcke himself placed the phenomenon of "mass suggestion" in the context of a general

[177] Fricke to Gehrcke, January 3, 1920, GN 76-A-9. [178] Gehrcke to Kraus, [1920], GN 71-G-4.
[179] The reviews of *Massensuggestion* come from 1924 and later and were not included in the original collection of newspaper articles, which, after all, provided the material for the book.

cultural degeneration that he thought he had recognized in post-war Germany.[180] In the afterword, however, he did refer to Freud, according to whom "'emotional relationships' [constituted] the essence of the soul of the masses," citing "the relationships of Zionism to the theory of relativity" as an example of such emotional relationships.[181]

A more decisive factor for the political reception of the booklet than this psychologized folk-nationalist sentimental nonsense was the fact that the advertising argument as such was structurally similar to a widespread anti-Semitic pattern of interpretation. "Propaganda and advertising, a real Jewish sham, instead of quietly working and researching."[182] The enthusiastic reception of Gehrcke's work in German nationalist, folk-nationalist, and anti-Semitic circles is thus not surprising. The nationalistic *Deutsche Zeitung* newspaper, for instance, stated in its review, "Yes, Mr. Einstein knows the soul of the masses indeed! Of course, Jews know this extremely well anyway; that's why they succeed in revolutions even up to the moment."[183] In the July 31, 1924, edition of *Rundschau für Literatur und Kunst*, the authors of the review went overboard in creating neologisms:

> They [Gehrcke's documents] simultaneously reveal the suggestion dictatorship of progress Jewry [*Fortschrittsjudentum*] that is hostile to obtrusive Jewry [*Aufdrangsjudentum*] as well as to upwardly humanizing Germanness [*Empormenschlichungsdeutschtum*] [...] Therefore, every German who is not nationalistically, but nationally disposed, should know this book, which simply allows the facts to speak.

In the August 9, 1924 issue of the *Mecklenburger Warte* newspaper, Gehrcke's work was praised as "mercilessly pointing out the revolting advertising that the small-minded Jews have used to lie about this minor luminary from Judea in order to elevate him into a shining star."[184] For the *Völkischer Kurier* newspaper of September 11, 1924, Gehrcke's book was actually not folk-nationalistic enough, but Gehrcke had nevertheless "done a service to the folk-nationalist cause, which has the same goal as true science in a certain sense, namely asserting the truth."

Critical reviews of *Massensuggestion* came from various quarters. It was said that Gehrcke had only compiled "unobjective elements from newspapers and journals

[180] Cf. Gehrcke 1924b, Preface. Lenard inquired of Gehrcke in advance of publication whether he couldn't send him the articles in his collection that related to "Einstein's political, Jew-connected and similar German-hostile activities," which in Lenard's view Gehrcke had not sufficiently utilized in *Massensuggestion*. Lenard wanted to "transmit" these articles "to the offices accessible to me for suitable exploitation." Whether this occurred is unknown. Lenard to Gehrcke, May 11, 1924, GN 6-I-2.

[181] Gehrcke 1924b, p. 100.

[182] Fischer, Franz. [Review of Gehrcke 1924b]. *Volksruf*, 11(43). All of the reviews quoted below are found in Gehrcke's papers, GN VII.

[183] Einsteins Massensuggestion [Einstein's mass suggestion], [Review of Gehrcke 1924b]. *Deutsche Zeitung* newspaper, June 10, 1924; a similar review is also found in *Der Tag*, July 18, 1924.

[184] Similarly anti-Semitically motivated reviews are found in *Deutscher Wille. Werkgemeinschaftliches Wochenblatt*, August 7, 1924; Roderich. (1924). [Review of Gehrcke 1924b]. *Deutsches Haus*, Issue 8–9, September, p. 3. A review with a folk-nationalist focus in *Hammer*, No. 532, August 1924.

from every corner of the Earth,"[185] and the mathematician Otmar Zettl stated that it might perhaps be possible to speak of suggestion with regard to the popular reception, but not with regard to acceptance of the theory of relativity by the "leading men of science."[186] Similarly, the *Berliner Börsenzeitung* newspaper stated under the headline "A pointless book," "[I]t is completely immaterial whether or not the public makes a great fuss out of such a theory."[187] As is apparent in these statements, critics of Gehrcke's book also shared his skepticism towards the sensationalist press and the great enthusiasm in the public for the theory of relativity, but they did not follow him in his assessment of the significance of this public euphoria for the scientific tenability of the theory of relativity. The most apt criticism came from Bernhard Bavink in the *Hannoverscher Kurier* newspaper, who stated that one could indeed clearly discern the public enthusiasm for Einstein from this booklet, but:

> Unfortunately, it is less possible to clearly discern that Einstein himself is completely removed from this external sham and even less possible that – apart from a few exceptions – the general recognition of Einstein's accomplishment in the circles of his professional colleagues was neither caused by that exaggerated sham nor has it been modified in any way to any considerable degree by the later waning of the 'relativity fuss'. Thus a layperson cannot be led to a fair judgment about the theory itself from this booklet.[188]

But this was precisely the intention of Gehrcke's excursion into mass psychology. He wanted to show that the theory of relativity was a suggestion that had seized the public *and* science ("If the rel[altivity] th [eory] were a special physical theory like the ether theory, for instance, then it never would have been able to cause such a sensation")[189] and was now "on the wane" not just in the public sphere, but also in science. As a logical consequence, one chapter of the book was dedicated to the "Waning of the movement," and the final chapter to the "Vanishing of the theory of relativity."

The advertising argument often included conspiratorial patterns of interpretation, since most of those using this argument assumed that Einstein had consciously triggered the advertising and thus also the mass suggestion. It thus represented a kind of preliminary stage compared to openly conspiratorial arguments, which will be examined in more detail in the next section. In a letter to Lenard as early as 1917, Gehrcke complained that others were not seeing through the "tireless advertising carried on by Einstein."[190] Reuterdahl drew parallels between Einstein and Phineas

185 [Review of Gehrcke 1924b]. *Breslauer Zeitung* newspaper, August 24, 1924.
186 Zettl, Otmar. (1924). [Review of Gehrcke 1924b]. *Augsburger Postzeitung* newspaper [no specific date].
187 Ein zweckloses Buch [A pointless book]. *Berliner Börsenzeitung* newspaper, entertainment supplement, November 29, 1924.
188 Bavink, Bernhard. [Review of Gehrcke 1924b] *Hannoverscher Kurier*, 77(67); cf. also Mie 1926.
189 Handwritten comment on the copy of *Die Relativitätstheorie, eine wissenschaftliche Massensuggestion [The Theory of Relativity, a Scientific Mass Suggestion]* used as a lecture script, GN 83-E-17 (see p. 11).
190 Gehrcke to Lenard, [draft], December 3, 1917, GN 84-E-4.

Taylor Barnum, a showman of the nineteenth century ("Einstein branded Barnum of science, Minnesota man calls relativity 'bunk.'"[191]) and stated in a letter to Gehrcke about the rumors of a planned trip by Einstein to Russia:[192]

It would seem that Einstein has developed what we jokingly call 'softening the brain.' Napoleon too must rush madly into Russia – let us hope that the parallelism will complete itself. It would be far better to invade Russia with 'bushels and bushels of wheat' instead of with a 'hat-full of Relativity.' He is *mad* for glory. Next he will invade Borneo, Tasmania, etc., etc.[193]

T. J. J. See stated in the *San Francisco Journal*, "[H]e [Einstein] has the habit of a promoter" and continued, "[T]he magazine and newspaper press for the last eight years has been so filled with systematic propaganda, undoubtedly organized and directed by Einstein and his agents, that the public has become familiar with the name of Einstein and with the phrase 'Theory of Relativity'."[194] However, there were also individuals among Einstein's opponents who did not use the advertising argument and stood up for Einstein against the accusation that he had initiated the advertising.[195]

Conspiratorial patterns of interpretation

The politicization and intensification of the debate about the theory of relativity can be explained in large degree by the relevance of conspiratorial patterns of interpretation in this dispute. Conspiracy theories, whether they relate to the "Protocols of the Elders of Zion" or the attacks on September 11, 2001, are political arguments with a specific logic. This means that adherents of conspiracy theories are not mentally ill, and conspiracy theories cannot be interpreted as paranoia in the clinical sense, because they are used strategically for political purposes. Richard Hofstadter has examined the political relevance of the use of conspiratorial arguments in politics and suggested the label "paranoid style." He uses the concept of paranoia to emphasize an important element of conspiracy theories. The conspiracy is an explanation for alleged threats and an expression of the feeling of being persecuted. Hofstadter defines paranoid style in the context of political psychology as "use of paranoid modes of expression by more or less normal people," as specific, apocalyptic political rhetoric that is independent from particular programs or content.[196]

[191] *The Minneapolis Sunday Tribune*, April 10, 1921.

[192] Cf. the contemporary press reports of August 1921 in Gehrcke's Papers, GN 4, XI, 47r to 49r.

[193] Reuterdahl to Gehrcke, August 11, 1921, GN 3-I-8. [194] See 1923a.

[195] Cf. e.g. Ruckhaber: "It must [...] however, not be forgotten that one cannot hold the author of the theory of relativity himself responsible for it [the advertising] [...] for him, as I have been told, it is supposed to be very embarrassing." Ruckhaber, Erich. (1920e). Zur Relativitätstheorie [On the theory of relativity]. *Neue Weltanschauung [New World View]*, **9**(9), 274–75 (see pp. 274f).

[196] Cf. Hofstadter, Richard. (1964). The Paranoid Style in American Politics. *Harper's Magazine*, November, 77–86, reprinted in idem (1979). *The Paranoid Style in American Politics and Other Essays*, Chicago, pp. 3–40 (see p. 4).

Graumann and Moscovici point out the apparently paradoxical situation that in the age of reason of all times, in the enlightened world, conspiracy theories have experienced an unforeseen boost, while in feudal society, it was actually true that a relatively small circle of regents held power in their hands, but conspiracy theories did not play a role to such a degree.[197] But precisely due to the complexity and sophistication of modern societies, where attributability and responsibility can hardly be determined unequivocally any more,[198] conspiracy theories flourish, as do counter-movements to secularization and enlightenment. Conspiracy theories are particularly effective at reducing complexity and are therefore attractive, particularly in uncertain and excessively complex situations. A single, irrefutable reason is behind each and every occurrence, and it is easy to find others who share the same view. Dieter Groh has most concisely expressed the question that conspiracy theories answer: "Why do bad things happen to good people?"[199] Conspiracy theories answer the "Why?" They are in great demand particularly in periods of political instability (the "scapegoat reaction"), such as during the French Revolution or in Germany in the 1920s – remember, for instance, the legend of the stab in the back (Germany betrayed by its own politicians) or the "Protocols of the Elders of Zion."[200]

The triumphal march of modern physics led to similarly extensive uncertainty among the various groups of Einstein's opponents, not merely with regard to their conception of science in the narrower sense, but also with regard to the world view significance that science had for them. In fact, the marginalized position and threatened situation that Einstein's opponents perceived themselves to be in was interpreted conspiratorially by most of them.

If the success of the theory of relativity was due to a conspiracy, then the question of who was behind it arose. Sometimes it was Einstein and his colleagues personally, but more often it was the Jews in general and, even more frequently, both, who were suspected of pulling the strings. Ziegler held the opinion that the theory of relativity had been entirely "invented" in order not to have to concede that a chemist has solved the mysteries of the universe. He said Planck and others had systematically promoted Einstein with this goal. "It [the theory of relativity] is thus not to be conceived of as anything but a means of warding off enlightenment, with the task of heading off the primordial light theory."[201] In fact, Ziegler was firmly convinced that Planck and other theoretical physicists spent the majority of their time suppressing primordial light theory, and that Planck's works "are all directed against an

197 Cf. Graumann, Carl F. and Moscovici, Serge (eds.). (1987). *Changing Conceptions of Conspiracy*. New York (Springer Series in Social Psychology).

198 Cf. e.g. Beck, Ulrich. (1992). *Risk Society: Towards a New Modernity*. London.

199 Cf. Groh, Dieter. (1992). Die verschwörungstheoretische Versuchung oder: [The temptation of conspiracy theories, or:] Why do bad things happen to good people? in *Anthropologische Dimensionen der Geschichte [Anthropological Dimensions of History]*, ed. Dieter Groh. Frankfurt (Main), pp. 267–304.

200 Cf. the articles in Graumann and Moscovici 1987; cf. Groh 1992.

201 Ziegler 1931, p. 119.

opponent who is never named" – namely him.[202] Fricke assumed a targeted "tactic by the Planck physicists" of representing ether theory as an anachronism that did not need to be taken seriously.[203] Even he, who had remained extremely objective for years, got himself into deep water in the mid 1930s by alluding to conspiracies. He complained to Gehrcke, "For 30 years I have been attempting to let common sense have its say in physics. My endeavors were always sabotaged."[204]

Mohorovičić is an interesting case; he attempted to pursue an academic career for a long time, but was unsuccessful. The exact reasons for his failure are obscure, but are most probably attributable to his less than convincing scientific accomplishments.[205] He himself interpreted the lack of success of his university career purely conspiratorially. In 1924, Mohorovičić attempted almost desperately to become Dvorak's successor to the chair for experimental physics at the University of Zagreb ("I will never again have such an opportunity") and was supported by Professors Vladimir Njegovan and Johannes Plotnikow. However, he was unsuccessful in obtaining the chair. Mohorovičić placed the blame for this on the influence of the professors at the University of Zagreb who were not favorably disposed towards him ("But I have too many enemies here"), specifically on Stanko Hondl:

> [T]he 'relativist' *Hondl* has remained true to his ideas, and he attempted to also apply the theory of relativity, which is merely a scientific justification of the basic com[m]unist concept, in practice. [...] We will now act very energetically against this cursed com[m]unist rabble (which the *Jews* are secretly supporting).[206]

He repeatedly emphasized to Gehrcke that the "communists" and "Jews" among the professors were attempting to block him as a candidate:

> There are a good many candidates who without exception have not accomplished anything scientific until now (but politics plays a powerful role in this case, unfortunately). On top of that, there is a contractual professor in the philos. faculty for physical chemistry, a Jew from St.-Petersburg (Leningrad!), who will strive to push in some 'Galician', etc. (*strictly confidential!*).[207]

202 Ziegler 1935, p. 36; cf. also the accusation of plagiarism against Planck in a letter to Gehrcke of February 11, 1931, GN 83-L-4: "I spoke with Plank already in 1902 at the meeting of the GDNÄ in Karlsbad & visited him in 1907 and 1908 at his villa in Grunewald. But he never mentioned my name, even though I had sent him my complete publications up to 1908. That proves sufficiently that even he regarded the results of my intellectual work as outlawed & suitable for plagiarizing." Ziegler speaks in this letter of a "plot" against his primordial light theory. Cf. Ziegler to Gehrcke, April 26, 1923, GN 6-G-4.

203 Fricke [ca. 1942]. 204 Fricke to Gehrcke, April 17, 1936, GN 38-I-3.

205 Cf. also pp. 167f. Kragh points out that Mohorovičić was the first to have postulated the existence of the positronium, although it was in the context of his works that deviated to a great extent from the state of physics of his time. Cf. Kragh, Helge. (1990). From 'Electrum' to Positronium. *Journal of Chemical Education*, **67**(3), 196–97.

206 This and the two preceding quotes from: Mohorovičić to Gehrcke, December 20, 1924, GN 29-F-1.

207 Ibid. Njegovan then wanted to arrange a chair in "cosmic physics" for Mohorovičić at the new College of Technology. Apparently that also did not work out, which Mohorovičić had already anticipated, since in his view the "Varicak-Hondl party is working against it" again. Mohorovičić to Gehrcke, January 28, 1925, GN 29-F-4.

Even 15 years later, he complained to Hugo Dingler, "Already as a student, since 1910, I fought against Einsteinery [...] Unfortunately I am fighting here entirely alone, supported by no one! What difficulties I have had to overcome!" He said he had now founded a "private station for cosmic physics" and asked Dingler to lend his support so that this private laboratory would be equipped with instruments. This time, Mohorovičić explained to Dingler that he did not obtain a position in the academic world because of the influence of the Freemasons in that world. It was only because he refused to become a Freemason that he was now forced to turn to Dingler with this request, "since everything possible was undertaken to prevent me from my scientific activity."[208]

Rejection by the professional press of manuscripts intended for publication was often interpreted conspiratorially. Charles Lane Poor, an astronomer at Columbia University, complained, for instance:

> [A]t the present moment relativists control all scientific societies, all scientific journals and it is impossible to get the truth before scientists. Further all scientific news in the public press is syndicated by Science Service in Washington, which is under control of the relativists. They are running the scientific world today just as a political boss runs a city, which is under his control.[209]

Often, however, it is unclear who is supposed to be behind the organized suppression that is assumed by the individual author. Reuterdahl, for instance, repeatedly indulged in hints:

> I am in possession of letters from eminent European Scientists [Einstein's European opponents] describing the deplorable methods employed to hinder and, if possible, completely prevent an unbiased and free discussion of the problem of relativity. In addition to this evidence, my own experience is proof conclusive that the known evil effects are not due to accidental causes, but arise from a well-defined and strongly organized plan.[210]

Only a few of Einstein's opponents, like Fricke, saw theoretical physics itself as "pulling the strings" behind the acceptance of the theory of relativity. It was obvious to many people that it had to primarily be a "Jewish conspiracy" for the benefit of the Jew Einstein.

Not all of Einstein's opponents thought anti-Semitically; there were opponents of Einstein who – themselves of Jewish extraction – were expressly not anti-Semites (Gilbert, Driesch, Kraus, Palagyi, and Israel, for instance). However, there were a large number of people (Gehrcke, Ruckhaber, Mohorovičić, and von Gleich, for

[208] Mohorovičić to Dingler, October 31, 1939, [copy], GN 22-C-7. Dingler sent this copy to Gehrcke with a request for help for Mohorovičić: "He [Mohorovičić] is also a victim of the international Einstein clique and has certainly earned it." Dingler to Gehrcke, November 21, 1939, GN 22-C-7.

[209] Poor to Redman, October 14, 1926, [copy], RP 5–7.

[210] Reuterdahl, Arvid. (1922a). The Academy of Nations – Its aims and hopes. *The Dearborn Independent*, January 7, p. 14.

instance) who shared the widespread anti-Semitic resentment,[211] but for whom it did not, however, play a role in their criticism of the theory of relativity, or only a subordinate one. Finally, there were confirmed anti-Semites, some of whom carried out their criticism of the theory of relativity unpolitically (Lenard, initially), and others for whom an anti-Semitic conspiracy theory played an important role, sometimes even the decisive role, in their argumentation against the theory of relativity (Mewes, Ziegler, and later Lenard, for instance).

Nevertheless, Einstein's opponents discussed in this book did not discuss the theory of relativity *because* they were anti-Semites, as did Theodor Fritsch or Arthur Trebitsch,[212] for example; instead, anti-Semitic or conspiratorial patterns of interpretation as a whole were a suitable explanation in their view, because they were seen as a plausible explanation for the lack of recognition of their own theories (or the classic-physical or philosophical objections) and the recognition of the theory of relativity.

It was a given for many of Einstein's opponents that the press was under "Jewish control" and therefore played up the theory of relativity by Einstein, a Jew. The fact that some of the leading newspapers and important scientific journals such as *Die Naturwissenschaften* under its editor Arnold Berliner were run by publishers and journalists of Jewish extraction served as a sweeping explanation for the rejection of Einstein's opponents in these media. Reuterdahl stated to Mohorovičić in May 1923, "Our trouble in America is that all scientific journals are closed to the Anti-Relativists through Jewish influence. The daily press is almost entirely under the control of the Jews."[213] Ruckhaber, who highly regarded Reuterdahl's theistic approach, wanted to disseminate his ideas from *The God of Science* in Germany – but where? "There is however the great difficulty that the newspapers are either – and mostly – in the hands of Jews and consequently Einsteinians (and I am of course on the blacklist) or antisemitic or little interested in philosophy."[214] Although he considered the "Jewish press" to be Einstein-friendly, it is interesting that publication in an anti-Semitic periodical was out of the question for Ruckhaber, not because

211 On anti-Semitism in the German Imperial period and the Weimar Republic, cf. Poliakov, León. (1985). *The History of Anti-Semitism*, Vol. 4: *Suicidal Europe, 1870–1933*, translated from the French by George Klin, Oxford; Hammerstein, Notker. (1995). *Antisemitismus und deutsche Universitäten 1871–1933 [Anti-Semitism and German Universities 1871–1933]*. Frankfurt (Main)/New York. The correspondence very clearly shows who had an anti-Semitic attitude and who did not. Cf. e.g. the statements by Mohorovičić on p. 229. Cf. also von Gleich to Gehrcke on the fact that "the *abuse* of mathematics among the Hebrews [was] so popular." February 10, 1926, GN 71-B-4.

212 Cf. Fritsch 1921; Trebitsch, Artur. (1921). *Deutscher Geist – oder Judentum! Ein Weg der Befreiung [German Spirit – or Jewishness! A Path to Liberation]*. Berlin [*et al.*], on the theory of relativity, pp. 252–57.

213 Reuterdahl to Mohorovičić, May 24, 1923, RP 4–35. Even years later, he said to William Fischer, "The force behind Einstein has excellent control over the press and scientific journals. They control our mathematical and scientific departments (indirectly) in our universities and colleges." Reuterdahl to Fischer, October 11, 1931, RP 4–14.

214 Ruckhaber to Reuterdahl, April 11, 1931, RP 4–17.

he rejected anti-Semitism; instead, it was because he was one of the people who consciously and consistently made a distinction between the refutation of the theory of relativity and political views.

As early as December 1920, Ruckhaber expressed his views in *Neue Weltanschauung* on accusations that the periodical was anti-Semitic due to its publication of critical articles on the theory of relativity. In fact, the articles in *Neue Weltanschauung* did not include any political attacks on Einstein or the theory of relativity. Ruckhaber defended himself as follows: "The strong anti-Semitism of our present time is in fact contrasted on the other side with an almost oversensitive inhibition from the mere suspicion of anti-Semitism. 'Either Einsteinian or anti-Semite' seems to be the slogan today." With this statement, Ruckhaber is suggesting that the "Einsteinians" had used the accusation of anti-Semitism to disqualify the opponents of the theory of relativity. This position generalizes the cases in which unjustified anti-Semitic motivations had actually been assumed, without taking into account the context that led to such appraisals. Ruckhaber was in a similar situation to Gehrcke, who had been sharply attacked by Einstein in 1920 and for whom, together with Weyland, anti-Semitic motivations for his criticism of the theory of relativity had been assumed.[215] Gehrcke, along with Lenard, who had likewise been criticized by Einstein, were able to indignantly point out that they had never undertaken any anti-Semitic or personally motivated attacks on the theory of relativity.[216] It was pointed out multiple times that in his article "My Answer," Einstein was the first to publicly make a topic of anti-Semitism in connection with the criticism of the theory of relativity. But is it actually possible to attribute a "certain complicity"[217] to Einstein for the politicization of the debate about the theory of relativity from this fact?

Einstein's statement reacted to existing anti-Semitic attitudes and did not trigger the anti-Semitism.[218] Anti-Semitic mindsets and consequential behavior (from hiring policies accompanied by anti-Semitic resentments up to violent attacks)

[215] Cf. Einstein, Albert. (1920c). Meine Antwort. Ueber die anti-relativistische G.m.b.H. [My response. On the anti-relativity company]. *Berliner Tageblatt* newspaper, August 27, morning edn., reprinted in English translation in Janssen *et al.* 2002b, Doc. 45.

[216] Cf. e.g. Lenard's statement, "I myself unfortunately doubt whether my presence at the meeting [of the DPG in Bad Nauheim on September 21, 1920] will be advisable. The cause is the strange attacks of a personal nature that Mr. Laue and Mr. Einstein have directed against me in full view of the public (Tägl. Rundschau of Aug. 11, evening edn., Berliner Tageblatt of Aug. 27, No. 402), although I have made no public statement relating to the relativity principle that is supported by these gentlemen other than the purely objective and extensively substantiated one in my pamphlet 'Ueber Relativitätsprinzip, Äther, Gravitation' [On the Relativity Principle, Ether, Gravitation]." Lenard to Gehrcke [and eleven other physicists], Einladung [zu einer Beratung über eine angestrebte Erneuerung der DPG] [Invitation to a consultation on an intended renewal of the DPG], September 14, 1920, GN 3-F-1.

[217] Cf. Könneker 2001, pp. 138f.: "A certain complicity was thus ascribable to Einstein for the fact that the controversy about the theory of relativity also irreversibly expanded into the political discourse in the summer of 1920 across all other positions." Ibid., p. 138; similarly Hentschel 1990, pp. 134f.

[218] Similarly Rowe 1993, p. 102.

were widespread in the German Imperial Period and the Weimar Republic. Einstein was already subject to vehement anti-Semitic and nationalist attacks in the early 1920s. Anti-Semitic resentments could also often be found among Einstein's opponents, which is clearly revealed particularly in the correspondence; it can be assumed that these personal attitudes did not remain hidden from their contemporaries. Lenard already drew parallels early on between "correct" science and national origin (in his polemics against the English, for instance) or extraction ("Judah always remains cunning and is always poison for the German spirit"[219]). After reading Hermann Weyl's description of the Bad Nauheim debate, which did not turn out to be very positive for Lenard, he remarked to Gehrcke, "Admittedly I am now wondering about Mr. Weyl's so very German appearance – as I believed."[220] It was not just in the correspondence with Gehrcke that questions like: "Incidentally, is Bohr a Jew?"[221] – letter dated November 18, 1920, or comments like: "What is an 'anti-Semite' actually? It is the disparaging Jewish term for a German who really is one and who does not want to be controlled by a foreign spirit," – letter dated September 9, 1921,[222] were standard fare. Even in 1920, people knew what Lenard thought of Jews and non-Germans, even though it was only in 1922 that he publicly polemecized against the "foreign spirit" that he saw embodied in the theory of relativity.[223]

While Ruckhaber thus argued calculatedly unpolitically against Einstein, Ziegler and Mewes developed elaborate anti-Semitic conspiracy theories early on. Sweeping anti-Semitic accusations against Einstein were customary from the beginning of the 1920s in Ziegler's printed works and in his newspaper articles written under a pseudonym. From 1931, Ziegler also expressly referred in his polemics to the "Protocols of the Elders of Zion" and Henry Ford's pamphlet *The International Jew*. The "Protocols of the Elders of Zion," which allegedly reflected the records of a Jewish world conspiracy, is probably the most influential work of conspiracy

219 Lenard to Gehrcke, January 8, 1920, GN 83-E-10.

220 Lenard to Gehrcke, November 6, 1920, GN 3-F-3; cf. Weyl 1920.

221 Lenard to Gehrcke, November 18, 1920, GN 3-F-6.

222 Lenard to Gehrcke, September 9, 1921, GN 3-F-16.

223 Cf. Lenard 1922a, p. 9. The public lecture series opposing the theory of relativity at the Berlin Philharmonic caused Einstein to seriously consider leaving Germany. Lenard stated his opinion in a letter to Gehrcke: "[T]he power of these people [around Einstein] has obviously climbed so high that one could blow up against it straightaway. How is it even possible that someone in Germany can successfully threaten (even to the Minister!) to want to leave Germany? Why don't they say to him, Oh, then you *aren't* a German at all; go to hell!" Lenard to Gehrcke, October 1, 1920, GN 83-E-17. Nevertheless, Lenard was in contact with Einstein's Jewish opponents, Kraus and Palagyi; Kraus even visited Lenard (cf. Lenard to Gehrcke, 1921 [lacking more specific date], GN 83-F-7). Presumably Lenard was not aware of Kraus' and Palagyi's Jewish ancestry; in a letter to Gehrcke, he thus first called on "truly German philosophers" to examine the theory of relativity, only to seamlessly expressly thanks in the next sentence for a "very high quality" (Lenard) letter from Palagyi that Gehrcke had sent him. Lenard to Gehrcke, November 30, 1920, GN 83-E-18; cf. also Lenard to Gehrcke, December 7, 1921, GN 83-F-5.

theorists.[224] The fact that it had been exposed as a forgery shortly after it appeared did no harm to its reception. The "Protocols" fit (and still fit) much too well into widespread anti-Semitic thought patterns, which only wanted to find their confirmation therein. Moreover, it is of no concern to an adherent of a conspiracy theory if his conspiracy, in this case the "Protocols," is exposed as false. In the conspiratorial conception of the world, this is of course only a tactic by the opposition.

For Ziegler, the important passages in the "Protocols" were those where "the Jews" (like Einstein) described the oppression of the "non-Jews" (like Ziegler). Ziegler quotes long passages from the "Protocols" about how control of public opinion will be taken over using the press, and the way in which the public sphere was to be misled by empty concepts and fantastic theories (like the theory of relativity). According to his view, "the Jews" were attempting "to snatch away" his universal theory from him, but they were unsuccessful because the theory was an "ancient, Aryan inheritance that the Aryans are no longer letting out of their hands." Ziegler's protection of his universal theory from the "world domination delusion" of the Jews was intended to be for the benefit of all mankind – here, he was being entirely philanthropic again.[225]

Ziegler's only essay dedicated exclusively to dealing with the theory of relativity, according to the title, was a pamphlet riddled with anti-Semitic stereotypes and conspiracy theories. According to him, it was not only academic physics that conspired against him; instead "world Jewry" was behind it.

> But they also had to support the spell [of the theory of relativity] with misleading speeches, or with the requisite advertising. And this action had to be all the stronger, the poorer the prospects that a theory that is difficult to understand would be able to reach the public in place of one that is easy to understand and then could stand its ground vis-à-vis the other. Einstein should not be lacking such advertising, and certainly all the more because the whole advertising system is, after all, an invention of his tribe; and this tribe, stronger in holding together than all other tribes of people in the world, was perceptive enough to correctly appreciate the great moral advantage that would accrue to it if it could be allowed to boast that the true renewal of the sciences had sprung forth from its intellect.[226]

For Ziegler it was proven "that the whole Einsteinery is primarily a question of Jewish ambition and only secondarily of scientific interests."[227] Mewes, a Pan-Germanist supporter of the Kaiser, also developed a conspiracy theory, which was

[224] Cf. Sammons, Jeffrey L. (1998). *Die Protokolle der Weisen von Zion. Die Grundlage des modernen Antisemitismus – eine Fälschung. Text und Kommentar [The Protocols of the Elders of Zion. The Foundation of Modern Anti-Semitism – A Forgery. Text and Commentary]*. Göttingen.

[225] All quotes in this paragraph from: Ziegler 1931, pp. 84ff.

[226] Ziegler 1923, p. 28. Cf. also Ziegler to Gehrcke: "Recently, since Einstein has returned from Japan, Palestine, & Spain, strong propaganda in his favor has now recently been carried out by the Zionists. Their intention is clear. Jewry should have intellectual hegemony." April 26, 1923, GN 6-G-4.

[227] Ziegler 1923, p. 29.

embedded in his deeply racist world view. Quite unusually for a turbine engineer, Mewes' magnum opus – based on its scope and the fact that he published new editions four times, sometimes extensively expanded – consisted of a treatise on historical philosophy that ultimately comprised 672 pages. The table of contents covered topics from Dürer's Christ, to the Flood, to Kinetic Gas Theory. It included discussions of the justification of the "Synchronization of the life of Wilhelm II in his youth with the sunspot curve and the biography of the author" using number theory, side-by-side with an "Assessment of Wilhelm II from a racial standpoint."[228] A section was inserted in the third edition of 1922 with the programmatic title: "War periods and intellectual periods in the life of peoples from the standpoint of racial and world development (relativity theory) explained using the life of the Kaiser." The explanation of the "Scientific basis for the discussion" begins with a paragraph about the "injustice to the Kaiser," moves to the "spacetime effects of racial forces," and arrives at the "connection between Einstein and the press of his racial comrades."[229] The anti-Semitic stereotype that "the Jew" could not build up anything concrete or content-based in science, but rather only develop abstract, unreal theories, was a fact for Mewes:

> In Einstein and Minkowski, one gets the impression of large hollow-sounding wordy pomp, not to say a mere sham or misleading bluff in place of significant progress and sound work. [...] The relationship between Mewes and Einstein can accordingly be indicated in brief using the following slogans: 'German versus Jew, knowledge creator versus knowledge leech, ownership versus theft, monarchist versus communist.'[230]

Mewes also took up the line of argument of the "Jewish press," stating "that my efforts to clarify the circumstances [that Einstein and Minkowski had plagiarized his spacetime theory] in periodicals and in the daily newspapers have failed completely. The influence of the Jewish manipulators is so great that such an undertaking is pointless today."[231] At the same time, however, Mewes had to admit that the "German-folk nationalist press," particularly the *Deutsche Zeitung*, did not want to take up his cause in the 1920s[232] – and he was left with no explanation. In this pattern of interpretation, the fact that he and Weber had been unable to gain acceptance for their electrodynamic interpretation of gravitation was not attributable to reasons of content, but rather to the "resistance from Berlin." That meant that the

[228] In his work, Mewes wanted to describe wars between peoples as a scientific event. Similarly to Wilhelm Fliess, who developed a comprehensive theory of human life using the periodicity of the numbers 28 and 23, Mewes was an adherent of the idea of a fundamental periodicity in world affairs, but did not interpret it like Fliess as a mechanism inherent in the living essence, but rather astronomically as consisting of "the effects of the periodically changing influences of the heavenly bodies." Mewes 1922a, p. 59. Mewes wanted to convince the world of his historical theory in order to help the German Empire recover its former position. Cf. also the discussion on pp. 53f. of this book.

[229] Mewes 1922a, pp. 388ff. [230] Mewes 1920a, p. 78. [231] Ibid., p. 70.

[232] An article about Mewes still appeared there in 1919. Cf. p. 209.

theoretical physicists were to blame, but also the "hostility of the Jewish press" towards the "truly German researchers."[233] Lenard also interpreted the fact that he was increasingly isolated in the professional world in the same way. In his correspondence with Gehrcke and others from the early 1920s, as well as publicly from 1922, Lenard repeatedly conjured up the "Jewish spirit" in physics and polemicized against the "depravity of the current rabble of scholars"[234] who were to blame for his marginalized position. Thus he declined to participate in a committee meeting of the Reich Institute of Physics and Technology in 1927 with the following justification: "I don't believe that I could help in the least, because merely the fact that I am not a Jew and also do not have the Jewish spirit, would isolate me from the start."[235]

While anti-Semitic conspiracy scenarios were developed in detail by Ziegler and Mewes, for other opponents of Einstein, anti-Semitism usually played a role as one of multiple dimensions of their world view. Johannes Riem, for instance, a member of the Kepler League, first of all fought against materialism; his anti-Semitic prejudices are apparent in the fact that he laid the blame for materialism as well as the theory of relativity on the Jews. "Both things, the theory of relativity and materialism, must be fought against in the same way, and it is very characteristic that both cases almost only involve Jews. In any event, the Monist League in Berlin consists almost only of Jews, and I have heard that it is the same in other places."[236]

It is hardly surprising that the conspiratorial argumentation intensified over the years, in the course of which Einstein's opponents increasingly styled themselves as unrecognized geniuses and an (as yet) unrecognized elite. Along with a firm belief in their own theories and conceptions, the fact that the success of the theory of relativity was often interpreted using conspiratorial patterns (caused by the press, the theoretical physicists, or the Jews) is an additional explanation for the fact that Einstein's opponents were largely immune to criticism or kindly intended explanations. Wanting to make the theory of relativity comprehensible to them didn't help any more than rational information helps against conspiracy theories. This resistance was an expression of a perceived threat within a conspiratorial world view immunized against criticism, not a lack of education.

However, Einstein's opponents were predominantly not content with styling themselves as victims. Their great intellectual role as prophets made them lonely. Fricke justified his appearance at the DPG conference in Bad Salzbrunn (which was harshly criticized by academic physicists) to Gehrcke by saying, "In any event, I opposed a hostile majority alone." In another letter, he complained, "Why do they have such a hostile and distrusting attitude towards me everywhere?"[237] In a letter to

[233] All quotes from Mewes 1920a, p. 93. [234] Lenard to Gehrcke, August 16, 1922, GN 83-F-10.
[235] Lenard to Gehrcke, March 4, 1927, GN 33-F-6. [236] Riem to Reuterdahl, October 7, 1921, RP 4–67.
[237] First quote: Fricke to Gehrcke, December 21, 1936, GN 33-H-3; second quote: Fricke to Gehrcke, May 10, 1948, GN 2-G-8.

the Swiss literary figure and patron Georg Reinhart in 1929, Ziegler described his "almost frightening isolatedness."[238] The role of a victim was therefore usually only a last resort to some kind of dramatic interpretation of their own situation at the end of a non-academic career. However, Einstein's opponents did not want to be isolated, lonely warriors; instead, they sought to remedy their marginalization by building networks among themselves, as the following section will show.

4.3 Network formation and protest organizations of Einstein's opponents

The network of Einstein's opponents was formed for two reasons. First, deviant science that is not formally organized tends to form networks. Alternative networks of world riddle solvers already existed, in which they pursued their scientific interests that diverged from academic physics, as was clearly shown in the example of the Breslau gravitation researchers (cf. pp. 126ff.).[239] Second, and this will be the focus of the following discussion, this networking was promoted by the shared perceived threatening situation that was described in Chapter 2, and by the marginalization that was described at the beginning of this chapter – the various groups of marginalized Einstein opponents put aside their content-based differences in order to form a visible countermovement of the marginalized.

4.3.1 The network of Einstein's opponents

Within the network of Einstein's opponents, some individuals, such as Gehrcke, Reuterdahl, or Mohorovičić, were more significant than others. Two factors were decisive and mutually reinforcing: social status (including recognition in academic science) and commitment and interest in reaching the broadest possible network of Einstein's opponents. Gehrcke, for example, as a university professor, an academically trained physicist, and an employee of the Reich Institute of Physics and Technology, was an important first point of contact for many of Einstein's non-academic opponents. They hoped to receive support from him in their search for professional scientific publication venues, as well as recognition and assistance in

[238] Ziegler to Reinhart, September 28, 1929, Winterthur City Archives.

[239] The networking efforts of Einstein's opponents fit into Dolby's differentiation of various degrees of alternative organization of deviant sciences, from very exclusive groupings that can be entered only through an initiation process, to open groupings that develop by the adherents of deviant science identifying each other, corresponding, exchanging their writings, and holding meetings. Cf. Dolby 1979. The network of Einstein's opponents was such an open grouping. Cf. e.g. Zacharias' statement, "The *independent* researchers must band together *just as firmly* as the science officials. Only then can they force the urgently necessary education. Newspapers, periodicals, book publishers who want to promote real progress must be won over." Zacharias 1920, p. 112.

their dispute with academic physics ("I only need to know one thing: Does the expert understand me, or am I incomprehensible?"[240]). It is true that Gehrcke was also confronted with rejection from the academic world, but he did have more opportunities to be heard than Einstein's non-academic opponents – not due to his convincing arguments, but rather because he was a member of academic science.

The social prestige of an opponent of Einstein, on its own, cannot explain his significance within the network of Einstein's opponents. This is apparent in the comparison between Gehrcke and Lenard; Lenard had a higher status as a Nobel Prize winner and received far more attention than Gehrcke as an opponent of Einstein, both in academic science as well as in the public sphere. However, he was less relevant to the network of Einstein's opponents. This was because Lenard was not willing to get too involved with the network and was harsher in his judgments about potential non-academic allies. While Gehrcke was accepted by Lenard as a confederate, like other academic physicists and sometimes also philosophers – "It is necessary, however, for this matter [of the opposing discussion on Weyl's report about the direction of the Bad Nauheim meeting] that I do not stand alone, and I would therefore very much welcome it if you and/or Mr. Palagyi would take the floor"[241] – most of the others did not satisfy his requirements. "Dr. Geissler is unfortunately too limited and too full of himself for him to be of much use. With Fricke it is unfortunately almost the same. There are too few *people* alive."[242] In another letter to Gehrcke, he complains about their isolation. "With Ziegler and Mewes, and also some others, it will not be possible to deal with the Einstein madness; I don't see anyone at all besides us two on the physics side who have been working publicly thus far."[243] Such statements and deep sighs ("But who will help me; I cannot help all of them if they are fools on whom an empire must founder"[244]) can be found frequently in Lenard's letters to Gehrcke. Lenard accordingly attempted to persuade physicists like Mie and Wiechert to work harder for the cause of Einstein's opponents:

I am still placing much hope in Mie, and I recently wrote to him again, also about the fact that it would be shameful, after all, if I were to remain alone with you [Gehrcke] in our

240 Gilbert to Gehrcke, February 1, 1930, GN 11-B-2. Mohorovičić also asked Gehrcke – following Debye's rejection of the article (cf. pp. 201f.) – to review the manuscript of "Die Folgerungen der allgemeinen Relativitätstheorie und die Newtonsche Physik [The implications of the general theory of relativity and Newtonian physics]" for any errors. Cf. Mohorovičić to Gehrcke, July 30, 1921, GN 3-G-4; Gimmerthal approached Gehrcke, who was supposed to judge his work: "Now with you, it will be subjected to the most competent judgment." Gimmerthal to Gehrcke, April 29, 1927, GN 76-A-4.

241 Lenard to Gehrcke, November 6, 1920, GN 3-F-3. Cf. also Lenard's statement in the same letter: "There are a great many people who want to 'help' me, but they cannot do it (letters opposing the rel[altivity] principle come to me almost daily from people who are entirely unacquainted with physics and mathematics)."

242 Lenard to Gehrcke, October 14, 1921, GN 3-F-20. Vogtherr he valued, however, cf. Lenard to Gehrcke, July 24, 1921, GN 83-F-2.

243 Lenard to Gehrcke, July 21, 1921, GN 83-F-1.

244 Lenard to Gehrcke, November 14, 1920, GN 3-F-5.

decisive rejection of such obvious intellectual nonsense as the genl. RTh. [theory of general relativity], together with the efforts to overturn the space-time concept.[245]

Lenard preferred to leave it to Gehrcke to deal directly with Geissler, Fricke, and Reuterdahl. This second factor of network development was much more important than social prestige. Gehrcke rarely rejected even those of Einstein's opponents who Lenard would not accept as allies.[246] On the one hand, Gehrcke himself had a very broad orientation in his scientific interests. One of his main arguments was that the theory of relativity affected the fields of physics, mathematics, and philosophy equally, which was why it was so difficult for an individual physicist, mathematician, or philosopher to criticize it, and he looked for support specifically in the field of philosophy.[247] Along with this openness, though, the decisive reason was that Gehrcke thought in extremely political ways. He was a strategist in the opposition to modern physics. Even though he was concerned about a certain level of quality in the content-based criticism, the visibility of the opposition was the focus of his interest, and he was not very selective with regard to the methods of achieving visibility, which could be the masses and volume, suitable alliances, or targeted campaigns. The fact that Fricke was defending ether was accordingly more important for Gehrcke than the fact that Fricke's ether theory was called into doubt even by ether proponents like Gehrcke and Lenard.

On the other side of the Atlantic, Reuterdahl was the second central figure in the network of Einstein's opponents. As Dean of the Department of Engineering and Architecture at the College of St. Thomas, and as President of the Ramsey Institute of Technology from 1922, an institution he founded that primarily offered industry-oriented post-graduate training, he was in a social position with a certain amount of prestige. In addition, he had a charismatic personality and enjoyed an excellent network in society and local politics in Minnesota.[248] He was not only convinced of his own theory of "Scientific Theism" and wanted to convince the rest of the world, but was also just as keen on collecting material about "the true story of

[245] Lenard to Gehrcke, November 30, 1920, GN 83-E-18.

[246] On the other hand, if someone was too close to the theory of relativity, that represented grounds for rejection. Cf. Gehrcke's rejection of Johann Radakovits: "The same item [Radakovits' manuscript] does provide ideas deviating from the theory of relativity, but in its approach and in its spirit, it is very close to this theory, and I therefore request of you that you refrain from taking up my time with this matter." March 31, 1921, GN 3-J-22.

[247] The most important thing for him was confirmation of his own philosophical views, not academic status. He therefore profited in content more from Melchior Palagyi and Lenore Kühn than from Oskar Kraus, which was already apparent in the discussion of the fundamentals of space and ether between Kraus and Gehrcke in Chapter 3 (cf. pp. 120f.). Lenard, on the other hand, did value the works of Kühn (cf. p. 247) and Palagyi (cf. this chapter, Note 223), but was not interested in an interdisciplinary countermovement and remarked to Gehrcke: "I myself, for me, doubt whether the philosophers could offer me anything that I don't already know?" Lenard to Gehrcke, November 30, 1920, GN 83-E-18.

[248] Reuterdahl was locally well known as the President of the Ramsey Institute of Technology and was socially and politically active. Among other things, he was a member of a Masonic group (Scottish Rite Masonry) and the founder of the Interchurch Theistic Alliance and the International Theistic Society, often giving popular lectures, particularly on religious topics.

Einsteinism."[249] In fact, his papers in St. Paul include a comprehensive collection of anti-Einstein material. While Gehrcke was more likely to limit himself to reacting, Reuterdahl actively created a collection of addresses, photographs, and opinions of Einstein's opponents,[250] and he was even less selective than Gehrcke. This combination of Reuterdahl's mission and Gehrcke's strategic interests brought Einstein's opponents – world riddle solvers, physicists, and philosophers – together for several years.

Beyond Weyland & Co.

The only organization of Einstein's opponents that has been mentioned previously in the Einstein research is the Arbeitsgemeinschaft deutscher Naturforscher zur Erhaltung reiner Wissenschaft e.V. [Association of German Natural Scientists for the Preservation of Pure Science, reg. assn.], which Paul Weyland (1888–1972) claimed to have founded. Weyland was one of the presenters at the first public protest event against the theory of relativity at the Berlin Philharmonic on August 24, 1920. In his lecture, he declared the theory of relativity to be "scientific dadaism" and established a connection to a general cultural and political degeneration that he perceived:

> Now that we have become financially poor, there is a movement afoot to take our intellectual property from us; from now on, we are to be prevented from independent thinking. In politics, they have succeeded. You see it daily and hourly in every news item about how a crazy group of uncritical people under the unscrupulous leadership of self-serving people are heading towards Bolshevism. Ethics and morality only exist as words anymore, and after people set to work killing everything in the German that has made him great, science is now also to be taken from him. [...] For the consequences and intentions of the general Einsteinian R.[elativity] T.[heory] relativity principle and its backers are more extensive and more difficult than the public can imagine.[251]

Who was Paul Weyland, who suspected backers behind the general theory of relativity and attributed dark intentions to them? Until his lecture, he was entirely unknown among scientists. He described himself as an engineer and chemist, among other things.[252] Weyland was a staunch anti-Semite and published the

[249] Cf. also Reuterdahl to See, May 17, 1923, RP 4–49. "Today I have letters, documents, and original papers which tell the story of the advent and development of Einsteinism. Some time in the near future I hope that a publisher can be found for this material compiled into book form. I believe that I alone have all the data. I have published item after item wherever an opportunity appeared. The facts have been transmitted back to Europe to those who were not familiar with them. From Jogoslavien to Sweden I have supplied this ammunition for the batteries of truth, wherever located."

[250] Cf. e.g. Fricke to Gehrcke: "Did Mr. Reuterdahl also ask for your picture?" August 5, 1921, GN 3-G-8.

[251] Typescript of the lecture given on August 24 at the Berlin Philharmonic, with handwritten corrections by Ernst Gehrcke, [1920], GN 72-E-6. The section quoted here is crossed out in the manuscript. The script contains many careless mistakes, such as the "R.T. relativity principle."

[252] Cf. Kleinert 1993.

Deutschvölkische Monatshefte [German Folk National Monthly] at the beginning of the 1920s. While older research did not cast any doubt on the existence of Weyland's Association,[253] it was primarily Andreas Kleinert who convincingly showed that this Association was in all probability only an idea that was never carried out.[254] In fact, Weyland only played a minor role in the network of Einstein's opponents, even though he achieved great effects in the public arena with his polemical campaigns and his political views, and therefore has justifiably been examined in the research on the anti-Einstein movement. On the other hand, this strong focus on Weyland has neglected aspects that are central to an understanding of the contexts of the widespread opposition to the theory of relativity. Why was Weyland unable to build up an organization of Einstein's opponents, and were Einstein's opponents perhaps differently networked than to Weyland's organization, which ultimately did not exist? The failure of Weyland's plan to build up an organization against Einstein and the success of an alternative organization conveys a great deal about the status of Weyland in the countermovement against the theory of relativity, and about the motives of Einstein's opponents. When the event at the Philharmonic and the relationships of Einstein's opponents to Weyland are placed in the context of the network of Einstein's opponents, it becomes apparent that they did not consider him one of their own.

The continuation of the lecture series at the Berlin Philharmonic with lectures by Oskar Kraus and Ludwig Glaser, which had been announced for September 2, 1920, did not occur as planned. Alarmed by newspaper reports about the anti-Semitic context of the event, Oskar Kraus had cancelled his attendance in a letter to Gehrcke dated August 30, 1920, although the trip had already been organized and the lecture had been written. Anti-Semitism in the anti-Einstein movement became a topic between Kraus and Gehrcke for the first (and only) time:

> You know that I have been struggling with the absurdity of the theory of relativity since 1913. […] Now, after Mr. Weyland's actions have made the matter into a political one, a personal one, and above all unfortunately a crudely anti-Semitic one, I cannot participate in it. I thought that it could not be unknown to you that I myself am of Jewish extraction.[255]

Gehrcke's answer of September 3 reinforces the assessment stated above that he certainly shared anti-Semitic resentments and thought in anti-Semitic stereotypes, but they played a marginal role in his rejection of the theory of relativity and by no means called into question his alliance with Einstein's Jewish opponents:

[253] Cf. e.g. Grundmann 1967; Hermann 1977. [254] Cf. Kleinert 1993.
[255] Kraus to Gehrcke, August 30, 1920, GN 71-G-1.

I am not accustomed to consider the question of whether someone is of Jewish extraction as such to be of special importance, since I make it a habit to assess people according to their deeds. Of course, I also have among my friends those of Jewish extraction.

I have not criticized Einstein *because* he is of Jewish extraction, but rather at most *although* he is, and believe myself to be fully in agreement with you on this matter. I see both the preference of a member of a race for their own racial comrades and the aversion to the different race as a humanly understandable, although objectively unjustified feeling, and would therefore like to plead for not all too strict an assessment of racial chauvinism as well as racial hatred.

In particular, I would like to recognize that there are mitigating circumstances for the pro-Semitic friends of Einstein as well as his anti-Semitic enemies, because both of them are under the influence of a feeling. On the other hand, I would not wish to make such feelings my own.[256]

Kraus was apparently satisfied with this response, but frequently asked Gehrcke about the further course of the connection with Weyland. "Does the 'Association' still exist? Has Weyland resigned? Who gave the next lectures there?"[257] Gehrcke's report to Kraus confirms the supposition that Weyland's Association was a one-man company:

Despite all enticements, I have not yet joined Weyland's Association, and you can imagine that I also have no inclination to join in the future. Incidentally, the future of this Association is a complete mystery to me, since I have not yet gotten to know any member besides Mr. Weyland.[258]

Gehrcke still wanted to organize a Kraus lecture, outside the framework of Weyland's series of events.

In connection with your lecture, the matter is such that, assuming your willingness, I would very much like for it to occur before the *largest* possible audience in the interest of the matter itself. I firmly believe that the hall will be full if your lecture is brought to the attention of the public beforehand in the correct manner.[259]

The dubious impression that Weyland's anti-Einstein activities made on Einstein's opponents is apparent in Kraus' reaction when Weyland absolutely wanted to visit him, a Jew, in Prague – for reasons that were unknown to Kraus. He wrote to Gehrcke, completely consternated:

I have *long since* disclosed my heritage to Weyland – also that I will not speak under his patronage. I also wrote to him that I had visitors (in fact, our colleague Kastil from

256 Gehrcke to Kraus, September 3, 1920, GN 71-G-2.

257 Kraus to Gehrcke, October 16, 1920, GN 71-G-5. Ludwig Glaser had given his lecture on September 2, 1920, as provided for in the original program.

258 Gehrcke to Kraus, [Fall 1920], GN 71-G-4.

259 Ibid. According to Gehrcke's idea, the lecture would take place at Berlin's Urania Society or before the Kant Society.

Innsbruck was at my house) – but he now wants to come later. Isn't this man a little bit abnormal? He gives me the impression of being a fanatical – exceedingly busy-unclear [*sic*] – daredevil.

What he writes about the Jews is grotesque. The 'Kol Nidre' prayer supposedly reveals the moral reprehensibility of the race. I would have to inquire of 'more knowing' Jews in order to understand what he means at all. The Kol Nidre prayer asks God to release the sinner from all vows to the Divinity! These gentlemen interpret that to mean that it releases Jews from all vows to other people.

You can see how ridiculous and simultaneously unjust this monstrous accusation is from the fact that the likes of us don't even know what he is accusing the Jews of! We have been raised by German culture and cannot do anything else than feel with the Germans, even if they may despise us and kick us. We, that is, just those people who have not become Zionists because of anti-Semitism. For my part, I have documented clearly enough with my conversion to Protestantism that I don't place any importance on membership in a synagogue.[260]

In the following years, Gehrcke frequently warned other Einstein opponents away from getting more deeply involved with Weyland or even just stating his name in connection with other opponents of the theory of relativity,[261] and distanced himself:

In summary, my perception is that Weyland made it possible for me to bring an end to being passed over by Einstein in silence, in that he created the circumstances making it possible to take a position against the theory of relativity in public. But with that, I am done with him: The Moor has done his duty [...] It was already clear to me before my Philharmonic lectures that Weyland must be dealt with extremely cautiously and is nothing more than a character from the morass of revolutionary post-war Berlin.[262]

This, at any rate, was Gehrcke's retrospective assessment in his report to Lenard. It is an open question whether he actually did just use Weyland and later drop him, as he described, or whether he worked out the matter in this way in retrospect. However, entering into such an alliance from a cost-benefit aspect certainly would have been in accordance with his strategic thinking; precisely due to that aspect, he

260 Kraus to Gehrcke, November 22, 1920, GN 71-G-6.

261 Cf. e.g. the statement to Reuterdahl: "In conclusion, I would like to warn you once again about Weyland. He does more harm than good." Gehrcke to Reuterdahl, August 29, 1921, RP 4–16.

262 Gehrcke to Lenard, [draft], [1921], GN 71-G-13. The slightly revised letter actually sent is dated February 3, 1921, cf. Kleinert and Schönbeck 1978, p. 327. On Lenard's contacts with Weyland, cf. ibid.; Rowe 1993; Schönbeck 2000. Although Lenard did not allow himself to be roped into the lecture series, he did see Weyland's "action" as a legitimate method against marginalization: "Shouldn't one [...] conclude that a Mr. W[eyland] is almost a necessity vis-à-vis the 'relativists'?" Lenard to Gehrcke, October 1, 1920, GN 83-E-17. In addition, he sympathized with Weyland's political activities. Cf. Lenard to Gehrcke, November 6, 1920, GN 3-F-3: "After all, it is good for spirits of power and air like Weyland to roam among the people now and then; it brings those together who belong together, even if some confusion is created. I always wished that Mr. W[eyland] would not come off too badly. What is he doing now?"; cf. also Lenard to Gehrcke, January 8, [1921], GN 83-E-10: "Mr. Weyland's versatility in the articles in his new paper [*Deutschvölkische Monatshefte*], which it seems to me are written skillfully and aptly, could now become fruitful. This paper's success would be very desirable; it touches the core of the matter."

would then have distanced himself from the alliance when it turned out to be more harmful than beneficial.[263]

Nonetheless, the desire to organize and to network in order to better assert themselves was strongly represented among Einstein's opponents, and other organizations were actually founded – but without Weyland, who was not to be counted as part of serious science in their eyes, in contrast to themselves. Shortly after the Philharmonic event, Gehrcke stated to Kraus:

> It would be something different [from Weyland's Association] if founding something new would be successful, that is, a preliminary league consisting of, say, you, Lenard, Palagyi, Dr. Glaser, and me, that takes as its general task to offer the people good scientific presentations in the form of lectures and generally comprehensible writings. So not a separate Corp[oration] against Einstein, but rather a scientific leadership league for the purpose of popularizing scientific ideas and discoveries.[264]

It is remarkable that Gehrcke was not planning a league of the "determined opponents of Einstein,"[265] instead placing the popularization of "good scientific presentations" in central position in the tradition of popular education in the nineteenth century. Along with this educational intent, his insight to create a "counter-public relations" dictated the plan – without public relations, they would not be visible and could not exercise any influence. The "scientific leadership league" that Gehrcke imagined did not materialize after all, presumably because his acquaintance with Reuterdahl opened up other options.

How did Einstein's opponents get to know each other?

This book has frequently discussed contacts among Einstein's opponents. But when and how did they actually get to know each other? There is no exhaustive answer to this question. How contacts occurred among Einstein's opponents cannot be entirely reconstructed and was moreover individually different. However, an important factor was the presence of Einstein criticism in the media and the personal references of Einstein's opponents to each other.

263 Max von Laue and Erich Regener severely criticized Gehrcke's alliance with Weyland, above all the fact that he had not expressly distanced himself from Weyland in retrospect. Cf. von Laue to Gehrcke, August 28, 1920, [copy], GN 4-B-12; Regener to Gehrcke, September 10, 1920, GN 4-B-13 and Regener to Gehrcke, September 15, 1920, GN 4-B-14. Several days before the lecture, the physicist Paul Spies called on Gehrcke at home in order to warn him against a connection with Weyland, whose anti-Semitic lectures were well known. It can be seen from Gehrcke's answer to Spies – reproduced in a letter to Lenard – that he was indifferent to any anti-Semitic tone of the event. "Spies may be right insofar as the anti-Semites were happy about the matter, but I could not influence any partisanship of the anti-Semites any more than Einstein could influence any partisanship of the pro-Semites for the Rel[altivity] th[eory]." Gehrcke to Lenard, [around August 20, 1920], GN 3-J-17.

264 Gehrcke to Kraus, [Fall 1920], GN 71-G-4.

265 This was the characterization of Einstein's opponents by the *Berliner Tageblatt* newspaper, cf. "Die verbissenen Gegner Einsteins …" In *Berliner Tageblatt*, November 17, morning edn., GN 4, I, 35r.

Lacking the mass media, there would not have been public enthusiasm for the theory of relativity, much less the network of Einstein's opponents. Many opponents of modern physics found out about world riddle solvers and Einstein's academic opponents through their "publicity work," that is, lectures, publications, and press reports. Gehrcke had sometimes underlined the names of Einstein's opponents in his newspaper clipping collection and sometimes also noted who it was, what profession that person was practicing, etc. ("valuable opponent of R[elativity] Th[eory] and clear logician, (formerly outside lecturer?) of physics").[266] They then simply contacted the people who had been mentioned in the article. This was also how the world riddle solvers Patschke and Wutke got to know each other, as Patschke tells retrospectively:

On June 24, 1920, I was visited in the evening at about 6 1/2 o'clock by the former pharmacy owner Georg Wutke, who is known because of his geophysical works 'Die Erde als Motor [The Earth as a Motor],' 'Kann die Erde erkalten [Can the Earth Grow Cold],' 'Über Ursachen und Wesen der Eiszeiten [On the Causes and Nature of the Ice Ages],' etc. 'Mr. Patschke,' he said, 'I have been looking for you for 15 years, but it was only from a review in today's Wilmersdorfer Zeitung newspaper that I found out your address, although I live quite close to you, in fact diagonally across from you.'[267]

From then on, lively exchanges occurred; what was important for both of them was a discussion of their own theories, and not primarily a criticism of modern physics, which neither of them thought highly of. Through Wutke, Patschke then learned about the non-academic researchers around Johannes Zacharias, who were very close to his own theory in their content, and gave Wutke Reuterdahl's address, which he himself had found in a newspaper report about Reuterdahl as an opponent of Einstein, probably written by Johannes Riem, who published several press reports about Reuterdahl.[268] Zacharias and Ziegler in turn knew each other through other connections,[269] possibly through Fricke, who had already become visible in the Weltwissen group around Zacharias.[270] Reuterdahl also probably became familiar with Zacharias and his periodical *Weltwissen* through Ziegler.[271] Gehrcke also

266 Note by Gehrcke on the edge of a photograph that had appeared in the Swedish press on the occasion of Nordenson's fiftieth birthday. Caption: "Direktören, fil. dr Harald Nordenson, Stockholm," [unknown newspaper], August 10, 1936, GN 4, (2), 32r.

267 Quoted from Patschke 1925, p. 55.

268 Cf. Patschke 1925, p. 57; cf. e.g. Riem, Johannes. (1921a). Zu Einsteins Amerikafahrt [On Einstein's trip to America]. *Deutsche Zeitung* newspaper, September 13, GN 4, XII, 26r; ibid. (1921b). Ein amerikanisches Weltanschauungsbuch [An American world view book]. *Der Reichsbote* newspaper, October 4, GN 4, XII, 55r.

269 Cf. Ziegler to Reuterdahl, December 29, 1923, RP 4–66.

270 Fricke published many items in Zacharias' periodical *Weltwissen*. Cf. e.g. Fricke, Hermann. (1914). Die allgemeine Gravitation. Ein neuer Weg zur Lösung des Problems [General gravitation. A new path to a solution of the problem]. *Weltwissen*, **2**(38), 177–80, Part 2, ibid. (39), 185–88; idem. Nochmals die 'Gravitation' ['Gravitation' again]. Ibid. (45), 239.

271 Reuterdahl quotes a contribution by Zacharias from *Weltwissen* in an article about Ziegler. Cf. Reuterdahl 1921c.

came in contact with Reuterdahl through Fricke, who moved in remarkably many sub-academic contexts.[272] Erich Ruckhaber, one of the editors of *100 Autoren gegen Einstein [100 Authors Against Einstein]*, learned of Reuterdahl's opposition to the theory of relativity through an article in the *Luzerner Neueste Nachrichten* newspaper written by Ziegler and therefore contacted Reuterdahl years later on the occasion of the publication of *100 Autoren gegen Einstein* (cf. pp. 288ff.).[273]

Even Gehrcke himself represented a publicly visible point of contact for many of Einstein's opponents, as already mentioned. His publications against the theory of relativity in the 1910s, including those in the periodical *Die Naturwissenschaften* that was addressed to scientists in various disciplines, and in the philosophical journal *Kant-Studien*, had made him known as an opponent of Einstein not only to his physics colleagues, but also to the interested public. Thus Kraus, for instance, had read Gehrcke's early critical articles on the theory of relativity, in which Gehrcke points out the philosophical dimension of the dispute with the theory of relativity. Reading this encouraged Kraus to approach Gehrcke in 1913 and the following years – "such an excellent expert on the matter and a physicist of some reputation."[274] His appearance at the Berlin Philharmonic in 1920 made Gehrcke known to wider circles.[275] Following Lenard's recommendation, Vogtherr approached Gehrcke in 1921, whom he "knows of and values from writings and lectures as a diligent and courageous defender of common sense against the absurd impressions of the theory of relativity."[276] At the end of 1921, Vogtherr asked Gehrcke to review the manuscript of his booklet *Wohin führt die Relativitätstheorie? [Where does the Theory of Relativity Lead?]* and to help him to look for a publisher for it – and the two of them remained in contact for the next

[272] Cf. Fricke to Reuterdahl, June 19, 1921, RP 4–15. Fricke reported to Reuterdahl that he had asked Gehrcke to send Reuterdahl his Philharmonic lecture. He had also given Johannes Riem a newspaper article by Reuterdahl that Riem was supposed to place in the German press.

[273] Cf. Ruckhaber to Reuterdahl, February 11, 1931, RP 4–48; cf. Ziegler, Johann Heinrich. (1921). Albertus Maximus und die Blamage der Schulweisheit [Albert the Great and the disgrace of book-learning]. *Luzerner Neueste Nachrichten* newspaper, September 20 (published under the pseudonym J. E. G. Hirzel).

[274] Kraus to Gehrcke, June 19, 1920, GN 76-A-18.

[275] Other opponents of Einstein who contacted Gehrcke in the 1920s included Gimmerthal, Geissler, and Orthner: "I would have gladly publicized the enclosed paper in scientific circles, but I have neither a name nor a title nor connections and don't know where to put it." Gimmerthal to Gehrcke, April 29, 1927, GN 76-A-4; cf. also Geissler to Gehrcke: "Other ones [of Gehrcke's writings] are already known to me from earlier, e.g. what you had published in Kantstudien." Geissler to Gehrcke, December 9, 1920, GN 76-A-7; the chemist Rudolf Orthner thanked Gehrcke for his review of the manuscript of *Über physikalische und mathematische Abhängigkeit [On Physical and Mathematical Dependency]*, cf. Orthner to Gehrcke, August 24, 1928, GN 33-F-1.

[276] Vogtherr to Gehrcke, July 28, 1921, GN 3-J-23; on the publication of Vogtherr 1923, cf. also Vogtherr to Gehrcke, December 27, 1921: "Since I know that you take a lively, active interest in fighting the peculiarities that have crept into modern physics, I may perhaps take the liberty of approaching you with a request that has the same interest as its foundation." In 1932, after a longer break, Vogtherr approached Gehrcke despairingly: "Hopefully you will not hold it against me if I [...] take the liberty of approaching you in a desperate situation." Vogtherr to Gehrcke, April 29, 1932, GN 33-F-8. The issue was where he would be able to publish *Das Problem der Gleichzeitigkeit [The Problem of Simultaneity]*. The book was then published by Reinhardt (Munich) in 1933.

ten years. For Vogtherr, a physician who had no connection to the academic field, this exchange with Gehrcke, a physicist, was extremely important. "I still think back gladly on the pleasant hours that I was allowed to spend at your home and in your company. Unfortunately, I almost completely lack such stimulating scientific contact here."[277]

The first contact between Gehrcke and Mohorovičić occurred on March 3, 1921. Mohorovičić, the youngest person in the network of Einstein's opponents at 31, approached Gehrcke upon the recommendation of Johannes Plotnikow, "since I will be giving a lecture on the *dark sides* of the Einsteinian theory on the occasion of a colloquium at Prof. Plotnikow's because I am a staunch opponent of the Einsteinian special and general theory of relativity."[278] Over the next several years, Mohorovičić became one of Gehrcke's most important discussion partners.

Several publications offered indications of potential allies. Reuterdahl's "Einstein and the New Science" was set up as a kind of collection of materials and was therefore rewarding for other opponents of Einstein who appreciated this information. "There [in "Einstein and the New Science"] I found quotations unknown to me, to which I shall refer."[279] References to interesting authors or potential correspondents that resulted from personal correspondence were particularly promising; inquiries about unknown letter writers were also often obtained from acquaintances. Thus Kraus received a "suggestion re. book purchases" from Gehrcke and for his part advised Gehrcke as to which philosophical journals he could publish in, but also asked what Gehrcke thought of Fricke, who had sent him some articles and about whom he wasn't entirely sure what he should think.[280] Lenard inquired of Gehrcke about Lenore Kühn, whose pamphlet *Kant contra Einstein* he considered so useful that he suggested to an association that they allow Kühn to give a lecture in his place.[281] Gehrcke repeatedly made Einstein's opponents who corresponded with him aware of each other and distributed addresses. For instance, Geissler thanked Gehrcke "for the list of addresses that you sent me back then for the relativity article. I got agreement from most of them, and also started more extensive correspondence."[282]

277 Vogtherr to Gehrcke, January 31, 1926, GN 76-A-2.

278 Mohorovičić to Gehrcke, March 3, 1921, GN 3-G-2.

279 Mohorovičić to Reuterdahl, February 25, 1922, RP 4–35.

280 Cf. Kraus to Gehrcke, October 17, 1913, GN 72-A-4 (on the periodicals); cf. Kraus to Gehrcke, January 29, 1921, GN 71-G-10 (on Fricke). The relevant literature or correspondence was also very frequently forwarded, sometimes as loans that were passed along from one to the next. Cf. Fricke: "I am providing you with some German newspapers that Mr. Mewes provided to me, which contain very vehement attacks against Einstein." Fricke to Gehrcke, January 14, 1920, GN 68-E-18; "Enclosed further Einstein literature." Fricke to Gehrcke, September 11, 1921, GN 3-G-10.

281 Lenard to Gehrcke, November 6, 1920, GN 3-F-3. The letter does not show which association he recommended Kühn to as a speaker.

282 Geissler to Gehrcke, January 26, 1921, GN 3-J-9; Glaser received the address of the engineer Giulio Alliata from Gehrcke. Cf. Glaser to Gehrcke, July 22, 1921, GN 3-J-1.

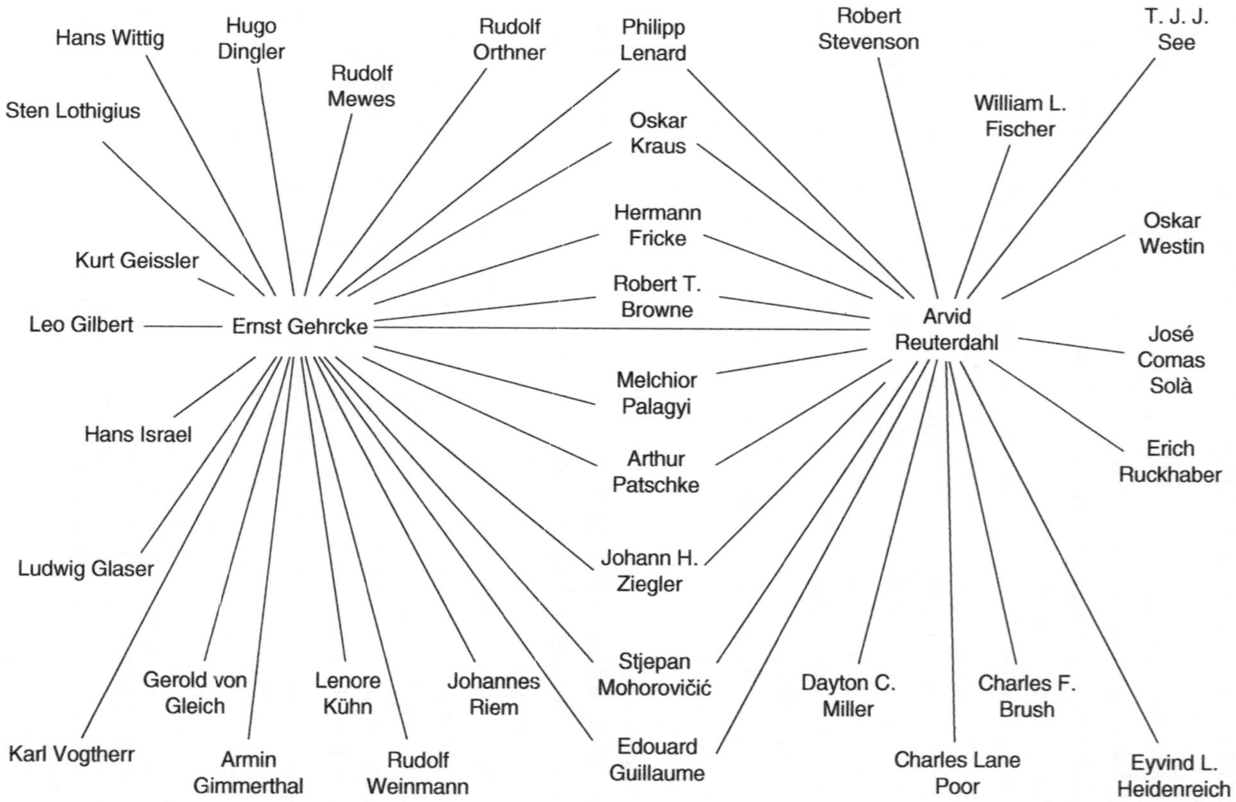

Fig. 4.4 The network of Einstein's opponents – based on the correspondence in Gehrcke's and Reuterdahl's papers[283]. Source: Milena Wazeck.

[283] The figure lists the most important correspondents; this also takes into account opponents of Einstein who are not the focus of this book and are therefore not listed in the table on page 13. In addition, both Gehrcke's and Reuterdahl's papers include letters (often only a single one) from opponents of Einstein not shown in the figure.

4.3.2 On the concept of the protest organization

Einstein's opponents contacted each other because they sought like-minded people who were conducting divergent science that was "correct" in their eyes, and in association with whom it would be possible to gain acceptance for this correct science against the current "aberrations" in physics. The protest against the theory of relativity became institutionalized. Until now, these protest organizations have been completely unknown and their role with regard to the protest against the theory of relativity is examined here for the first time. The fact that the institutionalization of the protest against the theory of relativity has escaped the attention of researchers until now has to do with the source material situation, which has considerably improved since Gehrcke's and Reuterdahl's papers have become available, and also to do with the concentration of Einstein research on Weyland (cf. pp. 3f., p. 240), which has prevented further inquiry about the opposition to the theory of relativity.

It is first necessary to clarify what should be understood as a "protest organization." Alongside the large scientific organizations, the specialized science organizations and the popular science associations, there were organizations that were characterized by their engagement in a non-academic program. Their founding was decisively determined by an attitude of protest not against other popular science or world view associations, but rather against academic science and by the effort to establish a counter-program. The popular science or world view associations, on the other hand, were committed to the idea of imparting academic knowledge. Thus the Christian Kepler League was indeed a "protest association" in a certain sense against the Monist League, which was critical of religion, but it specifically advocated "careful" popularization of academic knowledge and did not question acceptance of academic science as a producer of truth. In contrast, the renaming of the Breslau Physics Association to the Central Association for the Solution of the Problem of Attraction (cf. pp. 126ff.) represented its conversion into a protest association.

The source situation on these original organizations of "scientific outlaws" is extremely problematic. Sometimes founding manifests and articles of incorporation can be found in the world riddle solver publications, but it is often unclear whether they represent an intention that possibly never came into being at all, or at least not in the form that was announced. Thus Patschke's large-scale plans for implementing a "world exhibition of natural forces" were in all probability only a strongly utopian one-man enterprise that never got beyond the status of a plan.[284]

The following discussion is therefore limited to two organizations for which there are numerous documents in Gehrcke's and Reuterdahl's papers. The first organization assumed an outstanding position among these protest leagues because it was an

[284] References beyond an announcement in one of his publications were not found. Cf. Patschke 1925, pp. 134ff.

internationally focused network. It obtained its cohesiveness primarily from a consensus, not further specified in its content, that "correct" science had to be set against the theory of relativity. For the second, smaller organization, it was primarily specific content-focused interests that established cohesiveness. It is likely that additional protest organizations of this nature existed.[285]

4.3.3 The Academy of Nations as a protest organization of Einstein's opponents

The truth league...

The role of truth in Einstein's opponents' conception of science has been highlighted at various points in this book. This struggle for truth was not only reflected on paper, but also in targeted campaigns. The goal of this organization of the powers was just as much to establish a "true science," an "organization of the good people," as it was to combat the theory of relativity. This will become clear in the following example of the Academy of Nations (AoN). Both goals went hand in hand; the struggle against the theory of relativity was seen as the foundation for the reorganization of science.

In 1921, Reuterdahl and the esoteric Robert Browne (cf. pp. 253f. on Browne) decided to found an organization of "true scientists."[286] They had met in May 1921 when Browne approached Reuterdahl in a letter emphasizing the necessity of acting educationally in order to remedy the chaos in the area of knowledge for which the Einsteinian theory of relativity was particularly responsible in Reuterdahl's opinion. Browne's attention was probably drawn to Reuterdahl through a press report about one of Reuterdahl's lectures on religious topics. The AoN was intended to remedy the current complexity in the knowledge systems. "The Academy of Nations will function in the unification and co-ordination of systems of knowledge, thus procuring the development of a synthesized body of knowledge as against the highly specialized condition now existing." In addition, it was to be a world tribunal for deciding scientific and philosophical controversies and one of its prominent content goals was, "Discovery, investigation and dissemination of truth."[287] The struggle

[285] For instance, the "Weltbund" [World League] founded by Johannes Zacharias. However, there is only scanty documentation that it actually existed, in the form of two articles in the only preserved issue of *Weltwissen* from 1921. Cf. Rutan. (1921a). Was will der Weltbund [What does the Weltbund want]. *Weltwissen*, 5, 1–2, April 1, and Rutan. (1921b). Die Organisation des Weltbundes [The organization of the Weltbund]. Ibid., pp. 2–4. It can be seen, however, from the preserved volumes of the periodical *Weltwissen* that the protest organization "Gesellschaft Weltwissen" [Society of Universal Knowledge], likewise founded by Zacharias, existed at least for a few years.

[286] The founding of the AoN was announced for the first time in 1921 (without a more specific date). Cf. *Announcement. The Academy of Nations. Organized for the Advancement of Knowledge in all its Branches and for the Promotion of Amity, Unity of Purpose and Organized Cooperation among the Nations of the World. An Outline of its Purposes, Scope and Plan of Organization*, RP 3–5.

[287] Both quotes from Reuterdahl 1922a.

for truth and the defense of "correct" science and world view were at least as important for the members of the AoN as the struggle against the theory of relativity.[288] Gehrcke stated in his correspondence with Reuterdahl in advance of the AoN's foundation:

I see the primary point of your program in the term in §2 of the agenda: Truth! It must become a *Truth League*; this *positive* content alone will bring together the people for an effective defense against anti-cultural powers. Such *positive* content is far better than any merely *negative* one would be; this has been taken into account in your announcement booklet in a very clever way. What we want is an imperialism of *truth* – otherwise, no imperialism.[289]

The first point in a draft by Lenore Kühn on *What should the Academy of Nations be?* states, "It should be a collective upswing of the nations for united acquisition of ultimate truths." Point two states, "It should be an intellectual directive force for the path of civilized humanity, in the intellectual field, that republic of Plato's where the 'philosophers' are the rulers."[290] In this draft, there are many references to truth, but not a single one to the theory of relativity. In a letter to Reuterdahl, Palagyi called it "truth research," which united them against the "deep intellectual confusion of our day, the 'relativism'."[291] The fact that the AoN was understood as an exclusive circle and elite of the seekers for truth is also apparent in Dušan Grubić's letter to Reuterdahl, in which he expressed his "deepest thanks for the trust that you and the 'College of Fellows' of our Academy have shown me by electing me as a full member of this honorable body," emphasizing:

In association with our Dr. Mohorovičić and other members of our Academy, we will offer our best and do all that is possible to prepare the soil in our country for the noble tasks of the 'Academy of Nations' by creating our national institute in Yugoslavia (S.H.S.) [...] After all, we stand in the service of truth.[292]

Mohorovičić assured Reuterdahl that together they would be able to assert "correct" science: "I never doubted that our affair will succe[e]d. We shall write and fight so

[288] This is also shown not least by the fact that Gerold von Gleich, who played a special role in the network of Einstein's opponents as a non-academic opponent of the theory of relativity who did not argue non-academically, clearly showed his uneasiness with the universal claims of truth by the AoN members: "Frankly speaking, it [Kühn's draft of the goals of the AoN, cf. note 290 below] is somewhat too ambitious. Above all the 'acquisition of ultimate truth.' [...] Let's just start by saying that we simply want to combat the Rel.Th." Von Gleich to Gehrcke, April 4, 1925, GN 29-F-7.

[289] Gehrcke to Reuterdahl, October 17, 1921, RP 4–16. Gehrcke was referring to the goals of the AoN listed in Reuterdahl 1922a; cf. also Gehrcke to Browne: "With all my heart, I hope for beneficial and advantageous progress of the Academy of Nations in all your national institutions, as an international body for the dissemination of the truth for the benefit of all mankind." January 18, 1922, [draft], GN 82-G-5.

[290] [Kühn, Lenore]. *Was soll die Akademie der Nationen sein? [What should the Academy of Nations be?]*. [Undated draft], GN 29-F-10.

[291] Palagyi to Reuterdahl, October 12, 1922, RP 4–40.

[292] Grubić to Reuterdahl, January 17, 1924, RP 4–57.

long, until this clear scientifical truth conquers and I am sure that moment is not far."[293] He was encouraged in turn by Reuterdahl in phases of discouragement:

> Dear friend, do not lose courage! We must continue this fight for the preservation of 'free science.' The issue is greater than Einsteinism. These sinister forces are trying to stifle the truth. The few fearless men like ourselves, and the staunch Germans like Lenard, Gehrcke, Fricke, Riem, etc., etc., the Swiss – Guillaume, Ziegler etc. – constitute the truth phalanx which must battle on to victory. We will win eventually.[294]

The consciousness that the decisive battle for "true" science was imminent was also shared by Einstein's former Patent Office colleague Guillaume: "[I]l est urgent que toutes les bonnes volontés s'unissent pour lutter contre ces méthodes déplorables, qui ne pourraient que conduire à la mort de la Science."[295] The fact that there would have to be a "positive program" and that it would involve "the truth" was a goal that various opponents of academic physics could initially agree upon. So was the AoN just a league of noble seekers of the truth?

...and anti-Einstein academy

Along with this basic consensus on restoration of "true" science – whatever this was specifically understood to be – the struggle against the theory of relativity was the second unifying interest. The struggle against the theory of relativity was not understood as something different from the struggle for "true" science; instead, the theory of relativity was considered something that stood in the way of "true" science and should therefore first be removed. Half of the AoN's founding announcement accordingly polemicizes against the theory of relativity. The first sentences already leave no doubt as to the fact that the theory of relativity was perceived as a significant threat:

> We are emerging from a period of material and intellectual chaos. Nations have clashed in war. The intellectual world is still in conflict on the fields of knowledge. Never before has the demarcation between intellectual camps been so clearly defined. The meteoric rise of Einstein marks the beginning of this division in the modern kingdom of intellect. [...] Now the intellectual world is divided broadly into the Relativistic and the Anti-Relativistic schools. Einstein has served as a chemical reagent which has precipitated relativity from the present content of knowledge as a mass insoluble to the average man.[296]

Here, Reuterdahl and Browne clearly state that two "camps" existed in their view, which Fricke had also frequently pointed out (cf. p. 78). This "camp mentality" is a reason why the heterogeneity of content in the anti-Einstein movement moved

[293] Mohorovičić to Reuterdahl, July 1, 1923, RP 4–35.
[294] Reuterdahl to Mohorovičić, May 24, 1923, RP 4–35.
[295] "[I]t is imperative that all people of good will unite to fight against these deplorable methods, which could only lead to the death of Science." Guillaume to Reuterdahl, April 26, 1922, RP 4–19.
[296] Reuterdahl 1922a.

at least partially into the background. Anyone who rejected the theory of relativity and had the characteristics of a "proper" natural scientist (as demanded by Einstein's opponents) was in the right camp. In this strategic dispute, the terminology of Einstein's opponents was that of a battle, not that of science ("in our battle we have only *one* opponent and we must now quickly exploit those errors by our enemies who want to totally annihilate us;"[297] "We must use the same means as our enemies do"[298]). In Europe, opposition to the theory of relativity was the most important and perhaps even the only recruiting criterion for the AoN, a development that its Secretary General Browne, who had little interest in modern physics, observed with suspicion. The leading figures in the European national institutions presented in the following paragraphs – Gehrcke, Mohorovičić, Guillaume, and Ziegler – were all well-known opponents of Einstein and brought in their local networks.

Organizational structure

The AoN was run from New York, but was set up as an international federal organization that provided for the founding of national institutes that were able to make independent decisions (such as on accepting members). The AoN's "Committee on Organization" was composed primarily of persons who have already been discussed in this book: Arvid Reuterdahl, Robert T. Browne, Sidney T. Skidmore, Hermann Fricke, Thomas J. J. See, Eyvind Lee Heidenreich, Ludwig Glaser, Hudson Maxim, James F. Morton, Ernst Gehrcke, Sten Lothigius, Oskar Kraus, Johann Heinrich Ziegler, Lenore Kühn, Edouard Guillaume, J. G. A. Goedhart, Johannes Riem, José Comas Solà.[299] The AoN was organized like an academic academy. It had simple "members," "savants" who were at a level above the simple members, and a limited number of "fellows" who were above both in the hierarchy. Its origin in the theistic movement was apparent in that it was supposed to be led over the course of time by "archons," spiritual leaders.

The national institutes in the US and Europe

Along with Reuterdahl, Robert Tecumtha Browne (born 1882) was the central figure in the US portion of the AoN network. Browne functioned as the Secretary

297 Mohorovičić to Gehrcke, December 20, 1924, GN 29-F-1.

298 Mohorovičić to Reuterdahl, July 1, 1923, RP 4–35; cf. also Reuterdahl to See: "There is very little danger of overdoing the fight on Einstein, I have had the pulse both here and abroad on that matter for a long time. With the excellent control and censorship which the Einsteinians can exert, and because of the comparatively small number of Einstein's opponents, there is a hard fight ahead. Einstein and his cohorts are solidly intrenched and their Rock of Relativity can only be destroyed by continuous assault. I propose, for myself at least, to fight this charlatan until he is vanquished or I a[m] dead." May 17, 1923, RP 4–49.

299 "Committee on Organization", November 15, 1921, RP 3–5.

General of the AoN and thus as the head of the entire organization. His interest did not lie in the area of scientific knowledge, but rather occult knowledge. In 1919, he published *The Mystery of Space*,[300] a parapsychological book on the dimensions of space, and otherwise only published on occult topics. Presumably the AoN in the US developed into a theistic sect under Browne's influence. At any rate, Browne, who from 1925 called himself Mullah Hanaranda, is still seen as an authority in occult-esoteric circles today.[301] His goal was to integrate the AoN into a larger spiritual organizational structure.[302] Browne had little interest in Einstein criticism; he wanted to assert his theistic program using the AoN, meaning that Reuterdahl remained the contact person for Einstein's European and American opponents, who only shared Browne's program to a minor degree.

Reuterdahl's follower Heidenreich, together with Browne and Reuterdahl, constituted the "big three" at the top of the US AoN. The question of whether other US opponents of Einstein besides T. J. J. See who were in contact with Reuterdahl, such as Charles Francis Brush, Charles Lane Poor, or Dayton Clarence Miller, belonged to the AoN cannot be verified from the sources examined.

When Reuterdahl approached Gehrcke in fall 1921 with the idea of founding the AoN, through a referral by Fricke, Gehrcke immediately saw it as a potential forum for an institutionalization of anti-Einstein activities, which he had already suggested to Kraus the year before:

> Through your invitation to an Academy of Nations, a discussion took place at my home on October 23 among the following persons who are known to you from literary contributions to the Einstein affair: Fricke, Gehrcke, Glaser, Kühn, Riem. The discussion resulted in general consent to the plan and to participation in this Academy of Nations, whose founding by the American committee is greatly welcomed. […] The 5 persons listed have constituted themselves as the preliminary working group for the German section of the Academy of Nations.[303]

As for recruiting additional members, they initially wanted to wait "in consideration of the special circumstances"; the Americans were told that Kurt Geissler, Philipp Lenard, and Rudolf Mewes might join.[304] Gehrcke insisted that it would first be necessary to win over well-known scientists with widely appealing names before it

300 Browne, Robert T. (1919). *The Mystery of Space. A Study of the Hyperspace Movement in the Light of the Evolution of New Psychic Faculties and An Inquiry into the Genesis and the Essential Nature of Space.* New York.

301 Including, for instance, the Hermetic Society for World Service, which most recently prepared a new edition of Browne's writings.

302 Cf. "The Archons." *Occult Series*, 2, [no date], RP 03a-02.

303 Gehrcke to Reuterdahl, October 24, 1921, GN 79-A-4.

304 Ibid. Geissler joined in 1924 (cf. the membership form, GN 82-F-6), Lenard did not join, and it is unclear whether Mewes became a member.

Fig. 4.5 Robert T. Browne. Source: Reuterdahl papers; courtesy of the Department of Special Collections, University of St. Thomas, St. Paul, MN.

would be possible to go public; otherwise, they would not be acknowledged and "only serve as a target for attack by our opponents."[305]

In his function as "Acting Secretary General," Browne issued Gehrcke the official "Authority for the Organization of National Institutes" on November 15, 1921, thus authorizing him to establish a German national institute of the AoN. This semi-official impression given by the AoN was important for Einstein's opponents in order to establish what was, in their eyes, a serious counter-structure. At first glance, in fact, the letterheads from the US headquarters – the documents for authorizing the chairmen of the national institutes, the articles of incorporation, the bulletins, and not least the induction procedure for new members (detailed forms had to be filled out and a scientific paper submitted, in response to which an official membership certificate was issued) – do not show any difference from a normal scientific organization. From 1923 Reuterdahl even used a doctoral title awarded by the AoN.[306]

While the motivation for founding the AoN in the US was primarily Reuterdahl's and Browne's desire to assert a religious science, the question arises as to what Gehrcke actually expected from the German AoN and an alliance with the American theists. The anti-Einstein program appealed to him, and he hoped primarily for support for Einstein's German opponents – in terms of non-material and openly of material support.

In terms of non-material support, the honors that were frequently exchanged among the AoN members represented moral support. Thus Browne stated to Gehrcke, "We are deeply gratified to know that so eminent an authority as yourself will undertake the work of organizing The German National Institute of The Academy of Nations."[307] The financial side is discussed in the letter from Gehrcke to Reuterdahl, quoted above, on the constitution of the German section, under Point 6. "The [...] working group [of the German section of the AoN] assumes that, in consideration of the generally known difficult economic situation in Germany, funds will be made available by the American central committee in order to be able to form the German group of the Academy of Nations."[308]

The lack of funds was a constant topic among Einstein's German opponents. In the post-war period, the economic situation in Germany was strained, many people

[305] Gehrcke to Reuterdahl, October 24, 1921, RP 4–16. Cf. also Gehrcke to Browne: "We are of the opinion that the founding of a German National Institute under the local circumstances must best occur in that a larger group of famous and well-known researchers makes an appeal to the broad public sphere in order to recruit new members. Given the local conditions, it would be wrong if *such a small* group as we are thus far would enter the broad public sphere with an appeal! We have accordingly negotiated with a number of well-known scholars with an international reputation and initially explored opinions about founding the Academy of Nations *strictly in confidence*. We found a great deal of interest and encouragement, but so far all of them did *not yet immediately* want to decide, but rather to wait for the present to see how the whole matter and very specifically the financial side will turn out." June 20, 1922, [draft], GN 82-G-9.

[306] The award certificate, signed by Browne, can be found in Reuterdahl's papers, RP 3–1.

[307] Browne to Gehrcke, November 17, 1921, GN 82-G-4.

[308] Gehrcke to Reuterdahl, October 24, 1921, RP 4–16.

had no assets, and reparations payments and inflation characterized the post-war economy.[309] Now it is true that the reason Einstein's opponents lacked opportunities to publish had less to do with the paper shortage than with the academic system of quality assurance. The difficulty of placing their articles and books in established periodicals and publishing houses ever more frequently forced many of them to create their own publication venues, and the new foundation of periodicals or self-publishing cost money that was often not available.[310] Moreover, self-publication was linked to low prestige. On the other hand, a periodical belonging to the AoN and its own book series, financed from America if at all possible, were attractive. The Americans were also tempting them with a powerful ally, the industrialist Henry Ford (1863–1947).

On August 7, 1921, Reuterdahl wrote to Gehrcke, "Mr. Henry Ford has been a great help to my attack on Einstein. Most of the American scientific journals have been hoodwinked by Einstein – either by financial influences or by other means. I am enclosing a terrific attack on Einstein by Ford's paper *The Dearborn Independent*."[311] Gehrcke replied, "The fact that you have a connection with the industrialist Ford is very good and pleased me very much. Ford will be interested in unmasking Einsteinism. Einstein is the national saint of his race."[312] In a second letter, Gehrcke once again asked for financial support for the German section of the AoN.[313] Secretary General Browne denied it in a confidential letter, stating that the AoN would have a prospect for financing in America only when it had sufficient support and a membership base; the previous expenses had been paid by the American Fellows out of their own pockets.[314] The question of whether Henry Ford actually supported the American AoN financially is unclear, but rather unlikely. Apart from Reuterdahl's connection with Ford's newspaper, *The Dearborn Independent*, which was established through Stevenson,[315] no further contact is known. There is no correspondence with Henry or Edsel Ford in Reuterdahl's papers. Perhaps support from Ford could have been achieved through Stevenson, but it was no longer possible to implement this idea due to Stevenson's death in 1922.

309 On the economic situation in the post-war period, cf. the comprehensive treatment in Feldman, Gerald D. (1993). *The Great Disorder. Politics, Economics, and Society in the German Inflation, 1914–1924*. New York, NY/Oxford.

310 "If I had not lost all my savings to the inflation, it [the printing of Vogtherr's manuscript] would probably long since have occurred." Vogtherr to Gehrcke, May 5, 1926, GN 76-A-3. Einstein's US opponents also complained about the high costs of self-publication. Cf. e.g. See's statement, "My Aether work alone will have cost me close on to $2000.00." See to Reuterdahl, October 11, 1921, RP 4–49.

311 Reuterdahl to Gehrcke, August 7, 1921, GN 3-I-5.

312 Gehrcke to Reuterdahl, October 17, 1921, RP 4–16. This is one of the few direct anti-Semitic statements by Gehrcke that are expressed in the correspondence. On Gehrcke's attitude towards the topic of anti-Semitism, cf. pp. 241f.

313 Gehrcke to Browne, June 22, 1922, [draft], GN 79-A-5.

314 Browne to Gehrcke, July 27, 1922, GN 79-A-3.

315 Cf. Stevenson to Reuterdahl, December 30, 1921, RP 4–56.

Palagyi was also hoping for financial aid from America. He approached Reuterdahl with a request to publish his essays in American periodicals so that he could obtain foreign currency that would allow him to continue his physics experiments. "The current living conditions on this side and that side of the ocean are of such a strange nature that you would solve the problem of my opportunities for experimental work with the least bit of effort – at least that's how it seems to me!"[316] Experimental confirmation of his natural philosophy was of great significance to Palagyi, who considered himself both a physicist and a natural philosopher.[317] In another letter, he asked once again about the possibility of receiving an honorarium in foreign currency for an article. He was thinking of $100, to be delivered as quickly as possible, as long as the exchange rate was still so high.[318] Palagyi was betting on the hyperinflation that was just beginning. In January 1921, $1 cost 191.80 marks, but at the time of his letter in November 1922, $1 brought in 7,183.10 marks, while in January 1923 it was as high as 17,972 marks.[319]

Apparently the rumor that Henry Ford wanted to support their work had quickly made the rounds in the circles of Einstein's German opponents. Gehrcke was able to briefly get Lenard interested in the joint project with the Americans using the prospect of massive support by the industrialist Henry Ford, who was also a dedicated anti-Semite – which was likewise a recommendation for Lenard. In accordance with his strategy of winning over "big names" for the AoN, Gehrcke had initially attempted to harness Lenard as a crowd puller for the German section of the AoN. But Lenard paid more attention than Gehrcke to scientific quality in the choice of his associates, and was governed much more strongly than Gehrcke by nationalistic and anti-Semitic resentments. While Lenard favorably mentioned to Gehrcke the anti-Semitic publication *Einstein's Truglehre [Einstein's Fallacious Theory]* that had been published by the Hammer-Verlag publishing house, despite the weakness of the physics that he had discovered himself,[320] an alliance with an American like Reuterdahl was unthinkable for Lenard. He therefore wanted to have nothing to do with the AoN:

> To be honest about it, I do not like the Reuterdahl letter very much. His 'Acad. of N.' is – as you also more or less say yourself, incidentally – fooling around of the American type; nothing German. But only thoroughly German things can help us. I consider participating in any kind of international things (except for specialized scientific things in special cases) to be entirely wrong.[321]

316 Palagyi to Reuterdahl, October 12, 1922, RP 4–40. 317 Cf. ibid.
318 Palagyi to Reuterdahl, November 25, 1922, RP 4–40. 319 Cf. Feldman 1993, Table 1, p. 5.
320 Cf. Lenard to Gehrcke, May 7, 1921, GN 3-F-10.
321 Lenard to Gehrcke, October 14, 1921, GN 3-F-20. In a letter to Gehrcke dated July 21, 1921, he had already stated: "It could be that you are right to support him [Reuterdahl]; but for my part I would not like to attempt to extend my influence beyond Germany." GN 83-F-1. Major General (ret.) Gerold von Gleich also quarreled with the internationalism of the AoN. He protested resolutely against considerations of moving the headquarters of

The only international alliance that he would allow, he added, would be an alliance against the "Jewish spirit," but of course only if the "German spirit" would take on the leadership role.[322] The fact that the initiative for the foundation of the AoN emanated from the US did not suit him at all. "If Mr. R.[euterdahl] in America has difficulties with the Einstein sphinx, then he should first clean out *his* stall over there, instead of laughing and acting as if *we*, who see much more clearly, would be needy of American nannying." Despite these reservations, Lenard also thought in tactical terms and therefore continued his deliberations as follows:

This of course does not exclude the possibility that Mr. R.[euterdahl] can be *used* by *us*, which you surely can cleverly achieve. For example, Dr. Glaser, when he translates [Reuterdahl wanted to have his works translated into German], would for his part certainly have to put the *German* statements on the relativity that is misleading the people into the clearest of light, along with their authors, using suitable *additions*.

And it would be possible, moreover, to exploit the American freedom of speech for their own purposes. "Something like this – such as the relative national origin [of Einstein] – *must* be said quite publicly, and for that purpose, America is quite good." A year later, however, Lenard wrote a letter in fluent English to Reuterdahl. (Lenard had wanted to emigrate to England in 1890, but ultimately only worked there for six months.)[323] The letter deals with the various priority claims that were raised against Einstein, and particularly with the fact that Reuterdahl wanted to support the priority claim that Lenard had raised on behalf of Soldner (cf. pp. 277ff.). But those were preliminaries; the main focus for Lenard was now an international exchange on the characteristics of and defense against the alleged suppression of Einstein's opponents. There was no question in Lenard's mind that the Jews were behind everything:

Your remark that it is difficult, almost impossible, to get anything against Einstein published in America, was astonishing to me. The power[324] of the Jews is behind that, and it must accordingly be greater in your country even, than it is here. Indeed that may be seen too from Henry Fords very deserving book [*The International Jew*]. One would think that Henry Ford with his great means would be able to do something in America toward clarifying and saving

the AoN to Paris: "[A]s long as the scandalous Treaty of Versailles exists, [I am] completely incapable of commencing any contact whatsoever with French people, simply as a member of the German Nobility Society and the National Association of German Officers." Von Gleich to Gehrcke, April 4, 1925, GN 29-F-7.

322 All of the following quotes from: Lenard to Gehrcke, October 14, 1921, GN 3-F-20.

323 Cf. Lenard, Philipp. (2009). *Erinnerungen eines Naturforschers. Kritische annotierte Ausgabe des Originaltyposkriptes von 1931/1943 [Memories of a Natural Scientist. Critical Annotated Edition of the Original Typescript of 1931/1943]*, ed. Arne Schirrmacher. Berlin. Einstein's German opponents wrote in German, except for Ruckhaber and Mohorovičić who wrote in English, while the Americans wrote in English. Guillaume and Solà wrote in French, and the Scandinavians in Swedish. Reuterdahl stated to Mohorovičić that he had difficulties particularly with the German handwriting and that it took him a great deal of time to decipher the letters. Cf. Reuterdahl to Mohorovičić, May 24, 1923, RP 4–35.

324 In an earlier version: "influence."

the Aryan (non-Jew) spirit. But evidently there is a lack of self-sacrificing mental forces in sufficient numbers, including a lack of publishers and other men of the press.

Such assistance is still to be had in Germany. In case Henry Ford should actually wish to give support to auxiliary forces in the struggle to free the world, not only America, from the spirit of Jewry, he could do it most effectively and at the same time most economically through Germans. There is a considerable number of prominent German authors and scholars who are skillfully and successfully active for clarification from pure motives of self-sacrifice, (among them also Fritsch, the publisher of the German translation of Ford's book and G. Bartels, of Weimar, the historian of literature), and they experience so much opposition that it is difficult for them to make their living in their advancing age.

If such men had support, the frightful (tremendous) problem of saving the Aryan spirit would be as effectively advanced as it could be, and at the same time in a part of the Earth where men of strong and still pure Aryan spirit are yet to be found; for it is a problem of the fate of the entire world, a world-problem, as Henry Ford himself says. Perhaps it would be possible for you to direct Mr. Fords attention to this side of the question.[325]

The motivation for this letter is obvious and reflects the direction stated in the correspondence with Gehrcke of profiting as much as possible from cooperation with the Americans. Even Oscar Kraus, who had been confronted with the anti-Semitism in the anti-Einstein movement as early as 1920, turned to Reuterdahl in 1922 in order to request Ford's support. "Perhaps it is possible for you to do something for the intellectual life of the Germans here [in Prague] using your connections. – The economy measures of the government are primarily affecting German cultural interests. Would it be possible to forward the appeal to Ford?"[326]

Gehrcke attempted to recruit additional members for the AoN; among other things, he wanted to speak to potential members who seemed suitable to him at the hundredth anniversary celebration of the GDNÄ in Leipzig (cf. pp. 266ff.).[327] In the following years, the initial group of five members of the German section was expanded by an additional seven members, all of whom had publicly appeared as Einstein's opponents (except for the archaeologist Eckard Unger, for whom I am unaware of any position on the theory of relativity). While Karl Vogtherr, Gerold von Gleich, Kurt Geissler, and Hans Wittig were recruited as well-known opponents of Einstein, the forest zoologists Max Wolff and Anton Krausse of Eberswald did

[325] Lenard to Reuterdahl, October 22, 1922, RP 4–31. In a letter to Gehrcke dated December 21, 1921, Lenard had also suggested that a merger of the *Stark'sches Jahrbuch* with the *Physikalische Zeitschrift* (which was rejected by experimental physicists like Lenard himself) could possibly be countered by founding a new journal. Since the difficulties in finding a publisher in Germany were great, it would be possible "in the absolute worst case" to take advantage of the aid of the Americans and particularly of Ford. Lenard to Gehrcke, December 21, 1921, GN 83-F-6.

[326] Kraus to Reuterdahl, June 1922 [without date], with postscript dated July 17, 1922, RP 4–30.

[327] Cf. Gehrcke to Browne, June 22, 1922 [draft], GN 79-A-5. He additionally attempted to win over Palagyi and Mewes (cf. Gehrcke to Mewes, [undated draft], GN 3-I-12; cf. Palagyi to Gehrcke, July 22, 1921, GN 83-B-22; Palagyi to Gehrcke, August 21, 1921, GN 3-I-4). Neither probably joined, although Palagyi carried on an extensive correspondence with Reuterdahl (cf. RP 4–40).

not take a position against the theory of relativity with numerous publications;[328] however, both had signed the Leipzig call for action against the theory of relativity (cf. p. 269).

Differences arose as to the desire of Patschke, the world riddle solver, to join the AoN. As an engineer, Patschke had professional connections to Chicago and had also visited that city on the occasion of the 1893 World's Columbian Exposition. He had stumbled upon Reuterdahl and the AoN through a press report independently from the opponents of Einstein associated with Gehrcke. In accordance with the decision-making structures of the AoN, Browne handed over the decision on admitting Patschke to Gehrcke as the head of the German national institute.[329] But Gehrcke rejected Patschke's application:

With regard to Mr. Patschke, we decided not to recommend him as a member of the Academy of Nations for the time being. He does not appear to us to be a person who would be suitable at the current time to be useful to the German group, which is presently small. The question of whether he should be considered as a member at a later time is a separate question which is postponed for the time being.[330]

It was probably not only Patschke's ambitions as a world riddle solver that caused Gehrcke to refrain from an association in this case, but also the fact that he did not know Patschke personally – in contrast to the world riddle solvers Ziegler, Fricke, and Gilbert, who represented similarly undogmatic theories. Nevertheless, Patschke remained in contact with the Americans. Reuterdahl once again proved to be a networker, and Patschke proudly stated, "Through Reuterdahl, I was then put in contact with a larger number of first-class foreign natural scientists for the purpose of helping the truth in natural science to victory."[331] The "first class natural scientists" from America that Patschke is referring to are specifically Robert T. Browne, Dayton Clarence Miller, Sidney Tuthill Skidmore, and particularly Heidenreich, who was a professional colleague of Patschke. Patschke printed a portrait of Reuterdahl in his last work, *Weltlichttheorie [Universal Light Theory]*, and quoted Reuterdahl's praise of Patschke's work from the correspondence with him:

There are only a few men in the world who have nailed down the Einsteinian fiasco in the right place. I myself [Reuterdahl] have pointed out in America that the Einsteinian calculations on the Michelson experiment are based on a false conception of the nature of the light ether; but you deserve world priority. In a slide presentation at the local engineers' association, I pointed out your clear mathematical proofs and brought your works to the attention of

[328] But cf. Wolff, Max. (1922). Einige Bemerkungen zur Bewertung der Descendenz-Hypothese und neuerer Hypothesen der Physik [Some remarks on the assessment of the descendance hypothesis and more recent hypotheses in physics]. *Archiv für Naturgeschichte*, **88**(A)(7), 113–21.
[329] Browne to Patschke, May 10, 1922, [copy], GN 82-G-7.
[330] Gehrcke to Browne, June 20, 1922, GN 82-G-9. [331] Patschke 1925, p. 57.

first class movers here in America (Browne, Miller, Skidmore, See, and others), in Sweden (Westin), and in Yugoslavia (Mohorovičić).[332]

Due to the unstable political situation after World War I, time delays occurred in the founding of the Yugoslavian national institute. It is probable that Mohorovičić was made aware of Reuterdahl through Gehrcke's mediation. In a letter dated February 8, 1922, Browne stated, "We are endeavoring to extend the organization of the College of Fellows to all European countries. And we should esteem it a very great courtesy if you would kindly send us the names of persons in Poland and Russia or any other European country except Switzerland which is already represented."[333] Presumably Gehrcke recommended his acquaintance Mohorovičić in response. However, the first letter from Mohorovičić to Reuterdahl is not included in his papers, where the correspondence begins with a letter dated February 25, 1922, which refers to earlier correspondence. At any rate, Mohorovičić reported with satisfaction to St. Paul on October 18, 1923, that the Yugoslavian national institute of the AoN had now finally been founded.[334]

Its composition differed significantly from the German national institute. Along with Mohorovičić, the first three Fellows were Milan Marjanović, Dušan Grubić, and Dragiša Lapčević. Lapčević (1864–1939) was a co-founder and vice chairman of the Serbian social democratic party, and the head of the AoN institute; Marjanović (1879–1955), an author and publicist, did not appear as an opponent of Einstein or with ambitions in natural science. Mohorovičić pointed this out in a letter to Reuterdahl. "Our chairman is an literate and in a small degree interested in our strugg[le]; but he rejoice a great respect and he distinguishes himself with an extraordinary honesty."[335] Except for Patschke, Marjanović was probably the only European AoN member who visited the Americans in person. He dropped in on Browne in 1924 when he was in New York for a friend's art exhibition. Browne reported on this to Reuterdahl:

He is a man of spiritual inclinations and is a practical idealist, I should say. He does not impress me as being a strong man; nevertheless, I was pleased with his visit. Most of the time was spent in explaining to him some of the deeper denotations of the Pantelicon. He assured me, however, that there are not more than two or three men throughout the Kingdom of the Serbs, Croats and Slovenes who could measure up to the Principles of the Pantelicon, and inquired dubiously as to whether we thought there were any in America who were commensurate. He has evidently been disappointed with Americanism, somewhat.[336]

332 Ibid., p. 131. 333 Browne to Gehrcke, February 8, 1922, GN 82-G-6. See pp. 263ff. on the Swiss group.
334 Mohorovičić to Reuterdahl, October 18, 1923, RP 4–35.
335 Mohorovičić to Reuterdahl, November 6, 1923, RP 4–35.
336 Browne to Reuterdahl, August 19, 1924, RP 3–3.

The only member of the Yugoslavian national institute[337] besides Mohorovičić who had ambitions in natural science was Dušan Grubić. In 1917, he published *Das Universal-Kausalgesetz als unser oberstes Naturgesetz, ein Versuch des Aufbaus einer allumfassenden Naturphilosophie [The Universal-Causal Law as our Highest Natural Law, an Attempt at Constructing a Comprehensive Natural Philosophy]*.[338] However, I am also unaware of any positions taken by Grubić against the theory of relativity. Mohorovičić's strategy for building up the AoN in Yugoslavia apparently consisted of finding reputed and influential comrades-in-arms in order to ease the difficult status of being opponents of Einstein. In contrast to Gehrcke, who pursued the same strategy but limited himself to big names in science, Mohorovičić sought support among intellectuals in general. Using this strategy, he was also successful in repurposing the literary-popular science periodical *Univerzum* into the official publication of the Yugoslavian section of the AoN, for which financial support from the US was again planned in:

> [W]e have resolved to transform our scientific Review *Univerzum* in the Journal of our Institute; [...] In deed we have gr[e]at expenses by the publication of our own Journal, in consequence of what is now quite passive, but we trust that we shall find a great support on the part of our fraternal American National Institute, for we work now with our own very modest means.[339]

Starting with the May–August 1923 issue, the title page carried the addition "The Journal of the Jugo-Slavian Institute of the Academy of Nations" and an AoN logo. Mohorovičić had taken over editing the natural science section, and now requested contributions from Einstein's European and American opponents. Among other things, *Univerzum* printed critical articles on the theory of relativity by Gehrcke, Reuterdahl, and Fricke – translated into Croatian by Mohorovičić – as well as reports of critical contributions against modern physics by Gehrcke, Fricke, Reuterdahl, Heidenreich, Lenard, Ziegler, Dingler, and Wiechert, among others.[340] In contrast to Gehrcke, Mohorovičić was liberal in recruiting new members – by 1923, the Yugoslavian section already had 51 members, mostly artists, politicians, and literary figures.

A national institute was also supposed to be created in Switzerland; again, two of Einstein's opponents, Ziegler and Guillaume, were favored by the Americans as

337 No archive of Mohorovičić's papers could be found. The information on the internal structure of the Yugoslavian network comes from the material on this topic that exists in Gehrcke's and Reuterdahl's papers.

338 Grubić 1917. 339 Mohorovičić to Reuterdahl, October 18, 1923, RP 4–35.

340 Cf. e.g. Mohorovičić to Reuterdahl, June 27, 1923, RP 4–35; cf. Mohorovičić to Reuterdahl, July 1, 1923, RP 4–35: "In our scientifical review 'Univerzum' I shall publish under your name in English and in Croatian both your articles [...] So I shall make our learned circles better acquainted with your work and at the same time I shall rise the reputation of our scientifical review." A large number of critical articles on the theory of relativity can be found particularly in the 1922 and 1923 volumes. Cf. e.g. Gehrcke, Ernst. (1922–1923). Soldnerov efekat i teorija relativnosti. *Univerzum*, 1–2(4–5), 159; Fricke, Hermann. (1923). Novi rezultati Michelsonova pokusa. Ibid. 2(6–8), 191.

possible protagonists. Ziegler had been disappointed by the (at times) intensive contacts with other world riddle solvers such as Zacharias, since he thought that discussions only kept him from formulating the primordial light theory, and he therefore worked quite in isolation. For a brief period, he placed great hopes in the network. Ziegler thought that he had finally found a supporter in Reuterdahl and his Academy of Nations that highly valued the content of the primordial light theory and would bear it to the public.[341] It was not possible to reconstruct how the contact between Ziegler and the AoN was established. However, since Browne informed Gehrcke that they already had representatives for Switzerland (cf. p. 262), this contact must have existed earlier than the contact with Gehrcke. But the founding of a Swiss national institute of the AoN ultimately did not come to pass due to personal differences between Ziegler on the one side, and Guillaume and his comrade-in-arms René de Saussure on the other.[342]

Browne and Reuterdahl had apparently initially placed their hopes in Guillaume, for whom the AoN was truly a ray of hope in a period when people had to defend themselves against the "suppression" of "true" science. "Votre idée de créer une 'Academy' est très belle et a ma complète approbation. Vous pouvez compter sur moi comme un membre très dévoué. [Your idea of creating an 'Academy' is very nice and has my full approval. You can count on me as a very devoted member.]"[343] The AoN was expressly seen by Einstein's former colleague as a countermeasure justified by the marginalization of Einstein's opponents. Guillaume reports in this letter that he had spoken about the AoN with various people whom he considered suitable. In Switzerland, the AoN was based on the specific existing contexts and networks. Guillaume's closest comrade-in-arms, René de Saussure, a mathematician from the Institut de France, was also the publisher of an Esperanto journal. What interested him most about the AoN were the possibilities of working on translations; for instance, he wanted to translate the announcement of the founding of the AoN into Esperanto. Due to the difficult financial situation, however, Guillaume did not see himself in a position to found a national institute.[344] According to the Americans' conception, Ziegler, Guillaume, and de Saussure together would call a constituting session. However, they were not in agreement as to who should have which powers. Thus Browne wrote to Reuterdahl on November 8, 1923,[345] that Ziegler thought that Guillaume and de Saussure were acting behind his back, while de Saussure had reported as late as October 3, 1923, that he and Guillaume had not yet taken any steps whatsoever with regard to founding the Swiss AoN institute, for two reasons that Browne reported on from

[341] Cf. Ziegler to Reuterdahl, December 29, 1929, RP 4–66.
[342] Guillaume and Ziegler had been in contact since at least 1921. Cf. Guillaume to Reuterdahl, October 20, 1921, RP 4–19.
[343] Ibid. [344] Ibid. [345] All of the following quotes from: Browne to Reuterdahl, November 8, 1923, RP 3–3.

de Saussure's letter: "1st. that the Swiss scientists fear to link up in a fight against Einsteinism lest such action might imperil their official positions; 2d. that they have been busily engaged in the promotion of Esperantida."

While the first point likely applied primarily to Guillaume, the second point was clearly de Saussure's excuse. Browne, who could handle Ziegler's religious science significantly better than Guillaume's criticism of Einstein or de Saussure's interest in Esperanto, now suggested withdrawing Guillaume's and de Saussure's authority to found an AoN national institute ("holds no hope for future activities") and to concentrate it exclusively in Ziegler. It is true that Ziegler had initially tried to bring his small network into the AoN; he indicated interested parties to Reuterdahl that included the ophthalmologist Emil Hegg-Bitzius, who had written the document *Das Ewige im Zeitlichen: Eine naturwissenschaftliche Formulierung [The Eternal in the Temporal: A Scientific Formulation]*[346] on Ziegler's theory, and Amé Pictet (1857–1937), a professor of chemistry. On December 29, 1923, however, Ziegler told Reuterdahl that it was now too late to found a Swiss section in consideration of his advanced age.[347] In Sweden, there was also the intention of founding a national section of the AoN, specifically by Einstein's opponents Oskar Edvard Westen and Sten Lothigius.[348]

What became of the Academy of Nations?

Under Browne, the US Academy of Nations acquired an ever more spiritual orientation. This development was attributable to the fact that Browne and Reuterdahl focused increasingly on spiritual matters and classified the struggle against Einstein, which had taken up more than half of the founding announcement three years earlier, as less relevant. They studied the Bhagavad Gita, learned Sanskrit, and wrote about rebirth and karma yoga. The important thing for Browne was now to develop the "Pantelicon." Spiritual leaders, so-called "Archons," were supposed to come to the AoN from Germany, Japan, Italy, Russia, and America ("We are not to go out looking for them, however. They are coming to us, in time"), whereby the "Great Master of All" was expected from India.[349]

While the US AoN existed until at least 1930, it can be assumed that the German group was dissolved or petered out in the mid 1920s; at any rate, no correspondence can be documented between Gehrcke and the American AoN after 1925. This is possibly for the reason that at this point Gehrcke had gotten away from making

346 Hegg-Bitzius, Emil. (1914). *Das Ewige im Zeitlichen: Eine naturwissenschaftliche Formulierung [The Eternal in the Temporal: A Scientific Formulation]*. Bern.

347 Cf. Ziegler to Reuterdahl, December 29, 1923, RP 4–66.

348 In a letter to Reuterdahl, Westin mentions that he had received the authorization for himself, Lothigius, and others, to found a Swedish national institute of the AoN. Cf. Westin to Reuterdahl, July 10, 1922, RP 4–63.

349 Both quotes from Browne to Reuterdahl, August 19, 1924, RP 3–3.

public statements against Einstein – although he was a dedicated opponent of Einstein to his dying day – and had, moreover, shifted his interests increasingly into the medical area.[350] Riem was also no longer interested in sacrificing his rare free time on Sundays, which were dedicated to the meetings of Einstein's Berlin opponents, to the struggle against Einstein, wanting rather to spend this time with his family.[351] On January 27, 1927, Gehrcke reported on the fate of the AoN in a letter to Kühn:

> There is little news about the Ac. of Nations, it 'fulfilled its mission, since it brought together Einstein's opponents', as Lothigius wrote to me recently. (The Americans have become completely passive.) But we are occasionally using the enclosed card that I am including for you as a sample, as well as a very nice leaflet on my relativistic literature. It has become astonishingly quiet, however, in the entire scientific jungle, where things used to be so stormy. I am only receiving about 1/5 as many inquiries [...] as a year ago, almost exclusively purely physics and astrophysics matters.[352]

4.3.4 The joint protest campaigns of Einstein's opponents

Along with founding protest organizations (a second organization that was not created until 1933/34 will be presented below, following the chronology of the events), joint protest campaigns were carried out by Einstein's opponents in the 1920s and early 1930s. The protest campaigns discussed here are highly varied. Their common denominator was that they were campaigns jointly planned and carried out by Einstein's opponents that were intended to help "correct" science achieve a breakthrough, or to unmask the theory of relativity as an illusion, or both. I will first discuss the 1922 Leipzig protest statement of Einstein's opponents, and then present the strategically motivated campaigns for the benefit of alleged precursors of Einstein. At the beginning of the 1930s, the joint campaigns ended with Reuterdahl's application for the Nobel Prize and the publication of *100 Autoren gegen Einstein [100 Authors Against Einstein]*.

Leipzig 1922 – The protest statement of Einstein's opponents

On August 5, 1922, an article appeared in the *Leipziger Neueste Nachrichten* newspaper with an apparently innocuous headline, "Is Professor Einstein Coming to Leipzig?"[353] The GDNÄ wanted to celebrate its hundredth anniversary there with a festive assembly. A special highlight of the event was to be an address by Einstein on the theory of relativity. Therefore, the opponents of the theory of relativity had

[350] Gehrcke developed an artificial healthy climate (the "Gehrcke Climate"), founded Gehrcke Climate Institutes, and treated patients.

[351] Cf. Riem to Gehrcke, October 22, 1925, GN 77-C-2. [352] Gehrcke to Kühn, January 27, 1927, GN 38-I-8.

[353] Kommt Professor Einstein nach Leipzig? *Leipziger Neueste Nachrichten* newspaper, August 5, 1922, GN 4, XXI, 5r.

prepared well. They drafted a "protest statement" in order to prevent Einstein's address in Leipzig. This protest statement became particularly volatile due to the political events of the summer of 1922. After Foreign Minister Walter Rathenau had been murdered by members of the private paramilitary Freikorps groups,[354] the situation had become increasingly precarious for Einstein as a committed democrat and Jew. Einstein had been subjected to vehement nationalist criticism already on the occasion of his trip to France in spring 1922. At the time, the press stated that Einstein had "taken up the cause of the relativity of the national sense of honor."[355] Now, however, several months later, specific threats existed and Einstein decided to withdraw from public life for a period of time; rumors even circulated that he was considering leaving Berlin forever.[356]

On July 6, 1922, Einstein cancelled his promised lecture at the hundredth anniversary celebration in a letter to Max Planck, the chairman of the GDNÄ at the time, with the following justification: "I supposedly belong to the group against whom assassinations are planned by the folk-nationalists [...] Now nothing will help but patience and going out of town."[357] A little later, he reported to his friend Maurice Solovine, "There have been turbulent times here since the horrible murder of Rathenau. I am also being warned, have given up my course of lectures and am officially absent, but am in fact here after all. The anti-Semitism is very great."[358] Planck was outraged at the influence that right-wing radical circles could exercise with their threats, and wrote to Max von Laue on July 9, "Those rogues have actually taken it so far that they are able to block an event of historical significance to German science."[359] They agreed that von Laue would give the address on the theory of relativity in place of Einstein.

The circumstances for Einstein's canceling the address were discussed extensively in the public sphere. The course of this controversy is documented by numerous articles from the daily newspapers in summer 1922. Ever since the beginning of the excessive reporting on Einstein and the theory of relativity starting in 1919, it could be seen that the position of a newspaper for or against the theory of relativity corresponded conspicuously with its political orientation (cf. also pp. 269f.). Once the murder threats against Einstein became known, the

354 On this matter, cf. Sabrow, Martin. (1999). *Die verdrängte Verschwörung. Der Rathenau-Mord und die deutsche Gegenrevolution [The Repressed Conspiracy. The Rathenau Murder and the German Counterrevolution]*. Frankfurt (Main).

355 Der Film des physikalischen Nihilismus [The film of physical nihilism]. *Kinorat*, Nos. 9–10 (1922), GN 4, XVI, 67r.

356 Cf. Austritt Einsteins aus der Kaiser-Wilhelm-Akademie. Eine Folge der reaktionären Hetze [Einstein's resignation from the Kaiser Wilhelm Academy. A consequence of reactionary hounding]. *Neue Berliner 12 Uhr Zeitung* newspaper, September 29, 1922, GN 1, 1, 2r.

357 Einstein to Planck, July 6, 1922, MPIWG Archival Collection, 100-V.

358 Einstein to Solovine, July 16, 1922, reprinted in Einstein, Albert. (1960). *Briefe an Maurice Solovine [Letters to Maurice Solovine]*. Berlin, p. 42.

359 Planck to von Laue, July 9, 1922, quoted from Hermann 1994, p. 281.

democratic press immediately declared its solidarity with Einstein. The *Dresdener Volkszeitung* newspaper judged it this way: "It is a disgrace to all of Germany that a world-famous scholar can be put on the assassination list by a pack of unintellectual reactionary scoundrels and hounded out of the country,"[360] while the *Berliner Tageblatt* newspaper spoke of a "moral feralization [...] in large sections of right-wing radicalism."[361] In contrast, the anti-Semitic newspaper *Die Wahrheit [The Truth]* sneered that Einstein himself was at fault for losing the opportunity to give the address, "since the idea that there should actually be mad people who contemplate such murderous intentions is not credible."[362] In light of over 350 political murders that were committed by right-wing radicals in the period from 1919 to 1922, this can only be seen as an attempt to play down the real danger to Einstein.[363] The *Rheinisch-Westfälische Zeitung* newspaper even suggested to its readers to "interpret the flight produced" by Einstein "as advertising [...], which should allow his star, which had already noticeably faded, to shine with new lustre,"[364] thus insinuating that Einstein had stage-managed the threat to his life himself.

The idea that scientific acceptance and public enthusiasm for the theory of relativity was supposedly only thanks to "advertising" was also a firmly established fact for Einstein's opponents, who used the hundredth anniversary of the GDNÄ as a welcome opportunity to make a "protest statement" (see Fig. 4.6), despite the politically heated atmosphere.

The "protest statement" was presented as a scientific correction, but was clearly a strategically oriented statement that tried to derive its legitimacy from the presence of the theory of relativity in science and the public. The statement was leaked to the press and also distributed directly at the conference. Heisenberg, who was present as a student in Leipzig, reported that the leaflets were distributed at the entrance to the lecture hall and how appalled he was that a respected physicist like Lenard was participating in this campaign.[365] In fact, Lenard had publicly taken a position on the theory of relativity, also in political terms, since 1922. At the beginning of the second edition of *Äther und Uräther [Ether and Primordial Ether]* of 1922, he had placed "A Word of Admonishment to German Natural Scientists," which was

[360] Einsteins Absage. Der Gelehrte auf der Mörderliste [Einstein's cancellation. The scholar on the murder list]. *Dresdener Volkszeitung* newspaper, August 5, 1922, GN 4, XXI, 5r.

[361] Einsteins Absage an den Naturforschertag. Auf der Liste der Mörderorganisation [Einstein's cancellation at the scientists' conference. On the list of the murder organization]. *Berliner Tageblatt* newspaper, August 5, 1922, morning edn., GN 4, XXI, 6r.

[362] Einstein. *Die Wahrheit*, September 23, 1922, GN 1, I, 3r.

[363] Cf. Gumbel, Emil Julius. (1980 [1922]). *Vier Jahre politischer Mord [Four Years of Political Murder]*. Heidelberg.

[364] Die flüchtige Relativität [The flee(t)ing relativity]. *Rheinisch-Westfälische Zeitung* newspaper, August 5, 1922, GN XXI, 6r.

[365] Cf. Heisenberg, Werner. (1969). *Der Teil und das Ganze. Gespräche im Umkreis der Atomphysik [The Part and the Whole. Conversations Surrounding Atomic Physics]*. Munich, pp. 66ff., particularly p. 67 (published in English translation as *Physics and Beyond: Encounters and Conversations*).

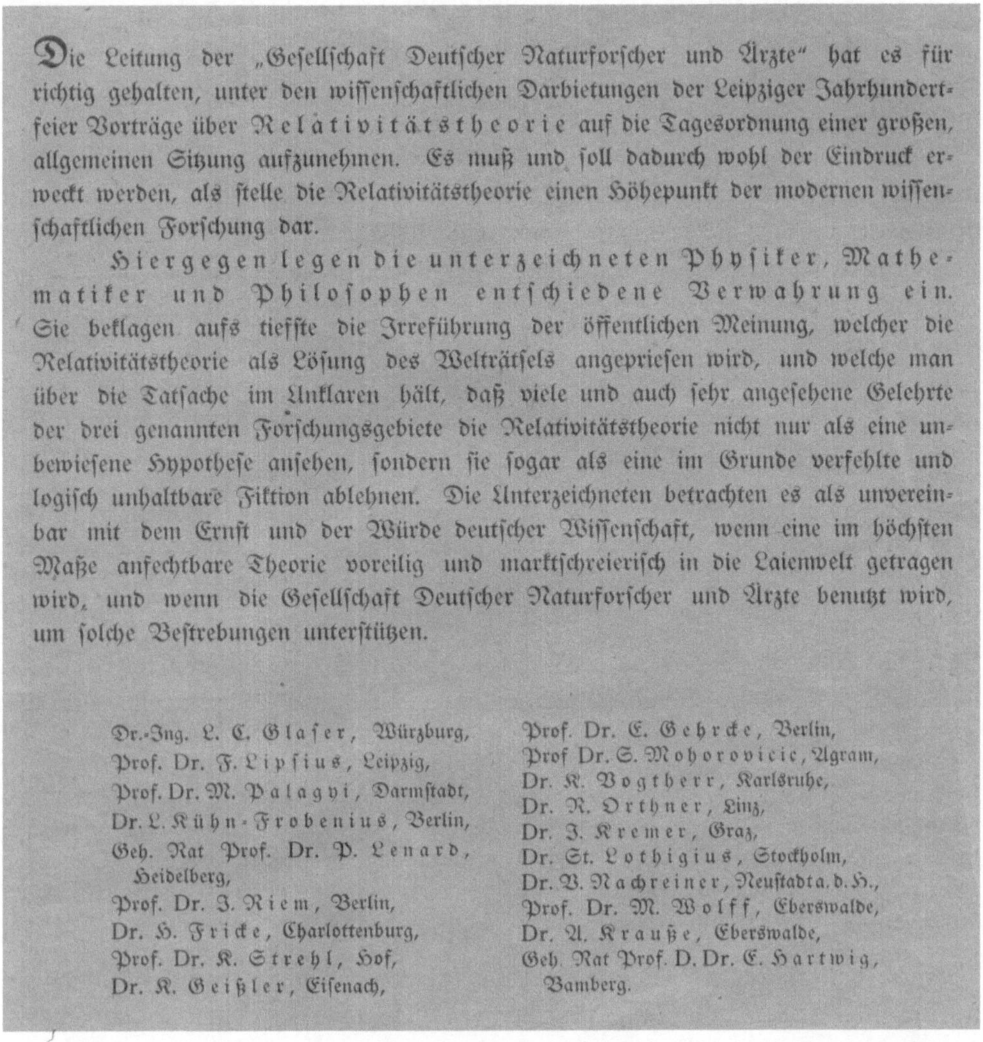

Die Leitung der „Gesellschaft Deutscher Naturforscher und Ärzte" hat es für richtig gehalten, unter den wissenschaftlichen Darbietungen der Leipziger Jahrhundertfeier Vorträge über *Relativitätstheorie* auf die Tagesordnung einer großen, allgemeinen Sitzung aufzunehmen. Es muß und soll dadurch wohl der Eindruck erweckt werden, als stelle die Relativitätstheorie einen Höhepunkt der modernen wissenschaftlichen Forschung dar.

Hiergegen legen die unterzeichneten Physiker, Mathematiker und Philosophen entschiedene Verwahrung ein. Sie beklagen aufs tiefste die Irreführung der öffentlichen Meinung, welcher die Relativitätstheorie als Lösung des Welträtsels angepriesen wird, und welche man über die Tatsache im Unklaren hält, daß viele und auch sehr angesehene Gelehrte der drei genannten Forschungsgebiete die Relativitätstheorie nicht nur als eine unbewiesene Hypothese ansehen, sondern sie sogar als eine im Grunde verfehlte und logisch unhaltbare Fiktion ablehnen. Die Unterzeichneten betrachten es als unvereinbar mit dem Ernst und der Würde deutscher Wissenschaft, wenn eine im höchsten Maße anfechtbare Theorie voreilig und marktschreierisch in die Laienwelt getragen wird, und wenn die Gesellschaft Deutscher Naturforscher und Ärzte benutzt wird, um solche Bestrebungen unterstützen.

Dr.-Ing. L. C. *Glaser*, Würzburg,
Prof. Dr. F. *Lipsius*, Leipzig,
Prof. Dr. M. *Palagyi*, Darmstadt,
Dr. L. *Kühn-Frobenius*, Berlin,
Geh. Rat Prof. Dr. P. *Lenard*, Heidelberg,
Prof. Dr. J. *Riem*, Berlin,
Dr. H. *Fricke*, Charlottenburg,
Prof. Dr. K. *Strehl*, Hof,
Dr. K. *Geißler*, Eisenach,
Prof. Dr. E. *Gehrcke*, Berlin,
Prof Dr. S. *Mohorovicic*, Agram,
Dr. K. *Vogtherr*, Karlsruhe,
Dr. R. *Orthner*, Linz,
Dr. J. *Kremer*, Graz,
Dr. St. *Lothigius*, Stockholm,
Dr. V. *Nachreiner*, Neustadt a. d. H.,
Prof. Dr. M. *Wolff*, Eberswalde,
Dr. A. *Krauße*, Eberswalde,
Geh. Rat Prof. D. Dr. E. *Hartwig*, Bamberg.

Fig. 4.6 Leaflet with the protest statement opposing the lectures on the theory of relativity.[366] Source: Max Planck Institute for the History of Science, Archival Collection.

366 "The leaders of the 'Society of German Natural Scientists and Physicians' considered it correct to include lectures on the *theory of relativity* among the scientific presentations at the anniversary celebration in Leipzig. This will certainly create the impression, intentionally, as if the theory of relativity would represent the pinnacle of modern scientific research.

The undersigned physicists, mathematicians, and philosophers firmly object to this. They deeply deplore this misleading of public opinion, to which the theory of relativity is praised as the solution to the world riddle, and which is kept in the dark as to the fact that many scholars, including very respected ones, in the stated research fields view the theory of relativity not just as an unproven hypothesis, but rather reject it as a fundamentally unsuccessful and logically untenable fiction. The undersigned consider it incompatible with the seriousness and dignity of German science when a theory that is extremely questionable is taken out into the world of laypersons prematurely and blatantly, and when the Society of German Natural Scientists and Physicians is used to support such efforts."

printed separately in the nationalistic daily newspapers.[367] In it, he polemicizes about the "foreign spirit, [...] which surfaces everywhere as a dark power and which also is so clearly developed in everything that belongs to the 'theory of relativity',"[368] among other things.

Many of Einstein's opponents came to Leipzig, either because they had arranged it in advance, or because after the conference in Bad Nauheim, this conference promised to become the next major event in the debate about the theory of relativity. Furthermore, Einstein's opponents, like other scientists, used the conference as an opportunity to get to know each other personally. Thus Rudolf Orthner and Gehrcke met for the first time in Leipzig.[369] Einstein's opponents were apparently satisfied with the effect of their protest. Gehrcke sent a copy of the protest statement to all of the participants on October 6 and stated in the accompanying letter that the protest had had a great effect both directly in Leipzig and in the press:

> Its effect was obviously that many doubted the correctness of the Einsteinian theory, which they had previously still considered accurate. I must also say that unfortunately it was no longer possible to include the signatures of Prof. Kraus – Prague and Dr. Mach – Munich (the son of E. Mach) because they were received too late.[370]

Einstein's American opponents were not involved in this protest campaign. A few months later, when T. J. J. See found out about it, he did write a press release for the Associated Press in which he more than doubled the number of protesters:

> In January of this year it was reported from Berlin that fifty German Physicists, Mathematicians and other philosophers were seriously grieved to see public opinion misled by the suggestion that the Theory of relativity is the solution of the problems of the Universe, and by the concealment of the fact that many savants, including the most distinguished, do not accept this Theory as a proved hypothesis, but look upon it as fiction.[371]

Vogtherr, who was unable to be present in Leipzig in person for reasons of time, was not satisfied with the press reporting, however, in contrast to other opponents of Einstein:

> As I saw from the 'Vossische' and 'Berliner Tagblatt' newspapers, the protest was taken up by that side with proper contempt; the daily press only knows about a political or tendentious attitude, which corresponds to their school of thought, and senses anti-Semitism behind it! No proper enlightenment can be expected from that side; the suggestion is already

367 Cf. Lenard 1922a; ibid. (1922). 'Ueber Aether und Uräther.' Ein Mahnwort an deutsche Naturforscher ['On ether and primordial ether.' A Word of Admonishment to German Natural Scientists]. *Deutsche Zeitung* newspaper, September 15, GN 4, I, 20r.

368 Lenard 1922a, p. 9. 369 Cf. Orthner to Gehrcke, May 10, 1928, GN 83-A-3.

370 Gehrcke to the signers of the protest statement, October 6, 1922, GN 4-B-3. On Ernst Mach's son as an opponent of Einstein, cf. Wolters, Gereon. (1987). *Mach I, Mach II, Einstein und die Relativitätstheorie: eine Fälschung und ihre Folgen [Mach I, Mach II, Einstein and the Theory of Relativity: A Forgery and its Consequences]*. Berlin/New York.

371 See, T. J. J. (April 18, 1923). To the Associated Press, RP 4–49.

seated too firmly for the people to still be able to see clearly. But this again shows how necessary a really proper enlightenment of the educated public is.[372]

This assessment by Vogtherr clearly expresses the pattern of interpreting the theory of relativity as mass suggestion, as outlined above. On the other hand, the practical-political significance of this manner of perception is apparent here. The suggestion that Einstein's opponents implied legitimated the protest. The protest was a legitimate and necessary means against the threat that the theory of relativity represented in their eyes, as well as against the alleged suppression by either the "Jewish press" or the "relativists." As a co-signer, Fricke countered press reports stating that the protesters had wanted to prevent a discussion about the theory of relativity as follows:

That is a misunderstanding. The signers naturally did not have the least objection to a discussion of this theory at such a conference. The protest was merely directed against the idea that the theory of relativity [...] had been placed on the agenda of a large general session where no discussion occurs, that is, that the opponents of this theory, who are becoming ever more numerous, do not get to speak.[373]

For Fricke, convinced as he was of the correctness of his mechanical ether theory, the theory of relativity was still an unconfirmed hypothesis that was greatly in need of discussion, even in 1922. Yet by this time the special theory of relativity was already a fixed component of academic physics, and the general theory of relativity was also generally accepted, so that an address on one of the most influential scientific theories of the twentieth century at the occasion of the GDNÄ's festive event almost went without saying. However, this was not accepted by Fricke and Einstein's other opponents, who vastly overestimated their own importance as critics of academic physics ("opponents [...] who are becoming ever more numerous"). Therefore, the mere fact that an address was to be given without a discussion on the principles of the theory of relativity represented suppression in his eyes. "It has the appearance as if any resistance against the theory was to be forcibly suppressed from the start."[374] Ziegler, who had not signed the call but had been present in Leipzig and sympathized with the call for action, took a positive report about the lectures on the theory of relativity that were given in Leipzig as an opportunity to make a correction that he saw as necessary:

372 Vogtherr to Gehrcke, September 25, 1922, GN 83-A-10. The *Berliner Tageblatt* newspaper had in fact placed the protest into the tradition of the public lecture series directed against Einstein at the Philharmonic. Considering the motivation and the participants, this was not incorrect. Cf. Die verbissenen Gegner Einsteins... [Einstein's determined opponents...]. *Berliner Tageblatt* newspaper, November 17, 1922, morning edn., GN 4, I, 35r.

373 Fricke, Hermann. (1922). Ueber den Gelehrtenprotest gegen die Relativitätstheorie [On the scholars' protest against the theory of relativity]. *Der Tag* newspaper, September 28, GN 4, I, 46r.

374 Ibid.

According to that [the positive report], it appears as if the so-called theory of relativity has finally arrived at general and unopposed recognition and has become a fixed component of science. However, that would be a serious error. On the contrary, that theory is continually losing ground, specifically to the same degree that those who were initially caught unawares by its excessive praise are coming to their senses again.[375]

Ziegler then constructed the legitimacy of the protest statement from a Jewish conspiracy and stated that "the large press media [...] in Germany" would be "almost exclusively in the hands of Einstein's national comrades" and would be fighting "a public discussion in the exclusive interest of Einsteinianism."[376] Mohorovičić also felt compelled to justify the protest. "A question had been put about what reasons might have suggested us in proceeding so rigorously against the theory [of relativity]."[377] For him, the protest likewise unequivocally represented a justified reaction:

[T]he manner ought to be mentioned in which Mr. Einstein himself and by means of his followers is endeavoring to work against us anti-relativists, furthermore the means they are making use of in consequence of which many men are obliged to continually bear such things on the part of those inconsiderate mobs! Up till now such a manner had never been used in science! If that theory had not been advanced by such manner its impossibility would have been proved already long ago and we never had proceeded so sharply. But now they had used different means in order to create that theory to a modern and dangerous religion. But we are not inclined to kneel down before the golden calf!

Einstein's opponents' opinion that their criticism of the theory of relativity was being deliberately suppressed was already rendered absurd by the fact that their protest statement was reprinted in many newspapers. Most of the newspapers reproduced the protest neutrally and, while some made critical statements, almost exactly the same number made positive ones. A newspaper like the anti-Semitic *Wahrheit*, which had just played down the murder threat against Einstein, presented the "protest [...] against the propaganda for the benefit of Professor Einstein" favorably in the same article.[378] The leaking of this protest statement, regardless of the circumstances of Einstein's cancelling his announced address, clearly shows that the protest against the theory of relativity in no way reflected the political context in which it was placed. Instead, the murder threats against Einstein were played down. Thus Reuterdahl inquired of Gehrcke, "What is the truth concerning the alleged threat against Einstein's life? Is this a form of Jewish sympathy propaganda?"[379]

[375] Ziegler 1922, Part 2. [376] Ibid. [377] This and the following quote from: Mohorovičić 1923b, p. 5.
[378] Einstein. *Die Wahrheit*, September 23, 1922, GN 1, I, 3r.
[379] Reuterdahl to Gehrcke, September 10, 1922, GN 4-B-2.

Since Einstein's opponents did not distance themselves from the attacks on Einstein as a person, instead taking advantage of them, they took a political position, even though their own motivation continued to be the scientific refutation of the theory of relativity. Since even entirely or mostly objective criticism of the theory of relativity represents political positioning by apolitical behavior in political contexts, where no position is also a position, Gehrcke's and particularly Fricke's behavior as opponents of Einstein in the National Socialist state will manifest itself even more blatantly (cf. pp. 296ff.).

The campaigns in favor of alleged precursors of Einstein

Along with the accusations of plagiarism, which referred to individuals' own theories and were therefore examined in the chapter on content-based criticism of the theory of relativity (cf. pp. 157ff.), Einstein's opponents also made accusations of plagiarism that were primarily of a strategic nature. Two of these accusations, which developed into full-blown campaigns, are interesting here; first, the Gerber case that involved the perihelion motion of Mercury, and second, the Soldner case relating to light deviation. The small deviation of Mercury's perihelion motion from the orbit required by Newtonian theory and the light deviation at the edge of the Sun are observations that can both be explained by the general theory of relativity. It was therefore of great importance to Einstein's opponents to be able to provide alternative explanations for these undisputed effects.

In his 1916 article "Zur Kritik und Geschichte der neueren Gravitationstheorien [On the criticism and history of more recent gravitation theories]," Ernst Gehrcke attributed the explanation of the phenomenon of the deviation in Mercury's perihelion motion to the senior primary school teacher Paul Gerber (1854–1909). In 1917, he published a reprint of Gerber's articles of 1898/1902 in *Annalen der Physik*.[380] Gerber's work was done in the context of the attempts in the latter third of the nineteenth century to apply the laws of electrodynamics to gravitation, as Zöllner had attempted, for example. Gerber wanted to replace the Newtonian assumption of instantly working gravitational force by the assumption based on electrodynamics that gravitation would propagate at the speed of light. He derived a deviation in the perihelion of Mercury from this assumption that corresponded well to the observed value.[381] However, the astronomer Hugo von Seeliger pointed out in response to the

[380] Cf. Gerber, Paul. (1898). Räumliche und zeitliche Ausbreitung der Gravitation [Spatial and temporal diffusion of gravitation]. *Zeitschrift für Mathematik und Physik*, **43**, 93–104; idem (1902). Die Fortpflanzungsgeschwindigkeit der Gravitation [The propagational speed of gravitation]. *Programmabhandlung des städtischen Realgymnasiums Stargard in Pommern [Program Transactions of Stargard Municipal High School in Pomerania]*. [Stargard] (also in *Annalen der Physik*, **357**(4), 415–41).

[381] On this discussion, cf. Seeliger, Hugo von. (1917). Bemerkung zu P. Gerbers Aufsatz: 'Die Fortpflanzungsgeschwindigkeit der Gravitation' [Comment on P. Gerber's article 'The propagational speed of gravitation']. *Annalen der Physik*, **358**(9), 31–32; ibid. (1917). Weitere Bemerkungen zur 'Fortpflanzungsgeschwindigkeit der Gravitation' [Further comments on 'The propagational speed of gravitation']. *Annalen der Physik*, **359**(17), 38–40; Laue, Max von. (1917). Die Fortpflanzungsgeschwindigkeit der Gravitation.

reprint of Gerber's work that he had been aware since its first publication that this derivation contained multiple errors.[382] Gehrcke did not accept this criticism from von Seeliger, but came under immense pressure when Lenard reported to him that the Heidelberg mathematician Paul Stäckel had confirmed an error discovered by von Seeliger.[383] Gehrcke attempted to evade replying to von Seeliger and suggested to Lenard to either let Stäckel reply or – a last desperate measure – to set a student to locate an earlier paper by von Seeliger in which Gehrcke thought he remembered that von Seeliger had mentioned Gerber without discussing an error in his calculations.[384]

Other opponents of Einstein such as Gerold von Gleich also criticized Gerber's work internally – "[I]t was entirely impossible to satisfy myself as to the validity of his [Gerber's] discussion. The more I dealt with it, the more untenable it seemed to me. Now it is actually a comfort to me that you write to me of Seeliger's negative attitude" – but wanted to appear as a united front to the outside. "I quoted Gerber in such a way that Prof. Seeliger can at least not oppose it,"[385] is how von Gleich reassured Gehrcke.

However, most of Einstein's opponents were not interested in either Gerber's gravitation theory or in the errors existing in it, but rather in the formula. The main issue was that a formula for the deviation in the perihelion motion of Mercury was possible outside of the general theory of relativity. For, according to Gehrcke's argument, if Einstein and Gerber had arrived at the same result, then Gerber's formula deserved preference, and he emphasized, "One may think what one wishes about Gerber's theory, but at any rate enough comes out of it that it is not necessary to employ considerations of relativity theory in order to derive Gerber's formula for the perihelion motion of Mercury."[386] Defending a formula about whose theoretical

Bemerkungen zur gleichnamigen Abhandlung von P. Gerber [The propagational speed of gravitation. Comments on the treatise of the same name by P. Gerber]. *Annalen der Physik*, **358**(11), 214–16; for secondary literature, cf. Schönbeck 2000, pp. 18ff.

382 Cf. Seeliger 1917a. Michele Besso also pointed out an error, cf. Besso to Einstein, December 5, 1916, in Schulmann *et al.* 1998, Doc. 283. Lenard also concedes in the foreword to the reprint of Soldner's paper (Soldner 1921), making reference to von Seeliger's and von Laue's objections, "that Gerber's calculation was not found to be without fault." Lenard (1921b), p. 594. Critical remarks on Gerber were already found in Osten, Hans. (1906). Über eine Erklärung der anomalen Bewegung des Merkurperihels [On an explanation of the anomalous motion of the perihelion of Mercury]. *Astronomische Nachrichten*, **171**(8), 119–20.

383 Lenard first attempted to appease Gehrcke: "Since I had strongly advised printing this paper in the Ann. at the time, I would like to say that it is very improbable to me to find a serious error in Gerber's work because the final result, the diffusion of grav. at the speed of light, is probably not to be doubted." Lenard to Gehrcke, October 30, 1917, GN 6-J-7. Then, however, Lenard received information from Stäckel that von Seeliger was correct in discovering a calculation error, but that Gerber's entire train of thought could possibly be rescued nevertheless. Cf. Stäckel to Lenard, November 14, 1917, [copy], GN 6-J-9, as well as Lenard to Gehrcke, November 22, 1917, GN 6-J-9.

384 Gehrcke to Lenard, November 24, 1917, GN 6-J-8.

385 Von Gleich to Gehrcke, February 3, 1923, GN 78-C-1.

386 Gehrcke 1916, p. 124.

Fig. 4.7 Paul Gerber. Source: Max Planck Institute for the History of Science, Archival Collection.

context one would rather not be nailed down, however, is an extremely defensive form of argumentation, yet Riem and Reuterdahl also used it. The latter stated that if Gerber's value for the perihelion deviation were correct, the theory of relativity would not be needed at all. How Gerber's value came about, on the other hand, was not interesting, or at least, only marginally.[387] However, Einstein's opponents were not only concerned with the priority of a formula ripped out of the context of its theory; instead, an accusation of plagiarism was raised from the beginning. In 1916, Gehrcke had already hinted that Einstein had plagiarized Gerber:

Thus Gerber already anticipated the Einsteinian formula; moreover, Gerber already explained the perihelion motion of Mercury quantitatively; he also pointed out the very small size of the perihelion motion that results for the other planets. One might think that this would be a great coincidence, and that Einstein would have arrived at the same result without knowledge of Gerber's paper. However, such an assumption is made more difficult by the fact that Gerber's treatise is discussed in the well-known mechanics by Mach, and that Einstein just recently indicated his precise knowledge of this book on the occasion of an obituary for Mach.[388]

In 1916 and 1917, this suspicion of plagiarism aroused little public sensation. Many of Einstein's opponents supported the priority claim asserted in favor of Gerber only in the 1920s in order to integrate it into a larger accusation of plagiarism together with the priority claim asserted for Soldner's explanation of light deviation (cf. pp. 277ff.). Oskar Kraus was one of the few who did not expressly insinuate that Einstein had plagiarized Gerber. In contrast to other opponents of Einstein, Kraus was remarkably fair and made a serious effort to understand the theory of relativity, although without success, since he discussed his problems understanding the physics primarily with Gehrcke, who himself had not understood them. He stated:

In my opinion, the question is whether Gerber had stated and utilized the thought that gravitation would propagate at the speed of light. The former is unquestionable – as for the calculation, Lenard has also defended it.[389] [...] But since Einstein arrives at the same quantum using other conditions, I have not wanted to believe in plagiarism on his part. In this way, Weyland [who had accused Einstein of plagiarizing Gerber's formula] put himself in the wrong.[390]

But in doing so, Kraus represented an absolute minority position, and the deceased Gerber was practically exploited for the purposes of Einstein's opponents. Here again, Reuterdahl played a central role. He wished to obtain a portrait of Gerber to illustrate a publication.[391] As Fricke reported to Gehrcke, Reuterdahl developed a

387 Cf. inter alia Reuterdahl 1921a, p. 9; cf. Mohorovičić 1923b, p. 2. Fricke thought that he would be able to provide a new justification for Gerber's formula based on the relationship between temperature and gravity that he postulated, and thus to block the attacks on Gerber's formula. Cf. Fricke to Gehrcke, May 2, 1920, GN 76-A-10.

388 Gehrcke 1916, p. 124. Cf. Mach, Ernst. (1919 [1883]). *The Science of Mechanics: A Critical and Historical Account of its Development*, translated by Thomas J. McCormack, Chicago [*et al.*], p. 535.

389 Here, Kraus makes a mistake; Lenard merely qualified the significance of the incorrect calculation for the overall value of the work. Cf. Notes 4 and 20 in Lenard 1918, Lenard 1920b, Lenard 1921a.

390 Kraus to Gehrcke, October 16, 1920, GN 71-G-5. Cf. Weyland, Paul. (1920a). Einsteins Relativitätstheorie – eine Massensuggestion [Einstein's theory of relativity – a mass suggestion]. *Tägliche Rundschau* newspaper, August 6.

391 Reuterdahl to Gehrcke, August 7, 1921, GN 3-I-6.

lively interest in Gerber's work and quotes from a letter from Reuterdahl addressed to him: "I am anxious to obtain the complete details of the work of the Oberlehrer Gerber [...] I would like very much to secure copies of the works of Dr. Gerber and photographs (2) for the purpose of framing and reproduction in my work."[392]

In fall 1921, through Gehrcke's agency, Reuterdahl located Gerber's widow in Stargard, Pomerania, where she was living in impoverished circumstances after the death of her husband. An acquaintance of the Gerbers from Stargard described Meta Gerber's difficult financial situation and suggested that Einstein's opponents should collect money in order to purchase the desired images from her.[393] Meta Gerber sent two photographs to Reuterdahl via Gehrcke and received in exchange two checks for 200 marks, again via Gehrcke.[394] Along with this support, she was thankful that her husband's work enjoyed such great interest among Einstein's opponents. "Deeply, from my heart, I thank you that you want to speak for my husband, then all of his work was not entirely for naught after all." She further indicated a request to Gehrcke "to explain to me how the Einstein matter is going, I know so little about it."[395] However, no further correspondence between Meta Gerber and Einstein's opponents is preserved. Meta Gerber was also in demand from other people. More than a year later, she gave the original manuscript of Gerber's article to Robert Braun, a private tutor in Cologne, who assured her that "not only Einstein's opponents, [...] but also Einstein's dedicated proponents – and I belong to them – will acknowledge without reservation that your husband may claim the right of priority for certain sequences of thoughts in the Einstein theory for himself." With him, according to Braun, it would be certain "that this valuable document has not fallen into unworthy hands."[396]

While Gehrcke had taken on the role of godparent for Gerber, so to speak, Lenard elevated the Bavarian geodesist and astronomer Johann Soldner (1776–1833) to new honors. In 1801, Soldner had published his treatise "Über die Ablenkung eines Lichtstrahls von seiner geradlinigen Bewegung durch die Attraktion eines Weltkörpers, an welchem er nahe vorbeigeht [On the deviation of a light ray from its straight-line motion due to the attraction of a heavenly body which it passes by closely]."[397] In the context of Newtonian emission theory, Soldner, who was primarily interested in whether this deviation has effects for astronomers'

[392] Reuterdahl, quoted from: Fricke to Gehrcke, August 5, 1921, GN 3-G-8.

[393] Kursch to Reuterdahl, August 13, 1921, RP 4–16.

[394] Gerber to Gehrcke, September 14, 1921, GN 3-K-8; Gerber to Reuterdahl, October 2, 1921, RP 4–16.

[395] Gerber to Reuterdahl, September 7, 1921, RP 4–16.

[396] Braun to Gerber, January 21, 1923, [copy], GN 83-B-24.

[397] Cf. Soldner, Johann. (1921). Über die Ablenkung eines Lichtstrahls von seiner geradlinigen Bewegung durch die Attraktion eines Weltkörpers, an welchem er nahe vorbeigeht [On the deviation of a light ray from its straight-line motion due to the attraction of a heavenly body which it passes by closely]. *Annalen der Physik*, **370**(15), 593–604. It first appeared in *Astronomisches Jahrbuch für das Jahr 1804 [Astronomical Yearbook for the Year 1804]*. [Berlin 1801].

observations, also calculated the values for the light deviation at the edge of the Moon and the Sun. The formula resulted in a light deviation at the edge of the Sun of 0.84 arcsec.[398]

Lenard initiated a reprint of Soldner's paper in *Annalen der Physik*, but in a strongly modified form. He had not only shortened, revised, and commented on the text, but sometimes even paraphrased it. Lenard also put an introduction before the text, which better deserves the name "Guide for Reading and Aid to Interpretation," since it constituted more than half of the article. Lenard's intention was to use Soldner's work strategically in the struggle against the theory of relativity, and not for any interest in the history of science or in a return to the physics of around 1800. In the introduction to Soldner's paper, Lenard hinted that Einstein had made use of Soldner's older work for his own involvement with light deviation. "The theory of relativity has attracted a great deal of attention; in contrast, Soldner's much older accomplishment also deserves generally accessible emphasis, all the more because – as always in such cases – no one can say to what extent the older accomplishment had been the cause and support for later involvement with the same object."[399]

Similarly to Gerber's case, the claim to priority or the suspicion of plagiarism raised in Soldner's case was rejected immediately. Max von Laue replied to Lenard's introduction in *Annalen der Physik*,[400] and in addition, von Laue as well as David Hilbert and Max Born took an article that appeared in the *Frankfurter Zeitung* newspaper on "Fame. Soldner and Einstein"[401] as an opportunity to correct the significance of Soldner's work in the daily press. After all, while the priority claim for Gerber in 1916/17 had been discussed in professional journals, largely excluding the general public, the Soldner debate of 1921 was greedily taken up by the daily newspapers.[402] Von Laue as well as Hilbert and Born emphasize in their

[398] This discussion is based on Jaki, Stanley L. (1978). Johann Georg von Soldner and the gravitational bending of light, with an English translation of his essay on it published in 1801. *Foundations of Physics*, **8**(11–12), 927–50.

[399] Lenard 1921b, p. 594.

[400] Laue, Max von. (1921a). Eine Erwiderung auf Hrn. Lenards Vorbemerkungen zur Soldnerschen Arbeit von 1801 [A reply to Mr. Lenard's introductory remarks on Soldner's article of 1801]. *Annalen der Physik*, **371**(20), 283–84.

[401] There, Baumgardt narrated the Soldner–Einstein case as a story in which the confirmation during his life for one – Einstein – who had predicted the light deviation, brought him happiness and fame, while the other – Soldner – had arrived at the same result more than 100 years earlier by other paths, but unfortunately had not enjoyed any confirmation and had therefore sunk into oblivion. Now if Soldner's prediction had been confirmed during his life, Baumgardt emphasizes in another article, then he would have earned the fame and Einstein would never have become so famous. Cf. Baumgardt, Ludwig. (1921a). Ruhm. Soldner und Einstein [Fame. Soldner and Einstein]. *Frankfurter Zeitung* newspaper, November 6, GN 4, XII, 72r; idem (1921b). Soldner und Einstein. *Frankfurter Zeitung* newspaper, December 2, evening edn., GN 4, XII, 75r.

[402] Cf. inter alia Riem 1921c; idem (1922b). Nachdenkliches zur Sonnenfinsternis im September [Contemplative thoughts on the solar eclipse in September]. *Rheinisch-Westfälische Zeitung* newspaper, July 29, GN 4, XX, 26r; idem (1922c). Von der deutschen Sonnenfinsternis-Expedition [On the German solar eclipse expedition]. *Deutsche Tageszeitung* newspaper, July 27, GN 4, XX, 27r; Die Relativitätstheorie bestätigt? [The theory of relativity confirmed?] *Hamburger Nachrichten* newspaper, May 25, 1923, GN 4, XX, 88r; Palagyi 1921b; Um die Priorität der Einstein-Theorie [On the priority of the Einstein theory]. *Magdeburgische Zeitung* newspaper, November 30, 1921, GN 4, XII, 74r; Nochmals Soldner und Einstein [Soldner and Einstein again]. *Neue Zürcher Zeitung* newspaper, November 30, 1921, second morning edn., GN 4, XII, 75r.

specific articles that the question of priority could not be asked in that way, since Einstein's and Soldner's values were calculated on completely different theoretical bases and were thus not comparable.[403] As in Gerber's case, however, this correction had no effects on the debate. Instead, the demand to grant Soldner historical priority spread rapidly among Einstein's opponents, who used the same argument as with Gerber. "But if Soldner's simple means are sufficient, then the great artificiality of the Einsteinian mathematical apparatus is superfluous, at least in this case."[404]

Particularly on the occasion of the solar eclipse in 1922, where more precise measurements of the light deviation were supposed to finally confirm the theory of relativity (the measurement results of 1919 had been insufficient for the astronomers[405]), the question of "Soldner or Einstein" was widely discussed. Johannes Riem published several articles in the daily press and stated on the deviation of light, "This effect must accordingly be designated the Soldner effect, even though Einstein derives the same assumption from his ideas more than 100 years later."[406]

The demand for "historical justice" for Soldner developed into a full-blown campaign and came to a boil in the US somewhat later than the 1921 debate in Germany. The network connections of Einstein's opponents played a decisive role in this – though belated – reception of Soldner in the US. It was Gehrcke who brought Soldner to Reuterdahl's attention, and it was Lenard who ensured that Reuterdahl received the reprint of the Soldner article from *Annalen der Physik*.[407] In their accusations, T. J. J. See and Reuterdahl always referred to the fact that "leading German physicists" – particularly Gehrcke and Lenard – had documented the plagiarism.[408] Reuterdahl sharpened the insinuations into a clear accusation of plagiarism and ensured that it was disseminated in the US. He held slide lectures intended to help Soldner achieve justice and published multiple newspaper articles that charged Einstein with plagiarizing Soldner's theory.[409] Reuterdahl also ensured the dissemination of the accusation of plagiarism among Einstein's American

403 Cf. Laue, Max von. (1921b). Soldner und Einstein. Die Entdeckung der Ablenkung des Lichts durch die Schwere [Soldner and Einstein. The discovery of the deviation of light by gravitation]. *Frankfurter Zeitung* newspaper, November 18, morning edn., GN 4, XII, 73r; Born, Max, and Hilbert, David. Soldner und Einstein. Ibid. Emanuel Beke also pointed out the fundamental dissimilarity of Soldner's and Einstein's explanations, cf. Beke, Emanuel. (1921). Lenard und Einstein. *Pester Lloyd* newspaper, November 29, GN 4, XII, 75r.

404 Palagyi 1921b. Riem said the same thing: "He [Soldner] had indicated a physical reason, while the theory of relativity is nothing more than a speculation that is built up on a highly dubious basis, is scientifically impossible, and philosophically just as impossible." Riem 1922b.

405 Cf. Note 4 in Chapter 2. 406 Riem 1922b.

407 The reprint from *Annalen der Physik* with a personal dedication from Lenard can be found in Reuterdahl's papers. It is unclear whether Lenard personally sent it on its way or sent it via Gehrcke or his publisher Hirzel, as occurred in the case of *Aether und Uraether*. In a letter to Gehrcke, he said that it would be desirable for Reuterdahl to receive the article, but that he could "not subject myself to a correspondence with the American." Lenard to Gehrcke, September 2, 1921, GN 3-F-16. But cf. Lenard's letter to Reuterdahl a year later, pp. 259f.

408 Cf. e.g. See 1923a; cf. also Crelinsten 2006, particularly pp. 213ff.

409 Cf. inter alia Reuterdahl 1921b; Reuterdahl 1921c; Reuterdahl 1922b; Reuterdahl 1923b; idem (1923c). Einstein's Predecessors. Dr. Arvid Reuterdahl Argues German Scientist Used Soldner's, Gerber's, and Other Formulae Without Acknowledgment. *The New York Times*, June 3, RP 4–4.

opponents that he knew of, including T. J. J. See, who in his position was able to achieve broader public resonance than Reuterdahl, and who even succeeded in getting the Soldner–Einstein controversy into the renowned professional journal *Science* (cf. pp. 202f.).[410] Just as in the German-speaking area, the accusation of plagiarism raised by See and Reuterdahl not only disregarded the differing theoretical contexts of the calculation of light deviation, but also referred very specifically to the formula and ultimately focused on the accusation that Einstein would have copied one of Soldner's errors. Reuterdahl stated:

Soldner, as Lenard noticed, made a slight mistake in his calculations in that he neglected the factor 2 in his formula. That was in the year 1801. In the year 1911 Einstein was overtaken by the same misfortune, that is, he also omitted the factor 2 in his formula [...] Through the courtesy of Dr. Lenard and Dr. Gehrcke, the author [Reuterdahl] has all evidence before him as he writes this article. This most unfortunate blunder, on the part of Einstein, in including the error of Soldner, was adroitly camouflaged in a subsequent article which appeared in the year 1915.[411]

The article mentioned, which was supposedly intended to hush up the plagiarism, deals with Einstein's general theory of relativity. Reuterdahl had hardly any knowledge about the differences between the special and the general theory of relativity, much less about differences between Einstein's papers on gravitation around 1911 and the general theory of relativity.[412] This lack of knowledge played a major role in the accusation of plagiarism. Einstein's paper "Über den Einfluß der Schwerkraft auf die Ausbreitung des Lichtes [On the influence of gravitation on the propagation of light]"[413] of 1911 did not yet include the additional deviation caused by the curvature of space and by exclusive application of the equivalence principle, arrived at a significantly lower value for light deviation, 0.83 arcsec, which roughly corresponded to Soldner's value. Since at first glance an error of the factor of 2

410 Cf. Reuterdahl's report to Mohorovičić: "I supplied many American scientists including Dr. T. J. J. See with this information. Dr. See's exposé of the Soldner–Einstein affair was based upon personal communication, and copies of my articles which I sent to him. His reference to Westin's part in the exposé was quoted verbatim from my article in the Dearborn Independent, issue of Jan. 6, 1923 [Reuterdahl 1923b]." Reuterdahl to Mohorovičić, [1924], RP 3–88; cf. also See to Reuterdahl: "I thank you very much for the data about Soldner's original paper, now reprinted by Lenard. It certainly proves Einstein to be a thief." May 22, 1923, RP 4–49.

411 Reuterdahl 1922b. On See, cf. Note 408 (this chapter), as well as various newspaper reports that reported on See's accusations published by a press release, e.g. Government scientist exposes Einstein trick. Declares German astronomer was detected in plagiarizing von Soldner in 1911. Newton theory held true. Says bending of stars' rays do not bear out claims which he terms crazy vagaries. *The Philadelphia Journal*, April 12, 1923.

412 On Einstein's path to developing the general theory of relativity, cf. Renn, Jürgen (ed.). (2007). *The Genesis of General Relativity*, 4 volumes, Dordrecht (Boston Studies in the Philosophy of Science Vol. 250), Vol. 1: Janssen, Michel, Norton, John, Renn, Jürgen, Sauer, Tilman, and Stachel, John. *Einstein's Zurich Notebook: Introduction and Source*; idem., Vol. 2: *Einstein's Zurich Notebook. Commentary and Essays*.

413 Einstein, Albert. (1911b). Über den Einfluß der Schwerkraft auf die Ausbreitung des Lichtes [On the influence of gravitation on the propagation of light]. *Annalen der Physik*, **340**(10), 898–908, reprinted in English translation in Stachel *et al.* 1994, Doc. 23.

seemed to exist in Soldner's calculation,[414] and according to the final theory of relativity (including the curvature of space) likewise resulted in a value that was about twice as great, this was seen as convincing proof of plagiarism, according to the following principle: having the same correct result is already suspicious, but if the errors are copied, too, then the situation is unequivocal. How irresistible the putative "calculation error" was is shown in the assessment of Harris A. Houghton in a letter to Reuterdahl:

Am I not right in believing that the best evidence of Einstein's duplicity is in the fact that Soldner corrected his own formula two years after his original article by inserting "2" in the denominator, a mistake which Einstein copied and afterwards corrected. It seems to me that this point is conclusive to the lay mind, of which type mine is. Such a thing could not have occurred without conscious plagiarism.[415]

In Germany, it was primarily Johannes Riem who conjured up the "factor of 2" in newspaper articles:

But now comes the odd thing about the matter. *The formulae of Soldner and Einstein are completely identical.* Lenard, who dug up the old treatise and published it, points out that Soldner commits a small oversight in his derivation. He neglects the factor of 2 and, after 110 years, it happens that exactly the same error happens to Einstein at exactly the same place![416]

Reuterdahl himself actually would not have used Soldner at all for his primary criticism of the "factor of 2." He already considered the fact that Einstein's calculations using the 1911 theory and using the final general theory of relativity resulted in different values to be an inconsistency in the theory of relativity. The truly bizarre core argument in Reuterdahl's criticism is that the theories of 1911 and 1915 are in principle the same, and therefore, if the results of the calculations of 1911 and 1915 are both mathematically correct, then there would be an obvious contradiction in the theory of relativity.[417] Many critics who, like Reuterdahl, hardly had any knowledge of the general theory of relativity and its development shared this view. Others argued in the firm conviction that if the theory of 1911 was a precursor of the theory of 1915, but the value of the light deviation resulting from it was declared to be false, then the theory of 1915, and any theory of relativity at all, would also be false. "[T]hus the relativity theoretician of 1911 would not just have refuted that of 1915,

414 However, Hans-Jürgen Treder and Gerhard Jackisch show that this factor of 2 was not an error, corresponding instead to the physics terminology of the eighteenth century. Cf. Treder, Hans-Jürgen, and Jackisch, Gerhard. (1981) On Soldner's value of Newtonian deflection of light. *Astronomische Nachrichten*, **302**(6), 275–77.

415 Houghton to Reuterdahl, July 15, 1923, RP 4–67. Houghton, a physician and military officer, played a central role in the English publication of the "Protocols of the Elders of Zion."

416 Riem 1923. Cf. also Lenard to Gehrcke, August 15, 1921, GN 3-F-15: "Both times [in Soldner and in Einstein], the factor of 2 is the error. Certainly, anyone can be mistaken with such things; but I hope that I have avoided that?"

417 Cf. Reuterdahl 1923d.

but would already have proven the impossibility of any relativity, both the general as well as the special, in 1911 by setting up an unsuccessful theory of relativity."[418]

Reuterdahl intensified the debate about Soldner using sharp wording, which in turn had repercussions for Einstein's European opponents. For instance, Mohorovičić's views on priority issues were considerably radicalized by his acquaintance with Reuterdahl. In his work *Die Einsteinsche Relativitätstheorie und ihr mathematischer, philosophischer und physikalischer Charakter [The Einsteinian Theory of Relativity and its Mathematical, Physical, and Philosophical Character]*, Mohorovičić argued that the Einsteinian value of 1.7 arcsec of light deviation would not necessarily have to prove the theory of relativity, but would also be explainable in the context of classical physics. This was not linked to an accusation of plagiarism; Mohorovičić only pointed out the possibility of interpreting this effect without the theory of relativity, stating "that one cannot speak of experimental confirmation of the general theory of relativity."[419] In his article "Worthlessness of Einstein's theory of relativity for the physics," which was written at Reuterdahl's request for the US market, Mohorovičić became more explicit. He downplayed the significance of the confirmation – "There had been no other thing but that a ray passing close by enormous masses of matter deviates extraordinarily little from its way and – nothing else" – and he now openly accused Einstein of plagiarism:

> [T]he measurements really confirmed the existence of s[o] c[alled] Soldner's effect which erroneously is called by modern relativists 'Einstein's' effect, whereas this was 'prophesied' more than 100 years ago (thus before Einstein) [...] Neither he [Einstein] was a faultless inquirer of natural things, because he did not mention anywhere his predecessors, although he had known them very well. That never could be pardoned to Mr. Einstein![420]

Even the church intervened in the US plagiarism debate with regard to Soldner. In 1929, Cardinal O'Connell made a statement about the theory of relativity. O'Connell had nothing good to say about Einstein anyway, since he had refused his wish for a personal discussion, transmitted via his wife, Elsa. "Frau Einstein said that Einstein did not wish to dispute with me about his theories and that my assertions left him cold."[421] These assertions will most likely have been accusations of plagiarism; the cardinal supported Reuterdahl's accusations of plagiarism, after all, stating that it would be typical behavior by Einstein to avoid disputes with the great scientists, with the "best minds in Germany," who had accused him, among other things, of plagiarizing Soldner and who had shown, he said, "that his so-called new theory of relativity is not new at all, but that whatever there is in it of scientific

[418] Nochmals Soldner und Einstein [Soldner and Einstein again]. *Neue Zürcher Zeitung* newspaper, November 30, 1921, second morning edn.

[419] Mohorovičić 1923a, p. 42. [420] Both quotes from Mohorovičić 1923b, p. 5.

[421] All of the following quotes from: O'Connell, William. (1929). Cardinal gives further view on Einstein theory. *Boston Evening American*, April 12.

value is nothing but a plagiarism of Von Soldner's system, explaining the deflection of light published as far back as about 1810."

The openness with which the cardinal presents his superficial knowledge is remarkable; even the year of Soldner's publication is incorrect. Without seeing any contradiction, the cardinal also approved of Reuterdahl's own claim to priority for the theory of relativity. "[T]his same theory [of relativity] was borrowed or filched from the works of Professor Reuterdahl." The cardinal considered the fact that God assumed a central role in Reuterdahl's theory a great advantage. According to this line of argument, Einstein not only plagiarized Reuterdahl's theory, but also simply let the existence of God fall by the wayside. For the cardinal it was clear: "[B]ehind the Einstein cloak lurks in very truth the ghastly spectre of atheism." This polemic by the cardinal shows particularly clearly that the campaign for Soldner was not about the question of historical priority, but rather that the Soldner argument was integrated into more comprehensive lines of argumentation against the theory of relativity that were shaped by world view.

In contrast to the accusations raised by Einstein's opponents that the theory of relativity plagiarized their own theories (cf. pp. 157ff.), the campaigns in favor of Soldner and Gerber were less content-based and more strategic in nature. Lenard did not want to promote the emission theory used by Soldner as a new basis for physics, nor did Gehrcke want to do the same for Gerber's electrodynamic gravitation theory.[422] For them, the main point was showing that classical physics explanations were possible in principle for the phenomena explained by the theory of relativity, in order to set something against such things as the persuasiveness of the connection between the special theory of relativity and the explanation of the outcome of the Michelson–Morley experiment. For this reason, they continually pointed out that there were alternative explanations. Initially, no fault can be found with doing this, but both Lenard and Gehrcke went further by making an accusation of plagiarism. If the classical explanation existed before Einstein's explanation, then it should be obvious, they said, that Einstein had committed plagiarism.

The argument that multiple individual explanations extracted from various theoretical contexts do not replace a theory that can explain all the effects was either ignored or simply not perceived by Einstein's opponents. Physicists like von Laue and Born insisted that the value of the theory of relativity lay in the explanatory power of the entire theory and not in an individual, isolated formula, and Frederick Bond stated vividly in the *New York Times*, "[I]f, 120 years ago, Archimedes,

[422] Although Lenard valued Gerber's electrodynamic explanation of gravitation very highly. Cf. Lenard to Gehrcke, December 1, 1916, GN 83-E-2. In a letter to Gehrcke dated December 10, 1920 (GN 83-E-19), however, Lenard conceded that if the explanation by Hugo von Seeliger stating that unknown masses near the sun were the cause of Mercury's perihelion motion, which he preferred, were accurate, then Gerber's formula would no longer have any value.

Galileo, Newton, Laplace and Benjamin Franklin had been assembled in some realm on Earth or among the isles of the blessed, they could not have worked out Einstein's theory. Gauss and Clark-Maxwell had to live first."[423] Einstein's opponents did not think of entering into this level of discussion. The campaigns for the "historical precursors" Soldner and Gerber were such a success not least because these individuals presented themselves to Einstein's opponents as persons to identify with, similarly to Robert Mayer and Eugen Dühring in the criticism of the professoriate (cf. pp. 58ff.). "Poor Soldner"[424] had to be helped to justice – this was a typical reaction, and one gets the impression that it was actually the "poor" opponent of Einstein himself who was meant.

The Nobel Prize for the unified theory!

Since Einstein's opponents were absolutely convinced that they had formulated a much better theory than Einstein, they sought authorities with whose help they could take steps against the non-acknowledgement of their theories. Some hopes were set on the government. Mewes, for example, turned to the Prussian Ministry of Culture in order to enforce his claims to priority, but received no answer from them. For a German nationalist like Mewes, this came as no surprise "considering the situation of the current direction there" in the Weimar democracy.[425] Mewes also registered his claim to priority with the Prussian Academy of Sciences "in relation to the Semitic professors A. Einstein and H. Minkowski."[426] However, the Academy did not take a position, initially stating that the letter had gotten lost. After Mewes lodged his accusations a second time, the Academy merely replied noncommittally that it was not their task to rule on priority issues.[427]

Ziegler set his hopes on private support, because he had gotten to know Ernst Solvay, a patron of science, in 1905 at a conference in Lüttich and pinned great expectations on this acquaintanceship. He had sent Solvay his works, and Solvay

423 Bond, Frederick. (1923a). Relating to relativity. *The New York Times*, May 13. Bond referred to Houghton, Harris A. (1923). A Newtonian duplication? Einstein's triumph a long distance ahead. *The New York Times*, April 21. In response, Reuterdahl entered the controversy (Houghton referred to Reuterdahl in his article and Bond had sharply attacked Reuterdahl's accusations of plagiarism published in *The Dearborn Independent* in his reply). Cf. Reuterdahl 1923c; Bond, Frederick (1923b). Reuterdahl and the Einstein theory. *The New York Times*, July 15; Reuterdahl, Arvid. (1923f). The origin of Einsteinism. Prof. Reuterdahl, replying to Mr. Bond, says that, in the absence of proof of innocence, Einstein stands convicted of plagiarism. *The New York Times*, August 12. All the articles, along with a letter from Bond to Reuterdahl, are in Reuterdahl's papers, RP 4–4.

424 Nochmals Soldner und Einstein [Soldner and Einstein again]. *Neue Zürcher Zeitung* newspaper, November 30, 1921, second morning edn.

425 Mewes 1922a, p. 418; cf. also Mewes 1920a, pp. 72ff. Cf. also Vogtherr: "In my opinion, it would actually be the task of the government or learned societies to come up with the printing costs for the works of authors without means, provided that they are not worthless." Vogtherr was thinking particularly of his own works that he was unable to publish due to a lack of funds. Cf. Vogtherr to Gehrcke, May 5, 1926, GN 76-A-3. In the National Socialist state as well, Fricke was still looking for an official authority that could provide him with a hearing: "But where in the new Germany is the office that reviews and promotes such discoveries [like his own]? Who is responsible? Who discovers the discoverer?" Fricke 1937, p. 195.

426 Mewes 1920a, p. 74. 427 Cf. ibid.

had replied kindly. Five years later, at the International Conference for Radiology and Electricity in Brussels, the opportunity for a second meeting arose. Solvay was not participating in the conference, but invited Ziegler to visit him at his company offices and accepted Ziegler's publications. However, this meeting did not go as Ziegler had hoped:

> Unfortunately, a strong difference in views was apparent at this discussion. Solvay considered reason alone to be insufficient to determine the fundamentals of a correct world view and still held the old view that wanted to solve this task using experiments and mathematical calculation.[428]

Ziegler did remain in contact with Solvay through mountain climbing. Solvay spent several weeks each year in the Engadine region of the Alps, lodging like Ziegler in Pontresina, and invited Ziegler several times to an evening glass of beer and once for a walk, "but he [Solvay] did not allow himself to be converted, despite the unified theory and the color sphere; instead, he persisted in his delusion that one could never determine the universal laws without mathematics."[429] Ziegler interpreted the following Solvay Conferences, financed by Solvay, which were dedicated to topics in modern physics,[430] from a conspiratorial perspective as assemblies that were deliberately directed against his primordial light theory. It was not just from Solvay that Ziegler attempted to obtain support; he also applied for several science prizes, including the Swiss Science Prize that was offered for the first time in 1920, the Marcel Benoit Prize, and – of course – to the Nobel Foundation for the Nobel Prize.

The Nobel Prize aroused particular desires in the world riddle solvers. Along with the prize money, what primarily lured them was the ideational support, the prestige that was linked to the Nobel Prize – finally, their own work would be granted the highest possible recognition, and they would be officially recognized as outstanding researchers. The award of the Nobel Prize to Einstein in 1921 proved to be a particular catalyst for these self-applications for the Nobel Prize. After all, the world riddle solvers thought that they had already refuted Einstein and were able to offer a much better and more useful theory. For this reason, the Nobel Committee was vehemently criticized. Ziegler – like other "unrecognized scientists" – accused the Nobel Committee of having perverted the prize from its original purpose of promoting promising scientists who really needed the support for their research (of course, he was primarily thinking of himself) into a reward for the luminaries.[431] At the same time, most opponents of Einstein knew nothing about the award procedure,

[428] Ziegler 1931, p. 66. [429] Ibid.

[430] The topics of the first three conferences were "Theory of radiation and quanta" (1911), "The structure of matter" (1913), and "Atoms and electrons" (1921).

[431] For a variety of reasons, the award policy for Nobel Prizes was then and continues to be frequently criticized. Cf. Ziegler 1931, pp. 133ff. On the award policy in general, cf. Friedman, Robert Marc. (2001). *The Politics of Excellence: Behind the Nobel Prize in Science*. New York.

for instance that self-nomination is prohibited, or they didn't care. Ziegler sent all of his works from the beginning to the Nobel Committee in Stockholm,[432] and the following dedication precedes Patschke's works which started in 1920:

Dedicated to the Academy of Sciences in Stockholm and to all friends of the natural sciences from astronomy to electronics and chemistry. The present work is part of my natural science works, which represent the product of many years of activity that were full of discoveries. These works will be publicized in suitable venues. They lead in a natural way, and therefore also in a simple, obvious way to correct knowledge and utilization of the natural forces, and thus to promotion of the general welfare. Based on the valuable results of these works for theory and practice, I am applying for the Nobel Prize for physics.[433]

The Nobel Prize was also particularly attractive to the world riddle solvers because they saw a role model in Nobel as a person who was inclined towards the practical successes of natural science. The inventor Nobel, according to their unanimous opinion, would never have wanted these "abstract," "peculiar" theories of modern physics to be recognized with his prize.[434] Along with such self-nominations and praise by non-academic colleagues ("This noble researcher and brave pioneer of true knowledge of nature, who has proven in his many works in various areas of knowledge to be a Galileo of natural science worthy of a Nobel Prize"[435]), Reuterdahl also carried out a targeted effort to receive the Nobel Prize. He was able to fall back on support from the network, and in contrast to the self-nominators Patschke and Ziegler, he familiarized himself with the structures of decision making in the Nobel Committee. The initiative came primarily from Reuterdahl's friend Heidenreich, who in 1915 had already attempted to convince academic science in the form of the Carnegie Institute of Technology (Carnegie Mellon University), as well as several other universities, of the value of Reuterdahl's works, but to no avail.[436]

Heidenreich and Reuterdahl had gotten to know each other on the occasion of lectures that they gave at the Kansas State College of Agriculture. Heidenreich spoke on reinforced concrete and Reuterdahl on his "space-time potential," and Heidenreich was fascinated with Reuterdahl's theory from then on. In August 1931, Heidenreich sent an application for Reuterdahl directly to Stockholm, including a list of signatures of supporters whom he had largely recruited from the network of Einstein's opponents. However, it was then pointed out to him that only former Nobel Prize winners and possibly the leading scientists in the world would have an influence on the selection of the Nobel Prize winners. Friends and acquaintances in

[432] Cf. Ziegler 1931, p. 74. [433] Cf. Patschke 1920; Patschke 1921; Patschke 1922; Patschke 1925.
[434] Cf. e.g. Ziegler 1931, pp. 133ff. [435] Patschke on Reuterdahl in Patschke 1925, p. 132.
[436] Cf. e.g. Heidenreich to Carnegie Institute, October 9, 1915, [copy], RP 4-24; cf. also Heidenreich to Kielland, December 15, 1931, [copy], RP 4-24.

Stockholm advised Heidenreich to try to get an endorsement of the application from the Massachusetts Institute of Technology (MIT) or Harvard University.[437]

Along with the attempt to win over a former Nobel Prize winner for the project with the assistance of the community of Scandinavian immigrants in the US,[438] Heidenreich placed great expectations in Edwin Pliny Seaver, the Superintendent of the Boston Primary Public Schools, who had studied at Harvard and who had been recommended to him for the Nobel Prize campaign by a mutual friend, Søren M. Kielland. Heidenreich warned Seaver that a difficult undertaking lay ahead. "Dr. Reuterdahl, for some reason or other, has been much feared, and therefore apparently deprecated by scientific institutions."[439] Seaver did approach Harvard University, but had to send a disappointing message to Heidenreich:

I have received a letter from Prof. Charles E. [H.] Grandgent as follows: 'Reuterdahl's credentials were examined by three of the most eminent experts in our Engineering School, and I regret to say they all agree in the opinion that he does not measure up to the standard of the Nobel Prize.' The men were H. E. Clifford, Dean of the School, G. W. Pierce, and E. L. Chaffee.[440]

Heidenreich insisted that Seaver should not give up so quickly, and Seaver then received information from Chaffee on February 4, 1932, that Harvard would only be willing to examine Reuterdahl's work in its entirety once again if an independent third-party opinion from among the physicists – he recommended contacting John C. Slater (1900–1976) of MIT or Edwin C. Kemble (1889–1984) at Harvard – arrived at something other than rejection as a result.[441] Heidenreich then approached Kemble, who became very blunt in response to Heidenreich's repeated inquiries, however:

My primary thesis is that the Nobel prizes in Physics have always been given to men whose work has been accepted by the great body of competent scientists and incorporated as an integral part into the structure of orthodox physical science. So far as I know in every case the prize has been given long after the quality of the work done has been fully recognized by practically all competent judges.

Just this qualification is, it seems to me, completely lacking in the case of Dr. Reuterdahl, although the collection of reviews which you have sent me shows that some men of scientific standing have had good words to say about his work. These reviews do not by any means constitute the kind of recommendation of which I speak.

437 According to Heidenreich in a letter to Kemble, February 12, 1932, [copy], RP 4-24.

438 Reuterdahl and Heidenreich received support from Erik Sjöstrand (the secretary of the Swedish Journalists Association), who wanted to attempt to win the support of the 1930 Nobel Peace Prize winner, Swedish archbishop Nathan Söderblom. Cf. Sjöstrand to Reuterdahl, November 30, 1930, RP 4-51.

439 Heidenreich to Seaver, January 5, 1932, [copy], RP 4-24.

440 Seaver to Heidenreich, January 18, 1932, [copy], RP 4-24.

441 Chaffee to Seaver, February 4, 1932, [copy], RP 4-24.

Certainly his work has not been incorporated into the body of accepted physical science and I suspect it to be true that there have been few if any papers by others of either theoretical or experimental nature based upon the work of Dr. Reuterdahl.[442]

Furthermore, Kemble had not understood Reuterdahl's works[443] – Reuterdahl was applying for the Nobel Prize in physics using his theistic work *The God of Science*, among other things. "I can make neither head nor tail of the articles of Dr. Reuterdahl that you have sent me." He said that he had not spent much time on the papers, "but the language which he uses makes no connection with my own equipment of scientific concepts and I see no reason for supposing that it conceals any important scientific truth."[444] The list of supporters with the names of Einstein's opponents – Heidenreich used a kind of chain letter that was initially sent to Mohorovičić, who then asked Gehrcke for his signature and to gather additional signatures[445] – was unable to convince someone like Kemble. During the Nobel Prize campaign, Heidenreich protested vehemently and repeatedly against "these professors and solidly established salaried scientists"[446] and reported about Kemble to Reuterdahl on March 3:

I received the [...] letter from Professor Kemble of Harvard, which shows he either does not understand your theory or will not understand it, claiming he never has seen or heard of it before, while naturally he is well acquainted with it and perfectly understands the Einstein conundrums. He advises you to send a convincing article of this kind to the Physical Review or some other reputable journal of physics and claims that thereby you will place yourself before the public. To my mind Professor Edwin C. Kemble must be either singularly dumb, stupid or a very lazy or slovenly reader of current discoveries of physical endeavors.[447]

After receiving rejections from all the institutions that were approached, and even from Reuterdahl's former employer, the University of Kansas, the Nobel Prize application fizzled out.[448]

100 Authors Against Einstein *goes to America*

A booklet with the ambitious title *100 Autoren gegen Einstein* appeared in 1931.[449] Even though the promised one hundred authors did not come together,

[442] Kemble to Heidenreich, February 29, 1932, [copy], RP 4-24.
[443] What was sent was: Reuterdahl 1923a, Reuterdahl 1928 and Reuterdahl 1929a. Cf. Heidenreich to Reuterdahl, February 16, 1932, RP 4-24.
[444] Kemble to Heidenreich, February 29, 1932, [copy], RP 4-24.
[445] Mohorovičić to Gehrcke, September 10, 1931, GN 33-B-5.
[446] Heidenreich to Reuterdahl, January 25, 1932, RP 4-24.
[447] Heidenreich to Reuterdahl, March 3, 1932, RP 4-24.
[448] The Chancellor of the University of Kansas, E. H. Lindley, diplomatically withdrew from the affair. "Our men have a high regard for Dr. Reuterdahl, but feel doubtful of his success in that competition. This is, of course, confidential. We have no desire to stand in the way of recognition of Dr. Reuterdahl's distinguished service in science." Lindley to Heidenreich, November 6, 1931, [copy], RP 4-24.
[449] Cf. also Goenner 1993b.

but rather only 28 original contributions beefed up with additional quotes from opposing documents that had already been published, it was the first published collective volume against the theory of relativity. Publishing such a collective volume had already been suggested in the 1920s, and the idea was pursued even after *100 Autoren gegen Einstein* came out.[450] The project of *100 Autoren gegen Einstein* was based on the initiative of the editors, Erich Ruckhaber, Hans Israel and Rudolf Weinmann, and was initially planned independently from Einstein's opponents in the narrower network around Gehrcke and Reuterdahl, who were, however, asked for contributions due to their fame as opponents of Einstein. The inquiries came from Rudolf Weinmann, who was managing the project as the chief editor. The letters were sent along with a two-page justification for the project and a list of the authors who were initially being asked.[451] However, Gehrcke rejected Weinmann:

For a variety of reasons, however, I would like to distance myself from involvement in this matter, not only because some of the points in the document don't appeal to me, but above all because I do not consider the entire campaign, as it lies before me, to be effective in the sense that its organizers apparently expect.[452]

He was then asked once again both by Weinmann and Israel,[453] but didn't allow himself to be persuaded. However, he did permit an excerpt from one of his older works against the theory of relativity to be printed. It is not known whether Gehrcke did not want to participate because the publicly conducted struggle against the theory of relativity had been over for him since the mid 1920s, or whether he had similar reservations to Mohorovičić's, who was skeptical about the campaign:

Today I received a letter from a Dr. Weinmann who is unknown to me, who surely had also sent you [Gehrcke] an identical letter. Please be so kind as to notify me as to

450 Lenard had already been thinking of a "little collective volume to combat the rel. nonsense" in 1920; cf. Lenard to Gehrcke, December 10, 1920, GN 83-E-19. Cf. p. 292 on Reuterdahl's planned collective volume. In 1942, Fricke wanted to publish a collective volume, "Critics of Modern Physics" with contributions by Gehrcke, Dingler, Thüring, Müller, Hilgenberg, Ruckhaber, Fricke, Stark, and Lenard, and possibly also by Glaser, Vogtherr, and Mey. Cf. Fricke to Gehrcke, November 9, 1942, GN 18-C-6. In another letter, he stated: "One would have to try to publish a competing encyclopedia. For years I have been planning an encyclopedia about aether and the modern conception of the world in which *everyone* should get a chance to speak. Fricke to Gehrcke, May 10, 1945, GN 3-G-8.

451 Weinmann to Gehrcke, January 12, 1930, enclosure: *Der Sturz der Relativitätstheorie [The Fall of the Theory of Relativity]*, GN 3-J-1.

452 Gehrcke to Weinmann, January 19, 1930, [carbon copy], GN 12-J-2. Gehrcke notes in the letter that Lenore Kühn is missing from the list of authors.

453 "We would certainly be particularly obliged to you if you would perhaps decide to participate after all. A change in the document could be discussed – for that matter, I would be thankful to hear about your reservations, also in *our* interest." Weinmann to Gehrcke, January 21, 1930, GN 12-J-3; cf. also similarly Israel to Gehrcke, January 23, 1930, GN 12-J-9.

how far I could pursue this invitation. A joint attack on the theory of relativity could hardly be led by people who do not have a significant name in science; for this purpose, the Hillmann publishing house does not seem to me to be sufficiently serious for such purposes.[454]

After *100 Autoren* came out, Weinmann turned again to Gehrcke. *100 Autoren* had been sharply criticized in a review by A. von Brunn in *Naturwissenschaften*.[455] Weinmann thereupon attempted to place an opposing opinion, but was thwarted by the editor, Arnold Berliner. As always, Gehrcke was interested in disputes with Berliner and was glad to have Weinmann send him the relevant correspondence. Weinmann further suggested that Gehrcke should do the following:

… strongly reject the *positively false* assertion that you and Lenard had 'long since come entirely to terms with the spec[ial] R[elativity] th[eory].' My request is now that you would authorize this rejection in a form that seems suitable to you, or, perhaps even better, simpler, and more effectively, cause the editorial office to withdraw this false allegation. It cannot in any way be permitted to remain unchallenged, I think. I am writing in the same vein to Prof. Lenard.[456]

However, Gehrcke did not wish to become actively entangled in a dispute with Berliner again. It was true, he said, that he had by no means come to terms with the special theory of relativity. "But approaching Dr. Berliner, whom I was only able to reach in an earlier case with the help of the Public Prosecutor, for the purpose of printing a factual correction in Naturwissenschaften […] does not seem advisable to me given the insignificance of the present matter."[457]

There are two reasons why *100 Autoren* is mentioned here. First, the volume was created in the context of marginalization and protest; its primary goal was to show by its mere scope that Einstein's opponents were not individual, marginalized people, but rather a large group to be taken seriously. The heterogeneity of the argumentation – proponents and opponents of ether stood side by side in *100 Autoren* – was sold by the editors as a strength: "for every argument on the Einstein side, the corresponding revealing counterargument will certainly be

454 Mohorovičić to Gehrcke, January 17, 1930, GN 12-J-7. In his cover letter, Weinmann had stated that Hillmann would probably publish the book, but it was ultimately R. Voigtländer that did so. Cf. Weinmann to Gehrcke, January 12, 1930, GN 12-J-1. Ruckhaber also reported to Reuterdahl about the difficulties in finding a publishing house for *100 Autoren*, even though they had deliberately put forward the academically regarded contributing philosophy professors Hans Driesch and Oskar Kraus. Cf. Ruckhaber to Reuterdahl, February 11, 1931, RP 4–48.

455 Cf. Brunn, A. von. (1931). [Review of Israel, Ruckhaber, and Weinmann 1931]. *Die Naturwissenschaften*, **19**(11), 254–56.

456 Weinmann to Gehrcke, May 8, 1931, GN 12-J-4.

457 Gehrcke to Weinmann, May 10, 1931, [carbon copy], GN 12-J-5. On Gehrcke's disputes with Berliner, cf. pp. 200f.

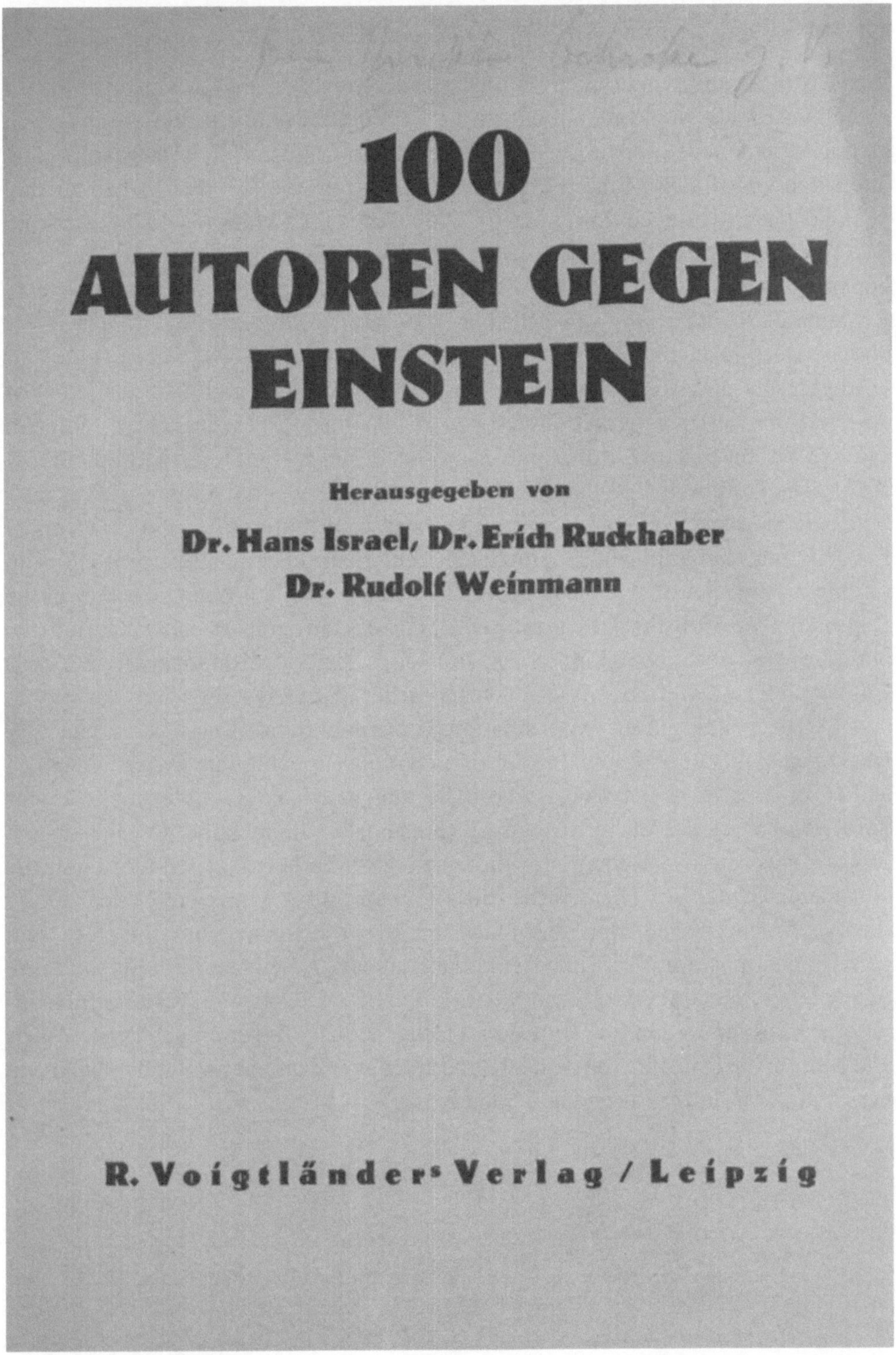
100 AUTOREN GEGEN EINSTEIN

Herausgegeben von

Dr. Hans Israel, Dr. Erich Ruckhaber

Dr. Rudolf Weinmann

R. Voigtländers Verlag / Leipzig

Fig. 4.8 An English translation of this collective volume (*100 Authors against Einstein*) opposing the theory of relativity was also planned. Source: Israel, Ruckhaber, and Weinmann 1931.

found."[458] Second, while Gehrcke was not interested in the project, Reuterdahl certainly was. "It would be a great stroke for truth if we could find the means of getting '100 Autoren gegen Einstein' published in English."[459] Reuterdahl's great interest in *100 Autoren* can be explained not least by the fact that he had already begun collecting brief contributions of 2,000 words for a volume with the working title *The Fallacies of Einstein* in 1921.[460] He wanted to use it to make Einstein's opponents known to the US public. In fact, it is probably only attributable to Reuterdahl's poor health – he had diabetes – that *Fallacies of Einstein* was not completed at the beginning of the 1920s.

Ruckhaber and Reuterdahl had planned to put an English translation of *100 Autoren* on the market in the US.[461] William L. Fischer (aka Wilhelm Ludwig Fischer) from Akron, Ohio, was also definitively interested in the book. He had been referred by Ruckhaber to Reuterdahl. Fischer not only wanted to have *100 Autoren* translated, but also expanded with contributions by US opponents of Einstein, including himself.[462] However, Reuterdahl foresaw great difficulties in finding a publisher. It would be necessary to deal with the financing and, in the Depression period, this was a problem that was difficult to solve. During the period of the world economic crisis, Einstein's American opponents were confronted with similar problems as ten years earlier when Einstein's German opponents had been sliding into hyperinflation.Reuterdahl himself was also financially weakened by his long illness. He considered it questionable that Dayton Clarence Miller or Charles Lane Poor, whom he suggested as possible sponsors for *100 Autoren*, would be in a position (presumably financially) to take on the project; according to Reuterdahl, though, only a certain L. J. Moore could be considered otherwise. There were still a few scientists who were critical of the theory of relativity, he said, "but many are afraid of losing scientific caste, and perhaps their position."[463] Thus Reuterdahl, at least, stated that he could not think of any other potential patrons for *100 Autoren*. Since Fischer knew Miller personally, he was supposed to give him a try. Fischer actually did approach Poor, Miller, and Moore, but it is quite obvious that nothing ever came of an English language version of *100 Autoren* – no such book exists.

458 Israel, Ruckhaber, and Weinmann 1931, foreword, p. 4.
459 Reuterdahl to Fischer, October 11, 1931, RP 4-14.
460 Cf. Prof. Reuterdahl to write book to show Einstein fallacies. *Evening Tribune*, April 18, 1921.Cf. also a letter from Guillaume to Reuterdahl announcing that he would write the requested text of 2,000 words, August 29, 1921, RP 4-19.
461 Cf. the correspondence between Ruckhaber and Reuterdahl in Reuterdahl's papers, RP 4–48.
462 Cf. Fischer to Reuterdahl, November 3, 1931, RP 4-14. He also asked Reuterdahl about other possible contributors in the US.
463 Reuterdahl to Fischer, October 11, 1931, RP 4-14.

4.3.5 *The German Society for Universal Ether Research and Comprehensible Physics as a protest organization of Einstein's opponents*

In the 1930s, after the decline of the AoN and following the joint protest campaigns, another protest organization of Einstein's opponents became established: The Deutsche Gesellschaft für Weltätherforschung und anschauliche Physik (DGWAP).

Organization and goals

The DGWAP was the result of a project that Fricke had long planned. He had wanted to found such an organization as early as 1920. Nothing initially came of this; he joined the AoN and belonged to the circle of Einstein's Berlin opponents around Gehrcke. He justified the formation of the DGWAP as part of this tradition of concerted action. Following the attempt at an uprising of the "true scientists" in Bad Nauheim in 1920, the protest at the GDNÄ meeting in Leipzig in 1922, and the AoN, he now felt called to continue the struggle for "true science" by founding an organization of his own.[464] In 1934, Fricke stated the following on the aims and objectives:

> The Society for Universal Ether Research intends to organize the errors and fallacies on which, among other things, the modern relativistic discussion is based in the books by Einstein, Planck, v. Laue, Born, Reichenbach, Heisenberg, Westphal, Jordan, Eddington, and many others, in an easily and clearly arranged way using calm, objective statements.[465]

Like the AoN, Fricke's organization had a double focus; first, against modern physics, and second, for the development of a new ether physics. The DGWAP appealed just as expressly to the orientation towards truth. "The Society [for Universal Ether Research] calls for help from everyone who cares about researching the truth!"[466] Fricke acted as the head of the society, a Dr. Kothe was listed as the managing director, and the society was registered at his home address. The other members remained largely obscure and were probably not very numerous ("the adherents of ether are terrible loners").[467] Fricke listed the physicists Karl Strehl and Lenore Kühn as sympathizers with the DGWAP. The founding members were Fricke, Erich Ruckhaber, and, from the subsequent generation of Einstein's opponents, the engineer Ott Christoph Hilgenberg (1896–1976).[468] When Gehrcke

[464] Cf. Fricke, Hermann. Die Notwendigkeit einer 'Gesellschaft für Weltätherforschung und anschauliche Physik' [The necessity of a 'Society for Universal Ether Research and Comprehensible Physics'], in Fricke 1939a, pp. 102–8.

[465] Fricke 1934e. [466] Ibid. [467] Fricke to Gehrcke, December 21, 1936, GN 33-H-3.

[468] By this time, there were multiple generations of Einstein's opponents, the new generations often referring back to the first generation of Einstein's opponents: Gehrcke's generation. In the 1950s, there were direct contacts between the first and second generations. Mohorovičić and Vogtherr were in contact with the group connected with *Wissen im Werden*, the periodical of the Gesellschaft für rationale Physik und Naturphilosophie [Society for Rational Physics and Natural Philosophy], in the 1950s; Mohorovičić was even an editor of the periodical along with Karl Sapper and Gotthard Barth. Gehrcke also published there. The dispute with the theory of relativity held central status in *Wissen im Werden*.

became a member is unknown, but in any event Fricke told him in 1948 in response to an inquiry that he still considered him a member.[469]

The meetings occurred on the third Tuesday of every month in the Reserve Officers' Casino at the Berlin Zoo and triggered a certain amount of press response. In a report on February 20, 1934, for instance, the *BZ* newspaper announced a lecture by Fricke for that evening. The report is uncritical and sensationalist – "'The Sun is cold – but eternal.' New space theory to be debated this evening" – and stated, "Nothing is so droll that one should not listen to it, if only the new idea is fruitful." For the *BZ*, incidentally, the fruitful droll ideas, although they were dwindling away, included both Einstein's theory of relativity ("today his meteor is dwindling away") and Darwin's theory of evolution ("it has been rejected again").[470] Lectures by other world riddle solvers were also given in conjunction with DGWAP events. Patschke's acquaintance, the world riddle solver Georg Wutke, gave a lecture on "The Motion of the Earth as the Cause for Major Earthly Natural Phenomena," and Major Carl Frederick Krafft one on "A New Ether Hypothesis."[471]

The struggle against the theory of relativity continued to play a major role among ether theoreticians in the 1930s.[472] After Reuterdahl and Gehrcke had turned to other areas, Reuterdahl suffered from poor health in the 1920s and ultimately died in 1933; it was then Fricke, still very active as an ether theoretician and opponent of Einstein, who maintained contacts with Einstein's opponents – and also with those who were not yet active in the 1920s. Thus he was in contact with Carl Frederick Krafft, who founded an "Ether Vortex Society" in Iowa together with Russel Moore.[473] The DGWAP's financial and staff capacity was extremely limited. Nevertheless, the society served as a point of contact for ether physicists. The DGWAP was only able to issue its publications as stapled typewritten booklets, but did so in great numbers, with more than 100 publications in the years from 1935 to 1947.[474] The topics ranged from *Die einfache Aufklärung des Einsteinstreits [The Simple Clarification of the Einstein Dispute]* to *Goethe als Entdecker einer neuen Meteorologie [Goethe as the Discoverer of a New Meteorology]* to *Plan zur schnellen Überwindung der europäischen Not [Plan for Rapidly Overcoming the European Distress]* by redirecting the Elbe river to Berlin.[475]

[469] Fricke to Gehrcke, July 21, 1948, GN 2-E-5.
[470] 'Die Sonne ist kalt – aber ewig'. Neue Weltalltheorie heute abend zur Debatte ['The sun is cold – but eternal.' New space theory to be debated this evening]. *BZ* newspaper, February 20, 1934, GN 29-F-11.
[471] Cf. Einladung zur Monatsversammlung der Deutschen Gesellschaft für Weltätherforschung [Invitation to the monthly meeting of the German Society for Universal Ether Research], 1934, GN 4-B-15.
[472] Cf. e.g. Fricke 1939b, p. 106. [473] Cf. ibid., p. 107.
[474] The numbering of the publications partly preserved in Gehrcke's papers extends up to No. 104 in 1947.
[475] Fricke 1934b; Fricke 1934c; Fricke 1947.

The "Literary Patent Office"

Along with the DGWAP, Ruckhaber and Fricke pursued an additional project that unmistakably bears traces of the preceding struggles for recognition, namely establishing a "Literary Patent Office." The fundamental principle of exchanging scientific recognition – the citation system of the scientific community – was unavailable to the world riddle solvers. References to their works in publications by other world riddle solvers were unable to offset this. An attempt to set up a "scientific assessment office" had already been made in conjunction with the AoN:

> An institution that, by a certain analogy to the patent office, serves to give everyone the opportunity to present his new scientific thoughts to a suitable body and have them reviewed objectively, has been lacking to date and would satisfy an existing need all the more in that the existing academies reject such a review as not part of their duties.[476]

At the beginning of the 1930s, the effort to codify the symbolic recognition procedures culminated in the demand for a patent on ideas and for the foundation of a "Literary Patent Office," as Ruckhaber demanded:

> The only thing that would help against these ignominious circumstances [of non-recognition of their work] would be an institution that the well-known gravity theorist, Gov't. Councilor Dr. Fricke, suggested years ago, namely an institute at which new scientific thoughts are registered similarly to inventions at the patent offices.[477]

Along with legally securing intellectual property against the alleged plagiarisms by modern physics that were distorting the world riddle solvers' research, the particularly important thing was access to the scientific discourse by institutionally securing the publication of the ideas. "[S]uch a protection of priority [could] be simultaneously linked with a fruitful thought and its author becoming known." Ruckhaber did not want to create some kind of additional institution, but rather to establish central management of all science. "A […] centralization of intellectual life, a forum before which there are no authorities, no arbitrary decisions by an editor, no partisan points of view." Ruckhaber imagined a privately founded institute with several departments under the head department of philosophy, which, contrary to the "partisan daily newspapers or periodical press," could ensure that an "unadulterated philosophical spirit" would be disseminated.

[476] [Confidential circular to the members of the German national institute of the AoN], [around 1921/1922], GN 82-G-12. Cf. also Fricke, Hermann. (1921). Die Akademie die uns fehlt. Ein praktischer Vorschlag zur Neu-Organisation der Wissenschaft [The academy we are lacking. A practical suggestion for re-organizing science]. *Weltwissen*, 5, April 1, pp. 9–10.

[477] All of the following quotes from: Ruckhaber, Erich. ([1931]). *Expansion und Unendlichkeit [Expansion and Infinity]*. Berlin, Section B. This document was already designed as a patent application for the patent office that was to be founded. Ruckhaber wanted to patent the thought of expansion of substance as the primordial reason for all events.

The DGWAP and national socialism

As a protest organization against the theory of relativity in the 1930s and 1940s, the DGWAP found itself in a completely different political context than that of the Academy of Nations at the beginning of the 1920s. The fact that the DGWAP was founded in 1933/34 was not happenstance; through a series of lectures at the Humboldt-Hochschule [college], Fricke had gathered a circle of adherents around him during these years that he wanted to put on a permanent basis. The Humboldt-Hochschule Berlin was an adult education college that was created in 1915 by a merger of the Humboldt-Akademie adult education center (that had been operating since 1878) with the Freie Hochschule [Free College] adult education center (founded in 1902), each of which had about 20,000 students at the time of the merger. According to the 1919 and 1920 lecture catalogs, the theory of relativity was an object of knowledge transfer.[478] This was followed by many years without events on the theory of relativity at the adult education center, and it was only in Fricke's lectures in the 1930s that Einstein's theory appeared in the catalogs again. Fricke debuted at the Humboldt-Hochschule on January 28, 1933, with a lecture on his solar theory, "Overturning our view of the Sun (with slides). The unsatisfying cooling theory. The Sun, dark and cold on the inside. Space atmosphere and cosmic meteorology. Gravity and universal ether. The nebulae ['star fog'] as real fog." This course was so popular that the adult education center set up repetitions. "The expositions by Dr. Fricke [...] on his new conception of the world have excited the greatest of sensations, so we are offering our students two additional opportunities to attend the lecture."[479]

Was it only by chance that Fricke succeeded in establishing his universal ether research at the Humboldt-Hochschule just at the time of the National Socialist

[478] This course was usually taught by Viktor Engelhardt. He lectured on: "From Aristotle to Einstein. The history of the development of the physical conception of the world." Lecture Catalog, 4th quarter of 1919, p. 80; "The world view of recent physics," Lecture Catalog, 1st quarter of 1920, p. 65; "The theory of relativity," Lecture Catalog, 2nd quarter of 1920, p. 33. In the 4th quarter, Engelhardt already offered the course at two different times, and Cl. von Horbath additionally offered "The foundations of Newtonian mechanics in light of the Einsteinian theory of relativity," and there was a course given by Max Apel in the Philosophy Department on "The theory of relativity and philosophy. A critical discussion of the fundamental theories of relativity in their significance for the philosophical world view," Lecture Catalog, 4th quarter of 1920. The Lecture Catalog for the 3rd quarter of 1920 was unavailable to me. The lecture catalogs can be found in Volkshochschule Gross-Berlin.

[479] Lecture Catalog, April–June 1933, p. 2. Fricke gave lectures on April 21 and 28. Then, according to the Lecture Catalog, Winter 1933, p. 4, as well as the Lecture Catalog, Oct.–Dec. 1933, p. 11, a regular series of lectures in the classification "Natural History." Next, an individual lecture on January 27, 1934 (Lecture Catalog, Jan.–March 1934, p. 3), as well as a regular course (ibid. p. 9), this time with the title "Sun, gravity, and ether. Universal problems in a new solution. The deliverance from relativity," with the following subsections: "1. & 2. The sun, similar to the Earth in its interior. Fixed stars and nebulae. Gravity. 3. New meteorology and gravity – a great discovery by Goethe. Glider flight. 4. Against Einstein and Eddington. 5. Pro and contra cosmic ice theory. Conclusion." Then individual lectures again (the topics were worded somewhat differently, but remained essentially the same) on December 12, 1934 (Lecture Catalog, Fall 1934, p. 9), on February 13, 1935 (Lecture Catalog, Jan.–March 1935, p. 4), on September 22, 1936 (Lecture Catalog, Sept.–Dec. 1936, p. 11), on February 27, 1937 (Lecture Catalog, Jan.–March 1937, p. 15), on April 14, 1937 (Lecture Catalog, April–June 1937, p. 11), and on October 23, 1937 (Lecture Catalog, Sept.–Dec. 1937, p. 15).

takeover? There are clear indications that this political context played a role. In the fall–winter semester of 1933, which was entirely under the influence of the National Socialist takeover – now, "German prehistory and early history, German history, racial studies, hereditary health theory, family studies, and similar questions of folkish renewal" were on the curriculum, and the adult education center stated that it now had "a new face, a new character, a new significance"[480] – Fricke even held a weekly lecture in the regular classification "Natural History" with the slightly adapted title "New views of the Sun and universal ether. Constructive criticism of the so-called 'modern' conception of the world." The individual parts were "1. The Sun, similar to the Earth in its interior. 2. Universal ether. 3. New meteorology. Why glider flight is possible. 4. Relativity. Reservations against Einstein and Eddington. 5. Cosmic ice theory and space atmosphere. Final observations."[481]

It is not possible to clearly reconstruct the extent to which the inclusion of Fricke's course in the curriculum was connected to his offering his solar theory to National Socialism, or to the new orientation of adult education[482] after the takeover, or both. However, such a connection seems reasonable given that the curriculum of the 1920s was very sound and did not contain any non-academic world riddle solver theories or ideological training.

Fricke's opposition to Einstein was now placed in a completely different political context – in 1933, Einstein did not return from a foreign trip to Germany which was now governed by the National Socialists – to the Einstein criticism of the 1920s. Politically, Fricke behaved as an opportunist and used the changed political circumstances to assert what was truly important to him – ether physics. Before the National Socialist takeover, he had not made any anti-Semitic insinuations against the theory of relativity and Einstein either in his publications or in his correspondence, but in 1933 he placed great hope in the idea that the new government would help ether research achieve a breakthrough. This hope was not unjustified. For many National Socialists, Einstein and modern physics were considered "Jewish physics;" so-called "German physics" initially experienced an upturn and National Socialists politicized the appeal to "common sense" for their own purposes. Healthy German thinking was set against "Talmudic"[483] Jewish thinking.

[480] An unsere Hörer! [To our students!] Lecture Catalog, Oct.–Dec. 1933, p. 2. [481] Ibid., p. 11.

[482] The Reich Working Group and the Reich Association of German Adult Education Centers were dissolved and their duties were taken over by the newly-founded Central Office for Adult Education Centers at the Central Institute for Education and Instruction in Berlin. In 1934, the adult education centers were affiliated with the Reich Education Office of the NSDAP (Nazi Party) and the German Labor Front as the German People's Education Institution. Cf. Feidel-Mertz, Hildegard. (1999). Erwachsenenbildung im Nationalsozialismus [Adult education in National Socialism]. *Handbuch Erwachsenenbildung, Weiterbildung [Handbook of Adult Education, Continuing Education]*, ed. Rudolf Tippelt, 2nd ed. Opladen, p. 42–53.

[483] Thüring, Bruno. (1941). *Albert Einsteins Umsturzversuch der Physik und seine inneren Möglichkeiten und Ursachen, Forschungen zur Judenfrage [Albert Einstein's Attempt to Overthrow Physics and its Internal Possibilities and Causes, Research on the Jewish Question]*, Vol. 4. Berlin, pp. 134–62, quoted here from the offprint, p. 6; for secondary literature, cf. Beyerchen 1977, pp. 123ff.

Even if "German physics" and similar efforts like those of Fricke were unable to assert themselves over the long term, they represented an attempt at a complete redesign of physics that the academic physicists had to deal with. Previously marginalized scientists such as Lenard, Stark, and Gehrcke, and world riddle solvers like Fricke now felt a new impetus. Lenard was already emeritus at the National Socialist takeover, but in a memorandum to Hitler, he made himself available as an advisor for appointment matters, among other things, and received multiple honors. In 1933, he attempted to achieve an appointment for Gehrcke to a chair in physics in Berlin through his political connections.[484] His work *Deutsche Physik [German Physics]* appeared in 1936. It was to be a new, "Aryan" textbook.[485] Stark became the director of the Reich Institute of Physics and Technology in 1933 although all the consulting scientists had rejected his candidacy. From 1934 to 1936, moreover, he was the president of the German Research Association.[486]

Shortly after the takeover by the National Socialists, Fricke published a contribution on *Nationalsozialistische Revolution und Weltätherforschung [National Socialist Revolution and Universal Ether Research]* in which he tendered his science to the new rulers.

> Due to the National Socialist revolution, a new world view has achieved a breakthrough in the German folk. This intellectual adjustment is strangely parallel to the endeavor that is linked to the idea of universal ether research. On January 28th, on the eve of the National Socialist takeover, so to speak, the physicist Dr. Fricke at the Humboldt Adult Education Center developed his great program for the overthrow of just that conception of the world that, as the result of fictitiously exact theoretical physics, had budded its last strange blossom in Einstein's theory of relativity. In this way, he started a 'Hitler movement in physics,' so to speak [...] Combating Einstein's ideas certainly fits entirely with the intent of the National Socialist train of thought. What has occurred in this direction in the past was hardly suitable to create clarity. Now, it is finally time to remedy the confusion that has arisen due to unilaterally negative, corrosive thinking.[487]

The DGWAP caused somewhat of a stir when Fricke approached Theodor Vahlen, head of the Post-Secondary Department at the Culture Ministry since 1934 and president of the Prussian Academy of Sciences since 1938, as well as Bernhard

[484] Cf. Lenard to Gehrcke: "On April 4th, I sent a memorandum to Minister Rust about the new appointment of a Berlin chair [in physics, a successor to Nernst]; it went through the referral of Governor Kühn and was personally delivered by him to the Minister following a correspondence that I had had with Kühn when he was a delegate about 'professors with Einstein complexes.' In the memorandum, I described you alone as the suitable candidate, with an extensive justification. Experimental research should again come to the fore; it should be a clear sign of moving away from Planck and Einstein; it should be an older, proven researcher familiar with all parts of physics." December 13, 1933, GN 79-G-9, cf. also Gehrcke to Lenard, December 20, 1933, with thanks for Lenard's efforts.

[485] Cf. Beyerchen 1977, pp. 124f. [486] Cf. ibid., pp. 115ff.

[487] Fricke, Hermann. ([1934]). *Nationalsozialistische Revolution und Weltätherforschung [National Socialist Revolution and Universal Ether Research]*. Berlin-Schmargendorf (a publication of the Deutsche Gesellschaft für Weltätherforschung).

Rust, Reich Education Minister from 1934, in order to familiarize them with the significance of his scientific discoveries. Within the new political framework, making such comments to the highest levels turned out to be successful. Rust asked the president of the Reich Institute of Physics and Technology, Johannes Stark, to state his opinion on Fricke's universal ether research;[488] Stark in turn passed the task on to his department head Gehrcke. In his expert opinion, Gehrcke stated:

The publications of the German Society for Universal Ether Research cover scientific questions of the day such as the theory of relativity, universal ether, Goethe as a natural scientist, and are therefore particularly remarkable because they combat the previously prevailing theories, particularly those of Einstein, Planck, and others.

When it is said with regard to Einstein's theory of relativity that it is 'monstrous nonsense,' then I would like to completely endorse this, even today, and likewise when it is explained that it was primarily Planck who suppressed 'the criticism by dissenters, the voice of common sense, the free opinion of science.' The fact that Dr. Berliner, the editor of the periodical 'Die Naturwissenschaften,' propagated Einstein's theory for decades and excluded any other opinion from the discussion is accurate and can be documented with individual examples.

The desire to achieve an 'orderly discussion' about Goethe's meteorological views must be agreed with in the interest of the 'freedom of science,' which was often demanded by Einstein, Planck, etc., but was just as often suppressed.[489]

One could almost believe that Gehrcke was trying to be funny – Goethe as a "scientific question of the day" and works that stand out because they are particularly incompatible with currently valid theories? But this expert opinion was meant entirely seriously. In all probability, Gehrcke and probably also Stark, who co-signed, saw it as a welcome opportunity to get even with their adversaries of many years. Thus the polemic against Arnold Berliner that is included in Gehrcke's expert opinion is without a doubt delayed revenge for the disputes with Berliner in 1914, 1919, and 1927/28 (cf. pp. 200f.). Planck sent Berliner the expert opinion for his information and Berliner answered with the ironic remark, "It hardly needs any affirmation that I am proud to stand next to you as a suppressor of the free opinion of science."[490]

488 "I sent the documents from the German Society for Universal Ether Research that were forwarded to me with the letter of May 2, 1934, to the president of the Reich Institute of Physics and Technology in Berlin-Charlottenburg for review and an opinion. A copy of the report he prepared on July 1, 1934 – IV No. 2059/34 – is enclosed. Signed on behalf of Schnoering." GN 38-D-19.

489 Reprinted in Fricke, Hermann. (1934). *Die im Innern erdähnliche Sonne [The Sun, Similar to the Earth in its Interior]*. Weimar, Appendix, pp. 3f.

490 Berliner to Planck, July 28, 1934, to Sommerfeld for information, in Sommerfeld, Arnold. (2004). *Wissenschaftlicher Briefwechsel [Scientific Correspondence]*. Vol. **2**: 1919–1951, ed. Michael Eckert and Karl Märker, Berlin [*et al.*], Doc. 188. During the National Socialist period, Arnold Berliner encountered great difficulty in continuing to edit *Naturwissenschaften* due to his Jewish origins. Cf. inter alia the correspondence between Berliner and Sommerfeld in this period. In 1935, Berliner was forced to resign from his office; in 1942, he committed suicide in Berlin. Cf. Beyerchen 1977, p. 65.

For Fricke, this semi-official recognition was very important; he quoted from the expert opinion in several subsequent publications. Fricke also sought a connection to the adherents of "German physics" and traveled to Munich in July 1942 in order to look up Wilhelm Müller (1880–1968) and Rudolf Tomaschek (1895–1966), both of whom were representatives of this racist politicized physics.[491] On July 10, 1942, Fricke gave his well-practiced slide lecture on the "Sun and Russell Diagram" at the Colloquium for Theoretical Physics at the University of Munich. However, this "Munich Affair," as Fricke labeled it, turned out to be disappointing, since even the "German physicists" did not appreciate his solar theory.[492]

Fricke repeatedly attempted to adapt politically and to achieve recognition for his scientific argumentation in this way. His view of "two camps" in physics was also modified accordingly for political reasons. In a criticism of the critical review of the book *Jüdische und deutsche Physik [Jewish and German Physics]*[493] by the physicist Walter Weizel of Bonn,[494] Fricke again emphasized the existence of two contrary methods of working in physics – ether physics focused on mechanical models on the one hand, and the methods of modern physics that were focused on mathematical formulations on the other. While the first method had been that of the nineteenth century, the second had become that of the twentieth century. This paradigm shift is then justified by Fricke using a political argument. "It now seems extraordinarily reasonable to draw a parallel between this change of heart and the Jewish mentality that pushed itself ever more into the foreground around the turn of the century. After all, it is only the Old Testament commandment 'Thou shalt not make unto thee any graven image, or any likeness of any thing' applied here to physics."[495]

For reasons of objectivity, Fricke preferred to speak of "Planck physics" rather than "Jewish physics," which by no means relativizes his parallelizing "Jewish thinking" and "mathematical, modern physics," showing instead only his opportunism. In fact, it was already apparent in the analysis of the marginalization process that it was not so much Einstein, but rather Planck and other physicists holding decisive positions that he considered to be his opponents, whom he now wanted to

[491] Cf. Litten 2000 on the Munich group around Müller.

[492] Fricke reported extensively on the course of his "Munich Affair" to Gehrcke. Cf. particularly the correspondence in GN 18-C.

[493] Müller, Wilhelm and Stark, Johannes (eds.). (1941). *Jüdische und deutsche Physik: Vorträge zur Eröffnung des Kolloquiums für theoretische Physik an der Universität München [Jewish and German Physics: Lectures on the Opening of the Colloquium for Theoretical Physics at the University of Munich]*. Leipzig.

[494] Weizel criticized the fact that Müller was using National Socialist ideology as a lead-in to his criticism of the theory of relativity and quantum theory, and had "apparently no great knowledge" of quantum theory. In his review, Weizel was concerned with emphasizing quantum theory as "an achievement of German science" and not of the Jews. Cf. Weizel, Walter. (1942). [Review of Müller and Stark 1941]. *Zeitschrift für technische Physik*, **23**(1), 25; cf. Fricke [ca. 1942].

[495] Ibid.

bring into play somehow, although they did not fit into the line of argument of "German" and "Jewish" physics:

I consider [...] Planck's influence on recent physics to be absolutely catastrophic and must most joyfully welcome a situation where men can be found who, like Lenard at the physicists' conference in Nauheim in 1920 or like the author of the book 'Jewish and German Physics,' finally energetically oppose this whole direction.[496]

But Fricke was unable to secure permanent success for his ether research even in the period between 1933 and 1945. Fricke, and Gehrcke as well, who had written expert opinions during the National Socialist period at the request of the Reich Education Ministry (with which he wanted to help opponents of Einstein like Mohorovičić and Glaser obtain a university chair or an honorary doctorate, respectively[497]), did not argue from a heartfelt conviction for the political goals of National Socialism against the theory of relativity. They were among those who accommodated and who themselves firmly believed that they had only pursued their scientific goals completely apolitically. They never reflected on the fact that there are political contexts in which it is by no means possible to behave apolitically any longer. This is apparent not least in the fact that they both continued their critical interpretations of the theory of relativity, which were riddled with politically opportune vocabulary, without a hitch after the collapse of the "Third Reich."

After 1945, Fricke and Gehrcke argued that the theory of relativity *and* National Socialism had been the two great aberrances of the twentieth century. Both of them used the analogy of the theory of relativity and National Socialism for their argument of the "blinding" of science and the public by modern physics. For Gehrcke, even at this time, the special theory of relativity was still a "scientific and political mass suggestion" that (at least) had been "greatly overshadowed by the Hitlerian mass suggestion of National Socialism."[498] In Fricke's *16 Thesen über und gegen*

[496] Ibid.

[497] On Mohorovičić, cf. the request for an expert opinion from Wilhelm Führer, Reich Education Ministry, to Gehrcke, August 14, 1939, GN 28-E-1, as well as Gehrcke's expert opinions: GN 20-K-2 on Glaser; GN 28-E-2 on Mohorovičić. In these expert opinions, Gehrcke strongly emphasized both individuals' opposition to the theory of relativity in the 1920s: "Mohorovičić usually had his works published in the German language and thus joined ranks with the scientific workers and fighters for *German* culture. This applies particularly to his critical works on the Einsteinian theories, to whose undermining he contributed significantly." In addition, Gehrcke pointed out that Mohorovičić signed the protest call of 1922. Gehrcke also attempted to help the German military with inventions. Thus he filed a patent for a "torpedo with target detonator" in 1940 with a note indicating that this would affect the "interests of national defense." The patent application was rejected with the justification that Gehrcke's procedure did not represent any significant innovation. Documentation on the patent process is found in GN 27-B-1 to 27-B-12. In 1942, Gehrcke applied for a patent for a "direction camouflager," again with a note indicating that the "subject matter of the application is possibly of interest to the armed forces." A review of the "direction camouflager" was carried out by the German navy, but showed that the invention was not practical; the reply from the Reich Patent Office said as much in June 1943, GN 28-F-1 to 28-F-13. Only a few weeks before the end of the war, Gehrcke offered his artificial desert climate, which had been developed in the 1930s, for use in German submarines. Gehrcke to an unknown recipient, ["Climate control in submarines"], March 9, 1945, GN 81-K-2.

[498] Gehrcke, Ernst, [note], [after 1945], GN 3-D-13.

Einstein [16 Theses on and against Einstein], written in 1948, his first thesis succinctly states, "In this century, Germany has fallen for 'leaders' twice; second for Hitler and first for Einstein."[499] Fricke frequently used this thought, which sounds like a variation on Kurt Tucholsky's satire "Hitler und Goethe. Ein Schulaufsatz [Hitler and Goethe. A class essay]," as an argument in the post-war years. While it was "atrocity propaganda"[500] that threatened his ether physics in the National Socialist period, new politically opportune terms can be found in the post-war terminology:

> Even today, Einstein's theory of relativity is considered 'complete nonsense' and a 'crime against humanity' by many physicists. […] The fanatical terrorism excluding any other opinion, which is exercised by opponents of ether, can only be compared with recently past political events.[501]

Later in this treatise he speaks of an "'iron curtain' that science would like to draw around my discovery."[502] A week before his death, he suggested establishing "relationships with the Russians" in a letter to Gehrcke. "Dialect, materialism & substantial aether should be a good combination!"[503] The DGWAP also continued to exist after the war; in 1947, Fricke mentioned that the DGWAP had been affiliated with the Gesellschaft Freunde der Philosophie und Kunst [Society of Friends of Philosophy and Art], which was licensed in the British sector of Berlin,[504] but the DGWAP presumably terminated all its activities no later than upon Fricke's death in 1949.

4.3.6 The strategic dimension of the anti-Einstein network

The fact that the countermovement to the theory of relativity had a strong strategic dimension is apparent in the fact that the alliances among Einstein's opponents broke up once the perceived threat (which the theory of relativity was perceived to be) no longer overshadowed their divergences in content. Before that time, however, the content differences among themselves were quite deliberately set aside in favor of a unified struggle for "true" science – whatever that was understood to be in particular.

In their mutual correspondence, for instance between Oskar Kraus and Ernst Gehrcke, the differences in their views on content certainly did play a role. Kraus attempted multiple times to give Gehrcke an understanding of Brentano ("*Please read the essay by Brentano*;" "But please, *study* the article by Brentano"),[505] which

[499] Fricke, Hermann. ([1948]). *16 Thesen über und gegen Einstein [16 Theses on and against Einstein]*, [typescript], GN 2-G-8.
[500] E.g. in Fricke 1934e. [501] Fricke to Gehrcke, May 10, 1948, GN 2-G-8. [502] Ibid.
[503] Fricke to Gehrcke, September 9, 1949, GN 59-E-3. [504] Fricke to Gehrcke, July 21, 1948, GN 2-E-5.
[505] Kraus to Gehrcke, December 29, 1920, GN 71-G-8; Kraus to Gehrcke, January 29, 1921, GN 71-G-10.

obviously did not bear fruit; at least, Gehrcke's philosophical work *Physik und Erkenntnistheorie [Physics and Epistemology]*, which was produced during this period, is clearly uninfluenced by Brentano and the positions that Kraus had taken in the correspondence. Gehrcke had an affinity to Kant and preferred to let Palagyi take on the assessment of his tentative efforts in the philosophy of science.

However, Kraus did not like anything to come between himself, Gehrcke, and Lenard. He remarked on the difference between his Brentanian view of motion as absolute and the classical-relativistic view of motion held by the physicists Gehrcke and Lenard, that "this contrast between me & you and Lenard should certainly not develop into a public dispute," and Gehrcke likewise stated, "The fact that Lenard, you, and I are not in agreement in all areas about the R[elativity]T[heory] does not disturb me, but to the outside we should not emphasize what separates us, but rather what we have in common."[506] Lenard and Gehrcke had little to criticize in each others' content, and Lenard assured Gehrcke in February 1918, "that – as far as I can see at all – no major difference exists in our works on the relativity principle, even if we perhaps see the origin of his deficiencies concentrated at somewhat different places in detail."[507] Following Kraus' devastating criticism of Lenore Kühn's superficial understanding of Kant[508], it was Gehrcke who in turn calmed things down using a strategic argument:

> I think that we can come to an understanding that the document by R[ipke] K[ühn] represents a valuable contribution in the intellectual struggle against relativism that will not fail to have its effect in practical terms, but that we do not concur entirely with R. K. in objective matters, and that we particularly disagree with Kant's phenomenalism exactly as with the phenomenalism of the theory of relativity.
>
> However, the common ground between R. K. and us seems to me to be more important for the coming period than what separates us, and simply for reasons of practical politics we should therefore move what separates us to the background.
>
> Even if the adherents of the rel.[ativity] theory would like to set about delighting in the points of difference between you, Lenard, Ripke-Kühn, and me, *we* would be smarter to let the unified front against relativism be the main thing.[509]

506 Kraus to Gehrcke, November 16, 1921, GN 71-G-5; Gehrcke to Kraus, [about 1920], GN 71-G-4.

507 Lenard to Gehrcke, February 12, 1918, GN 83-E-5.

508 Kraus to Gehrcke, November 11, 1920. Among other things, Kraus criticized the fact that Kühn refers both to pre-critical Kant and to critical Kant without discussing this. He also made negative statements to Reuterdahl about Palagyi, whose theory he found of little significance and furthermore full of "a large quantity of errors." Palagyi's opposition to Einstein, he said, would not be a reason "to describe him as an important thinker. [...] In my opinion, Prof. Gehrcke goes too far in his estimation of P[alagyi]." Cf. Kraus to Reuterdahl, October 28, 1921, RP 4–30. On the other hand, he valued Kurt Geissler. Cf. Kraus to Gehrcke, December 29, 1920, GN 71-G-8.

509 Gehrcke to Kraus, [about 1920], GN 3-J-18. Kühn was also attacked by Major General von Gleich, who was exclusively interested in practical questions of physics, not in philosophy; cf. von Gleich to Gehrcke, April 4, 1925, GN 29-F-7. Von Gleich also thought nothing of Fricke's ether vortex theory: "Unfortunately, I had to write and tell him [Fricke] about what were already serious concerns about his powerful ether vortex: above all, that *today* solar rotation can no longer turn the planets using interposition of the aether. It is much too slow *now* for that purpose." Von Gleich to Gehrcke, February 10, 1926, GN 71-B-4.

Above all, the connection to Reuterdahl was of a strategic nature for Gehrcke; he stated to him, "I have read a good bit further in your work Theism adversus Materialism [what is meant is Reuterdahl's work *Scientific Theism vs. Materialism*] with great interest. I see that we differ on individual physical points, such as in the question of the ether. But these differences will not separate us; after all, we otherwise run in parallel on too many points."[510] It is remarkable that the issue of whether one assumes the existence of the ether is not a central point for Gehrcke as a classical physicist, but is instead devalued as an "individual physical point," apart from which they shared a great deal of common ground. What is equally remarkable is that Reuterdahl's theistic publications did not cause him to rethink the selection of his ally. In fact, Gehrcke had studied the voluminous work *Scientific Theism vs. Materialism* in detail, as the many underlines and marginal notes in his personal copy attest. It is probable that Gehrcke did not take Reuterdahl's attempt to unify science and religion seriously. Gehrcke's conception of "correct" science was located within the framework of classical physics and was thus significantly different than Reuterdahl's – a difference that was never discussed between them. The topics of their correspondence were limited merely to practical questions with regard to the struggle against the theory of relativity and the exchange of generally worded statements of high regard. Things were different for Mohorovičić, who borrowed *Scientific Theism vs. Materialism* from Gehrcke and stated to Reuterdahl, "I agree with you in all the principal points," and even planned to move to St. Paul where Reuterdahl was living.[511]

For other allies against modern physics, though, the content certainly did play a role. After all, they understood themselves as the actual elite of physics, sought out confederates in this regard, and were concerned about quality assurance. Gehrcke had asked Hugo von Seeliger, his adversary in the Gerber affair, about the value of the work by Charles Lane Poor on redshift. Von Seeliger considered it false, which fact Gehrcke immediately forwarded to Mohorovičić. But Gehrcke was also not truly satisfied with Mohorovičić's works against the theory of relativity. He highlighted his dissatisfaction in a letter dated January 13, 1925, in which he told Mohorovičić that he considered his works from two points of view; one of them was the "objective scientific standpoint" from which he welcomed Mohorovičić's interest in demonstrating many additional transformation equations, since in this way the Lorentz transformation would be demoted to only one of many (unreal) systems of physics. But that was not all. "The second point of view by which I consider your works, is best made clear when we consider the *tactical* significance

[510] Gehrcke to Reuterdahl, August 29, 1921, RP 4-16. [511] Mohorovičić to Reuterdahl, April 10, 1923, RP 4-35.

of these works in the dispute about the theory of relativity."[512] From this tactical point of view, Gehrcke notes that despite his good intentions, Mohorovičić was using the same methods as Einstein and, like Einstein, wanted to derive real consequences from them, and admonished him, "Some physicists reject this method as pathological physics, while others admire it and perceive in it a method of research in physics. If one does the latter, then one puts oneself on Einstein's side, and you are doing it as well."

Gehrcke asked Mohorovičić to consider that with his work, he was making it possible for the adversaries of Einstein's opponents to make the argument that even the opponents of the theory of relativity would make use of Einstein's method, and advised him to emphasize the following fact in his works:

> … that the method is not new, but rather old. So old, in fact, as any logical speculation that is built up on concepts. If you emphasize this, you will contribute to the spread of further education on the internal nature of the theory of relativity and prevent the relativists from taking support, which they so urgently need, particularly from your recent works.

Einstein's opponents represented a group that was constituted both actively – people sought out their allies – as well as passively – anyone who made any kind of critical statement on the theory of relativity and fit the picture of a "real" natural scientist could be included or be claimed for Einstein's opponents.[513] Their doubts as to some of their confederates were not stated in the publications, but were openly discussed in the correspondence.[514] Many of Einstein's opponents were elitist in that, either as academics like Kraus or Lenard, or from the lofty standpoint of having solved the world riddles, they did not want to make common cause with the mass of Einstein's critics. Kraus sneered to Gehrcke, "The anti-Einstein publications are now shooting up like mushrooms; I came across a booklet by Gartelmann, but it is more than weak. Such booklets strengthen the opposing position – 'God protect us from our friends.'"[515] As a primordial light theorist, Ziegler would have agreed with Kraus without reservation. He thanked Reuterdahl for mentioning Ziegler in an article in the *New York Times*;[516] however, his pleasure was clouded by the company in which he again found himself:

512 This and the following quotes from: Gehrcke to Mohorovičić, January 13, 1925, GN 29-F-3.

513 Thus Reuterdahl stated to Gehrcke, "The great American astronomers, Dr. William Wallace Campbell and Dr. George Ellery Hale […] have joined the anti-Einstein ranks with the pronunciamento 'Not proved'." Reuterdahl to Gehrcke, September 9, 1921, GN 3-I-10.

514 In addition to the examples listed above, Gehrcke also expressed criticism of Rudolf Orthner, cf. Gehrcke to Orthner, December 24, 1929, GN 33-F-4, as well as of Gilbert, whose works he considered speculative, cf. Gehrcke to Gilbert, April 10, 1914, GN 83-C-23; Gehrcke to Gilbert, January 28, 1931, GN 11-B-3.

515 Kraus to Gehrcke, November 22, 1920, GN 71-G-6. Vogtherr said the same: "…since I see that inferior material is still being printed, also against Einstein." Vogtherr to Gehrcke, January 31, 1926, GN 76-A-2. Lamberty, the world riddle solver, was criticized by Johannes Riem, who considered Lamberty's works to be worthless. Cf. Lamberty to Gervai, [copy], May 11, 1925, GN 38-D-6.

516 Cf. Reuterdahl 1923c.

Being named with Palagyi, whom I got to know a few years ago in Bad Nauheim, whose philosophy has very little or nothing at all in common with mine, being named particularly together with him, I perceived as a certain weakening of that honor. In my view, P[alagyi] is conceited, vain & unimportant, and also not very honest.[517]

This content-based disdain was often mutual. Thus Lenard thought little of Fricke (cf. p. 238, for example), and vice versa.[518] There was no trace of respect or even an orientation towards the "great names" among Einstein's opponents – the criticism by the world riddle solvers of academic science could be directed against Lenard just as much as against Einstein.

Ultimately, the widespread network of Einstein's opponents only held together as long as the threat that the theory of relativity represented was greater than their content-based differences. The local networks, which also collaborated on content, turned out to be relatively stable. They were reinforced by personal meetings, such as the link between Reuterdahl, Browne, and Heidenreich with their theistic science, the cooperation between adherents of non-academic mechanics such as Patschke and Wutke, and the ether vortex theorists Fricke and Ruckhaber. In these forms of exchange and cooperation, however, which were focused more on the subject matter, opposition to the theory of relativity was no longer the focal point, which was the case in the network of Einstein's opponents around Gehrcke and Reuterdahl, which, as this chapter has shown, understood itself above all as a strategic alliance.

517 Ziegler to Reuterdahl, December 29, 1923, RP 4–66.

518 Cf. e.g. Fricke's statements: "But the manner in which Lenard, the Soc[iety] f[or] Techn[ical] Physics, Kerkhoff and many others treat my ideas is in no way better than Einstein, Planck, and many others." Fricke to Gehrcke, January 7, 1935, GN 81-A-3. "particularly the most important thing, the ether mechanics, is not permitted to be discussed either with Planck or with Lenard." Fricke [ca. 1942]. "The anti-Einstein struggle has been run into the ground, in my opinion; Lenard was clueless, and Glaser was too spiteful." Fricke to Gehrcke, May 10, 1948, GN 2-G-8. For Ziegler, Lenard was "not thorough enough." Ziegler 1923, p. 20.

Conclusion

This book has examined the contexts of a phenomenon that has largely been a mystery until now. Numerous publications appeared in the 1920s to refute the theory of relativity; their authors were engineers, physicians, businessmen, chemists, or writers who believed that they had refuted Albert Einstein for all time, frequently in their free time and in only a few pages, and self-confidently demanded recognition for this refutation. Why did a physics theory provoke so many and such vehement attacks from individuals who were not at home in academic science? Where did this self-confidence come from, to be able to go head-to-head with leading academic physicists?

Chapter 1 showed that in the course of the popularization of science in the nineteenth century, another role was established alongside that of the scientifically interested citizen satisfied with the role of an amateur, namely an independent, offensively anti-academic natural scientist. The world riddle solvers constructed their own universal theories, which they directed confrontationally against the theories prevailing in academic science. They accused academic scientists of not seeking the ultimate substance of reality and absolute certainty about the nature of natural phenomena. The world riddle solvers came from various world view contexts at the turn of the century (occultism, life reform, monism), that influenced the content and goals of their own theories. What they had in common was that their theories were strongly characterized by metaphysics and were often immunized similarly to religious systems, as well as having extensive significance to world view. The most important protagonists of the criticism of the theory of relativity from the non-academic sphere were the world riddle solvers.

This answers a central question of this book, namely on what basis the popular criticism of the theory of relativity was carried out: it was the non-academic bodies of knowledge that represented this basis. The analysis of the content-based dispute with the theory of relativity therefore had to take into account the fact that the theory of relativity was not only investigated with a bias by the experimental physicists from classical "comprehensible" physics and by the philosophers, who remained closely attached to their views of the nature of central philosophical concepts. As the second chapter emphasized, in the non-academic space as well, the theory of relativity was received as a threat to specific bodies of knowledge, conceptions of science, and world views. In their dispute with the theory of relativity, Einstein's opponents, academic as well as non-academic, thus aimed from the start not at understanding or agreement; instead, their mission was to protect their own views, about whose correctness they were deeply convinced. The fact that the theory of relativity was euphorically celebrated by the public and widely accepted by science represented an attack for them and in their eyes not only legitimized a counterattack, but practically forced them to make one. Science and the public had to be educated and brought back to the "correct" science of the non-academic researcher or the academic scientist. This educational motivation runs throughout the pamphlets against the theory of relativity.

However, the refutations of the theory of relativity did not encounter the acceptance that their authors expected, but rather rejection or disregard on the part of academic physics. This was an inevitable effect. Within a scientific discipline like physics, boundaries exist between scientifically compatible and scientifically incompatible knowledge, as well as a fundamental consensus about the state of research. This also means that generally acknowledged criteria exist for differentiating between outdated and current knowledge as well as between scientific and non-scientific knowledge. In the development of modern physics, which meant a further step away from comprehensible concepts and mechanical models, such boundaries were readjusted and a new consensus was established as to the status of physics research.

The world riddle solvers, who largely argued against modern physics outside of academically accepted standards, were affected by the fundamental restructuring of the knowledge system of academic physics insofar as, in their view, it represented the pinnacle of a whole series of mistakes in academic science and a renewed shift in the boundaries of what was considered scientifically compatible.

Their attempts at correcting these perceived mistakes were destined to fail since, in their conceptions of physical phenomena, they were often located outside of what could be discussed in academic physics. Their arguments were frequently based on concepts that were no longer being used in physics and, insofar as they were characterized by metaphysics or religion, were moreover often not even compatible with science at all. The latter primarily affected world riddle solvers from the contexts of occultism and vitalism.

Einstein's academic opponents who did not want to or were unable to follow this restructuring of the knowledge system were marginalized in that their approaches were now considered "outdated" explanations and they were thus pushed to the margin of the discourse in academic physics. From this margin, the views of Einstein's non-academic opponents were sometimes closer to the views of the experimental physicists than the views of their academic colleagues from theoretical physics. This was shown in the third chapter in the discussion of the alternative concepts of Einstein's opponents on time, light, space, gravitation, and ether – using the shared different conception, that is, the popular conception of science. In the discussion of the content-based alternatives that were asserted against the theory of relativity, it was initially apparent that Einstein's opponents criticized the theory of relativity using various emphases. For the vitalists, it was his time concept, and for the adherents of a mechanical theory, it was his gravitation or ether concept that made a refutation of the theory of relativity necessary in their view. For the experimental physicists, it was the demand for classical physical explanations, and for the philosophers, it was the philosophical statements attributed to the theory of relativity that made its criticism necessary. Due to their various theoretical assumptions and convictions based on world view, Einstein's opponents constituted a very heterogeneous group in terms of content. Nevertheless, content-based points of contact did occur between Einstein's academic opponents and world riddle solvers from the monist context, for instance in their conceptions about ether, since the mechanical theories of the world riddle solvers showed intersections with classical physics. For the topics of space and time, the spontaneous reactions of rejection were particularly similar. For almost all of Einstein's opponents, these concepts had only become virulent due to the theory of relativity, and the criticism emanated essentially from an everyday understanding. Here, the world riddle solvers from the vitalist context were exceptions who asserted a fundamentally different conception of time from that of science. On the concept of space, it was particularly the philosophers and those physicists who were interested in the foundations of their science, who likewise already had well-developed theoretical concepts prior to the confrontation with the theory of relativity. The greatest differences existed with regard to the nature of gravity; here, along with a few occult and vitalist explanations, the mechanical theory of gravitation was represented most frequently in the non-academic sphere, while the academic physicists subscribed to electrodynamic explanations or stuck with gravitation as an instantaneous force.

However, all Einstein's opponents shared a particular conception of science. They demanded comprehensible and non-complex descriptions of physical phenomena and interpreted comprehensibility and simplicity as decisive criteria for closeness to reality, while the mathematical complexity of modern physics stood for remoteness from reality. They saw natural truth – not constructed by science – as the goal of research, which as non-human-made truth had great significance for the metaphysical and ontological anchoring of reality and thus of world view.

This significance of science to world view in connection with the defensive attitude towards the theory of relativity, which was perceived as the greatest threat to central components of the respective world views, led to vehement and permanent attempts to dissuade academic science from modern physics. This occurred by friendly attempts at convincing as well as by vehement polemics – both were unsuccessful due to their theoretical bases and methods of argumentation, which either differed strongly from the state of academic physics or lay completely outside it; instead, their claims to truth were marginalized and excluded by academic science.

The marginalization and exclusion process examined in the fourth chapter was expressed in various forms and affected Einstein's academic and non-academic opponents in different ways. For his non-academic opponents, what often occurred was a direct rejection of their publications, their lectures, or their views expressed in personal conversations. Often, however, they were not even perceived in the academic world in the first place, and were subject to the most efficient type of exclusion: disregard. Einstein's academic opponents also encountered publication difficulties, but they had access to the academic space and were therefore marginalized more by devastating criticism than by disregard. Appointments to committees and policies for filling academic positions were also oriented to the current state of academic physics and made Einstein's academic opponents aware of their marginal position. Despite all the differences, Einstein's academic and non-academic opponents ultimately found themselves structurally in a similar position, namely that of marginalization.

This also answers the question of why an experimental physicist like Gehrcke allied himself with the world riddle solvers for the struggle against the theory of relativity. A discourse against academic physics arose as a countermovement against the marginalization and exclusion process by academic science that comprised both a content-based dimension in the narrower sense of heterogeneous, marginalized bodies of knowledge, as well as a strategic dimension of asserting claims to truth for these bodies of knowledge. In the course of the marginalization process, the focus of the discourse against the theory of relativity shifted ever more strongly towards the issue of access to the discourse of academic science. This strategic dimension of the counter-discourse was apparent on the one hand in the efforts to convince the academic physicists to accept "correct" science. On the other hand, the attempt was made to withdraw recognition from academic physics as the discourse producing truth, and to transfer it to the counter-discourse instead. Modern physics was disqualified by its opponents as "non-physics," "Einsteinism" was classified as an unscientific aberration, the success of the theory of relativity in science and the public sphere was attributed to unscientific "advertising" and "suggestion," and the "relativists" were excluded from "correct" science as "conjurers," "mathematicians" and "speculators," and as "non-scientists." On the one hand, the strategic dimension of the criticism of the theory of relativity showed itself to be argumentative and was reflected in political argumentation strategies against academic physics that disputed their status of being scientific, and on the other hand, it expressed itself practically by constructing social networks and planning targeted joint campaigns. A real "camp mentality" prevailed in the early 1920s. Whoever rejected the theory of relativity and whoever was rejected by modern physics – whether he was structurally excluded due to a lack of academic connections and training, or because he represented knowledge that was incompatible with science (the world riddle solvers, and partly also the philosophers) or outdated (the experimental physicists) – was automatically a potential comrade insofar as he additionally fit what Einstein's opponents understood as a "real" natural scientist. It was only in this way that a broad network against the theory of relativity could be created, and that the counter-discourse of marginalized bodies of knowledge – from the knowledge of the experimental physicists to the knowledge of the occultists – could even be placed on an institutional basis in the form of the Academy of Nations or other organizations.

At the same time, the dispute about the theory of relativity analyzed in this book points beyond a phenomenon to be delimited temporally to the 1920s and thematically to the dispute about a paradigm shift in physics. In the public dispute about scientific theories that reorganize the foundations of the understanding of the world and the self – which in modern times is strongly characterized by science, and to which far-reaching consequences for world view are attributed (the theory of evolution could be cited as having a comparable effect to the theory of relativity) – the issue is never just about the pure content of these theories, but also about their status as a description of reality and about the perception of these theories as a threat to other conceptions of reality. The same can also be determined for public controversies about individual results or developments in science that have an effect on this understanding of the world and the self, such as in the areas of climate research and medicine. Such controversies extend beyond the scope of scientific disputes with regard to the group of individuals involved as well as with regard to the levels of the dispute. They are therefore obviously not exclusively of a specialized scientific nature, but also not merely of a political (nationalistic, anti-Semitic) nature in the narrower sense, as was often assumed with regard to Einstein's opponents. Instead, as this book has shown, they have both socio-political and epistemological dimensions.

It was particularly apparent, in the analysis of the conspiratorial patterns of interpretation that were brought to bear by Einstein's opponents to explain the acceptance of the theory of relativity in science and the public sphere, that it was not possible to cure Einstein's opponents of their misunderstandings of the theory of relativity using education and more discriminating popularization, and to bring them to their senses. Where the legitimacy of an interpretation of reality is at stake, the issue is not one of finding a level of agreement, but rather of tactics to assert claims to truth. Where the issue is ultimately conversion to a counter-science and not a dispute within the scientific discourse, research approaches that deal with criticism of the physical theory of relativity within the narrowly defined framework of scientific adequacy miss the actual problem.

The controversy about the theory of relativity continues to exist even today. As this book has shown, it was not possible to resolve the controversy because almost all of Einstein's opponents were unable or unwilling to move beyond the level of their own argumentation due to their conviction that they themselves possessed infallible knowledge, and thus "true" understanding, which meant that it was impossible to deal with their own views while acknowledging a status of research and an area of validity of academic science that is historically not arbitrary. However, this makes it clear that controversies of this nature fundamentally cannot be resolved. Instead, they are inherent in the existence of science as a disputed, truth-producing discourse in modern societies. They constantly arise at the margins and in the space outside of science, but rarely develop such a widespread impact and polarization as did the dispute about the theory of relativity in the 1920s.

References

Secondary literature

Albrecht, Helmuth (ed.). (1993). *Naturwissenschaft und Technik in der Geschichte. 25 Jahre Lehrstuhl für Geschichte der Naturwissenschaft und Technik am Historischen Institut der Universität Stuttgart [Science and Technology in History. 25 Years of a Chair for the History of Science and Technology at the Historical Institute of the University of Stuttgart]*. Stuttgart.

Allen, David Elliston. (1994). *The Naturalist in Britain. A Social History*. Princeton, NJ.

Ashmore, Malcolm. (1993). The Theatre of the Blind: Starring a Promethean Prankster, a Phoney Phenomenon, a Prism, a Pocket, and a Piece of Wood. *Social Studies of Science*, **23**(1), 67–106.

Ashtekar, Abhay (ed.). (2003). *Revisiting the Foundations of Relativistic Physics. Festschrift in Honor of John Stachel*. Dordrecht [*et al.*] (Boston Studies in the Philosophy of Science Vol. 234).

Banse, Gerhard and Müller, Hans-Peter (eds.). (2001). *Johann Beckmann und die Folgen: Erfindungen – Versuch der historischen, theoretischen und empirischen Annäherung an einen vielschichtigen Begriff [Johann Beckmann and the Consequences: Inventions – An Attempt at an Historical, Theoretical, and Empirical Approach to a Complex Concept]*. Münster [*et al.*].

Barlösius, Eva. (1997). *Naturgemäße Lebensführung. Zur Geschichte der Lebensreform um die Jahrhundertwende [Natural Conduct of Life. On the History of Life Reform around the Turn of the Century]*. Frankfurt (Main)/New York.

Baudson, Michel (ed.). (1985). *Zeit. Die vierte Dimension in der Kunst [Time. The Fourth Dimension in Art]*, [exhibit catalog]. Weinheim.

Bayertz, Kurt, Gerhard, Myriam, and Jaeschke, Walter (eds.). (2007). *Weltanschauung, Philosophie und Naturwissenschaft im 19. Jahrhundert [Ideology, Philosophy, and Natural Science in the 19th Century]*, 3 vols. Hamburg.

Beck, Ulrich. (1992). *Risk Society: Towards a New Modernity*. London [*et al.*].

Ben-Yehuda, Nachman. (1985). *Deviance and Moral Boundaries: Witchcraft, the Occult, Science Fiction, Deviant Sciences and Scientists*. Chicago.

Bensaude-Vincent, Bernadette and Blondel, Christine. (2002). *Des Savants Face à l'Occulte, 1870–1940 [Scholars Face the Occult, 1870–1940]*. Paris.

Benz, Ulrich and Hermann, Armin. (1972). Quanten- und Relativitätstheorie im Spiegel der Naturforscherversammlungen 1906–1920 [Quantum and Relativity Theory as Reflected in the Annual Meetings of the Society of German Natural Scientists and Physicians, 1906–1920], in Querner and Schipperges 1972, pp. 125–37.

Beyerchen, Alan D. (1977). *Scientists Under Hitler: Politics and the Physics Community in the Third Reich*. New Haven.

Biezunski, Michel. (1991). *Einstein à Paris: Le Temps n'est Plus ...* Paris.

Blaschke, Olaf and Kuhlemann, Frank-Michael (eds.). (1996). *Religion im Kaiserreich. Milieus – Mentalitäten – Krisen [Religion in the German Empire. Milieus – Mentalities – Crises]*. Gütersloh (Religiöse Kulturen der Moderne Vol. 2).

Bothe, Detlef. (1991). *Neue Deutsche Heilkunde 1933–1945 [New German Medicine 1933–1945]*. Husum.

Bourdieu, Pierre. (1991). *Language and Symbolic Power [French original: Ce que parler veut dire: l'économie des échanges linguistiques]*. Cambridge, MA.

Bourdieu, Pierre. (1992). *Homo academicus*. Frankfurt (Main).

Brauchle, Alfred. (1951). *Die Geschichte der Naturheilkunde in Lebensbildern [The History of Naturopathy in Life Vignettes]*, [2nd expanded edn. by Brauchle, Alfred. *Great Naturopaths*]. Stuttgart.

Braun, Traute. (1980). *Die Entwicklung der Physik im Spiegel der Presse: Berichterstattung des Berliner Tageblattes 1919–1923* [*The Development of Physics as Reflected by the Press: The Reporting of the Berliner Tageblatt 1919–1923*], Vol. 1: *Text*, Vol. 2: *Dokumentation*, [Master's Thesis, University of Stuttgart]. [Stuttgart].

Bruce, Steve (ed.). (1992). *Religion and Modernization: Sociologists and Historians Debate the Secularization Thesis*. Oxford.

Bruce, Steve and Wallis, Roy. (1992). Secularization: The orthodox model, in Bruce 1992, pp. 8–30.

Brush, Stephen G. (1989). Prediction and theory evaluation: The case of light bending. *Science*, **246**(4934), 1124–29.

Brush, Stephen G. (1999). Why was relativity accepted? *Physics in Perspective*, **1**, 184–214.

Bry, Carl Christian. ([1964/1924]). *Verkappte Religionen [Closet Religions]*, 3rd edn. Lochham [*et al.*].

Busset, Thomas, Rosenbusch, Andrea, and Simon, Christian (eds.). (1997). *Chemie in der Schweiz: Geschichte der Forschung und der Industrie [Chemistry in Switzerland: History of Research and the Industry]*. Basel.

Canales, Jimena. (2001). Exit the frog, enter the human: Physiology and experimental psychology in nineteenth-century astronomy. *The British Journal for the History of Science*, **34**, 173–97.

Cancik, Hubert (ed.). (1981). *Religions- und Geistesgeschichte der Weimarer Republik [Religious and Spiritual History of the Weimar Republic]*. Düsseldorf.

Cantor, Geoffrey N., and Hodge, Michael J. S. (eds.). (1981). *Conceptions of Ether: Studies in the History of Ether Theories 1740–1900*. Cambridge [*et al.*].

Cantor, Geoffrey N. (1981). The theological significance of ethers, in Cantor and Hodge 1981, pp. 135–55.

Čapek, Milič. (1971). *Bergson and Modern Physics: A Re-interpretation and Re-evaluation*. Dordrecht (Boston Studies in the Philosophy of Science Vol. 7).

Čapek, Milič. (1991). *The New Aspects of Time: Its Continuities and Novelties*. Dordrecht (Boston Studies in the Philosophy of Science Vol. 125).

Cassidy, David. (2004). *Einstein and Our World*, 2nd edn. New York, NY.

Cassirer, Ernst. (1957/1929). *Philosophy of Symbolic Forms*], 3 volumes, Vol. 3: *The Phenomenology of Knowledge*. New Haven.

Cassirer, Ernst. (1923/1910). *Substance and Function*. Chicago.

Conze, Werner and Kocka, Jürgen (eds.). (1992). *Bildungsbürgertum im 19. Jahrhundert [The Educated Middle Class in the 19th Century]*, Vol. 1: *Bildungssystem und Professionalisierung in internationalen Vergleichen [Educational Systems and Professionalization in International Comparisons]*. Stuttgart (Industrielle Welt Vol. 38).

Cooter, Roger (1984). *The Cultural Meaning of Popular Science: Phrenology and the Organization of Consent in Nineteenth-Century Britain*. Cambridge [*et al.*].

Cooter, Roger and Pumfrey, Stephen. (1994). Separate spheres and public places: Reflections on the history of science popularization and science in popular culture. *History of Science* **32**(3), 237–67.

Crelinsten, Jeffrey. (2006). *Einstein's Jury: The Race to Test Relativity*. Princeton [*et al.*].

Crelinsten, Jeffrey. (1980). Physicists receive relativity: Revolution and reaction. *Physics Teacher*, **18**, 187–93.

Darnton, Robert. (1968). *Mesmerism and the End of the Enlightenment in France*. Cambridge, MA. [*et al.*].

Daston, Lorraine and Galison, Peter. (2007). *Objectivity*. New York, NY.

Daum, Andreas. (2002). *Wissenschaftspopularisierung im 19. Jahrhundert. Bürgerliche Kultur, naturwissenschaftliche Bildung und die deutsche Öffentlichkeit 1848–1914 [Popularization of Science in the 19th Century: Middle-class Culture, Scientific Education, and the German Public 1848–1914]*, 2nd edn. Munich.

Di Gregorio, Mario A. (2005). *From Here to Eternity: Ernst Haeckel and Scientific Faith*. Göttingen.

Dienel, Hans-Liudger. (1992). *Herrschaft über die Natur? Naturvorstellungen deutscher Ingenieure 1871–1914 [Control Over Nature? German Engineers' Conceptions of Nature, 1871–1914]*. Stuttgart.

Doerry, Martin. (1986). *Übergangsmenschen. Die Mentalität der Wilhelminer und die Krise des Kaiserreichs [Transitional Humans. The Mentality of the Wilhelminians and the Crisis of the German Empire]*, 2 volumes. Weinheim/Munich.

Dolby, R. G. A. (1979). Reflections on deviant science, in Wallis 1979, pp. 9–49.

Donley, Carol C. and Friedman, Alan J. (1985). *Einstein as Myth and Muse*. Cambridge [*et al.*].

Drehsen, Volker and Sparn, Walter (eds.). (1996). *Vom Weltbildwandel zur Weltanschauungsanalyse. Krisenwahrnehmung und Krisenbewältigung um 1900 [From a Change in World View to an Analysis of Ideology. Crisis Perception and Crisis Management around 1900]*. Berlin.

Drehsen, Volker and Zander, Helmut. (1996). Rationale Weltveränderung durch naturwissenschaftliche Weltinterpretation? Der Monistenbund – eine Religion der Fortschrittsgläubigkeit [Rational change in the world through natural scientific interpretation of the world? The Monist League – A religion of faith in progress?], in Drehsen and Sparn (eds.) 1996, pp. 217–38.

Dussel, Konrad. (2004). *Deutsche Tagespresse im 19. und 20. Jahrhundert [The German Daily Press in the 19th and 20th Centuries]*. Münster [*et al.*].

Earman, John, Janssen, Michel, and Norton, John (eds.). (1993). *The Attraction of Gravitation: New Studies in the History of General Relativity*. Boston [*et al.*] (Einstein Studies Vol. 5).

Eisenstaed, Jean and Kox, Anne J. (eds.). (1992). *Studies in the History of General Relativity. Based on the Proceedings of the Second International Conference on the History of General Relativity, Luminy, France, 6–9 September 1988*. Boston [*et al.*] (Einstein Studies Vol. 3).

Elias, Norbert. (2004/1984). *Gesammelte Schriften [Collected Works]*, Vol. 9: *Über die Zeit [On Time]*, ed. Michael Schröter. Frankfurt (Main).

Elm, Veit. (2001). *Die Moderne und der Kirchenstaat. Aufklärung und römisch-katholische Staatlichkeit im Urteil der Geschichtsschreibung vom 18. Jahrhundert bis zur Postmoderne [Modernism and the Papal States. Enlightenment and Roman Catholic Statehood in the Judgment of Historiography from the 18th Century to the Postmodern Era]*. Berlin (Historische Forschungen Vol. 72).

Elton, Lewis. (1986). Einstein, general relativity, and the German press, 1919–1920. *Isis*, **77**, 95–103.

Engelhardt, Dietrich von (ed.). (1997). *Forschung und Fortschritt: Festschrift zum 175jährigen Jubiläum der Gesellschaft Deutscher Naturforscher und Ärzte [Research and Progress: Festshrift for the 175th Anniversary of the Society of German Natural Scientists and Physicians]*. Stuttgart.

Feidel-Mertz, Hildegard. (1999). *Erwachsenenbildung im Nationalsozialismus [Adult Education in National Socialism]*, in Tippelt 1999, pp. 42–53.

Feldman, Gerald D. (1993). *The Great Disorder: Politics, Economics, and Society in the German Inflation, 1914–1924*. New York, NY/Oxford.

Fischer, Ernst Peter. (1996). *Einstein. Ein Genie und sein überfordertes Publikum [Einstein. A Genius and his Overburdened Public]*. Berlin [*et al.*].

Flasch, Kurt. (2000). *Die geistige Mobilmachung. Die deutschen Intellektuellen und der Erste Weltkrieg [Spiritual Mobilization. German Intellectuals and the First World War]*. Berlin.

Fleck, Ludwik. (1979/1935). *Genesis and Development of a Scientific Fact*. Chicago.

Fölsing, Albrecht. (1997). *Albert Einstein. A Biography*. New York, NY.

Forman, Paul. (1971). Weimar culture, causality, and quantum theory: adaption by German physicists and mathematicians to a hostile environment. *Historical Studies in the Physical Sciences*, **3**, 1–115.

Forman, Paul. (1984). Kausalität, Anschaulichkeit and Individualität, or How cultural values prescribed the character and lessons ascribed to quantum mechanics, in Stehr and Meja 1984, pp. 333–47.

Foucault, Michel. (2003). *"Society Must be Defended": Lectures at the Collège de France, 1975–1976*. New York, NY.

Friedman, Robert Marc. (2001). *The Politics of Excellence: Behind the Nobel Prize in Science*. New York, NY.

Galison, Peter. (2003). *Einstein's Clocks, Poincaré's Maps. Empires of Time*. New York, NY.

Gay, Peter. (1968). *Die Republik der Außenseiter. Geist und Kultur in der Weimarer Zeit 1918–1933 [The Republic of Outsiders. Spirit and Culture in the Weimar Period 1918–1933]*. Frankfurt (Main).

Gimpl, Georg. (2001). *Weil der Boden selbst hier brennt ...: aus dem Prager Salon der Berta Fanta (1865–1918) [Because the Ground Itself Burns Here ...: From the Prague Salon of Berta Fanta]*. Furth im Wald [*et al.*].

Glick, Thomas F. (ed.). (1987). *The Comparative Reception of Relativity*. Dordrecht [*et al.*] (Boston Studies in the Philosophy of Science, Vol. 103).

Glowka, Hans-Jürgen. (1981). *Deutsche Okkultgruppen 1875–1937 [German Occult Groups, 1875–1937]*. Munich (Hiram Edn. Vol. 12).

Goenner, Hubert. (1992). The reception of the theory of relativity in Germany as reflected by books published between 1908 and 1945, in Eisenstaed and Kox 1992, pp. 15–38.

Goenner, Hubert. (1993a). The reaction to relativity theory I: The anti-Einstein campaign in Germany in 1920. *Science in Context*, **6**, 107–13.

Goenner, Hubert. (1993b). The reaction to relativity theory in Germany, III: 'A hundred authors against Einstein', in Earman, Janssen, and Norton (eds.) 1993, pp. 248–73.

Goenner, Hubert. (2005). *Einstein in Berlin, 1914–1933*. Munich.

Goschler, Constantin (ed.). (2000). *Wissenschaft und Öffentlichkeit in Berlin, 1870–1930 [Science and Publicity in Berlin, 1870–1930]*. Stuttgart.

Graumann, Carl F. and Moscovici, Serge (eds.). (1987). *Changing Conceptions of Conspiracy*. New York [*et al.*] (Springer Series in Social Psychology).

Gregory, Jane and Miller, Steve. (1998). *Science in Public: Communication, Culture, and Credibility*. New York.

Groh, Dieter (ed.). (1992). Die verschwörungstheoretische Versuchung oder: [The temptation of conspiracy theories, or:] Why do bad things happen to good people? in *Anthropologische Dimensionen der Geschichte [Anthropological Dimensions of History]*, ed. Dieter Groh. Frankfurt (Main), pp. 267–304.

Grundmann, Siegfried. (1967). Das moralische Antlitz der Anti-Einstein-Liga [The moral countenance of the Anti-Einstein-League]. *Wissenschaftliche Zeitschrift der TU Dresden*, **16**(5), 1623–26.

Grundmann, Siegfried. (2005). *The Einstein Dossiers: Science and Politics – Einstein's Berlin Period*, Berlin [*et al.*].

Gumbel, Emil Julius. (1980/1922). *Vier Jahre politischer Mord [Four Years of Political Murder]*. Heidelberg.

Hammerstein, Notker. (1995). *Antisemitismus und deutsche Universitäten 1871–1933 [Anti-Semitism and German Universities 1871–1933]*. Frankfurt (Main)/New York.

Heilbron, John L. (2006). *Max Planck: ein Leben für die Wissenschaft, 1858–1947 [Max Planck: A Life for Science]*, 2nd, corrected and supplemented edn. Stuttgart.

Henderson, Linda Dalrymple. (1985). Theo van Doesburg. Die 'vierte Dimension' und die Relativitätstheorie in den Zwanziger Jahren [Theo van Doesburg. The 'fourth dimension' and the theory of relativity in the Twenties], in Baudson 1985, pp. 195–205.

Hendricks, Vincent F. (*et al.*) (eds.). (2006). *Interactions: Mathematics, Physics and Philosophy, 1860–1930*. Dordrecht (Boston Studies in the Philosophy of Science Vol. 251).

Hentschel, Klaus. (1990). *Interpretationen und Fehlinterpretationen der speziellen und der allgemeinen Relativitätstheorie durch Zeitgenossen Albert Einsteins [Interpretations and Misinterpretations of the Special and General Theory of Relativity by Albert Einstein's Contemporaries]*. Basel [*et al.*]

Hermann, Armin. (1977). Der Kampf um die Relativitätstheorie [The battle about the theory of relativity]. *Bild der Wissenschaft*, 109–16.

Hermann, Armin. (1994). *Einstein: der Weltweise und sein Jahrhundert. Eine Biographie [Einstein: The Wise Man of the World and his Century. A Biography]*. Munich [*et al.*].

Hessenbruch, Arne. (2000). Science as public sphere. X-rays between spiritualism and physics, in Goschler 2000, pp. 89–126.

Hilgartner, Stephen. (1990). The dominant view of popularization: Conceptual problems, political uses. *Social Studies of Science*, **20**(3), 519–39.

Hillermann, Horst. (1976). *Der vereinsmäßige Zusammenschluß bürgerlich-weltanschaulicher Reformvernunft in der Monismusbewegung des 19. Jahrhunderts [The Association-based Merger of Middle-class Ideological Reform Reasoning in the Monist Movement of the 19th Century]*. Kastellaun [*et al.*] (Schriften zur Geschichte und Politischen Bildung Vol. 16).

Hines, Terence M. (1998). Comprehensive review of biorhythm theory. *Psychological Reports*, **83**, 19–64.

Hoffmann, Dieter, and Swinne, Edgar. (1994). *Über die Geschichte der 'technischen Physik' in Deutschland und den Begründer ihrer wissenschaftlichen Gesellschaft Georg Gehlhoff [On the History of 'Technical Physics' in Germany and the Founder of its Scientific Society, Georg Gehlhoff]*, 2nd, revised and supplemented edn.

Berlin (Berliner Beiträge zur Geschichte der Naturwissenschaften und der Technik. Vol. 1).

Hoffmann, Dieter, Laitko, Hubert, and Müller-Wille, Staffan (eds.). (2004). *Lexikon der bedeutenden Naturwissenschaftler [Lexicon of Important Scientists]*, 3 volumes. Heidelberg [*et al.*].

Hoffmann, Dieter. (2004a). Philipp Lenard, in Hoffmann, Laitko, and Müller-Wille 2004, Vol. 2, pp. 393–94.

Hoffmann, Dieter. (2004b). Max Planck, in Hoffmann, Laitko, and Müller-Wille 2004, Vol. 3, pp. 150–55.

Hoffmann, Dieter. (2008). *Max Planck. Die Entstehung der modernen Physik [Max Planck. The Emergence of Modern Physics]*. Munich.

Hofstadter, Richard. (1964/1979). The paranoid style in American politics. *Harper's Magazine*, November, 77–86, reprinted in Ibid. *The Paranoid Style in American Politics and Other Essays*, Chicago, pp. 3–40.

Holton, Gerald and Blanpied, William A. (eds.). (1976). *Science and Its Public: The Changing Relationship*. Dordrecht/Boston.

Hoormann, Anne. (2003). *Lichtspiele. Zur Medienreflexion der Avantgarde in der Weimarer Republik [Moving Pictures. The Media Reflection of the Avantgarde in the Weimar Republic]*. Munich.

Howard, Don, and Stachel, John (eds.). (1989). *Einstein and the History of General Relativity.* Boston [*et al.*] (Einstein Studies, Vol. 1).

Jaki, Stanley L. (1978). Johann Georg von Soldner and the gravitational bending of light, with an English translation of his essay on it published in 1801. *Foundations of Physics*, **8**(11–12), 927–50.

Jammer, Max. (1969). *Concepts of Space: The History of Theories of Space in Physics*, 2nd edn. Cambridge, MA.

Janssen, Michel, Schulmann, Robert, Illy, József, Lehner, Christoph, and Kormos Buchwald, Diana (eds.) (2002a). *The Collected Papers of Albert Einstein*, Vol. 7: *The Berlin Years: Writings, 1918–1921*. Princeton.

Janssen, Michel, Schulmann, Robert, Illy, József, Lehner, Christoph, and Kormos Buchwald, Diana (eds.). (2002b). *The Collected Papers of Albert Einstein*, Vol. 7: *The Berlin Years: Writings, 1918–1921*. (English translation of selected texts), translated by Alfred Engel. Princeton.

Jungnickel, Christa, and McCormmach, Russell. (1986). *Intellectual Mastery of Nature. Theoretical Physics from Ohm to Einstein*, 2 volumes, Vol. 1: *The Torch of Mathematics 1800–1870*, Vol. 2: *The Now Mighty Theoretical Physics 1870–1925*. Chicago, Ill. [*et al.*]

Kerbs, Diethard, and Reulecke, Jürgen (eds.). (1998). *Handbuch der deutschen Reformbewegungen 1880–1933 [Handbook of German Reform Movements, 1880–1933]*. Wuppertal.

Kertz, Walter. (2002). *Biographisches Lexikon zur Geschichte der Geophysik [Biographical Lexicon of the History of Geophysics]*, ed. Karl-Heinz Glassmeier and Ruth Kertz. Braunschweig.

Kleeberg, Bernhard. (2005). *Theophysis: Ernst Haeckels Philosophie des Naturganzen [Theophysics: Ernst Haeckel's Philosophy of the Cosmos]*. Cologne [*et al.*].

Klein, Martin J., Kox, A. J., and Schulmann, Robert (eds.). (1995). *The Collected Papers of Albert Einstein*, Vol. 5: *The Swiss Years: Correspondence, 1902–1914*. (English translation supplement), translated by Anna Beck. Princeton.

Kleinert, Andreas, and Schönbeck, Charlotte. (1978). Lenard und Einstein. Ihr Briefwechsel und ihr Verhältnis vor der Nauheimer Diskussion von 1920 [Lenard and Einstein. Their correspondence and their relationship before the Nauheim discussion of 1920]. *Gesnerus* **35**(3–4), 318–33.

Kleinert, Andreas. (1993). Paul Weyland, der Berliner Einstein-Töter [Paul Weyland, the Einstein Killer of Berlin], in Albrecht 1993, pp. 198–232.

Kleinert, Andreas. (2005). Philipp Lenard and Johannes Stark: Two Nobel laureates against Einstein, in Renn 2005a, pp. 226–29.

Klimke, Friedrich. (1911). *Der Monismus und seine philosophischen Grundlagen [Monism and its Philosophical Foundations]*. Freiburg im Breisgau.

Klueting, Harm. (1986) 'Vernunftrepublikanismus' und 'Vertrauensdiktatur': Friedrich Meinecke in der Weimarer Republik ['Practical republicanism' and a 'dictatorship of trust': Friedrich Meinecke in the Weimar Republic]. *Historische Zeitschrift* **242**, 69–98.

Kocka, Jürgen (ed.). (1989). *Bildungsbürgertum im 19. Jahrhundert [The Educated Middle Class in the 19th Century]*, Vol. 4: *Bildungssystem und Professionalisierung in internationalen Vergleichen [Political Influence and Social Formation]*. Stuttgart (Industrielle Welt Vol. 48).

Kohlstedt, Sally Gregory. (1976). The nineteenth-century amateur tradition: The case of the Boston Society of Natural History, in Holton and Blanpied 1976, pp. 173–90.

Könneker, Carsten. (2001). *Auflösung der Natur, Auflösung der Geschichte. Moderner Roman und NS-'Weltanschauung' im Zeichen der theoretischen Physik [Dissolving Nature, Dissolving History. The Modern Novel and National Socialist 'World View' Under the Sign of Theoretical Physics]*. Stuttgart/Weimar.

Kormos Buchwald, Diana, Schulmann, Robert, Illy, Josef, Kennefick, Daniel J., and Sauer, Tilman (eds.). (2004). *The Collected Papers of Albert Einstein*, Vol. 9: *The Berlin Years: Correspondence, January 1919–April 1920*. (English translation of selected texts), translated by Ann Hentschel. Princeton.

Kormos Buchwald, Diana, Sauer, Tilman, Rosenkranz, Ze'ev, Illy, Josef, and Holmes, Virginia Iris (eds.). (2006a). *The Collected Papers of Albert Einstein*, Vol. 10: *The Berlin Years: Correspondence, May–December 1920, and Supplementary Correspondence, 1909–1920*. Princeton.

Kormos Buchwald, Diana, Sauer, Tilman, Rosenkranz, Ze'ev, Illy, Josef, and Holmes, Virginia Iris (eds.). (2006b). *The Collected Papers of Albert Einstein*, Vol. 10: *The Berlin Years: Correspondence, May-December 1920, and Supplementary Correspondence, 1909–1920*. (English translation supplement), translated by Ann Hentschel. Princeton.

Koselleck, Reinhart (ed.). (1990). *Bildungsbürgertum im 19. Jahrhundert [The Educated Middle Class in the 19th Century]*, Vol. 2: *Bildungsgüter und Bildungswissen [Educational Goods and Educated Knowledge]*. Stuttgart (Industrielle Welt Vol. 41).

Kox, A. J., Klein, Martin J., and Schulmann, Robert (eds.). (1997). *The Collected Papers of Albert Einstein*, Vol. 6: *The Berlin Years: Writings, 1914–1917*. (English translation supplement), translated by Alfred Engel. Princeton.

Krabbe, Wolfgang. (1998a). Lebensreform/Selbstreform [Life Reform/Self-Reform], in Kerbs and Reulecke 1998, pp. 73–75.

Krabbe, Wolfgang. (1998b). Naturheilbewegung [Naturopathic movement], in Kerbs and Reulecke 1998, pp. 77–85.

Kracauer, Siegfried. (1971/1930). *Die Angestellten [The Employees]*. Frankfurt (Main).

Kragh, Helge. (1990). From 'Electrum' to Positronium. *Journal of Chemical Education*, **67**(3), 196–97.

Krenzlin, Norbert (ed.). (1992). *Zwischen Angstmetapher und Terminus: Theorien der Massenkultur seit Nietzsche [Between Fear Metaphor and Term: Theories of Mass Culture since Nietzsche]*. Berlin.

Kuhn, Thomas S. (2012/1962). *The Structure of Scientific Revolutions*, 4th edn. Chicago.

Kühn, Detlef. (2010). *Lenore Kühn. Eine nationale Mitstreiterin der Frauenbewegung [Lenore Kühn. A National Campaigner for Women's Rights]*. Plaidt.

Lange, Friedrich Albert. (1974/1866). *Geschichte des Materialismus und Kritik seiner Bedeutung in der Gegenwart [The History of Materialism and a Criticism of its Significance in the Present]*, 2 volumes. ed. Alfred Schmidt. Frankfurt (Main).

Lepsius, M. Rainer (ed.). (1992). *Bildungsbürgertum im 19. Jahrhundert [The Educated Middle Class in the 19th Century]*, Vol. 3: *Lebensführung und ständige Vergesellschaftung [Conduct of Life and Class-based Socialization]*. Stuttgart (Industrielle Welt Vol. 47).

Linse, Ulrich. (1983). *Barfüßige Propheten. Erlöser der zwanziger Jahre [Barefoot Prophets. Redeemers of the Twenties]*. Berlin.

Litten, Freddy. (2000). *Mechanik und Antisemitismus. Wilhelm Müller (1880–1968) [Mechanics and Anti-Semitism. Wilhelm Müller (1880–1968)]*. Munich (Algorithmus Vol. 34).

Lübbe, Heinrich. (1974). *Politische Philosophie in Deutschland [Political Philosophy in Germany]*. Munich.

Maase, Kaspar. (1997). *Grenzenloses Vergnügen. Der Aufstieg der Massenkultur 1850–1970 [Unlimited Amusement. The Rise of Mass Culture 1850–1970]*. Frankfurt (Main).

Mariafai, Judith M. (1982). *Das Leben des Sanitätsrats Dr. med. Franz Kleinschrod [The Life of Medical Counselor Dr. Franz Kleinschrod]*. Bad Wörishofen.

Mauersberger, Klaus. (2001). Kolbendampfmaschine kontra Rotationsdampfmaschine [Piston steam engine contra rotating steam engine], in Banse and Müller 2001, pp. 65–88.

McCormmach, Russell. (1970). H.A. Lorentz and the electromagnetic view of nature. *Isis*, **61**(4), 459–97.

McLeod, Hugh. (1992). Secular cities? Berlin, London, and New York in the later nineteenth and early twentieth centuries, in Bruce 1992, pp. 59–89.

Meinel, Christoph. (1991). *Karl Friedrich Zöllner und die Wissenschaftskultur der Gründerzeit: Eine Fallstudie zur Genese konservativer Zivilisationskritik [Karl Friedrich Zöllner and the Scientific Culture of the Founders' Years: A Case Study on the Genesis of Conservative Criticism of Civilization]*. Berlin (Berliner Beiträge zur Geschichte der Naturwissenschaften und der Technik Vol. 13).

Merton, Robert. (1957). Priority disputes in scientific discovery: A chapter in the sociology of science. *American Sociological Review*, **22**(6), 635–59.

Miller, Arthur I. (2001). *Einstein, Picasso. Space, Time, and the Beauty that Causes Havoc*. New York, NY.

Mittelstrass, Jürgen (ed.). (2005). *Enzyklopädie Philosophie und Wissenschaftstheorie [Encyclopedia of Philosophy and the Philosophy of Science]*, Vol. 2: (*C–F*), 2nd edn. Stuttgart/Weimar.

Mohler, Armin. (1999/1950). *Die konservative Revolution in Deutschland 1918–1932. Ein Handbuch [The Conserviative Revolution in Germany 1918–1932. A Handbook]*, 5th edn. Graz [*et al.*].

Mommsen, Wolfgang J. (ed.). (1996). *Kultur und Krieg. Die Rolle der Intellektuellen, Künstler und Schriftsteller im Ersten Weltkrieg [Culture and War. The Role of the Intellectuals, Artists, and Authors in the First World War]*. Munich.

Moores, Karen M. (1997). *Presse und Meinungsklima in der Weimarer Republik [The Press and the Climate of Opinion in the Weimar Republic]*. [Dissertation, University of Mainz].

Müller, Gisela. (1992). Der Massencharakter des Lebens und das rastlose Ich. Eine Lesart zu Georg Simmels Moderne-Bild [The mass character of life and the restless ego. A reading of Georg Simmel's picture of the modern], in Krenzlin 1992, pp. 43–72.

Nowotny, Helga. (1979). Science and its critics: reflections on anti-science, in Nowotny and Rose 1979, pp. 1–26.

Nowotny, Helga and Rose, Hilary. (1979). *Counter-movements in the Sciences*. Dordrecht [*et al.*] (Sociology of Science Yearbook Vol. 3).

Owen, David S., Strong, Tracy B., and Livingstone, Rodney (eds.) (2004). *The Vocation Lectures: "Science as a Vocation", "Politics as a Vocation"*. Indianapolis.

Pais, Abraham. (1994). *Einstein Lived Here*. Oxford [*et al.*].

Peukert, Detlev. (1987). *Die Weimarer Republik: Krisenjahre der Klassischen Moderne [The Weimar Republic: Crisis Years of the Classic Modern Period]*. Frankfurt (Main).

Poliakov, León. (1985). *The History of Anti-Semitism*, Vol. 4: *Suicidal Europe, 1870–1933*, translated from the French by George Klin. Oxford.

Querner, Hans and Schipperges, Heinrich (eds.). (1972). *Wege der Naturforschung 1822–1972 im Spiegel der Versammlungen Deutscher Naturforscher und Ärzte [Paths of Science Research 1822–1972 as Reflected in the Meetings of German Natural Scientists and Physicians]*. Berlin [*et al.*].

Renn, Jürgen and Sauer, Tilman. (2003). Eclipses of the stars – Mandl, Einstein, and the early history of gravitational lensing, in Ashtekar 2003, pp. 69–92 (MPIWG Preprint No. 160).

Renn, Jürgen (ed.) (2005a). *Albert Einstein. Chief Engineer of the Universe*, 3 volumes, Weinheim, Vol. 1: *One Hundred Authors for Einstein*.

Renn, Jürgen (ed.). (2005b). Vol. 2: *Einstein's Life and Work in Context*.

Renn, Jürgen (ed.). (2005c). Vol. 3: *Documents of a Life's Pathway*.

Renn, Jürgen (2006). *Auf den Schultern von Riesen und Zwergen: Einsteins unvollendete Revolution [On the Shoulders of Giants and Dwarves: Einstein's Unfinished Revolution]*. Weinheim.

Renn, Jürgen (ed.). (2007a). *The Genesis of General Relativity*, 4 volumes, Dordrecht (Boston Studies in the Philosophy of Science Vol. 250), Vol. 1: *Einstein's Zurich Notebook. Introduction and Source*, eds. Michel Janssen (*et al.*).

Renn, Jürgen (ed.). (2007b). Vol. 2: *Einstein's Zurich Notebook. Commentary and Essays*, eds. Michel Janssen (*et al.*).

Renn, Jürgen (ed.). (2007c). Vol. 3: *Gravitation in the Twilight of Classical Physics. Between Mechanics, Field Theory, and Astronomy*, eds. Jürgen Renn and Matthias Schemmel.

Renn, Jürgen (ed.). (2007d). Vol. 4: *Gravitation in the Twilight of Classical Physics. The Promise of Mathematics*, eds. Jürgen Renn and Matthias Schemmel.

Reschke, Renate. (1992). 'Pöbel-Mischmasch' oder vom notwendigen Niedergang aller Kultur. Friedrich Nietzsches Ansätze zu einer Kulturkritik der Masse ['Rabble-hodgepodge' or the necessary decline of all culture. Friedrich Nietzsche's approaches to a cultural criticism of the masses], in Krenzlin 1992, pp. 14–42.

Rothschuh, Karl. (1983). *Naturheilbewegung, Reformbewegung, Alternativbewegung [Naturopathic Movement, Reform Movement, Alternative Movement]*. Stuttgart.

Rowe, David. (1993). Editorial note: Einstein's encounters with German anti-relativists, in Janssen et al. 2002a, pp. 101–13.

Rowe, David. (2006). Einstein's allies and enemies: Debating relativity in Germany, 1916–1920, in Hendricks 2006, pp. 231–80.

Sabrow, Martin. (1999). *Die verdrängte Verschwörung. Der Rathenau-Mord und die deutsche Gegenrevolution [The Repressed Conspiracy. The Rathenau Murder and the German Counterrevolution]*. Frankfurt (Main).

Sammons, Jeffrey L. (1998). *Die Protokolle der Weisen von Zion. Die Grundlage des modernen Antisemitismus – eine Fälschung. Text und Kommentar [The Protocols of the Elders of Zion. The Foundation of Modern Anti-Semitism – A Forgery. Text and Commentary]*. Göttingen.

Schneider, Ludwig Wilhelm. (1942). *Die erste Periode im philosophischen Schaffen Melchior Palágyis [The First Period in the Philosophical Work of Melchior Palágyi]*. Würzburg.

Schaffner, Kenneth F. (1972). *Nineteenth-Century Aether Theories*. Oxford [*et al.*].

Schönbeck, Charlotte. (2000). *Albert Einstein und Philipp Lenard: Antipoden im Spannungsfeld von Physik und Zeitgeschichte [Albert Einstein und Philipp Lenard: Adversaries in the Conflict of Physics and Contemporary History]*. Berlin [*et al.*] (Schriften der Mathematisch-Naturwissenschaftlichen Klasse der Heidelberger Akademie der Wissenschaften No. 8).

Schottenloher, Karl, and Binkowski, Johannes (eds.). (1985). *Flugblatt und Zeitung. Ein Wegweiser durch das gedruckte Tagesschrifttum [Leaflet and Newspaper. A Guide to the Printed Daily Literature]*, Vol. 2: *Von 1848 bis zur Gegenwart [From 1848 to the Present]*. Munich (Bibliothek für Kunst- und Antiquitätenfreunde Vol. 21/2).

Schroeder-Heister, Peter. (2005). Certismus. In Mittelstrass 2005, pp. 35–36.

Schulmann, Robert, Kox, A. J., Janssen, Michel, and Illy, József (eds.). (1998). *The Collected Papers of Albert Einstein*, Vol. 8: *The Berlin Years: Correspondence, 1914–1918*. (English supplement translation), translated by Ann Hentschel. Princeton.

Shapin, Steven and Schaffer, Simon. (1985). *Leviathan and the Air-Pump: Hobbes, Boyle, and the Experimental Life*. Princeton.

Sheets-Pyenson, Susan. (1985). Popular science periodicals in Paris and London: The emergence of a low scientific culture, 1820–1825. *Annals of Science*, **42**, 549–72.

Sherrill, Thomas J. (1999). A Career of Controversy: The Anomaly of T. J. J. See. *Journal for the History of Astronomy*, **30**(98), Part 1, 25–50.

Siegel, Daniel M. (1981). Thomson, Maxwell and the universal ether in Victorian physics, in Cantor and Hodge 1981, pp. 239–68.

Simon, Christian. (1997). *Chemie in der Schweiz [Chemistry in Switzerland]*, in Busset, Rosenbusch, and Simon 1997, pp. 25–57.

Simon-Ritz, Frank. (1996). Kulturelle Modernisierung und Krise des religiösen Bewußtseins. Freireligiöse, Freidenker und Monisten im Kaiserreich [Cultural modernization and the crisis of religious conciousness. Non-denominationalists, freethinkers, and monists in the German Empire], in Blaschke and Kuhlemann 1996, pp. 457–73.

Skirbekk, Gunnar (ed.). (1996). *Wahrheitstheorien. Eine Auswahl aus den Diskussionen über Wahrheit im 20. Jahrhundert [Theories of Truth. A Selection from the Discussions about Truth in the 20th Century]*, 7th edn. Frankfurt (Main).

Späth, Manfred. (1992). Die Professionalisierung von Ingenieuren in Deutschland und Rußland 1800–1914 [The professionalization of engineers in Germany and Russia 1800–1914], in Conze and Kocka 1992, pp. 561–88.

Stachel, John, Cassidy, David C., Renn, Jürgen, and Schulmann, Robert (eds.). (1989). *The Collected Papers of Albert Einstein*, Vol. 2: *The Swiss Years: Writings, 1900–1909*. (English translation supplement), tanslated by Anna Beck. Princeton.

Stachel, John, Cassidy, David C., Renn, Jürgen, and Schulmann, Robert (eds.). (1994). *The Collected Papers of Albert Einstein*, Vol. 3: *The Swiss Years: Writings, 1909–1911*. (English translation supplement), translated by Anna Beck. Princeton.

Staley, Richard. (2008a). World Views and Physicists' Experience of Disciplinary Change: On the Uses of 'Classical' Physics. *Studies in the History and Philosophy of Science*, **39**, 298–311.

Staley, Richard. (2008b). *Einstein's Generation: The Origins of the Relativity Revolution*. Chicago.

Stehr, Nico and Meja, Volker (eds.). (1984). *Society and Knowledge: Contemporary Perspectives in the Sociology of Knowledge*. New Brunswick.

Stichweh, Rudolf. (1984). *Zur Entstehung des modernen Systems wissenschaftlicher Disziplinen. Physik in Deutschland 1740–1890 [The Development of the Modern System of Scientific Disciplines. Physics in Germany 1740–1890]*. Frankfurt (Main).

Studer, Tobias. (1997). Wirtschaftliche Rahmenbedingungen als Determinanten des Berufsbildes des Chemikers [Structural Economic Factors as Determinants for the Professional Profile of the Chemist], in Busset, Rosenbusch, and Simon 1997, pp. 233–52.

Swenson, Lloyd S. (1972). *The Ethereal Aether: A History of the Michelson–Morley–Miller Aether-drift Experiments, 1880–1930*. Austin.

Tanner, Albert. (1995). *Arbeitsame Patrioten – wohlanständige Damen. Bürgertum und Bürgerlichkeit in der Schweiz 1830–1914 [Industrious Patriots – Respectable Ladies. Bourgeoisie and the Middle-Class Way of Life in Switzerland, 1830–1914]*. Zürich.

Teitel, Corinna. (2000). The culture of knowledge in the metropolis of science. Spiritualism and liberalism in Fin-de-Siècle Berlin, in Goschler 2000, pp. 127–54.

Tetens, Holm. (1999). Die Grenze [The Boundary]. *Die Zeit* newspaper, September 9.

Tippelt, Rudolf (ed.). (1999). *Handbuch Erwachsenenbildung/Weiterbildung [Handbook of Adult Education, Continuing Education]*, 2nd edn. Opladen.

Tiryakian, Edward A. (1974). *On the Margin of the Visible: Sociology, the Esoteric, and the Occult*. New York, NY [*et al.*].

Toellner, Richard. (1986). Liebhaber und Wissenschaft. Zur Rolle des Amateurs in der Geschichte der Wissenschaften. Ansprache zur Einführung in das Symposium [Hobbyists and science. The role of the amateur in the history of the sciences. Address introducing the symposium]. *Berichte zur Wissenschaftsgeschichte*, **9**, 137–45.

Treder, Hans-Jürgen and Jackisch, Gerhard. (1981). On Soldner's value of Newtonian deflection of light. *Astronomische Nachrichten*, **302**(6), 275–77.

Truzzi, Marcello. (1974). Definition and dimensions of the occult: Towards a sociological perspective, in Tiryakian 1974, pp. 243–55.

Ulbricht, Justus H. (1998). Religiosität und Spiritualität [Religiosity and spirituality], in Kerbs and Reulecke 1998, pp. 495–98.

Van Dongen, Jeroen. (2007). Reactionaries and Einstein's fame: "German Scientists for the Preservation of Pure Science," relativity and the Bad Nauheim conference. *Physics in Perspective*, **9**, 212–30.

Van Lunteren, Frans Herbert. (1991). *Framing Hypotheses. Conceptions of Gravity in the 18th and 19th Centuries*. Utrecht.

Vidoni, Ferdinando. (1991). *Ignorabimus! Emil du Bois-Reymond und die Debatte über die Grenzen wissenschaftlicher Erkenntnis im 19. Jahrhundert [Ignorabimus! Emil du Bois-Reymond and the Debate about the Limits of Scientific Knowledge in the 19th Century]*. Frankfurt (Main) [*et al.*].

Vom Bruch, Rüdiger, Graf, Friedrich W., and Hübinger, Gangolf. (1989). *Kultur und Kulturwissenschaften um 1900 [Culture and Cultural Sciences around 1900]*, Vol. 1: *Krise der Moderne und Glaube an die Wissenschaft [Crisis of Modernity and Faith in Science]*. Stuttgart.

Vom Bruch, Rüdiger, Graf, Friedrich W., and Hübinger, Gangolf. (1997). Vol. 2: *Idealismus und Positivismus [Idealism and Positivism]*. Stuttgart.

Wallis, Roy. (1979). *On the Margins of Science: The Social Construction of Rejected Knowledge*. Keele (Sociological Review Monograph Vol. 27).

Wazeck, Milena. (2005). 'Einstein on the murder list!' The attacks on Einstein and the theory of relativity in 1922, in Renn 2005a, pp. 222–25.

Wazeck, Milena. (2010). The 1922 Einstein Film: Cinematic Innovation and Public Controversy. *Physics in Perspective*, **12**(2), 163–79.

Wazeck, Milena. (2013). Marginalization processes in science: The controversy about the theory of relativity in the 1920s. *Social Studies of Science*, **43**(2), 163–90.

Weber, Heiko. (2000). *Monistische und antimonistische Weltanschauung: Eine Auswahlbibliographie [Monistic and Anti-monistic Ideology: A Selected Bibliography]*. Berlin (Ernst-Haeckel-Haus-Studien: Monographien zur Geschichte der Biowissenschaften und der Medizin Vol. 1).

Weber, Max. (1920). *Gesammelte Aufsätze zur Religionssoziologie [Collected Essays on the Sociology of Religion]*, Vol. 1. Tübingen.

Weber, Max. (2004/1917). *Science as a Vocation*, in Owen, Strong and Livingstone 2004.

Wendt, Herbert and Loacker, Norbert (eds.). (1983). *Soziales und geschichtliches Verhalten des Menschen [Social and Historical Behavior of Humans]*. Zürich (Kindlers Enzyklopädie *Der Mensch [Humans]* Vol. 5).

Whittaker, Edmund Taylor. (1951/1910). *A History of the Theories of Aether and Electricity*, **2** volumes, Vol. 1: *The Classical Theories*, 2nd revised and expanded edn. London [*et al.*].

Wimmer, Franz Martin. (1983). Zufall oder Notwendigkeit? Die Frage nach dem Grundgesetz des Weltlaufs [Chance or necessity? The question of the fundamental law of the course of the world], in Wendt and Loacker 1983, pp. 416–34.

Winter, Alison. (1998). *Mesmerized. Powers of Mind in Victorian Britain*. Chicago.

Wittkau-Horgby, Annette. (1998). *Materialismus [Materialism]*. Göttingen.

Wolters, Gereon. (1987). *Mach I, Mach II, Einstein und die Relativitätstheorie: eine Fälschung und ihre Folgen [Mach I, Mach II, Einstein and the Theory of Relativity: A Forgery and its Consequences]*. Berlin/New York.

Wolters, Gereon (ed.). (1988). *Franz Anton Mesmer und der Mesmerismus. Wissenschaft, Scharlatanerie, Poesie [Franz Anton Mesmer and Mesmerism. Science, Charlatanry, Poetry]*. Konstanz.

Zehe, Horst. (1980). *Die Gravitationstheorie des Nicolas Fatio de Duillier [The Gravitational Theory of Nicolas Fatio de Duillier]*. Hildesheim (Arbor scientiarum, Series A: Abhandlungen Vol. 7).

Ziche, Paul (ed.). (2000). *Monismus um 1900: Wissenschaftskultur und Weltanschauung [Monism around 1900. Scientific Culture and Worldview]*. Berlin.

Sources

Anderssohn, Aurel, (1876a). Zur Lösung des Problems der Anziehung [On the solution of the problem of attraction]. *Die Natur*, New Series, **2**(33), 357.

Anderssohn, Aurel. (1876b). Ein Apparat zum praktischen Beweise der Naturnotwendigkeit des Gravitationsgesetzes [An apparatus for the practical proof of the natural necessity of the law of gravitation]. *Die Natur*, New Series, **2**(35), 383–84.

Anderssohn, Aurel. (1877a). Zur Lösung des Problems der Anziehung [On the solution of the problem of attraction]. *Die Natur*, New Series, **3**(4), 53–54.

Anderssohn, Aurel. (1877b). Ebbe und Flut nach der Lehre 'vom Druck der Massen aus der Ferne' [Low and high tide according to the theory 'of the pressure of masses from a distance']. *Die Natur*, New Series, **3**(15), 207–8.

Anderssohn, Aurel. (1878). Attraktion oder Massendruck? [Attraction or mass pressure]. *Die Natur*, New Series, **4**(16), 581.

Anderssohn, Aurel. (1879). Aufruf an die deutschen Naturforscher [Appeal to German natural scientists]. *Die Natur*, New Series, **5**(47), 591.

Anderssohn, Aurel. (1880). *Die Theorie vom Massendruck aus der Ferne in ihren Umrissen [The Theory of Mass Pressure from a Distance in Outline Form]*. Breslau.

Anderssohn, Aurel. (1894). *Physikalische Prinzipien der Naturlehre [Physical Principles of Science]*. Halle (Saale).

Baege, Max. (1911). *Der Keplerbund und seine Gelehrten [The Kepler League and its Scholars]*. Frankfurt (Main).

Baumgardt, Ludwig. (1921a). Ruhm. Soldner und Einstein [Fame. Soldner and Einstein]. *Frankfurter Zeitung* newspaper, November 6, GN 4, XII, 72r.

Baumgardt, Ludwig. (1921b). Soldner und Einstein [Soldner and Einstein]. *Frankfurter Zeitung* newspaper, December 2, GN 4, XII, 75r.

Bavink, Bernhard. (1920). Philosophische Folgerungen der Einsteinschen Relativitätstheorie [Philosophical conclusions from the Einsteinian theory of relativity]. *Unsere Welt*, **12**(4), 131–38.

Bavink, Bernhard. (1925a). [Review of Gehrcke 1924b]. *Hannoverscher Kurier* newspaper, **77**(67), GN VII.

Bavink, Bernhard. (1925b). Die Quantentheorie [The quantum theory], Part 1. *Unsere Welt*, **17**(10), 253–61, Part 2. Ibid. (11), 286–89.

Beckenhaupt, Carl. (1912a). Über den Wert der Theorie und den Ersatz der Anziehungslehre [On the value of the theory and the replacement of attraction theory]. *Weltwissen*, **1**(1), 4–6.

Beckenhaupt, Carl. (1912b). Über die objektiven Grundlagen der Physik [On the objective foundations of physics]. *Neue Weltanschauung*, **5**(7), 250–61.

Beckenhaupt, Carl. (1914). Über die Notwendigkeit geordneter Ausgleiche der Wissenschaft und Lebenspraxis. Ein Versuch zur Organisation der freien Forschung [On the necessity of an orderly balancing of science and life experience. An attempt at organizing independent research]. Part 1 in *Weltwissen*, **2**(42), 205–7, Part 2, Ibid. (43), 215–18, Part 3. Ibid. (45), 236–38.

Beke, Emanuel. (1921). Lenard und Einstein. *Pester Lloyd*, newspaper, November 29, GN 4, XII, 75r.

Becker. (1920). [Review of Fricke 1919]. *Naturwissenschaftliche Wochenschrift*, New Series, **19**(29), 463–64.

Bergson, Henri. (1922). *Durée et simultanéité à propos de la théorie d'Einstein*. Paris (Bibliothèque de philosophie contemporaine).

Blavatsky, Helena Petrovna. (1877). *Isis Unveiled: A Master-Key to the Mysteries of Ancient and Modern Science and Theology*, Vol. 1: *Science*. New York.

Block, Heinrich. (1912). *Die erkenntnistheoretische Rolle des Äthers in der Entwicklung des Elektromagnetismus [The Epistemological Role of the Ether in the Development of Electromagnetism]*. Bonn.

Bond, Frederick. (1923a). Relating to Relativity. *The New York Times*, May 13, RP 4–4.

Bond, Frederick. (1923b). Reuterdahl And The Einstein Theory. *The New York Times*, July 15, RP 4–4.

Born, Max. (1913). Zum Relativitätsprinzip: Entgegnung auf Herrn Gehrckes Artikel 'Die gegen die Relativitätstheorie erhobenen Einwände' [On the relativity principle: Reply to Mr. Gehrcke's article]. *Die Naturwissenschaften*, **1**(4), 92–94.

Born, Max. (1919). Vom mechanischen Äther zur elektrischen Materie [From mechanical ether to electrical material]. *Die Naturwissenschaften*, **7**(9), 136–41.

Born, Max. and Hilbert, David. (1921). Soldner und Einstein. *Frankfurter Zeitung* newspaper, November 18, morning edn., GN 4, XII, 73.

Born, Max. and Einstein, Albert. (1991). *Briefwechsel [Correspondence] 1916–1955*. Munich.

Bouasse, Henri. (1923). *La Question Préalable Contre la Théorie d'Einstein*. Paris.

Brandhoff, Alfred. (1921). Newtons Gravitations-Mechanik. Eine Laien-Betrachtung [Newton's gravitational mechanics. A lay examination]. *Neue Weltanschauung*, **10** (10), 65–72.

Breslauer physikalischer Verein zur Begründung der Lehre vom Druck der Massen in der Ferne. (1877). Neues zur Lösung des Problems der Anziehung [New information on the solution of the problem of attraction]. *Die Natur*, New Series, **3**(10), 136–37.

Breslauer Physikalischer Verein. (1884). *Physikalisches Jahrbuch*, vol. 1. Breslau.

Browne, Robert T. (1919). *The Mystery of Space. A Study of the Hyperspace Movement in the Light of the Evolution of New Physic Faculties and An Inquiry into the Genesis and the Essential Nature of Space*. New York.

Browne, Robert T. (no date). The Archons. *Occult Series*, 2, RP 03a-02.

Brunn, A. von. (1931). [Review of Israel, Ruckhaber, and Weinmann 1931]. *Die Naturwissenschaften*, **19**(11), 254–56.

Brush, Charles Francis. (1911). A kinetic theory of gravitation. *Science*, **33**(845), 381–86.

Brush, Charles Francis. (1928a). Some experimental evidence supporting the kinetic theory of gravitation. *Journal of the Franklin Institute*, **206**(2), 143–50.

Brush, Charles Francis. (1928b). Gravitation. *Proceedings of the American Philosophical Society*, 55–68.

Büchler, Robert. (1921). *Lehrsätze über das Weltall mit Beweis in Form e. offenen Briefes an Prof. Einstein [Propositions About Space with Proof in the Form of an Open Letter to Prof. Einstein]*. Aachen.

Case, Thomas. (1921a). From Euclid to Einstein. Is space physical? The test of relativity. To the editor of *The Times*. *The Times* (London), December 3, GN 4, XV, 24r.

Case, Thomas. (1921b). From Euclid To Einstein. Infinite body of the universe. Time in connexion with space. To the editor of *The Times*. *The Times* (London), December 24, GN 4, XV, 22r.

Chamberlain, Houston Stewart. (1904/1898). *Die Grundlagen des Neunzehnten Jahrhunderts, 1. Hälfte [The Foundations of the Nineteenth Century, 1st half]*. 5th edn. Munich.

Christiansen, Hans. (1915). *Meine Lösung der Welträtsel [My Solution of the World Riddles]*, 2 volumes, 3rd corrected edn., Wiesbaden.

Christiansen, Hans. (1920). *Absolut und relativ! Eine Ablehnung des 'Relativitäts-Prinzips' Einsteins auf Grunde einer reinen Begriffs-Mathematik [Absolute and Relative! A Rejection of Einstein's 'Relativity Principle' based on a new Conceptual Mathematics]*. Wiesbaden.

Christiansen, Hans. (1930). *Der Gleichberechtigungswahn [The Delusion of Equal Rights]*. Wiesbaden.

Christiansen, Hans. (1931). *Offener Brief an Seine Heiligkeit Papst Pius XI [An Open Letter to His Holiness Pope Pius XI]*. Wiesbaden.

Dellinghausen, Nicolai. (1880). *Das Räthsel der Gravitation [The Riddle of Gravitation]*. Heidelberg.

Driesch, Hans. (1924). *Relativitätstheorie und Philosophie [Relativity Theory and Philosophy]*. Karlsruhe (Wissen und Wirken: Einzelschriften zu den Grundfragen des Erkennens und Schaffens Vol. 14).

Drude, Paul. (1912). *Lehrbuch der Optik [Textbook of Optics]*, ed. Ernst Gehrcke, 3rd edn. Leipzig.

Du Bois-Reymond, Emil. (1882). *Über die Grenzen des Naturerkennens. Die sieben Welträthsel. Zwei Vorträge [On the Limits to Understanding Nature. The Seven Riddles of the Universe. Two Lectures]*. Leipzig.

Du Bois-Reymond, Paul. (1888). Über die Unbegreiflichkeit der Schwerkraft [On the inconceivability of gravity]. *Naturwissenschaftliche Rundschau*, **3**(14), 170–78.

Dühring, Eugen. (1882). *Sache, Leben und Feinde. Als Hauptwerk und Schlüssel zu seinen sämmtlichen Schriften. Mit seinem Bildniss [Matter, Life, and Enemies. As Magnum Opus and Key to his Entire Works. With a Portrait]*. Karlsruhe/Leipzig.

Dühring, Eugen. (1905). *Logik und Wissenschaftstheorie: denkerisches Gesammtsystem verstandessouveräner Geisteshaltung [Logic and Philosophy of Science: A Total Intellectual System of Masterfully Reasonable Mentality]*, 2nd revised and expanded edn. Leipzig.

Eddington, Arthur Stanley. (1920). *Space, Time and Gravitation: An Outline of the General Relativity Theory.* Cambridge.

Einstein, Albert. (1903). Über die Theorie der elektromagnetischen Wellen [On the theory of electromagnetic waves]. [Lecture before the Natural Science Society of Berne], listed in the minutes of the 992nd meeting, 1903, Burger Bibliothek Bern, GA BNG – Minutes of the Meetings, Vol. 10.

Einstein, Albert. (1905a). Über einen die Erzeugung und Verwandlung des Lichtes betreffenden heuristischen Gesichtspunkt [On a heuristic point of view concerning the production and transformation of light]. *Annalen der Physik*, **322**(6), 132–48, reprinted in English translation in Stachel *et al.* 1989, Doc. 14.

Einstein, Albert. (1905b). Zur Elektrodynamik bewegter Körper [On the electrodynamics of moving bodies]. *Annalen der Physik*, **322**(10), 891–921, reprinted in English translation in Stachel *et al.* 1989, Doc. 23.

Einstein, Albert. (1907). Ueber die Natur der Bewegungen mikroskopisch kleiner, in Flüssigkeiten suspendierter Teilchen [On the nature of the movements of microscopically small particles suspended in liquids]. [Lecture before the Natural Science Society of Berne], printed as an abstract in English translation in Stachel *et al.* 1989, Doc. 43.

Einstein, Albert. (1909). Über die Entwickelung unserer Anschauungen über das Wesen und die Konstitution der Strahlung [On the development of our views concerning the nature and constitution of radiation]. *Physikalische Zeitschrift*, **10**, 817–25, reprinted in English translation in Stachel *et al.* 1989, Vol. 2, Doc. 60.

Einstein, Albert. (1910). Über das Boltzmann'sche Prinzip und einige unmittelbar aus demselben fliessende Forderungen [On Boltzmann's principle and some requirements directly derived from it], reprinted in Renn 2005c, pp. 76–89.

Einstein, Albert. (1911a). Die Relativitäts-Theorie [The theory of relativity]. *Vierteljahresschrift der Naturforschenden Gesellschaft Zürich*, **56**, 1–14, reprinted in English translation in Stachel *et al.* 1994, Doc. 17.

Einstein, Albert. (1911b). Über den Einfluß der Schwerkraft auf die Ausbreitung des Lichtes [On the influence of gravitation on the propagation of light]. *Annalen der Physik*, **340** (10), 898–908, reprinted in English translation in Stachel *et al.* 1994, Doc. 23.

Einstein, Albert. (1915a). Zur allgemeinen Relativitätstheorie [On the general theory of relativity]. *Sitzungsberichte der Königlich Preußischen Akademie der Wissenschaften*, 778–86, reprinted in English translation in Kox *et al.* 1997, Doc. 21.

Einstein, Albert. (1915b). Erklärung der Perihelbewegung des Merkur aus der allgemeinen Relativitätstheorie [Explanation of the perihelion motion of Mercury from the general theory of relativity]. *Sitzungsberichte der Königlich Preußischen Akademie der Wissenschaften*, pp. 831–39, reprinted in English translation in Kox *et al.* 1997, Doc. 24.

Einstein, Albert. (1916). Die Grundlage der allgemeinen Relativitätstheorie [The foundation of the general theory of relativity]. *Annalen der Physik*, **354**(7), 769–822, reprinted in English translation in Kox *et al.* 1997, Doc. 30.

Einstein, Albert. (1917a). *Über die spezielle und die allgemeine Relativitätstheorie. Gemeinverständlich [On the Special and General Theory of Relativity (A Popular Account)]*, Braunschweig (Sammlung Vieweg Vol. 38). English translation in Kox *et al.* 1997, Vol. 6, Doc. 42.

Einstein, Albert. (1917b). Kosmologische Betrachtungen zur allgemeinen Relativitätstheorie [Cosmological considerations in the general theory of relativity]. *Sitzungsberichte der Königlich Preußischen Akademie der Wissenschaften*, [Berlin], pp. 142–52, reprinted in English translation in Kox *et al.* 1997, Doc. 43.

Einstein, Albert. (1918a). Bemerkung zu Ernst Gehrckes Notiz 'Über den Äther' [Comment on Ernst Gehrcke's note 'On the ether']. *Verhandlungen der Deutschen Physikalischen Gesellschaft*, **20**, 261, reprinted in English translation in Janssen *et al.* 2002b, Doc. 15.

Einstein, Albert. (1918b). Dialog über Einwände gegen die Relativitätstheorie [Dialogue about objections to the theory of relativity]. *Die Naturwissenschaften*, **6**(48), 697–702, reprinted in English translation in Janssen *et al.* 2002b, Doc. 13.

Einstein, Albert. (1918c). Motive des Forschens [Motives for Research], in *Zu Max Plancks sechzigstem Geburtstag. Ansprachen, gehalten am 26. April 1918 in der Deutschen Physikalischen Gesellschaft von E. Warburg, M. v. Laue, A. Sommerfeld und A. Einstein [On Max Planck's Sixtieth Birthday. Speeches held on April 26, 1918, at the German Physics Society by E. Warburg, M. v. Laue, A. Sommerfeld and A. Einstein]*, Karlsruhe, pp. 29–32, reprinted in English translation in Janssen *et al.* 2002b, Doc. 7.

Einstein, Albert. (1920a). *Äther und Relativitätstheorie, Rede gehalten am 5. Mai 1920 an der Reichs-Universität zu Leiden [Ether and relativity theory, a speech made on May 5, 1920, at the Imperial University of Leiden]*. Berlin, reprinted in English translation in Janssen *et al.* 2002b, Doc. 38.

Einstein, Albert. (1920b). *Relativity: The Special and the General Theory*. New York.

Einstein, Albert. (1920c). Meine Antwort. Ueber die anti-relativistische G.m.b.H. [My response. On the anti-relativity company]. *Berliner Tageblatt* newspaper, August 27, morning edn., reprinted in English translation in Janssen *et al.* 2002b, Doc. 45.

Einstein, Albert. (1960). *Briefe an Maurice Solovine [Letters to Maurice Solovine]*. Berlin.

Eisenhart, Luther P. (1923). Einstein and Soldner. *Science*, **58**(1512), 516–17.

Erdmenger, Otto. (1881). Bericht über eine Sitzung des Breslauer physikalischen Vereines [Report on a meeting of the Breslau Physics Association]. *Die Natur*, New Series, **6** (52), 628.

Feigl, Franz. (1932). *Neue Wege zur Lösung der Welträtsel [New Paths to Solving the Mysteries of the Universe]*. Leipzig.

Fierz-David, Hans Eduard. (1926). *Technologie der Textilfasern [Technology of Textile Fibers]*, 3 volumes, Vol. 3: *Künstliche organische Farbstoffe [Artifical Organic Dyes]*. Berlin.

Fincke, Ewalt. (1913). Empiriker und Philosoph [Empiricist and Philosopher]. *Weltwissen*, **1**(7), 101–4.

Fischer, Franz. (1924). [Review of Gehrcke 1924b]. *Volksruf*, **11**(43).

Ford, Henry. (1922 [1920–1922]). *The International Jew. The World's Foremost Problem*. Dearborn.

Frank, Philipp. (1928). Über die 'Anschaulichkeit' physikalischer Theorien [On the 'comprehensibility' of physical theories]. *Die Naturwissenschaften*, **16**(8), 121–28.

Frank, Philipp. (1979). *Einstein: sein Leben und seine Zeit [Einstein: His Life and His Time]*. Braunschweig [*et al.*].

Fraser, Bodie A. (1921). From Euclid To Einstein. To the editor of *The Times*. *The Times* (London), December 27, GN XV, 23r.

Freud, Sigmund. (1922/1921). *Group Psychology and the Analysis of the Ego*, translated from the German by James Strachey. New York.

Fricke, Hermann Heinrich Karl. (1906). Über einen Versuch der Gebrüder Weber, transversale Luftschwingungen betr. [On an experiment by the Weber brothers relating to transverse air vibrations]. *Verhandlungen der Deutschen Physikalischen Gesellschaft*, **8**(12), 248–50.

Fricke, Hermann Heinrich Karl. (1914a). Die allgemeine Gravitation. Ein neuer Weg zur Lösung des Problems [General gravitation. A new path to a solution of the problem]. *Weltwissen*, **2**(38), 177–80, Part 2. Ibid. (39), 185–88.

Fricke, Hermann Heinrich Karl. (1914b). Nochmals die 'Gravitation' ['Gravitation' once again]. *Weltwissen*, **2**(45), 239.

Fricke, Hermann Heinrich Karl. (1918). *Eine neue und einfache Deutung der Schwerkraft [A New and Simple Interpretation of Gravity]*. Berlin [printed as a manuscript].

Fricke, Hermann Heinrich Karl. (1919). *Eine neue und einfache Deutung der Schwerkraft und eine anschauliche Erklärung der Physik des Raumes [A New and Simple Interpretation of Gravity and a Comprehensible Explanation of the Physics of Space]*. Wolfenbüttel.

Fricke, Hermann Heinrich Karl. (1920a). *Der Fehler in Einsteins Relativitätstheorie [The Error in Einstein's Theory of Relativity]*. Wolfenbüttel.

Fricke, Hermann Heinrich Karl. (1920b). Eine neue Erklärung der Schwerkraft [A new explanation of gravity]. *Neue Weltanschauung*, **9**(1), 9–18.

Fricke, Hermann Heinrich Karl. (1920c). *Über eine neue Form des Wirkungsgesetzes für bewegte Massen [On a New Form of the Law for Moving Masses]*, [manuscript], GN 76-A-10.

Fricke, Hermann Heinrich Karl. (1920d). Eine neue und anschauliche Physik des Aethers [A new and comprehensible physics of the aether]. *Glasers Annalen*, **86**(12) (entire series No. 1032), 95–96.

Fricke, Hermann Heinrich Karl. (1920e). *Protest des Mitgliedes der Deutschen Physikalischen Gesellschaft Reg. Rat Dr. Fricke gegen die ihm vom Vorstande der Gesellschaft zu Teil gewordene Behandlung [Protest by the Member of the Deutsche Physikalische Gesellschaft, Gov't Councillor Dr. Fricke, Against his Treatment by the Executive Board of the Society]*, Fricke to the executive committee of the DPG, [September], [carbon copy], GN 71-B-5.

Fricke, Hermann Heinrich Karl. (1921a). Wind und Wetter als Feldwirkung der Schwerkraft [Wind and weather as a field effect of gravity]. *Naturwissenschaftliche Wochenschrift*, New Series, **20**(7), 97–102.

Fricke, Hermann Heinrich Karl. (1921b). Die Akademie die uns fehlt. Ein praktischer Vorschlag zur Neu-Organisation der Wissenschaft [The academy we are lacking. A practical suggestion for re-organizing science]. *Weltwissen*, **5**(1), 9–10.

Fricke, Hermann Heinrich Karl. (1921c). [Review article on von Laue 1921c *et al.*]. *Naturwissenschaftliche Wochenschrift*, New Series, **20**(25), 373–76.

Fricke, Hermann Heinrich Karl. (1922a). Review of Karollus 1921]. *Zeitschrift für Sauerstoff- und Stickstoff-Industrie*, **14**(1), 18.

Fricke, Hermann Heinrich Karl. (1922b). Eine neue Erklärung für die Ursache der Barometerschwankungen und des Windes [A new explanation for the cause of barometer fluctuations and wind]. *Zeitschrift für Sauerstoff- und Stickstoff-Industrie*, **11**(13–14), 49–53.

Fricke, Hermann Heinrich Karl. (1922c). [Review of Wien 1921]. *Naturwissenschaftliche Wochenschrift*, New Series, **21**(12), 168.

Fricke, Hermann Heinrich Karl. (1922d). Zur Klärung des Ätherproblems [On clarifying the ether problem]. *Naturwissenschaftliche Wochenschrift*, New Series, **21**(13), 169–73.

Fricke, Hermann Heinrich Karl. (1922e). Ueber den Gelehrtenprotest gegen die Relativitätstheorie [On the scholars' protest against the theory of relativity]. *Der Tag* newspaper, September 28, GN 4, I, 46r.

Fricke, Hermann Heinrich Karl. (1923). Novi rezultati Michelsonova pokusa. *Univerzum*, **2** (6–8), 191.

Fricke, Hermann Heinrich Karl. (1933). *Eine neue Sonnentheorie [A New Solar Theory]*. Berlin-Westend.

Fricke, Hermann Heinrich Karl. (1934a). *Die im Innern erdähnliche Sonne [The Sun, Similar to the Earth in its Interior]*. Weimar.

Fricke, Hermann Heinrich Karl. (1934b). *Die einfache Aufklärung des Einsteinstreits [The Simple Clarification of the Einstein Dispute]*. Berlin-Schmargendorf (Deutsche Gesellschaft für Weltätherforschung, Publication No. 4).

Fricke, Hermann Heinrich Karl. (1934c). *Goethe als Entdecker einer neuen Meteorologie [Goethe as the Discoverer of a new Meteorology]*. Berlin-Schmargendorf (Deutsche Gesellschaft für Weltätherforschung, Publication No. 5).

Fricke, Hermann Heinrich Karl. (1934d). *Nationalsozialistische Revolution u. Weltätherforschung [National Socialist Revolution and Universal Ether Research]*. Berlin-Schmargendorf (Deutsche Gesellschaft für Weltätherforschung, Publication No. 6).

Fricke, Hermann Heinrich Karl. (1934e). Die moderne theoretische (relativistische) Physik – eine Wissenschaft, die nicht stimmt (Der Justizmord am Weltäther) [Modern theoretical (relativistic) physics – a science that is not right (The judicial murder of universal ether)]. *Docentra*, **55**(38), September 23, 1934, reprinted in Fricke 1939a, pp. 9–17.

Fricke, Hermann Heinrich Karl. (1936). *Luftfahrt und Wetterkunde – gehemmt durch die 'moderne' Physik [Air Travel and Meteorology – Inhibited by 'Modern' Physics]*. Berlin-Schmargendorf (Deutsche Gesellschaft für Weltätherforschung, Publication No. 33), GN 86-A.

Fricke, Hermann Heinrich Karl. (1937). Neue Strahlungslehre als Weg zur Verständigung in der Physik. Schwerkraftstrahlung und Physik des 'Als ob' aphorisch dargestellt [New radiation theory as a path to understanding in physics. Gravity radiation and physics of the 'as if' presented aphoristically]. *Docentra*, **58**(12), 190–95.

Fricke, Hermann Heinrich Karl. (about 1938). *Eine Bitte an alle Goethe-Freunde [A Request to all Friends of Goethe]*. Berlin-Schmargendorf (Deutsche Gesellschaft für Weltätherforschung, Publication No. 30), GN 86-A.

Fricke, Hermann Heinrich Karl. (1939a). *Weltätherforschung. Ein Aufbauprogramm nach dem Umsturz in der Physik [Universal Ether Research. A Development Program following the Revolution in Physics]*. Weimar.

Fricke, Hermann Heinrich Karl. (1939b). *Die Notwendigkeit einer 'Gesellschaft für Weltätherforschung und anschauliche Physik'* [The necessity of a 'Society for Universal Ether Research and Comprehensible Physics'], in Fricke 1939a, pp. 102–8.

Fricke, Hermann Heinrich Karl. (ca. 1942). *Ein Protest gegen die Planck-Physik [A Protest Against Planck Physics]*. Berlin-Schmargendorf (Deutsche Gesellschaft für Weltätherforschung, Publication No. 57), GN 18-C-11.

Fricke, Hermann Heinrich Karl. (1945). *Die Meerwassertheorie der Sonne [The Seawater Theory of the Sun]* (Original Work No. 2), [undated typescript], GN 86-B.

Fricke, Hermann Heinrich Karl. (1947). *Plan zur schnellen Überwindung der europäischen Not. Manuskript-Entwurf zur Veröffentlichung Nr. 103 zu Fricke's Urheberschriften [Plan for Rapidly Overcoming the European Distress. Manuscript Draft for Publication No. 103 of Fricke's Original Works]*, GN 86-A.

Fricke, Hermann Heinrich Karl. (1948a). *16 Thesen über und gegen Einstein [16 Theses on and against Einstein]*, [typescript], GN 2-G-8.

Fricke, Hermann Heinrich Karl. (1948b). Die Fricke-Wienersche Äthervorstellung [The Fricke-Wiener ether conception]. *Docentra*, **65**(16), [offprint], GN 86-B.

Fricke, Hermann Heinrich Karl. (no date). *Neue Weltdeutung [New Interpretation of the World]*, (Original Works No. 7), GN 86-B.

Fricke, Hermann Heinrich Karl. (no date). *Rettet den Weltäther! Ein Aufruf an alle, besonders an Physiker und Astronomen [Rescue Universal Ether! An Appeal to All, Particularly Physicists and Astronomers]*, [typescript], GN 86-C.

Fricke, Hermann Heinrich Karl. (no date). *Sonne, Wind und Aether [Sun, Wind, and Aether]*, [typescript], GN 18-C-9.

Fritsch, Theodor. (1921). *Einstein's Truglehre [Einstein's Fallacious Theory]*. Leipzig (Hammer-Schriften No. 29) (published under the pseudonym Ferdinand Roderich-Stoltheim).

Gartelmann, Henri. (1920a). Zur Philosophie Kants [On Kant's philosophy]. *Neue Weltanschauung*, **9**(2), 33–41.

Gartelmann, Henri. (1920b). Zur Relativitätslehre [On the relativity doctrine], Part 1. *Neue Weltanschauung*, **9**(6), 161–72, Part 2. Ibid. **9**(7), 198–214.

Gartelmann, Henri. (1920c). Zur Relativitätstheorie [On the theory of relativity]. *Neue Weltanschauung*, **9**(11), 341–42.

Gartelmann, Henri. (1920d). Grundsätzliches zur Relativitätstheorie [Fundamental ideas on the theory of relativity]. *Neue Weltanschauung*, **9**(12), 353–56.

Gehrcke, Ernst. (1901). *Über den Geschwindigkeitsverlust, welchen die Kathodenstrahlen bei der Reflexion erleiden [On the Loss of Velocity that Cathode Rays suffer upon Reflection]*. [Dissertation]. Berlin.

Gehrcke, Ernst. (1911). Nochmals über die Grenzen des Relativitätsprinzips [Once again on the limits of the relativity principle]. *Verhandlungen der Deutschen Physikalische Gesellschaft*, **13**(21), 990–1000, reprinted in Gehrcke 1924a, pp. 4–11.

Gehrcke, Ernst. (1912). Über den Sinn der absoluten Bewegung von Körpern [On the meaning of the absolute motion of bodies]. *Sitzungsberichte der Königlich Bayerischen Akademie der Wissenschaften*, **12**, 209–22, reprinted in Gehrcke 1924a, pp. 12–20.

Gehrcke, Ernst. (1913). Die gegen die Relativitätstheorie erhobenen Einwände [The objections raised against the theory of relativity]. *Die Naturwissenschaften*, **1**(3), 62–66, reprinted in Gehrcke 1924a, pp. 20–28.

Gehrcke, Ernst. (1914a). *Zur Diskussion der Einsteinschen Relativitätstheorie [On the discussion of the Einsteinian theory of relativity]*, reprinted in Gehrcke 1924a, pp. 34f.

Gehrcke, Ernst. (1914b). Die erkenntnistheoretischen Grundlagen der verschiedenen physikalischen Relativitätstheorien [The epistemological foundations of the various physical theories of relativity]. *Kant-Studien*, **19**(4), 481–87, reprinted in Gehrcke 1924a, pp. 35–40 (also as an offprint).

Gehrcke, Ernst. (1916). Zur Kritik und Geschichte der neueren Gravitationstheorien [On the criticism and history of more recent graviation theories]. *Annalen der Physik*, **356**(17), 119–24, reprinted in Gehrcke 1924a, pp. 40–44.

Gehrcke, Ernst. (1918). Über den Äther [On ether]. *Verhandlungen der Deutschen Physikalischen Gesellschaft*, **20**, 165–69, reprinted in Gehrcke 1924a, pp. 44–47.

Gehrcke, Ernst. (1919a). Zur Diskussion über den Äther [On the discussion about the ether]. *Verhandlungen der Deutschen Physikalischen Gesellschaft*, **21**, 67–68, reprinted in Gehrcke 1924a, pp. 47–48.

Gehrcke, Ernst. (1919b). Berichtigung zum Dialog über die Relativitätstheorien [Correction of the dialogue about the theories of relativity]. *Die Naturwissenschaften*, **7**(9), 147–48.

Gehrcke, Ernst. (1920), *Die Relativitätstheorie, eine wissenschaftliche Massensuggestion [The Theory of Relativity, a scientific Mass Suggestion]*. Berlin, reprinted in Gehrcke 1924a, pp. 54–68.

Gehrcke, Ernst. (about 1921). Einsteins Relativitätstheorie – ein Bluff! [Einstein's Theory of Relativity – a Bluff!], [unpublished typescript], GN 4-B-4.

Gehrcke, Ernst. (1921a). Über das Uhrenparadoxon in der Relativitätstheorie [On the clock paradox in the theory of relativity]. *Die Naturwissenschaften*, **9**(24), 482.

Gehrcke, Ernst. (1921b). Die Erörterung des Uhrenparadoxons in der Relativitätstheorie [The discussion of the clock paradox in the theory of relativity]. *Die Naturwissenschaften*, **9**(29), 550–51.

Gehrcke, Ernst. (1921c). *Physik und Erkenntnistheorie [Physics and Epistemology]*. Leipzig/Berlin (Wissenschaft und Hypothese Vol. 22).

Gehrcke, Ernst. (1921d). [Reply to Weyl 1920]. *Die Umschau*, **25**(8), 99.

Gehrcke, Ernst. (1921e). Zur Relativitätsfrage [On the question of relativity]. *Die Umschau*, **25**(17), 227.

Gehrcke, Ernst. (1922). [Review of Kraus, Lipsius, Linke, and Petzoldt 1921]. *Glasers Annalen*, **88**, (entire series. No. 1053), 80.

Gehrcke, Ernst. (1922–1923). Soldnerov efekat i teorija relativnosti. *Univerzum*, **1**–2(4–5), 159.

Gehrcke, Ernst. (1923). Die Gegensätze zwischen der Äthertheorie und Relativitätstheorie und ihre experimentelle Prüfung [The contrasts between the ether theory and the theory of relativity and their experimental testing]. *Zeitschrift für technische Physik*, **4**(9), 292–99, reprinted in Gehrcke 1924a, pp. 83–95.

Gehrcke, Ernst. (1924a). *Kritik der Relativitätstheorie [Criticism of the Theory of Relativity]*. Berlin.

Gehrcke, Ernst. (1924b). *Die Massensuggestion der Relativitätstheorie [The Mass Suggestion of the Theory of Relativity]*. Berlin.

Gehrcke, Ernst. (ed.). (1927–1928). *Handbuch der physikalischen Optik [Handbook of Physical Optics]*, 2 volumes, Leipzig.

Gehrcke, Ernst. (1946). *Lebenslauf [Curriculum Vitae]*, October 3, GN 3-A-1.

Geissler, Friedrich Kurt. (1921). *Gemeinverständliche Widerlegung des formalen Relativismus (von Einstein und verwandten) und zusammenhängende Darstellung einer grundwissenschaftlichen Relativität [Generally Comprehensible Refutation of Formal Relativity (Einstein's and Related Approaches) and a Coherent Presentation of a Basic Scientific Relativity]*. Leipzig.

Geissler, Friedrich Kurt. (1931). Schluss mit der Einstein-Irrung [An End to the Einstein Error], in Israel, Ruckhaber, and Weinmann 1931, pp. 10–12.

Gerber, Paul. (1898). Räumliche und zeitliche Ausbreitung der Gravitation [Spatial and temporal diffusion of gravitation]. *Zeitschrift für Mathematik und Physik*, **43**, 93–104.

Gerber, Paul. (1902). Die Fortpflanzungsgeschwindigkeit der Gravitation [The propagational speed of gravitation]. *Programmabhandlung des städtischen Realgymnasiums Stargard in Pommern*. Stargard.

Gerber, Paul. (1917). Die Fortpflanzungsgeschwindigkeit der Gravitation [The propagational speed of gravitation]. *Annalen der Physik*, **357**(4), 415–41 (reprint of Gerber 1902).

Gerlach, Walther. (1927). [Review of Gehrcke 1927]. *Die Naturwissenschaften*, **15**(3), 410–12.

Gilbert, Leo. (1912). *Fundamente des exakten Wissens [Foundations of Exact Knowledge]*, Vol. 1: *Neue Energetik [New Energetics]*. Dresden.

Gilbert, Leo. (1914). *Das Relativitätsprinzip die jüngste Modenarrheit der Wissenschaft [The Relativity Principle, the Latest Fashionable Folly of Science]*. Brackwede i. W.

Glaser, Ludwig. (1920). Ueber Versuche zur Bestätigung der Relativitätstheorie an der Beobachtung [On attempts to confirm the theory of relativity using observation], Part 1. *Glasers Annalen*, **87**(4) (entire series No. 1036), 29–33, Part 2. (1921). Ibid. **88**(2) (entire series No. 1048), 30–32, and (3) (entire series No. 1049), 42–43.

Gleich, Gerold von. (1923). Die allgemeine Relativitätstheorie und das Merkurperihel [The general theory of relativity and the perhelion of Mercury]. *Annalen der Physik*, **377** (19), 221–35.

Gleich, Gerold von. (1930). *Einsteins Relativitätstheorien und physikalische Wirklichkeit [Einstein's Theories of Relativity and Physical Reality]*. Leipzig.

Goethe, Johann Wolfgang von. (1887/1808). *Goethes Werke [Goethe's Works]*, Part 1: *Werke [Works]*, Vol. 14: *Faust 1. Theil [Faust Part 1]*, ed. Grand Duchess Sophie von Sachsen. Weimar.

Goldzier, Hans. (1906). *Einige Weltprobleme [Some Universal Problems]*, 4 volumes, Vol. 2: *Gegen die Wahnvorstellung vom heißen Erdinnern: populär-wissenschaftliche Abhandlung [Against the Delusion of a Hot Interior of the Earth: A Popular-Science Treatise]*. Vienna (published under the pseudonym Thomas Newest).

Goldzier, Hans. (1921). *Gegen Einstein. Die Erfahrung im Weltall [Against Einstein. The Experience in Space]*. 2nd edn. Vienna.

Grossmann, Marcel. (1920). Mise au point mathématique. *Archives des sciences physiques et naturelles*, **2**, 497–99.

Grote, August. (1922). [Reply to Lessing 1922]. *Der Forscher*, 8–10, 14.

Grubić, Dušan. (1917). *Universal-Kausalprozess als unser oberstes Naturgesetz [The Universal-Causal Process as our Highest Natural Law]*. Zagreb.

Hack, Karl. (1930). *Eine neue mechanische Universal-Theorie [A New Mechanical Universal Theory]*. Würzburg.

Haeckel, Ernst. (1901/1899). *The Riddle of the Universe at the Close of the Nineteenth Century*. New York [*et al.*].

Haedicke, Johannes. (1902). *Die Lösung des Rätsels von der Schwerkraft durch die Versuche von Huyghens: ein Beitrag zur wissenschaftlichen Weltanschauung [The Solution of the Mystery of Gravity by Huyghens' Experiments: A Contribution to Scientific Ideology]*. Leipzig.

Haedicke, Johannes. (1932). *Die physikalische Unhaltbarkeit der Relativitäts-Theorie Einsteins [The Physical Indefensibility of Einstein's Theory of Relativity]*. Leipzig.

Hegg-Bitzius, Emil. (1914). *Das Ewige im Zeitlichen: Eine naturwissenschaftliche Formulierung [The Eternal in the Temporal: A Scientific Formulation]*. Bern.

Heidenreich, Eyvind Lee. (1921). Calls Einstein's statements irreconcilable. *The Minneapolis Morning Tribune*, May 16.

Heidenreich, Eyvind Lee. (1922). Reuterdahl vs. Einstein: Nailing a fallacy. *The Dearborn Independent*, March 25.

Heidenreich, Eyvind Lee. (1929). Reuterdahl's light quantum theory [Prologue to Reuterdahl 1929], in Reuterdahl 1929, pp. 1–6.

Heisenberg, Werner. (1969). *Der Teil und das Ganze. Gespräche im Umkreis der Atomphysik [The Part and the Whole. Conversations Surrounding Atomic Physics]*, Munich (published in English translation as *Physics and Beyond: Encounters and Conversations*).

Heldt, A. (1921). Wirtschaftliche und politische Gegenwartsfragen in Amerika [Current economic and political questions in America]. *Esslinger Zeitung* newspaper, September 3, GN 4, XII, 36r.

Holz, A. L. (1904). *Neueste Resultate über Weltkraft und Radialströmung [Latest Results about Universal Force and Radial Current]*. Hof (Saale).

Horst, Emil. (1913). Erkenntnistheoretische Begründung der Drucktheorie [Epistemological grounds for pressure theory], Part 1. *Weltwissen*, **1**(9), 121–25, Part 2. Ibid. **1**(10), 133–35, Part 3. Ibid. **1**(11), 141–43.

Horst, Emil. (1921). [Review of Oltmanns 1920]. *Neue Weltanschauung*, **10**(1), 16–19.

Houghton, Harris A. (1923). A Newtonian duplication? Einstein's triumph a long distance ahead. *The New York Times*, April 21, RP 4–4.

Huygens, Christiaan. (1690/1920). *Traité de la lumière*. Paris.

Isenkrahe, Caspar. (1879). *Das Räthsel von der Schwerkraft. Kritik der bisherigen Lösungen des Gravitationsproblems und Versuch einer neuen auf rein mechanischer Grundlage [The Mystery of Gravity: Criticism of Previous Solutions to the Gravitation Problem and Attempt at a New One on a Purely Mechanical Basis]*. Braunschweig.

Israel, Hans, Ruckhaber, Erich, and Weinmann, Rudolf (eds.). (1931). *100 Autoren gegen Einstein [100 Authors Against Einstein]*. Leipzig.

Kant, Immanuel. (1922/1787). *Kritik der reinen Vernunft [Critique of Pure Reason]*, ed. Albert Görland, Berlin.

Karollus, Franz. (1921). *Wo irrt und was übersieht Einstein? Ein neuer Versuch zur Lösung einer strittigen physikalischen Frage [Where does Einstein Err and What does he Overlook? A New Attempt to Solve a Disputed Question of Physics]*. Brünn.

Kleinschrod, Franz. (1911a). *Wissenschaftliche Begründung der Naturheilkunde. Eine prinzipielle Untersuchung [Scientific Justification of Naturopathy. An Investigation of Principles]*. Wörishofen.

Kleinschrod, Franz. (1911b). *Denkschrift zur Errichtung eines Lehrstuhles für Erforschung der Gesetze und Vorgänge der Naturheilung, verbunden mit dem Lehrauftrag für sozialethische Volksmedizin [Memorandum on the Establishment of a Chair for Research of the Laws and Processes of Naturopathy, combined with a Special Teaching Position for Social-Ethical Folk Medicine]*, in Kleinschrod 1911a, pp. 121–75.

Kleinschrod, Franz. (1912). *Das Rätsel des Lebens: Besitzt das Leben gegenüber dem Leblosen eine Eigengesetzlichkeit oder nicht? Eine biologische Studie über die übermechanische Natur des Lebensprinzips. Nach einem Vortrage, gehalten im Keplerbund, Ortsgruppe München [The Mystery of Life: Does Life Possess Autonomy vis-à-vis Lifeless Things, or Not? A Biological Study of the Metamechanical Nature of the Life Principle. Based on a Lecture Given at the Kepler League, Munich Group]*. Wörishofen.

Kleinschrod, Franz. (1920). *Das Lebensproblem und das Positivitätsprinzip in Zeit und Raum und das Einsteinsche Relativitätsprinzip in Raum und Zeit. Eine prinzipielle Untersuchung, zugleich ein neuer Weg zur Lösung des Lebensproblems [The Problem of Life and the Positivity Principle in Time and Space and the Einsteinian Relativity Principle in Space and Time. An Investigation of Principles, and Simultaneously a New Path to Solving the Life Problem]*. Hamm-Westfalen (Frankfurter zeitgemäße Broschüren Vol. 40, No. 1–3).

Kleinschrod, Franz. (1921). *Sebastian Kneipps Lehre. Wissenschaftliche Begründung [Sebastian Kneipp's Theory. Scientific Justification]*. Kempten.

Kleinschrod, Franz. (1922). Sebastian Kneipp und die Wissenschaft [Sebastian Kneipp and science], Part 1. *Kneipp-Blätter [Kneipp Pamphlets]*, **3**, 58–62. Part 2. Ibid. **4**, 286–89.

Kleinschrod, Franz. (1928). *Die Übermechanik des Lebens [The Meta-mechanics of Life]*, Vol. 1: *Die Herrschaft des Lebens über die tote Welt. Das tierisch-pflanzliche Leben als Leibseeleproblem [The Power of Life over the Inanimate World. Animal and Plant Life as a Body-Soul Problem]*. Berlin.

Kleinschrod, Franz. (1929). *Einführung in die Übermechanik des Lebens. Ein neuer Weg zur Lösung des Lebensproblems als eines Leibseele-Problems [Introduction to the Meta-mechanics of Life. A New Path to Solving the Life Problem as a Body-Soul Problem]*. Munich [*et al.*].

Kleinschrod, Franz. (1931). *Konnersreuth: Natürliches oder übernatürliches Geschehen? [Konnersreuth: Natural or Supernatural Event?]*. Waldsassen.

Kniepf, Albert. (1904). Neue Energien im Sonnensystem [New energies in the solar system]. *Die Gegenwart, Wochenschrift für Literatur, Kunst und öffentliches Leben*, **65**(24), 377–80.

Koller-Aeby, Hermann. (1931). *Der Grundirrtum Newton's als Ursache des Einstein'schen Grundirrtums [Newton's Fundamental Error as a Cause for Einstein's Fundamental Error]*. Leipzig.

Korff. (1921). Relativitätstheorie und Praxis [Relativity theory and practice], *Der (Schwarze) Tag [The (Black) Day]*, June 15, GN 4, X, 47r.

Kraus, Oskar, (1921). Zur Relativitätstheorie [On the Theory of Relativity]. *Annalen der Philosophie*, **2**(3). Issue 3.

Kraus, Oskar. (1921). *Fiktion und Hypothese in der Einsteinschen Relativitätstheorie. Erkenntnistheoretische Betrachtungen* [Fiction and Hypothesis in the Einsteinian Theory of Relativity. Epistemological Considerations], in Kraus, Lipsius, Lincke, and Petzoldt 1921, pp. 335–96.

Kraus, Oskar. (1925). *Offene Briefe an Albert Einstein und Max von Laue über die gedanklichen Grundlagen der speziellen und allgemeinen Relativitätstheorie [Open Letters to Albert Einstein and Max von Laue on the Intellectual Foundations of the Special and General Theory of Relativity]*. Vienna [*et al.*].

Kühn, Lenore. (1921). Kant contra Einstein. *Hellweg: Wochenschrift für deutsche Kunst*, August 17, GN 4, XII, 16r.

Kühn, Lenore. (no date). Der Raum als angebliche Ursache oder Wirkung physikalischer Zustände (Frage des inhomogenen Raumes und Frage der Erzeugung des geometrischen Raumzustandes durch Zustand der Materie) [Space as the Alleged Cause or Effect of Physical States (the Question of Unhomogeneous Space and the Question of the generation of the Geometrical State of Space by the State of Matter], [unpublished typescript], GN 38-I-9.

Kühn, Lenore. (no date). Was soll die Akademie der Nationen sein? [What should the Academy of Nations be?], [unpublished manuscript], GN 29-F-10.

Kühn, Lenore. (1983). *Erinnerungen an livländisches Landleben [Memories of Livonian Rural Life]*, ed. Detlef Kühn. Lüneburg (Series: Lüneburger ostdeutsche Dokumentationen Vol. 2).

Kupffer, Karl Reinhold. (1925). 'Materialismus, Vitalismus und Relativitätstheorie'. Vortrag gehalten auf dem Jahresaktus der Herder-Gesellschaft zu Riga am 7. September 1924 ['Materialism, vitalism, and the theory of relativity.' Lecture held at the annual meeting of the Herder Society of Riga on September 7, 1924]. *Abhandlungen des Herder-Institutes zu Riga*, **1**(1), 23–49.

Lamberty, Paul. (1926). *Die neue Physik [The New Physics]*. Haag.

Lämmel, Rudolf. (1924). [Review of Gehrcke 1924b]. *Leipziger Tageblatt* newspaper, September 14, GN VII.

Langevin, Paul. (1911). L'évolution de l'espace et du temps. *Scientia*, **10**, 31–54.

Lau, Ernst. (1924). [Review of Gehrcke 1924a and Gehrcke 1924b]. *Annalen der Philosophie*, **4**(8).

Laue, Max von. (1917). Die Fortpflanzungsgeschwindigkeit der Gravitation. Bemerkungen zur gleichnamigen Abhandlung von P. Gerber [The propagational speed of gravitation. Comments on the treatise of the same name by P. Gerber]. *Annalen der Physik*, **358** (11), 214–16.

Laue, Max von. (1921a). Eine Erwiderung auf Hrn. Lenards Vorbemerkungen zur Soldnerschen Arbeit von 1801 [A reply to Mr. Lenard's introductory remarks on Soldner's article of 1801]. *Annalen der Physik*, **371**(20), 283–84.

Laue, Max von. (1921b). Soldner und Einstein. Die Entdeckung der Ablenkung des Lichts durch die Schwere [Soldner and Einstein. The discovery of the deviation of light by gravitation]. *Frankfurter Zeitung* newspaper, November 18, morning edn., GN 4, XII, 73r.

Laue, Max von. (1921c). *Die Relativitätstheorie [The Theory of Relativity]*. Braunschweig.

LeBon, Gustave. (1895). *Psychologie des foules*. Paris (Bibliothèque de philosophie contemporaine Vol. 145).

Lenard, Philipp. (1911/1910). *Über Äther und Materie: Vortrag, gehalten in der Gesamtsitzung der Heidelberger Akademie der Wissenschaften am 4. Juni 1910 [On Ether and Matter: Lecture held at the Plenary Meeting of the Heidelberg Academy of Sciences on June 4, 1910]*, 2nd, more extensive edn. with supplements. Heidelberg.

Lenard, Philipp. (1918). *Über Relativitätsprinzip, Äther, Gravitation [On the Relativity Principle, Ether, Gravitation]*, 1st edn. Leipzig.

Lenard, Philipp. (1920a). Über Äther und Uräther [On ether and primordial ether]. *Jahrbuch der Radioaktivität und Elektronik*, **17**, 307–56.

Lenard, Philipp. (1920b). *Über Relativitätsprinzip, Äther, Gravitation [On the Relativity Principle, Ether, Gravitation]*, 2nd expanded edn. Leipzig.

Lenard, Philipp. (1921a). *Über Relativitätsprinzip, Äther, Gravitation [On the Relativity Principle, Ether, Gravitation]*, 3rd edn. with a supplement related to the Nauheim discussion. Leipzig.

Lenard, Philipp. (1921b). [Introductory remarks on Soldner 1921/1801]. *Annalen der Physik*, **370**(15), 593–600.

Lenard, Philipp. (1921c). Fragen der Lichtgeschwindigkeit [Questions of the speed of light]. *Astronomische Nachrichten*, **213**, 303–8.

Lenard, Philipp. (1922a). *Über Äther und Uräther [On ether and primordial ether]*. 2nd expanded edn. with a word of admonishment to German natural scientists, Leipzig.

Lenard, Philipp. (1922b). 'Ueber Aether und Uräther.' Ein Mahnwort an deutsche Naturforscher ['On ether and primordial ether.' A Word of Admonishment to German Natural Scientists]. *Deutsche Zeitung* newspaper, September 15, GN 4, I, 20r.

Lenard, Philipp. (2009). *Erinnerungen eines Naturforschers. Kritische annotierte Ausgabe des Originaltyposkriptes von 1931/1943 [Memories of a Natural Scientist. Critical Annotated Edition of the Original Typescript of 1931/1943]*, ed. Arne Schirrmacher. Berlin.

Lessing, Theodor. (1922). Zu 'Der Irrtum Einsteins' in Nr. 6/7 des Forschers [On 'Einstein's error' in No. 6/7 of the Forscher]. *Der Forscher*, 8–10, 13–14.

Lodge, Oliver. (1921). From Euclid to Einstein. Sir O. Lodge and relativity. The 'interim tool of exploration.' To the editor of *The Times*. *The Times* (London), December 10, 1921, GN 4, XII, 32r.

Lodge, Oliver. (1933). *My Philosophy. Representing My Views on the Many Functions of the Ether of Space*. London.

Lorentz, Hendrik Antoon. (1909). *The Theory of Electrons and its Applications to the Phenomena of Light and Radiant Heat. A Course of Lectures Delivered in Columbia University, New York, in March and April, 1906*. (Publication Ernest Kempton Adams Fund for Physical Research Vol. 2).

Mach, Ernst. (1969/1872). *Die Geschichte und die Wurzel des Satzes von der Erhaltung der Arbeit [The History and the Root of the Theorem of Conservation of Work]*, ed. Joachim Thiele, photomechanical reprint of the Prague edition, Calve, 1872, Amsterdam.

Mach, Ernst. (1919/1883). *The Science of Mechanics: A Critical and Historical Account of its Development*, translated from the German by Thomas J. McCormack, Chicago/ London.

Maxwell, James Clerk. (1879). Æther. *Encyclopedia Britannica*, Vol. 8, pp. 568–72.

Maxwell, James Clerk. (1890/1861–62). On Physical Lines of Force. *Philosophical Magazine*, **21**, reprinted in Idem. *The Scientific Papers*, Vol. 1, pp. 451–513, ed. W. D. Niven, Cambridge.

Mewes, Rudolf. (1919). Unterschiede zwischen der Relativitätstheorie von Mewes (1892/ 1893) und Lorentz (1895) [Differences between the theory of relativity of Mewes (1892/1893) and Lorentz (1895)], Part 1. *Zeitschrift für Sauerstoff- und Stickstoff-Industrie*, **11**(17–18), 70, Part 2. Ibid. **11**(19–20), 75–76.

Mewes, Rudolf. (1920a). *Wissenschaftliche Begründung der Raumzeitlehre oder Relativitätstheorie (1884–1894) mit einem geschichtlichen Anhang [Scientific Justification of the Spacetime Theory or Theory of Relativity] (1884–1894) with an Historical Appendix*. Berlin.

Mewes, Rudolf. (1920b). *Kaiser Wilhelm II., seine Zeit und sein Recht [Kaiser Wilhelm II, His Time and His Law]*. Berlin.

Mewes, Rudolf. (1920c). Bemerkung zur Ableitung der Grundformeln der Relativitätstheorie [Comment on the derivation of the basic formulae of the theory of relativity]. *Zeitschrift für Sauerstoff- und Stickstoff-Industrie*, **12**(1–2), 6.

Mewes, Rudolf. (1920d). Bestimmung der Fortpflanzungsgeschwindigkeit der Schwerkraft nach Gerber und Einstein [Determination of the speed of propagation of gravity according to Gerber and Einstein]. *Zeitschrift für Sauerstoff- und Stickstoff-Industrie*, **12**(3–4), 13–14.

Mewes, Rudolf. (1920e). Die Versuchsformel von Fizeau und die Relativitätsgleichung von Einstein [The experimental formula of Fizeau and the relativity equation of Einstein]. *Zeitschrift für Sauerstoff- und Stickstoff-Industrie*, **12**(9–10), 38.

Mewes, Rudolf. (1920f). Elementarer Weg zur Ableitung der Einsteinschen Grundgleichungen [Elementary path to derivation of the Einsteinian basic equations]. *Zeitschrift für Sauerstoff- und Stickstoff-Industrie*, **12**(13–14), 65.

Mewes, Rudolf. (1921a). Lichtablenkung in bewegten Mitteln [Light deviation in moving media]. *Zeitschrift für Sauerstoff- und Stickstoff-Industrie*, **13**(1–2), 1–4.

Mewes, Rudolf. (1921b). Zur Beurteilung der abendländischen Naturerkenntnis nach Spengler (Entropie, Gas- und Relativitätstheorie) [On assessing occidental knowledge of nature after Spengler (entropy, gas theory, and relativity theory]. *Zeitschrift für Sauerstoff- und Stickstoff-Industrie*, **13**(11–12), 42–44.

Mewes, Rudolf. (1921c). Lenards und Reuterdahls Stellungnahmen zur Relativitätstheorie [Lenard's and Reuterdahl's opinions on the theory of relativity]. *Zeitschrift für Sauerstoff- und Stickstoff-Industrie*, **13**(17–18), 77–78.

Mewes, Rudolf. (1922a/1896). *Kriegs- und Geistesperioden im Völkerleben und Verkündigung des nächsten Weltkrieges [War Periods and Spiritual Periods in the Life of Peoples, and Proclamation of the Next World War]*, 3rd and 4th expanded edns. Leipzig.

Mewes, Rudolf. (1922b). Reuterdahl vs. Einstein. Festnagelung einer Täuschung [Reuterdahl vs. Einstein. Nailing a fallacy]. *Zeitschrift für Sauerstoff- und Stickstoff-Industrie*, **14**(5), 55.

Mewes, Rudolf. (1922c). Vergleichende Gegenüberstellung der Exponentialformeln von Mewes, Planck, Einstein und Nernst für die spezifische Wärme [Contrastive comparison of the exponential formulae of Mewes, Planck, Einstein, and Nernst for specific heat]. *Zeitschrift für Sauerstoff- und Stickstoff-Industrie*, **14**(10), 121–23.

Mewes, Rudolf. (1922d). Stimmen gegen die Relativitätstheorie [Voices against the theory of relativity]. *Zeitschrift für Sauerstoff- und Stickstoff-Industrie*, **14**(11), 134–35.

Meyer, Carl Wilhelm. (1914). Professoren-Weisheit [Professorial wisdom]. *Weltwissen*, **3**(51–52), 17–21.

Mie, Gustav. (1926). [Review of Gehrcke 1924a and Gehrcke 1924b]. *Deutsche Literaturzeitung*, **14**, 678–82.

Minkowski, Hermann. (1909). Raum und Zeit [Space and Time], [Lecture held at the annual meeting of the Society of German Natural Scientists and Physicians (GDNÄ) in Cologne in 1908]. *Physikalische Zeitschrift*, **10**(3), 104–11.

Mohorovičić, Stjepan. (1916–1918). Über die räumliche und zeitliche Translation [On spatial and temporal translation], Part 1. *Académie des Sciences et des Arts des Slaves du Sud de Zagreb, Croatie, Bulletin des Traveaux de la Classe des Sciences Mathématiques et Naturelles*, 6–7, 46–72, Part 2. Ibid., 9–10 (1918), 21–33.

Mohorovičić, Stjepan. (1921). Die Folgerungen der allgemeinen Relativitätstheorie und die Newtonsche Physik [The implications of the general theory of relativity and Newtonian physics]. *Naturwissenschaftlichen Wochenschrift*, New Series, **20**(52), 737–39.

Mohorovičić, Stjepan. (1922). Eine elementare Theorie der Gravitation [An elementary theory of gravitation]. *Naturwissenschaftliche Wochenschrift*, New Series, **21**(11), 145–53.

Mohorovičić, Stjepan. (1923a). *Die Einsteinsche Relativitätstheorie und ihr mathematischer, physikalischer und philosophischer Charakter [The Einsteinian Theory of Relativity and its Mathematical, Physical, and Philosophical Character]*. Berlin/Leipzig.

Mohorovičić, Stjepan. (1923b). *Worthlessness of Einstein's Theory of Relativity for the Physics*. [Unpublished, undated typescript], RP 3–27.

Morgan, Robert. (1923). Einstein theory declared colossal humbug by U. S. naval astronomer. *The Dearborn Independent*, July 21.

Müller, Johannes Paul. (1913). *Neues vom Weltall [New Information from the Cosmos]*. Vienna [*et al.*].

Müller, Karl. (1880). Eine drastische Kritik der Anziehungskraft [A drastic criticism of attractive force]. *Die Natur*, New Series, **6**(15), 194.

Müller, Wilhelm and Stark, Johannes (eds.). (1941). *Jüdische und deutsche Physik: Vorträge zur Eröffnung des Kolloquiums für theoretische Physik an der Universität München [Jewish and German Physics: Lectures on the Opening of the Colloquium for Theoretical Physics at the University of Munich]*. Leipzig.

Natonek, Hans. (1921). Relativismus [Relativism]. *Weser-Zeitung* newspaper, July 22, GN 4, XI, 40r.

Neubert, Hermann. (1926). [Review of Gehrcke 1924b]. *Grundwissenschaft*, **6**(3–4), GN VII.

Newton, Isaac. (1729/1687). *Philosophia Naturalis Principia Mathematica*, English edn. *The Mathematical Principles of Natural Philosophy*, translated by Andrew Motte. London.

Newton, Isaac. (1704). *Opticks or a Treatise of the Reflections, Refractions, Inflections and Colours of Light*. London.

Nordmann, Charles. (1933). Aux dernières nouvelles le soleil serait froid et entouré d'eau [The latest news is, the Sun is cold and surrounded by water]. *Le Matin*, July 7.

O'Connell, William. (1929). Cardinal gives further view on Einstein theory. *Boston Evening American*, April 12.

Oltmanns, Johannes. (1920). *Die Mechanik des Weltalls: Grundlagen einer einheitlichen mechanistischen Weltanschauung [The Mechanics of Space: Foundations of a Uniform Mechanistic Ideology]*. Hamburg.

Oppell-Patschke, Margarete von. (1934). *Urkraft. Anpassung der vor 30 Jahren von Ing. A. Patschke aufgestellten Lichtäthermechanik an d. Ergebnisse d. heut. Experimentalphysik [Primordial Force. Adaptation of the Light Ether Mechanics set forth by A. Patschke, Engineer, 30 years ago to the Results of Modern Experimental Physics]*. Berlin.

Orthner, Rudolf (1928). *Über physikalische und mathematische Abhängigkeit: Umrisse eines neuen Verfahrens zur Aufstellung von Gleichungen zwischen geometrischen und physikalischen Größen [On Physical and Mathematical Dependency: Outlines of a New Procedure for Setting Up Equations Between Geometrical and Physics Factors]*. Leipzig/Vienna.

Osten, Hans. (1906). Über eine Erklärung der anomalen Bewegung des Merkurperihels [On an explanation of the anomalous movement of the perihelion of Mercury]. *Astronomische Nachrichten*, **171**(8), 119–20.

Palagyi, Melchior. (1901). *Neue Theorie des Raumes und der Zeit. Die Grundbegriffe einer Metageometrie [A New Theory of Space and Time. The Basic Concepts of a Meta-Geometry]*. Darmstadt.

Palagyi, Melchior. (1921a). Physik und Erkenntnistheorie [Physics and epistemology]. *Frankfurter Nachrichten* newspaper, August 7, morning edn., GN 4, XII, 87r.

Palagyi, Melchior. (1921b). Lenard und Einstein. *Frankfurter Zeitung* newspaper, December 12, GN 4, XII, 75r.

Palagyi, Melchior. (1922). *Kurze und populäre Darstellung der Einheits- oder Zusammengehörigkeitslehre von Raum und Zeit [A Short and Popular Presentation of the Uniform or Connectedness Theory of Space and Time]*. [Unpublished manuscript], RP 4–4.

Passarge, Hans. (1928). *Die Gravitation. Wesen und Ursprung [Gravitation. Nature and Origin]*. Leipzig.

Patschke, Arthur. (1904). *Transversal-Dampfturbinen für elastische Kraftmittel [Transverse Steam Turbines for Elastic Force Media]*. Mühlheim-Ruhr.

Patschke, Arthur. (1905a). *Lösung der Welträtsel durch das einheitliche Weltgesetz der Kraft [Solution of the Universal Mysteries by the Uniform Natural Law of Force]*. Munich.

Patschke, Arthur. (1905b). *Der Sturz des Anziehungsgesetzes und die Entdeckung des einheitlichen Weltgesetzes der Kraft [The Fall of the Law of Attraction and the Discovery of the Uniform Universal Law of Force]*. Munich.

Patschke, Arthur. (1906). *Vom Stern zum Atom. Beitrag zum Weltgesetz [From the Star to the Atom. A Contribution to the Universal Law]*. Munich.

Patschke, Arthur. (1920). *Umsturz der Einsteinschen Relativitätstheorie. Einführung in die einheitliche Erklärung und Mechanik der Naturkräfte [Overthrow of the Einsteinian Theory of Relativity. Introduction to the Uniform Explanation and Mechanics of Natural Forces]*. Berlin-Wilmersdorf.

Patschke, Arthur. (1921). *Elektromechanik: Einheitliche Erklärung und Mechanik der Naturkräfte [Electromechanics: Uniform Explanation and Mechanics of Natural Forces]*. Berlin-Wilmersdorf.

Patschke, Arthur. (1922). *Umsturz der Einsteinschen Relativitätstheorie: 4 mathematische Geburtsfehler. Einführung in die einheitliche Erklärung und Mechanik der Naturkräfte. Kreuzigung und Auferstehung des Lichtäthers [Overthrow of the Einsteinian Theory of Relativity. Introduction to the Uniform Explanation and Mechanics of Natural Forces. Crucifixion and Reserrection of the Light Ether]*. [2nd expanded edn. of Patschke 1920]. Berlin-Wilmersdorf.

Patschke, Arthur. (1925). *Weltlichttheorie. Welttheorie der Naturkräfte. Lösung der Naturkraftprobleme. Erfindungen, Entdeckungen und Ausnutzung der Naturkräfte [Universal Light Theory. A Universal Theory of Natural Forces. Solution of the Natural Force Problems. Inventions, Discoveries, and Utilization of the Natural Forces]*. Diessen near Munich.

Philosophische Fakultät der Kgl. Universität zu Berlin. (1877). *Aktenstücke in der Angelegenheit des Privatdozenten Dr. Dühring [Documents on the Matter of Private Lecturer Professor Dr. Dühring]*. Berlin.

Pick, Leopold. (1922). *Die vierte Dimension als Grundlage des transzendentalen Idealismus [The Fourth Dimension as the Foundation of Transcendental Idealism]*. Leipzig.

Planck, Max. (1910). *Die Stellung der neueren Physik zur mechanischen Naturanschauung, Vortrag gehalten am 23. September 1910 auf der 82. Versammlung der GDNÄ in Königsberg [The Position of More Recent Physics on the Mechanical View of Nature, Lecture held on September 23, 1910, at the 82nd meeting of the GDNÄ in Königsberg]*, in Planck 1958, pp. 30–46.

Planck, Max. (1937). Zum 25jährigen Jubiläum der von W. Friedrich, P. Knipping und Max von Laue gemachten Entdeckung [On the 25th anniversary of the discovery made by W. Friedrich, P. Knipping, and Max von Laue]. *Verhandlungen der Deutschen Physikalischen Gesellschaft*, 3rd series, Vol. 18, 77–80.

Planck, Max. (1958). *Vorträge und Reden [Lectures and Speeches]*, ed. Max-Planck-Gesellschaft zur Förderung der Wissenschaften/Verband Deutscher Physikalischer Gesellschaften. Braunschweig.

Planck, Max. (1982/1942). *Selbstdarstellung [Self-Portrait]*. Berlin.

Prochnow, Oskar (ed.). (1927). *Erdball und Weltall [Globe and Space]*. Berlin-Lichterfelde.

Rauschenberger, Walther. (1921). Die philosophische Bedeutung der Relativitätstheorie [The philosophical significance of the theory of relativity]. *Frankfurter Zeitung* newspaper, June 21, 1st morning edn., GN 4, XI, 36r–37r1.

Reichenbach, Karl. (1850). *Physikalisch-physiologische Untersuchungen über die Dynamide des Magnetismus, der Elektricität, der Wärme, des Lichtes, der Krystallisation, des Chemismus in ihren Beziehungen zur Lebenskraft [Physical-physiological Investigations of the Dynamides of Magnetism, Electricity, Heat, Light, Crystallization, Chemism in their Relationships to Life Force]*. 2 volumes, 2nd corrected edn. Braunschweig.

Reinke, Johannes. (1911/1901). *Einleitung in die theoretische Biologie [Introduction to Theoretical Biology]*, 2nd revised edn. Berlin.

Reinke, Johannes. (1926). *Das dynamische Weltbild: Physik und Biologie [The Dynamic Conception of the World: Physics and Biology]*. Leipzig.

Reuterdahl, Arvid. (1920). *Scientific Theism Versus Materialism. The Space-Time Potential*. New York.

Reuterdahl, Arvid. (1921a). Einstein and the New Science. *Bi-monthly Journal of the College of St. Thomas*, **9**(3), 3–26 (also as an offprint).

Reuterdahl, Arvid. (1921b). 'Kinertia' versus Einstein. *The Dearborn Independent*, April 30, 1921.

Reuterdahl, Arvid. (1921c). Science's 'baby-guy' was simple child till Einstein adopted it. *Minneapolis Sunday Tribune*, May 22.

Reuterdahl, Arvid. (1922a). The Academy of Nations – Its aims and hopes. *The Dearborn Independent*, January 7.

Reuterdahl, Arvid. (1922b). What the Einstein test can and cannot prove. Fact that he 'forgot' to credit others before him can't be overlooked. *The Dearborn Independent*, October 28.

Reuterdahl, Arvid. (1923a). *A Synthesis of Number, Space, Time and Energy, And A Physical Basis for Planck's and Rydberg's Constants*. Brooklyn, NY.

Reuterdahl, Arvid. (1923b). Einstein and the Nobel premium. *The Dearborn Independent*, January 6.

Reuterdahl, Arvid. (1923c). Einstein's predecessors. Dr. Arvid Reuterdahl argues German scientist used Soldner's, Gerber's, and other formulae without acknowledgment. *The New York Times*, June 3, RP 4–4.

Reuterdahl, Arvid. (1923d). Einstein's Predecessors. [Unpublished typescript], RP 3–88.

Reuterdahl, Arvid. (1923e). The Fischer tide predicting machine. *The Dearborn Independent*, January 20.

Reuterdahl, Arvid. (1923f). The origin of Einsteinism. Prof. Reuterdahl, replying to Mr. Bond, says that, in the absence of proof of innocence, Einstein stands convicted of plagiarism. *The New York Times*, August 12, RP 4–4.

Reuterdahl, Arvid. (1923g). The origin and motion of planetary systems. *The Dearborn Independent*, September 15.

Reuterdahl, Arvid. (1923h). Protoplasm and heredity. *The Dearborn Independent*, November 10.

Reuterdahl, Arvid. (1923i). First attempt of the unification of electricity and gravitation. *Univerzum*, **2**(9–12), 292–94.

Reuterdahl, Arvid. (1924). The Einstein film and the debacle of Einsteinism. *The Dearborn Independent*, March 22.

Reuterdahl, Arvid. (1928). *The God of Science*. Minneapolis.

Reuterdahl, Arvid. (1929a). *The Physical Basis of the Constant Velocity of Light*. Minneapolis.

Reuterdahl, Arvid. (1929b). *Communication by Dr. Arvid Reuterdahl of St. Paul, in reference to the new theory of Dr. Albert Einstein*. [Typescript], RP.

Reuterdahl, Arvid. (1931a). *Einsteinism. Its Fallacies and Frauds*. [Typescript of Reuterdahl 1931b], RP 1–28.

Reuterdahl, Arvid. (1931b). Der Einsteinismus. Seine Trugschlüsse und Täuschungen [Einsteinism. Its fallacies and frauds], in Israel, Ruckhaber, and Weinmann 1931, pp. 40–45.

Riem, Johannes. (1913). Hat die Welt einen Zweck? [Does the world have a purpose?]. *Unsere Welt*, **5**(11), 774–78.

Riem, Johannes. (1920). [Review of Fricke 1919]. *Naturwissenschaftliche Wochenschrift*, New Series, **19**(10), 158–59.

Riem, Johannes. (1921a). Zu Einsteins Amerikafahrt [On Einstein's trip to America]. *Deutsche Zeitung* newspaper, September 13, GN 4, XII, 26r.

Riem, Johannes. (1921b). Ein amerikanisches Weltanschauungsbuch [An American World View Book]. *Der Reichsbote* newspaper, October 4, GN 4, XII, 55r.

Riem, Johannes. (1921c). Um Einsteins Relativitätstheorie [About Einstein's theory of relativity]. *Deutsche Zeitung*, November 18.

Riem, Johannes. (1921d). Keine Bestätigung der Relativitätstheorie [No confirmation of the theory of relativity]. *Naturwissenschaftliche Wochenschrift*, New Series, **20**(28), 420.

Riem, Johannes. (1922a). Um die Relativitätstheorie [About the theory of relativity]. *Deutsche Zeitung* newspaper, January 7, GN 4, XVI, 9r.

Riem, Johannes. (1922b). Nachdenkliches zur Sonnenfinsternis im September [Contemplative thoughts on the solar eclipse in September]. *Rheinisch-Westfälische Zeitung* newspaper, July 29, GN 4, XX, 26r.

Riem, Johannes. (1922c). Von der deutschen Sonnenfinsternis-Expedition [On the German solar eclipse expedition]. *Deutsche Tageszeitung* newspaper, July 27, GN 4, XX, 27r.

Riem, Johannes. (1922d). Neues zur Relativitätstheorie [New information on the theory of relativity]. *Naturwissenschaftliche Wochenschrift*, New Series, **21**(1), 13–14.

Riem, Johannes. (1922e). Rotverschiebung und Michelsonscher Versuch [Red shift and Michelson experiment]. *Naturwissenschaftliche Wochenschrift*, New Series, **21**(52), 717.

Riem, Johannes. (1923). Die Bestätigung des Soldner-Effekts [The confirmation of the Soldner effect]. *Rheinisch-Westfälische Zeitung* newspaper, April 23, evening edn., GN 4, XX, 85r.

Riem, Johannes. (1924). Astronomie und Religion [Astronomy and religion]. *Unsere Welt*, **15**(1), 1–3.

Riem, Johannes. (1927). Welten-Werden [Becoming worlds], in Prochnow 1927, pp. 202–68.

Roderich. (1924). [Review of Gehrcke 1924b]. *Deutsches Haus*, Issue 8–9 (September), p. 3.

Rousseau, Jean Jacques. (2009/1755). *Discourse on the Origin of Inequality*. Oxford [*et al.*].

Rousseau, Jean Jacques. (1993/1762). *Émile*. London [*et al.*].

Ruckhaber, Erich Hermann. (1913). Die zwingende Logik des Monismus [The conclusive logic of Monism]. *Neue Weltanschauung*, **6**(4), 125–28.

Ruckhaber, Erich Hermann. (1920a). Das Denkproblem und seine Lösung [The thought problem and its solution]. *Neue Weltanschauung*, **9**(1), 1–9.

Ruckhaber, Erich Hermann. (1920b). Schwerkraft und Temperatur [Gravity and temperature]. [Comment on Fricke 1920b]. *Neue Weltanschauung*, **9**(4), 118–19.

Ruckhaber, Erich Hermann. (1920c). Geld, Sozialismus und Intelligenz [Money, socialism, and intelligence]. *Neue Weltanschauung*, **9**(5), 141–45.

Ruckhaber, Erich Hermann. (1920d). Zur Relativitätstheorie [On the theory of relativity]. *Neue Weltanschauung*, **9**(7), 219–20.

Ruckhaber, Erich Hermann. (1920e). Zur Relativitätstheorie [On the theory of relativity]. *Neue Weltanschauung*, **9**(9), 274–75.

Ruckhaber, Erich Hermann. (1920f). Eine neue Energiegewinnung [New production of energy]. *Neue Weltanschauung*, **9**(10), 303–4.

Ruckhaber, Erich Hermann. (1920g). Noch einmal die Einsteintheorie [The Einstein theory once again]. *Neue Weltanschauung*, **9**(12), 356–60.

Ruckhaber, Erich Hermann. (1921a). [Review of Fricke 1920a]. *Neue Weltanschauung*, **10**(5), 159–60.

Ruckhaber, Erich Hermann. (1921b). Die Umwälzung in den Grundanschauungen der Naturwissenschaft [The upheaval in the fundamental views of science] [Review of Ziegler 1914]. *Neue Weltanschauung*, **10**(7), 215–18.

Ruckhaber, Erich Hermann. (1927). *Untersuchung über das Prinzip des Widerspruchs [Investigation of the Principle of Contradiction]*. Berlin.

Ruckhaber, Erich Hermann. (1928). *Die Relativitätstheorie widerlegt durch das Widerspruchsprinzip und die natürliche Erklärung des Michelsonversuchs [The Theory of Relativity Refuted by the Contradiction Principle and the Natural Explanation of the Michelson Experiment: The Three-Dimensional Space-Time-System]*. Leipzig.

Ruckhaber, Erich Hermann. (1929). *Relativia. Der Roman eines Propheten [Relativia. The Novel of a Prophet]*. Berlin-Spandau.

Ruckhaber, Erich Hermann. (1931). *Expansion und Unendlichkeit [Expansion and Infinity]*. Berlin.

Ruckhaber, Erich Hermann and Hellweg, Lothar Wilfried. (1940). *Persönlichkeiten der Gegenwart: Luftfahrt, Wissenschaft, Kunst [Personalities of the Present: Air Travel, Science, Art]*. Berlin.

Ruckhaber, Erich Hermann. (1941). *Die Ätherdynamik des Sonnensystems [The Ether Dynamics of the Solar System]*. Berlin.

Ruckhaber, Erich Hermann. (1952/1938). *Biomechanik: das gelöste Lebensproblem, die Ursachen des Alterns und ihre alleinige Bekämpfung [Biomechanics: The Solved Problem of Life, the Causes of Ageing and their Sole Remedy]*, 3rd edn. Berlin.

Ruckhaber, Erich Hermann. (1959). *Das Urprinzip der Welt [The Primordial Principle of the World]*. Berlin.

Rutan. (1921a). Was will der Weltbund [What does the Weltbund want]. *Weltwissen*, **5**, 1–2, April 1.

Rutan. (1921b). Die Organisation des Weltbundes [The organization of the Weltbund]. *Weltwissen*, **5**, 2–4, April 1.

Sahulka, Johann. (1908). *Über die bisherigen Versuche, die Gravitation aus dem Stoße der Ätherteilchen zu erklären [On the Previous Attempts to Explain Gravitation from the Impact of the Ether Particles]*. Vienna.

Saager, Adolf. (1930). *Der Winterthurer Naturphilosoph Johann Heinrich Ziegler [Johann Heinrich Ziegler, the Natural Philosopher of Winterthur]*. Winterthur.

Schlick, Moritz. (1921). [Review of Gehrcke 1921c]. *Die Naturwissenschaften*, **9**, 779.

Schögel, W. (1922). Philosophie und Okkultismus [Philosophy and occultism]. *Die Übersinnliche Welt*, **2**, 17–25.

Schöning, Max. (1920). Ziel und Zweck des Kepler-Bundes in der Gegenwart [The object and purpose of the Kepler League in the present]. *Unsere Welt*, **12**(1), 17–28.

Schmidt, Harry. (1920). Wege zum Verständnis der Einsteinschen Relativitätstheorie [Paths to understanding the Einsteinian theory of relativity], Part 1. *Monistische Monatshefte*, **5**(6), 254–60; Part 2. Ibid. **5**(7–8), 305–11; Part 3. Ibid. **5**(9–10), 359–62; Part 4. Ibid. **5**(11–12), 404–8.

Schwarzschild, Karl. (1903). Über Himmelsmechanik [On heavenly mechanics]. *Verhandlungen der Gesellschaft Deutscher Naturforscher und Ärzte*. General part, 1–14, reprinted in Ibid. *Gesammelte Werke [Collected Works]*, Vol. 1, ed. Hans H. Voigt, Berlin [*et al.*], pp. 84–95.

See, Thomas Jefferson Jackson. (1910). *Researches on the Evolution of the Stellar System*, Vol. 2: *The Capture Theory of Cosmical Evolution*. Lynn, MA.

See, Thomas Jefferson Jackson. (1922). *New Theory of the Aether,* [Reprint of articles that appeared 1920–1922 in *Astronomische Nachrichten*]. Kiel (Electrodynamic Wave-theory of physical forces Vol. 2).

See, Thomas Jefferson Jackson. (1923a). Objections to relativity theory, Part 1. *The San Francisco Journal*, May 13, Part 2. Ibid., May 20, Part 3. Ibid., May 27.

See, Thomas Jefferson Jackson. (1923b). To the editor of The London Times, March 21, RP 4–49.

See, Thomas Jefferson Jackson. (1923c). Soldner, Foucault, and Einstein. *Science*, **58** (1506), 372.

See, Thomas Jefferson Jackson. (1923d). To the Associated Press, April 18, 1923, RP 4–49.

See, Thomas Jefferson Jackson. (1923e). To the editor of The London Times, April 19, RP 4–49.

See, Thomas Jefferson Jackson. (1924). *New Discoveries showing the complete triumph of Newton over Einstein.* [Press release], October 12, RP 4–49.

See, Thomas Jefferson Jackson. (1925). Travelling at speed of 294.375 miles per second. Scientist believes etherons provide pathway for light and radio waves. *The Dearborn Independent*, January 10.

See, Thomas Jefferson Jackson. (1938). *Wave–Theory! Discovery of the Cause of Gravitation!* London [*et al.*].

Seeliger, Hugo von. (1917a). Bemerkung zu P. Gerbers Aufsatz: 'Die Fortpflanzungsgeschwindigkeit der Gravitation' [Comment on P. Gerber's article 'The propagational speed of gravitation']. *Annalen der Physik*, **358**(9), 31–32.

Seeliger, Hugo von. (1917b). Weitere Bemerkungen zur 'Fortpflanzungsgeschwindigkeit der Gravitation' [Further comments on 'The propagational speed of gravitation']. *Annalen der Physik*, **359**(17), 38–40.

Seiling, Max. (1915/1904). *Das Professorenthum, 'der Stolz der Nation?' [The Professoriate, 'The Pride of the Nation'?]*, 3rd revised edn. Leipzig.

Sellien, Ewald. (1924a). Kant und Einstein. *Unsere Welt*, **16**(5), 107–12.

Sellien, Ewald. (1924b). Die Stellung des Aethers im Weltbild der Physik [The position of aether in the world view of physics]. *Unsere Welt*, **16**(10), 217–20.

Siemens, Werner. (1997/1886). Das naturwissenschaftliche Zeitalter [The age of science]. Lecture given at the 59th meeting of the GDNÄ in Berlin, reprinted in Engelhardt 1997, pp. 167–74.

Soldner, Johann. (1921/1801). Ueber die Ablenkung eines Lichtstrahls von seiner gradlinigen Bewegung, durch die Attraktion eines Weltkörpers, an welchem er nahe vorbeigeht [On the deviation of a light ray from its straight-line motion due to the attraction of a heavenly body which it passes by closely], in Bode, J. E., *Astronomisches Jahrbuch für das Jahr 1804 [Astronomical Yearbook for the Year 1804]*. Berlin, pp. 161–72 [shortened and commented reprint in *Annalen der Physik*, **370**(15), 593–604].

Sommerfeld, Arnold. (2004). *Wissenschaftlicher Briefwechsel [Scientific Correspondence]*. Vol. 2: *1919–1951*, eds. Michael Eckert and Karl Märker. Berlin [*et al.*].

Stark, Johannes. (1922). *Die gegenwärtige Krisis in der deutschen Physik [The Current Crisis in German Physics]*. Leipzig.

Staub, Julius Bernhard. (1898). *Die thatsächliche Widerlegung der Newtonschen Hypothese von der allgemeinen Anziehungskraft durch den naturgemäßen Ersatz derselben als Grundlage einer neuen monistischen Weltanschauung [The Actual Refutation of the Newtonian Hypothesis of the General Force of Attraction by Natural Replacement thereof as a Basis for a New Monistic Ideology]*. Leipzig-Lindenau.

Stauff, Philipp. (1913). *Semi-Kürschner oder literarisches Lexikon der Schriftsteller, Dichter, Bankiers, Geldleute, Ärzte, Schauspieler, Künstler, Musiker, Offiziere, Rechtsanwälte,*

Revolutionäre, Frauenrechtlerinnen, Sozialdemokraten usw. jüdischer Rasse und Versippung, die von 1813–1913 in Deutschland tätig oder bekannt waren [Semi-Kürschner or Literary Lexicon of the Authors, Poets, Bankers, Financiers, Physicians, Actors, Artists, Musicians, Officers, Lawyers, Revolutionaries, Suffragists, Social Democrats, etc. of the Jewish Race and Kinship who were active or known in Germany from 1813–1913]. Berlin-Lichterfelde.

Steiner, Rudolf. (1982). *Texte zur Relativitätstheorie [Texts on the Theory of Relativity]*, ed. Wim Viersen, Dornach (Mathematisch-astronomische Blätter, Neue Folge, No. 10).

Stevenson, Robert. (1914). Do Bodies Fall?, Part 1. *Harper's Weekly*, August 22, 174, Part 2. Ibid., August 29, 210, Part 3. Ibid., September 5, 234, Part 4. Ibid., September 12, 254, Part 5. Ibid., September 19, 285, Part 6. Ibid., September 26, 309, Part 7. Ibid., October 3, 332, Part 8. Ibid., October 10, 357–59, Part 9. Ibid., October 17, 382–84, Part 10. Ibid., October 24, 405–7, Part 11. Ibid., October 31, 429–30 (published under the pseudonym Kinertia).

Stewart, Balfour, and Tait, Peter Guthrie. (1882). *The Unseen Universe or Physical Speculations on a Future State*. London.

Stromer-Reichenbach, Friedrich von. (1924). *Historionomie, ihr Wesen und ihre Bedeutung. Ein wissenschaftlicher Versuch [Historionomy, its Essence and its Significance. A Scientific Experiment]*. Constance.

Thedinga, Eddo. (1922). *Einstein und wir Laien: Wege zu einer neuen Licht-Theorie [Einstein and We Laypeople: Pathways to a New Light Theory]*. Leipzig.

Thirring, Hans. (1921a). Über das Uhrenparadoxon in der Relativitätstheorie [On the clock paradox in the theory of relativity]. *Die Naturwissenschaften*, **9**(13), 209–12.

Thirring, Hans. (1921b). Erwiderung hierzu [Reply thereto (to Gehrcke 1921a)]. *Die Naturwissenschaften*, **9**(24), 482–83.

Thirring, Hans. (1921c). Erwiderung hierzu [Reply thereto (to Gehrcke 1921b)]. *Die Naturwissenschaften*, **9**(28), 551.

Thüring, Bruno. (1941). Albert Einsteins Umsturzversuch der Physik und seine inneren Möglichkeiten und Ursachen [Albert Einstein's Attempt to Overthrow Physics and its Internal Possibilities and Causes], in *Forschungen zur Judenfrage [Research on the Jewish Question]*, Vol. 4, pp. 134–62 (also as an offprint).

Toller, Ernst. (1921). *Masse-Mensch [Masses and Man]*. Potsdam.

Tölle, Oswald. (1913). Die Triebkraft im Weltall [The driving force in space]. *Neue Weltanschauung*, **6**(4), 142–44.

Trebitsch, Artur. (1921). *Deutscher Geist – oder Judentum! Ein Weg der Befreiung [German Spirit – or Jewishness! A Path to Liberation]*. Berlin [*et al.*].

Trumpler, Robert. (1923). Historical note on the problem of light deflection in the Sun's gravitational field. *Science*, **58**(1496), 161–63.

Vogt, Johann Gustav. (1912a). Weltwissen [Universal Knowledge]. *Weltwissen*. **1**(1), 1–4.

Vogt, Johann Gustav. (1912b). Der neueste Stern [The newest star]. *Neue Weltanschauung*, **5**(6), 209–12.

Vogtherr, Karl. (1922a). Über Fragen der Aberration und Lichtausbreitung [On questions of aberration and propagation of light]. *Naturwissenschaftliche Wochenschrift*, New Series, **21**(2), 20–25.

Vogtherr, Karl. (1922b). Über Aberration und Michelsonversuch [On aberration and the Michelson experiment]. *Astronomische Nachrichten*, **217**(5203), 381–96.

Vogtherr, Karl. (1923). *Wohin führt die Relativitätstheorie? Kritische Betrachtungen vom physikalischen und erkenntnistheoretischen Standpunkt aus [Where does the Theory of Relativity Lead? Critical Reflections from the Standpoint of Physics and Epistemology]*. Leipzig.

Vogtherr, Karl. (1933). *Das Problem der Gleichzeitigkeit [The Problem of Simultaneity]*. Munich.

Voigt, Woldemar. (1915). Über das Dopplersche Prinzip [On the Doppler principle]. *Physikalische Zeitschrift*, **16**, 381–86.

Volkmann, Paul. (1924). [Review of Gehrcke 1924a and Gehrcke 1924b]. *Annalen der Philosophie*, **4**(6).

Volkshochschule Gross-Berlin. (1919–1934). *Arbeitsplan [Work Plan]* from 1929–1933, *Lehrplan [Curriculum]*. Berlin.

Weinmann, Rudolf. (1926). *Versuch einer endgültigen Widerlegung der speziellen Relativitätstheorie [An Attempt at a Final Refutation of the Special Theory of Relativity]*. Leipzig.

Weizel, Walter. (1942). [Review of Müller and Stark 1941]. *Zeitschrift für technische Physik*, **23**(1), 25.

Westland, Ernst. (1901). *Universität, Politik und Dummheit [The University, Politics, and Stupidity]*. Berlin.

Weyl, Hermann. (1920). Die Diskussion über die Relativitätstheorie auf der Naturforscherversammlung [The discussion of the theory of relativity at the annual meeting of the Society of German Natural Scientists and Physicians]. *Die Umschau*, **24**(42), 609–11.

Weyl, Hermann. (1921). Antwort an Herrn Prof. Gehrcke [Reply to Prof. Gehrcke]. *Die Umschau*, **25**(10), 123–24.

Weyland, Paul. (1920a). Einsteins Relativitätstheorie – eine Massensuggestion [Einstein's theory of relativity – a mass suggestion]. *Tägliche Rundschau* newspaper, August 6.

Weyland, Paul. (1920b). [Typescript of the lecture given on August 24 at the Berlin Philharmonic], GN 72-E-6.

Wiechert, Emil. (1921). Anmerkungen zur Theorie der Gravitation und über das Schicksal der Gestirne [Comments on the theory of gravitation and on the fate of the stars]. *Vierteljahresschrift der Astronomischen Gesellschaft*, **56**(3), 171–91.

Wien, Wilhelm. (1921). *Die Relativitätstheorie vom Standpunkte der Physik und Erkenntnislehre. Vortrag gehalten im Verwaltungsgebäude der Firma Siemens & Halske in Siemensstadt bei Berlin am 18. März 1921 [The Theory of Relativity from the Standpoint of Physics and Epistemology. Lecture given in the Adminstration Building of the Siemens & Halke Company in Siemensstadt near Berlin on March 18, 1921]*. Leipzig.

Wittig, Hans. (1921). *Die Geltung der Relativitätstheorie. Eine Untersuchung ihrer naturwiss. Bedeutung [The Validity of the Theory of Relativity: An Investigation of its Scientific Significance]*. Berlin.

Wolff, Max. (1922). Einige Bemerkungen zur Bewertung der Descendenz-Hypothese und neuerer Hypothesen der Physik [Some remarks on the assessment of the descendance hypothesis and more recent hypotheses in physics]. *Archiv für Naturgeschichte*, **88**(A) (7), 113–21.

Wutke, Georg. (1909). *Kann die Erde erkalten? Die Gestirne als Kraftquelle und die Ursachen der Schwerkraft [Can the Earth Go Cold? The Stars as a Source of Power and the Causes of Gravity]*. Berlin.

Wutke, Georg. (1920). *Die Erde als Motor [The Earth as a Motor]*. Berlin.

Zacharias, Johannes. (1912a). Zur Mechanik der Gravitation [On the mechanics of gravitation]. *Weltwissen*, **1**(1), 9–10.

Zacharias, Johannes. (1912b). Die sechsteilige Kugel zur Erklärung der Gravitation [The six-part sphere for explaining gravitation]. *Weltwissen*, **1**(2), 33–34.

Zacharias, Johannes. (1912c). Gibt es ein Monopol auf Wissen und Denken? [Is there a monopoly on knowledge and thinking?]. *Weltwissen*, **1**(6), 91–93.

Zacharias, Johannes. (1914a). 'Geniale' Offenbarungen der 'exakten' Forschung ['Brilliant' revelations from 'exact' research]. *Weltwissen*, **2**(35), 125–29.

Zacharias, Johannes. (1914b). Akademische Politik [Academic Politics]. *Weltwissen*, **3**(51–52), 25–26.

Zacharias, Johannes. (1914c). Ein Vorschlag zur Güte (offenes Schreiben an die führenden Herren Gelehrten) [A suggestion of amicability (open letter to the leading scholars)]. *Weltwissen* **3**(53–54), 32–33.

Zacharias, Johannes. (1920). *Rätsel der Natur und Totengräber der Wahrheit [Mysteries of Nature and Gravediggers of Truth]*. Munich.

Zacharias, Johannes. (1921). Ein Pionier der Physik als Brückenbauer zur Metaphysik. Autobiographisches vom Ingenieur Johannes Zacharias [A pioneer of physics as a bridge builder to metaphysics. Autobiographical information on the engineer Johannes Zacharias]. *Der Forscher*, **2**, 4–6.

Zacharias, Johannes. (1922). *Verborgene Gewalten im Weltgeschehen: Eine neue Raum-Kraft-Lehre [Hidden Powers in the Events of the World: A New Space-Force Theory]*. Munich.

Zehnder, Ludwig. (1922). *Der Aufbau der Atome aus Uratomen: Vortrag, gehalten in der Kantgesellschaft, Ortsgruppe Basel am 12. Dezember 1921 [The Construction of the Atoms from the Primordial Atoms: Lecture held at the Kant Society, Basel Group, on December 12, 1921]*. Tübingen.

Zehnder, Ludwig. (1927). Ueber die Grundlagen der Naturwissenschaften [On the foundations of the sciences]. *Verhandlungen der Naturforschenden Gesellschaft in Basel*, **38**, 84–100.

Zehnder, Ludwig. (1933). *Der Aether im Lichte der klassischen Zeit und der Neuzeit [The Aether in the Light of the Classical Period and the Modern Period]*. Tübingen.

Zehnder, Ludwig. (1938). *Die tiefsten Grundlagen der Physik und Chemie. Aus den Universitätsvorlesungen in den Jahren 1935 bis 1938 [The Deepest Foundations of Physics and Chemistry. From the University Lectures in the years 1935 to 1938]*. Zurich/Leipzig.

Zettl, Otmar. (1924). [Review of Gehrcke 1924b]. *Augsburger Postzeitung* newspaper [no specific date], GN 4, VII.

Ziegler, Johann Heinrich. (1902). *Die Universelle Weltformel und ihre Bedeutung für die wahre Erkenntnis aller Dinge: Erster Vortrag, gehalten an der Versammlung der Schweizerischen Naturforschenden Gesellschaft vom 7.-10. September in Genf [The Universal Unified Theory and its Significance for True Insight into All Things: First Lecture, Given at the Assembly of the Swiss Natural Research Society on September 7–10 in Geneva]*. Zürich.

Ziegler, Johann Heinrich. (1903). *Die Universelle Weltformel und ihre Bedeutung für die wahre Erkenntnis aller Dinge: Zweiter Vortrag [The Universal Unified Theory and its Significance for True Insight into All Things: Second Lecture]*. Zürich.

Ziegler, Johann Heinrich. (1904a). *Die wahre Einheit von Religion und Wissenschaft. Vier Abhandlungen [The True Unity of Religion and Science. Four Treatises]*. Zürich.

Ziegler, Johann Heinrich. (1904b). Über das wahre Wesen der sog. Schwerkraft [On the True Nature of So-Called Gravity], in Ziegler 1904a, pp. 60–85.

Ziegler, Johann Heinrich. (1904c). Über den Sonnengott von Sippar [On the sun god of Sippar], in Ziegler 1904a, pp. 175–93.

Ziegler, Johann Heinrich. (1905). Die wahre Ursache der hellen Lichtstrahlung des Radiums [The true cause of the bright light radiation of radium. *Psychische Studien*, **32**, 449–55.

Ziegler, Johann Heinrich. (1908a). *Die Struktur der Materie und das Welträtsel [The Structure of Matter and the Riddle of the Universe]*. Bern.

Ziegler, Johann Heinrich. (1908b). *Konstitution und Komplementät der Elemente [Constitution and Complementarity of the Elements]*. Bern.

Ziegler, Johann Heinrich. (1914). *Die Umwälzung in den Grundanschauungen der Naturwissenschaft: Acht kritische Betrachtungen [The Upheaval in the Fundamental Views of Science: Eight Critical Reflections]*. Bern.

Ziegler, Johann Heinrich. (1921). Albertus Maximus und die Blamage der Schulweisheit [Albert the Great and the disgrace of book-learning]. *Luzerner Neueste Nachrichten* newspaper, September 20, (published under the pseudonym J. E. G. Hirzel).

Ziegler, Johann Heinrich. (1922). Ostwalds Relativitätstheorie und Farbenlehre an der Leipziger Zentenarfeier. Ein Vergleich mit Zieglers Urlichtlehre [Ostwald's theory of relativity and color theory at the Leipzig centenary celebration. A comparison with Ziegler's primordial light theory]. Part 1. *Luzerner Neueste Nachrichten* newspaper, October 21. Part 2. Ibid., October 28. Part 3. Ibid., November 4 (published under the pseudonym J. E. G. Hirzel), GN 4, I, 40r, II, 36r, I, 43r.

Ziegler, Johann Heinrich. (1923). *Das Ding an sich. Das Ende der sog. Relativitätstheorie [The Thing in Itself. The End of the So-Called Theory of Relativity]*. Zürich.

Ziegler, Johann Heinrich. (1928). *Die Hauptaufgabe der Naturwissenschaft und ihre einfache Lösung [The Primary Task of Natural Science and its Simple Solution]*. Zürich.

Ziegler, Johann Heinrich. (1929). Was verstehen wir unter physikalischen und chemischen Elementen und wodurch unterscheiden sie sich hauptsächlich? [What do we understand by physical and chemical elements and how are they primarily distinguished?]. *Das freie Wort*, November 22.

Ziegler, Johann Heinrich. (1931). *Der grosse Welt- und Selbstbetrug der Physiker: Ein Dokument aus der Gegenwart für die Zukunft [The Physicists' Great Deception of the World and of Themselves: A Document from the Present for the Future]*. Zürich.

Ziegler, Johann Heinrich. (1932). *Das Urlicht und das Hauptgerüst richtiger Weltanschauung: Ueber das Absolute und das Relative [Primordial Light and the Basic Framework of Correct Ideology: On the Absolute and the Relative]*. Zürich.

Ziegler, Johann Heinrich. (1935). *Das Geheimnis des Selbstverständlichen: der Schlüssel zum Welträtsel [The Secret of the Self-Evident: The Key to the World Riddle]*. Zürich.

Zöllner, Karl Friedrich. (1879). *Wissenschaftliche Abhandlungen [Scientific Transactions]*, Vol. 3: *Die transcendentale Physik und die sogenannte Philosophie: eine deutsche Antwort auf eine 'sogenannte wissenschaftliche Frage' [Transcendental Physics and So-Called Philosophy: A German Answer to a 'So-Called Scientific Question']*. Leipzig.

Zöllner, Karl Friedrich. (1882). *Erklärung der universellen Gravitation aus den statischen Wirkungen der Elektricität und die allgemeine Bedeutung des Weber'schen Gesetzes [Explanation of Universal Gravitation from the Static Effects of Electricity and the General Meaning of Weber's Law]*. Leipzig.

Quoted newspaper and periodical articles

(1906). Was will der Deutsche Monistenbund? [What does the German Monist League want?]. *Blätter des Deutschen Monistenbundes [Gazette of the German Monist League]*. **1** (July), 7–9.

(1909). Diskussion. *Physikalische Zeitschrift*, **10**, 825–26, reprinted in Stachel *et al.*, 1989, Doc. 61.

(1913). Ist der Weltraum leer? [Is space empty?]. *Unsere Welt*, **5**(5), 335.

(1913). [Comment by the editors on Tölle 1913]. *Neue Weltanschauung*, **6**, 142–44.

(1919). Eine neue Größe der Weltgeschichte: Albert Einstein, dessen Erkenntnisse eine völlige Umwälzung unserer Naturbetrachtung bedeuten und den Erkenntnissen eines Kopernikus, Kepler und Newton gleichwertig sind [A new great man of world history: Albert Einstein, whose discoveries mean a complete upheaval of our consideration of nature and are equal to the discoveries of Copernicus, Kepler, and Newton]. *Berliner Illustrirte Zeitung* newspaper, December 14.

(1919). Lights all askew in the heavens. Men of science more or less agog over results of eclipse observations. Einstein theory triumphs. Stars not where they seemed or were calculated to be. But nobody need worry. *The New York Times*, November 10.

(1919). Massensuggestion [mass suggestion]. *Mitteilungen des deutschen Monistenbundes [Newsletter of the German Monist League]*, **4**(3), 33–34.

(1919). Wissenschaftsraub und Bluff [Scientific theft and bluff]. *Deutsche Zeitung* newspaper, December 19, quoted from Mewes 1920a, pp. 74–76.

(1920). Allgemeine Diskussion über Relativitätstheorie [General discussion about relativity theory]. *Physikalische Zeitschrift*, **21**, 650–51, 662, 666–68, reprinted Janssen *et al.* 2002a, Doc. 46.

(1920). Authorship of the famous theory of light bending contested by St. Paul scientist; lost paper is key. *St. Paul Sunday Pioneer Press*, February 15.

(1920). Die Relativitätstheorie auf der Naturforscher-Versammlung in Nauheim [The theory of relativity at the scientists' conference in Nauheim]. *Neue Weltanschauung*, **9**(11), 342–44.

(1920). Diskussion [Discussion]. *Physikalische Zeitschrift*, **21**, 650–51, 662, 666–68, reprinted in Janssen *et al.* 2002a, Doc. 46.

(1920). Diskussionen om relativitetstheorien [Discussion about relativity theory]. *Aftenposten*, June 18, morning edn.

(1920). Einstein theory causes scientific war. *The Dearborn Independent*, October 30.

(1921). Die Berliner Bonifatiustagung. Minister Dr. Brauns über die nationalen Ideale des Katholizismus [The Berlin Bonifatius Conference. Minister Dr. Brauns about the national ideals of Catholicism]. *Kölnische Volkszeitung* newspaper, June 18, evening edn., GN 4, IX, 9r.

(1921). Dr. J. H. Ziegler. *Zürcher Chronik* newspaper, July 6.

(1921). Einstein als Violinist. *Vossische Zeitung* newspaper, May 20, morning edn., GN 4, IX, 30r.

(1921). Einstein branded Barnum of science, Minnesota man calls relativity 'bunk.' *The Minneapolis Sunday Tribune*, April 10.

(1921). Einstein foes prove theory false, claim. Twin cities mathematical association hears talk on relativity. Former exponents are now sorry, says St. Thomas engineering dean. *The Minneapolis Sunday Tribune*, November 20.

(1921). Einstein sails. U.S. women 'lovely'. *The Chicago Tribune*, June [no specific date], GN 4, XVII, 129r.

(1921). Einsteins – Bezügigkeitsanschauungstheorie [Einstein's relational observation Theory]. *12 Uhr Mittagszeitung]* newspaper, July 8, GN 4, XI, 26r.

(1921). Einsteins Theorie im Ballsaale. 'Walzer nach den Gesetzen der Relativitätstheorie' [Einstein's theory in the ballroom. 'Waltzes according to the laws of the theory of relativity']. *National-Zeitung* newspaper, August 24, GN 4, XI, 62r.

(1921). Nochmals Soldner und Einstein [Soldner and Einstein again]. *Neue Zürcher Zeitung* newspaper, November 30, second morning edn., GN XII, 75r.

(1921). Prof. Einsteins Gehrock [Prof. Einstein's frock coat]. *Leipziger Neueste Nachrichten* newspaper, July 1, morning edn., GN 4, XI, 27r.

(1921). Prof. Reuterdahl to write book to show Einstein fallacies. *Evening Tribune*, April 18.
(1921). Relativität und Weltanschauung [Relativity and ideology]. *Baseler Nationalzeitung* newspaper, November 30, GN 4, XII, 64r.
(1921). Um die Priorität der Einstein-Theorie [On the priority of the Einstein theory]. *Magdeburgische Zeitung*, newspaper, November 30, GN 4, XII, 74r.
(1921). [Report about Reuterdahl]. *Hamburger Woche* newspaper, June 9.
(1922). Albert Einstein über die deutsche Not [Albert Einstein on German distress]. *Boersen-Courier Berlin* newspaper, November 6, 1922, GN 4, I, 68r.
(1922). Au Collège de France. M. Guillaume, de Genève, contre M. Einstein. *L'Œuvre*, June 6, GN 4, XVII, 130r.
(1922). Austritt Einsteins aus der Kaiser-Wilhelm-Akademie. Eine Folge der reaktionären Hetze [Einstein's resignation from the Kaiser Wilhelm Academy. A consequence of reactionary hounding]. *Neue Berliner 12 Uhr Zeitung* newspaper, September 29, GN 1, I, 2r.
(1922). Der Film des physikalischen Nihilismus [The film of physical nihilism]. *Kinorat*, 9–10, GN 4, XVI, 67r.
(1922). Die flüchtige Relativität [The flee(t)ing relativity]. *Rheinisch-Westfälische Zeitung* newspaper, August 5, GN XXI, 6r.
(1922). Die Krisis in der Lichttheorie [The crisis in light theory]. *Vossische Zeitung* newspaper, February 8, GN 4, XVI, 31r.
(1922). Die verbissenen Gegner Einsteins… [Einstein's determined opponents…]. *Berliner Tageblatt* newspaper, November 17, morning edn., GN 4, I, 35r.
(1922). Ein graphologisches Portrait Einsteins [A graphological portrait of Einstein]. *Vossische Zeitung* newspaper, February 18, GN 4, XVI, 4r.
(1922). Einstein wrong, Swiss declares. *Chicago Tribune*, March 31, GN 4, XVII, 129r.
(1922). Einstein, der Feind des Telephons [Einstein, the enemy of the telephone]. *Hamborner General-Anzeiger* newspaper, April 11, GN 4, XVII, 153r.
(1922). Einstein. *Die Wahrheit*, September 23, GN 1, I, 3r.
(1922). Einsteins Absage an den Naturforschertag. Auf der Liste der Mörderorganisation [Einstein's cancellation at the scientists' conference. On the list of the murder organization]. *Berliner Tageblatt* newspaper, August 5, morning edn., GN 4, XXI, 6r.
(1922). Einsteins Absage. Der Gelehrte auf der Mörderliste [Einstein's cancellation. The scholar on the murder list]. *Dresdener Volkszeitung* newspaper, August 5, GN 4, XXI, 5r.
(1922). Hints on Einstein. *The Times* (London), April 3, GN 4, XVII, 102r.
(1922). Kommt Professor Einstein nach Leipzig? [Is professor Einstein coming to Leipzig?]. *Leipziger Neueste Nachrichten* newspaper, August 5, GN 4, XXI, 5r.
(1922). Les Théories d'Einstein? *Intransigeant*, April 3, GN 4, XVII, 129r.
(1922). Un adversaire des Théories d'Einstein. *Les Messin*, April 1, GN XVII, 129r.
(1922). [Questions from the Psychological Society of Berlin]. *Die Übersinnliche Welt*, 3–4, 61.
(1923). Die Relativitätstheorie bestätigt? [The theory of relativity confirmed?]. *Hamburger Nachrichten* newspaper, May 25, GN 4, XX, 88r.
(1923). Einstein in Japan. Ein Mißverständnis der Japanerinnen über die Relativitätstheorie [Einstein in Japan. A misunderstanding by the Japanese women about the theory of relativity]. *Berliner Tageblatt* newspaper, [no specific date], GN 4, I, 54r.
(1923). *Einstein Theory Is 'Old Stuff.' U.S. Astronomer Says German Scientist Is Plagiarist.* [unknown newspaper], April [no specific date], GN 4, XX, 84r.
(1923). Government scientist exposes Einstein trick. Declares German astronomer was detected in plagiarizing von Soldner in 1911. Newton theory held true. Says bending of stars' rays do not bear out claims which he terms crazy vagaries. *The Philadelphia Journal*, April 12.

(1924). Ein zweckloses Buch [A pointless book]. *Berliner Börsenzeitung* newspaper, entertainment supplement, November 29, 1924.

(1924). Einsteins Massensuggestion [Einstein's mass suggestion] [Review of Gehrcke 1924b]. *Deutsche Zeitung* newspaper, June 10, GN VII.

(1924). [Review of Gehrcke 1924b]. *Völkischer Kurier* newspaper, September 11, GN VII.

(1924). [Review of Gehrcke 1924b]. *Breslauer Zeitung* newspaper, August 24, GN VII.

(1924). [Review of Gehrcke 1924b]. *Der Tag* newspaper, July 18, GN VII.

(1924). [Review of Gehrcke 1924b]. *Deutscher Wille. Werkgemeinschaftliches Wochenblatt* newspaper, August 7, GN VII.

(1924). [Review of Gehrcke 1924b]. *Hammer* newspaper, No. 532, GN VII.

(1924). [Review of Gehrcke 1924b]. *Mecklenburger Warte* newspaper, August 9, GN VII.

(1924). [Review of Gehrcke 1924b]. *Rundschau für Literatur und Kunst*, July 31, GN VII.

(1925). [Review of Gehrcke 1924b]. *Psychiatrisch-neurologische Wochenschrift*, **27**(19).

(1929). Reuterdahl asserts he is real originator of Einstein's new theory. *The St. Paul Daily News*, January 27, RP 3–26.

(1929). St. Paul Physicist concludes that relativity booster and his admirers are suffering from 'delirium mathematicum', scoffs at idea of curved space. *St. Paul Dispatch*, January 28.

(1930). Einstein's cosmic religion. *Literary Digest*, November 29.

(1931). Scientific opinion turning against Einsteinism, says noted scientist. *The Kansas City Star*, March 24.

(1934). 'Die Sonne ist kalt – aber ewig'. Neue Weltalltheorie heute abend zur Debatte ['The Sun is cold – but eternal.' New space theory to be debated this evening]. *BZ*, February 20, GN 29-F-11.

(1936). Direktören fil. dr Harald Nordenson Stockholm, August 10 [unknown newspaper], GN 4, (2), 32r.

(1936). Hermann Fricke's 30jähriger Krieg [Hermann Fricke's 30-year war]. *Docentra*, **57**(25), 421–22.

Unpublished sources

Ernst Gehrcke Papers, Archival Collection, Max Planck Institute for the History of Science, Berlin

Georg Reinhart Papers, Winterthur City Archives

Patschke Student File, Technikum Mittweida

Reuterdahl Papers, Department of Special Collections, University of St. Thomas, St. Paul, MN

Index